Nonlinear Acoustics

Mark F. Hamilton • David T. Blackstock
Editors

Nonlinear Acoustics

Third Edition

Editors

Mark F. Hamilton
W. R. Woolrich Professor in Engineering
and Research Professor at Applied Research Laboratories
The University of Texas at Austin
Austin, TX, USA

David T. Blackstock
Late E. P. Schoch Professor Emeritus in Engineering
and Research Professor at Applied Research Laboratories
The University of Texas at Austin
Austin, TX, USA

ISBN 978-3-031-58962-1   ISBN 978-3-031-58963-8   (eBook)
https://doi.org/10.1007/978-3-031-58963-8

1st edition: © Academic Press 1998
2nd edition: © The Editor(s) (if applicable) and The Author(s) 2008
3rd edition: © The Editor(s) (if applicable) and The Author(s) 2024. This book is an open access publication.

This Springer imprint is published by the registered company Springer Nature Switzerland AG
The registered company address is: Gewerbestrasse 11, 6330 Cham, Switzerland

If disposing of this product, please recycle the paper.

The ASA Press

The Acoustical Society of America

On 27 December 1928 a group of scientists and engineers met at Bell Telephone Laboratories in New York City to discuss organizing a society dedicated to the field of acoustics. Plans developed rapidly, and the Acoustical Society of America (ASA) held its first meeting on 10–11 May 1929 with a charter membership of about 450. Today, ASA has a worldwide membership of about 7000.

The scope of this new society incorporated a broad range of technical areas that continues to be reflected in ASA's present-day endeavors. Today, ASA serves the interests of its members and the acoustics community in all branches of acoustics, both theoretical and applied. To achieve this goal, ASA has established Technical Committees charged with keeping abreast of the developments and needs of membership in specialized fields, as well as identifying new ones as they develop.

The Technical Committees include acoustical oceanography, animal bioacoustics, architectural acoustics, biomedical acoustics, computational acoustics, engineering acoustics, musical acoustics, noise, physical acoustics, psychological and physiological acoustics, signal processing in acoustics, speech communication, structural acoustics and vibration, and underwater acoustics. This diversity is one of the Society's unique and strongest assets since it so strongly fosters and encourages cross-disciplinary learning, collaboration, and interactions.

ASA publications and meetings incorporate the diversity of these Technical Committees. In particular, publications play a major role in the Society. *The Journal of the Acoustical Society of America* (JASA) includes contributed papers and patent reviews. *JASA Express Letters* (JASA-EL) and *Proceedings of Meetings on Acoustics* (POMA) are online, open-access publications, offering rapid publication. *Acoustics Today*, published quarterly, is a popular open-access magazine. Other key features of ASA's publishing program include books, reprints of classic acoustics texts, and videos. ASA's biannual meetings offer opportunities for attendees to share information, with strong support throughout the career continuum, from students to retirees. Meetings incorporate many opportunities for professional and social interactions, and attendees find the personal contacts a rewarding experience. These experiences result in building a robust network of fellow scientists and engineers, many of whom become lifelong friends and colleagues.

From the Society's inception, members recognized the importance of developing acoustical standards with a focus on terminology, measurement procedures, and criteria for determining the effects of noise and vibration. The ASA Standards Program serves as the Secretariat for four American National Standards Institute Committees and provides administrative support for several international standards committees.

Throughout its history to present day, ASA's strength resides in attracting the interest and commitment of scholars devoted to promoting the knowledge and practical applications of acoustics. The unselfish activity of these individuals in the development of the Society is largely responsible for ASA's growth and present stature.

To the memory of Francis H. Fenlon.

Preface to the Third Edition

Published originally by Academic Press in 1998, *Nonlinear Acoustics* continues to prove useful both as a resource for established researchers and as a textbook for graduate students. Its sustained utility over the past quarter century is due to its emphasis on basic theory, and as noted in the Preface to the First Edition, individual chapters were written by experts on their respective topics, the notation is largely consistent throughout the book, and there is extensive cross-referencing between chapters. The level of presentation is oriented toward graduate students, with the first eight chapters providing the foundation for a semester course on the fundamentals of nonlinear acoustics.

Apart from minor corrections, the text has remained unaltered in both the second edition, published by Acoustical Society of America in 2008, and in the present edition, published by Springer. Publication of the second edition by ASA, a nonprofit organization, permitted the book to be sold at a substantially reduced cost. Thanks to support from Applied Research Laboratories, University of Texas at Austin, Springer has made the electronic version of the third edition open access, permitting free downloads worldwide.

As an addendum to the Preface to the First Edition, a list of the currently triennial International Symposia on Nonlinear Acoustics (ISNA) covering the first 50 years of this series, from 1968 to 2018, may be found in the ASA publication *Acoustics Today*.[1]

Austin, TX, USA   Mark F. Hamilton
January, 2024

---

[1] Sapozhnikov, O. A., Khokhlova, V. A., Cleveland, R. O., Blanc-Benon, P., and Hamilton, M. F. (2019). Nonlinear acoustics today. *Acoustics Today* **15**, 55–64; see p. 62. At the time of publication, the anticipated years of the symposia in Oxford and Nanjing that appear in Fig. 8 were postponed by one year because of COVID-19.

Preface to the Second Edition

On this 10th anniversary of the publication of *Nonlinear Acoustics* by Academic Press, we are happy that the book has proven useful not only as a reference but also as a graduate level textbook at various universities. As a result of a merger, in recent years the book has been distributed by Elsevier. We are indebted to Elsevier for its release of the copyright in order for the Acoustical Society of America, as a nonprofit organization, to offer the present edition at a fraction of the previous cost and thus make it more accessible to students. We are also grateful to the Society for providing us with the opportunity to correct errors that were discovered in the first edition. Except for these minor corrections, the present edition is identical to the first.

Austin, TX, USA

May, 2008

Mark F. Hamilton

David T. Blackstock

Preface to the First Edition

This book is an introductory text on the theory and applications of nonlinear acoustics. For nearly 30 years the editors have, between them, taught a graduate course on this topic at the University of Texas at Austin. During this period the field has grown enormously. Many of the advances in theory have been inspired by measurements and applications. Because nonlinear acoustics now encompasses so many diverse areas, we decided that the best approach for a new book would be for individual chapters to be written by experts on their respective subjects. The book is meant to be a useful resource and reference for scientists and engineers and at the same time serve as a text for a graduate level course on nonlinear acoustics. Each chapter is written at a level and in a style that is oriented toward classroom instruction. Moreover, consistent notation, insofar as practicable, and extensive cross-referencing between chapters have been used. Although we had graduate students in engineering and physics in mind, the material is accessible to anyone who is well grounded in the concepts of linear physical acoustics.

As chronicled in Chap. 1, the seeds of theoretical nonlinear acoustics were planted in the eighteenth and nineteenth centuries by the mathematicians and physicists who laid the foundations for fluid mechanics and wave motion. The principal contributions during this era are summarized in a benchmark article published by Lord Rayleigh at the beginning of the twentieth century.[1] The current era of nonlinear acoustics may be traced to an innovative application referred to as the parametric array (Chap. 8), the theory for which was presented in 1960 by P. J. Westervelt at the 59th meeting of the Acoustical Society of America (ASA).[2] Experimental confirmation of the parametric array was described by Bellin and Beyer in the very next paper.[3] Concurrently in the former Soviet Union, substantial progress in modeling the propagation of finite amplitude sound by using the

---

[1] Rayleigh, Lord (1910). Aerial plane waves of finite amplitude. *Proc. Roy. Soc.* **A 84**, 247–284.

[2] Westervelt, P. J. (1960). Parametric end-fire array. *J. Acoust. Soc. Am.* **32**(A), 934–935.

[3] Bellin, J. L. S., and Beyer, R. T. (1960). Experimental investigation of the parametric end-fire array. *J. Acoust Soc. Am.* **32**(A), 935.

Burgers equation was made by R. V. Khokhlov and coworkers.[4] These developments spawned a surge of related research in the 1960s, in response to which a symposium devoted exclusively to nonlinear acoustics was held in 1968 at the Navy Underwater Sound Laboratory, New London, Connecticut. The New London meeting has since been labeled the 1st International Symposium on Nonlinear Acoustics (ISNA). Fourteen symposia in this series have been held, the last five in Japan (1984), the former USSR (1987), the US (1990), Norway (1993), and China (1996).[5] At the same time in the semi-annual ASA gatherings, papers on nonlinear acoustics have grown from just a handful at each meeting to enough for one or two full sessions per meeting.

The following guidance is suggested for those who wish to use the book as a course text. The first half of the book, Chaps. 1–7, develops the physical concepts, mathematical models, and classical methods of solution that form the principal theoretical framework of nonlinear acoustics. Benchmark experiments are also described. These chapters, or at least portions of them, are appropriate as the core for an introductory course. The material is largely self-contained. In the interest of brevity, some equations are presented without derivation because they are derived in introductory texts on fluid mechanics (e.g., general conservation equations for thermoviscous fluids) and physical acoustics (e.g., various linear relations between acoustical quantities). For a text on fluid mechanics see the one by Landau and Lifshitz.[6] For acoustics see Pierce's book.[7]

The second half of the book, Chaps. 8–15, covers special topics and applications, both theory and experiment. These chapters may be read in greater or lesser depth depending on the time available in a course and the interests of the students and instructor. Although independent of each other, they build upon the basic material presented in the first half of this book. Because of the complexity or breadth of various subjects that are covered, a number of sections in the second half are written as reviews. The chapter topics are indicative of current research areas in nonlinear acoustics.

The editors relied upon contributions and efforts from many individuals for the successful completion of this book. Allan Pierce is gratefully acknowledged for originally inviting the editors to prepare this book as a volume in the now discontinued series *Physical Acoustics* published by Academic Press. The authors are thanked for writing their chapters with student readers in mind, and for their

---

[4] Soluyan, S. I., and Khokhlov, R. V. (1961). Propagation of acoustic waves of finite amplitude in a dissipative medium. *Vestn. Mosk. Univ. (Series III), Fiz. Astron.* **3**, 52–61. See also Rudenko, O. V., and Soluyan, S. I. (1977). *Theoretical Foundations of Nonlinear Acoustics* (Plenum, New York). For earlier work on the Burgers equation, see Sect. 4.5.1 of the present book.

[5] For a listing and description up through the 12th ISNA, see Hamilton, M. F., and Blackstock, D. T., eds. (1990). *Frontiers of Nonlinear Acoustics: 12th ISNA* (Elsevier, London).

[6] Landau, L. D., and Lifshitz, E. M. (1987). *Fluid Mechanics*, 2nd edition (Pergamon Press, New York).

[7] Pierce, A. D. (1989). *Acoustics—An Introduction to Its Physical Principles and Applications* (Acoustical Society of America, New York).

patience with editorial changes requested for consistency and completeness. Yurii Il'inskii and Christopher Morfey provided invaluable advice on technical content. Former and current graduate students of the editors are in many ways responsible for the philosophy of this book through their responses to classroom lectures and questions associated with their research. Peggy Dickens was instrumental in preparing many of the figures. Finally, the tone of the book has certainly been influenced by the basic research in nonlinear acoustics performed by the editors and their students, most of which was supported by contracts and grants provided by the Air Force Office of Scientific Research (DTB), the National Aeronautics and Space Administration (DTB), the Office of Naval Research (DTB and MFH), the National Science Foundation (MFH), and the David and Lucile Packard Foundation (MFH).

Austin, TX, USA   Mark F. Hamilton
May, 1997   David T. Blackstock

Contents

About the Editors

Mark F. Hamilton is the W. R. Woolrich Professor in the Walker Department of Mechanical Engineering, and Research Professor at Applied Research Laboratories, at The University of Texas at Austin. He received his B.S. in Electrical Engineering from Columbia University, his M.S. and Ph.D. in Acoustics from Penn State, and then spent a year in the Department of Mathematics at University of Bergen in Norway before becoming a faculty member at UT Austin in 1985. His career has been devoted to basic research in nonlinear acoustics. He served as President of both the Acoustical Society of America and the International Commission for Acoustics. The ASA awarded him its Helmholtz-Rayleigh Silver Medal in Physical Acoustics and Biomedical Acoustics, and subsequently its Gold Medal.

David T. Blackstock is the late E. P. Schoch Professor Emeritus in the Walker Department of Mechanical Engineering at The University of Texas at Austin, where he was also Research Professor at Applied Research Laboratories. He received his B.S and M.A. in Physics from UT Austin, and his Ph.D. in Applied Physics from Harvard University, after which he worked for General Dynamics, and was a faculty member in Electrical Engineering at University of Rochester, before joining UT Austin in 1970. Throughout his career he made many fundamental contributions to nonlinear acoustics. He served as President of both the Acoustical Society of America and the International Commission for Acoustics. The ASA awarded him its Silver Medal in Physical Acoustics, and subsequently its Gold Medal.

List of Symbols

Only the main symbols used most frequently throughout the book are listed below. Special symbols used only in a single chapter are generally not included, for example, much of the special notation used in Chap. 9 for elastic waves in solids. Occasionally, some of the symbols listed below are assigned alternative definitions (e.g., P is at times used for dimensionless sound pressure rather than total pressure), but in these instances the meaning is made clear in the neighboring text.

a	= material (Lagrangian) coordinate
$A(x)$	= cross-sectional area
A, B, C	= coefficients in expansion of equation of state for liquids
$\mathcal{A}, \mathcal{B}, \mathcal{C}$	= third-order constants for an isotropic solid (Landau notation)
B/A	= parameter of nonlinearity
c	= sound speed
c_0	= small-signal sound speed
c_p, c_v	= specific-heat coefficients at constant pressure, volume
D/Dt	= $\partial/\partial t + \mathbf{u} \cdot \nabla$, material derivative
e	= internal energy per unit mass
f	= frequency
h	= $e + P/\rho$, enthalpy per unit mass
i, j	= $\sqrt{-1}$ (time convention $e^{j\omega t}$ or $e^{-i\omega t}$)
$\mathbf{I}$	= acoustic intensity
I_n	= modified Bessel function
J_n	= Bessel function
J_+, J_-	= Riemann invariants
k	= wave number
$\mathcal{L}$	= $\frac{1}{2}\rho_0 u^2 - p^2/2\rho_0 c_0^2$, Lagrangian density
$p; p_0$	= $P - P_0$, acoustic pressure; amplitude of p
P	= total pressure
P_0	= ambient pressure
Pr	= $\mu c_p/\kappa$, Prandtl number
r	= radial distance in spherical or cylindrical coordinates

R $= c_p - c_v$, gas constant

s = specific entropy (per unit mass)

s_0 = ambient specific entropy

s' $= s - s_0$, acoustic fluctuation in specific entropy

t = time

T = temperature

T_0 = ambient temperature

T' $= T - T_0$, acoustic fluctuation in temperature

$\mathbf{u}$ = particle velocity vector

$u;\, u_0$ = particle velocity component; amplitude of u

$\bar{x}$ = shock formation distance for a plane wave

α = absorption coefficient (nepers per unit distance)

β = coefficient of nonlinearity, $1 + B/2A$ for liquids, $\frac{1}{2}(\gamma + 1)$ for gases

γ $= c_p/c_v$, ratio of specific heats

Γ $= \beta \varepsilon k/\alpha$, Gol'dberg number

δ $= \rho_0^{-1}[\frac{4}{3}\mu + \mu_B + \kappa(c_v^{-1} - c_p^{-1})] = \nu[\frac{4}{3} + \mu_B/\mu + (\gamma - 1)/\text{Pr}]$, sound diffusivity

δ_{ij} = Kronecker delta (1 for $i = j$, 0 for $i \neq j$)

$\delta(x),\, \delta(t)$ = Dirac delta functions

ε $= u_0/c_0 \simeq p_0/\rho_0 c_0^2$, acoustic Mach number

κ = thermal conductivity

λ = wavelength

              $= \int_{\rho_0}^{\rho}(c/\rho)\, d\rho$, thermodynamic quantity

μ = shear viscosity

μ_B = bulk viscosity

ν $= \mu/\rho_0$, kinematic viscosity

$\boldsymbol{\xi}$ = particle displacement vector

ρ = mass density

ρ_0 = ambient density

ρ' $= \rho - \rho_0$, acoustic fluctuation in density

σ = dimensionless distance

τ = retarded time

ϕ = scalar velocity potential

              = Earnshaw phase variable

ω $= 2\pi f$, angular frequency

∇^2 = Laplacian operator

$\nabla_\perp^2$ $= \nabla^2 - \partial^2/\partial z^2$, transverse Laplacian operator

∇ = gradient operator

$\nabla\cdot$ = divergence operator

$\nabla\times$ = curl operator

$\Box^2$ $= \nabla^2 - c_0^{-2}(\partial^2/\partial t^2)$, d'Alembertian operator

Chapter 1
History of Nonlinear Acoustics: 1750s–1930s

David T. Blackstock

Contents

1.1   Introduction

Although a book on nonlinear acoustics need not begin with the history of the subject, the chronology of actual discoveries and advances provides a logical study plan. Moreover, seeing the difficulties, mistakes, and triumphs as they actually developed offers motivation and appreciation that might otherwise be hard to convey.

A very detailed understanding of linear acoustics has developed from experiments and theories dating back to antiquity. We know the properties of small-signal sound waves in great detail: propagation, reflection from and transmission through interfaces, standing-wave fields, refraction, diffraction, absorption and dispersion, and so on. By comparison, our understanding of nonlinear acoustics is exceedingly limited.

Although nonlinear acoustics has expanded enormously in the last 40 years, the field itself is very old. The wave equation for finite-amplitude sound in fluids was developed at about the same time as the wave equation for small signals. For the first 200 years, however, progress was very slow. Two reasons are offered. First, because ordinary linear acoustics does an outstanding job of explaining most

D. T. Blackstock
Department of Mechanical Engineering and Applied Research Laboratories, The University of Texas at Austin, Austin, TX, USA

© The Author(s) 2024
M. F. Hamilton, D. T. Blackstock (eds.), *Nonlinear Acoustics*,
https://doi.org/10.1007/978-3-031-58963-8_1

acoustical phenomena, the need to understand finite-amplitude waves has not, until recently, been great. Second, the nonlinear mathematics necessary to describe finite-amplitude sound has been a very difficult lock to pick.

This chapter is about pioneering lock pickers, from Euler, who started it all, to Fay and Fubini at the brink of the modern era. Almost all their efforts were devoted to unraveling the mysteries of propagation of plane waves of finite amplitude in gases. Later work on propagation, described in several subsequent chapters of this book, expands our knowledge to cover more complicated geometries, media other than gases, and modern applications. The modern era has its own lock pickers. Because of the efforts of Burgers, Hopf, Cole, and Lighthill, the once very difficult problem of finite-amplitude propagation in dissipative fluids is no longer a mystery. And thanks to Westervelt, Khokhlov, and Zabolotskaya, the diffraction door is now open.

1.2   1759–1860: The Classical Era

The theoretical description of finite-amplitude sound has its origins in Euler's formulation of the equations that bear his name (Euler, 1755),

$$\text{Continuity:} \quad \frac{D\rho}{Dt} + \rho \nabla \cdot \mathbf{u} = 0, \tag{1.1}$$

$$\text{Momentum:} \quad \rho \frac{D\mathbf{u}}{Dt} + \nabla P = \mathbf{F}, \tag{1.2}$$

where ρ is density, $\mathbf{u}$ is particle velocity, P is (total) pressure, $\mathbf{F}$ is an external body force per unit volume, and t is time. The material derivative D/Dt represents the combination $\partial/\partial t + \mathbf{u} \cdot \nabla$. (If a wave equation is to be obtained, Euler's equations must be supplemented by a pressure–density relation. More about this below.) The era ended approximately a century later with the beautiful results of Earnshaw (1860) and Riemann (1860). The scope of the work was limited to plane waves in lossless gases. Moreover, with the notable exception of Riemann, investigators confined their attention to progressive waves. Nevertheless, the era is very significant: It produced the basic propagation laws for continuous disturbances of finite amplitude.

1.2.1   Preliminaries: Gas Laws and the Speed of Sound

Certain gas laws, along with the sound speeds they imply, figure prominently in our story. Boyle's law,

$$P/P_0 = \rho/\rho_0, \tag{1.3}$$

where P_0 and ρ_0 are the ambient pressure and ambient density, respectively, was well known in the time of Newton (seventeenth century) and Euler (eighteenth century). Its use in connection with sound propagation leads to the following formula for sound speed:

$$b = \sqrt{P_0/\rho_0}. \tag{1.4}$$

We now know this to be the sound speed for an isothermal gas; the symbol b denotes its association with Boyle's law. The conventional symbol c, reserved for sound speed in an isentropic fluid, is defined as

$$c^2 = (\partial P/\partial \rho)_s, \tag{1.5}$$

where s is entropy per unit mass. Thus for a so-called adiabatic gas

$$P/P_0 = (\rho/\rho_0)^\gamma, \tag{1.6}$$

where γ is the ratio of specific heats, the sound speed is

$$c = \sqrt{\gamma P/\rho}, \qquad c_0 = \sqrt{\gamma P_0/\rho_0}. \tag{1.7}$$

The first form is for waves of any amplitude; the second form is restricted to small signals. Notice that no distinction is necessary for an isothermal gas; Eq. (1.4) gives the speed regardless of the strength of the wave.

The first calculation of the speed of aerial waves was done by Newton (1686), who obtained Eq. (1.4).[1] Newton's prediction was about 16% lower than measured values, and the discrepancy was one of the mysteries of physics for well over a century. Indeed, one motivation for work on finite-amplitude waves was the suspicion that the discrepancy might be due to the small-signal assumption. Could the measured sound speed perhaps be explained as a finite-amplitude effect?

The correct explanation for the measured value of the speed of sound was finally given more than a century later. Laplace (1816) argued that heat does not flow when sound propagates. Instead, the local temperature changes in accordance with the compressions and expansions of the air. This makes the "elasticity of the air" greater than Newton assumed. As a result, Newton's formula should be corrected by multiplying it by $\sqrt{\gamma}$. Laplace thus obtained the second form of Eq. (1.7). Although Laplace's formula is now universally accepted (and interpreted as the *adiabatic* speed of sound), old mysteries die hard. Some investigators in the first half of the nineteenth century dismissed Laplace's argument and continued to look to finite-amplitude waves for the explanation.

---

[1] Newton is often erroneously criticized for using Boyle's law and thus assuming that sound propagation is isothermal rather than adiabatic. In the first place, Newton did not invoke Boyle's law; he was apparently unaware of the connection between his prediction and Boyle's law (Truesdell, 1956). More important, the concepts of isothermal and adiabatic thermodynamic pressure were not even formulated until long after the time of Newton (Hunt, 1978).

One final preliminary item is introduced here. For an arbitrary (isentropic) pressure–density relation, it is convenient to represent the thermodynamic state of the fluid by the quantity λ, defined by[2]

$$\lambda = \int_{\rho_0}^{\rho} (c/\rho)\, d\rho. \tag{1.8}$$

Two special cases are as follows:

$$\text{Isothermal gas:} \quad \lambda = b\ln(\rho/\rho_0), \tag{1.9}$$

$$\text{Adiabatic gas:} \quad \lambda = \frac{2}{\gamma - 1}(c - c_0). \tag{1.10}$$

If λ is introduced in the plane-wave form of Euler's equations and body forces are neglected, the following symmetric pair of equations results:

$$\text{Continuity:} \quad \frac{\partial \lambda}{\partial t} + u\frac{\partial \lambda}{\partial x} + c\frac{\partial u}{\partial x} = 0, \tag{1.11}$$

$$\text{Momentum:} \quad \frac{\partial u}{\partial t} + u\frac{\partial u}{\partial x} + c\frac{\partial \lambda}{\partial x} = 0. \tag{1.12}$$

A form of these equations was introduced by Riemann (1860), and a similar approach, although in much different notation, can be found in the article by Earnshaw (1860).

1.2.2 From Euler to Poisson

As advertised, our story begins with Euler. Besides establishing the foundation of classical hydrodynamics and contributing much to linear acoustics, Euler (1759) also flirted briefly with an equation for finite-amplitude waves. Using material coordinates a, t (today commonly called "Lagrangian" coordinates), he derived from first principles the following equation for aerial plane waves:

$$b^2\frac{\partial^2 \xi}{\partial a^2} - \left(1 + \frac{\partial \xi}{\partial a}\right)^2 \frac{\partial^2 \xi}{\partial t^2} = 0, \tag{1.13}$$

where ξ is particle displacement and a is particle rest position. The coefficient b^2 identifies the air as having the properties of Boyle's law.[3] In his 1759 derivation

---

[2] In this chapter, λ does not represent wavelength.

[3] Euler's notation for the coefficient of $\partial^2 \xi / \partial a^2$ is $2Gh$, where G is the distance a mass falls in 1 s (his $2G$ is thus the acceleration due to gravity g), $h = P_0/\rho_0 g$ is the height of a uniform

Euler made a slip and had $1 + (\partial \xi / \partial a)^2$ in place of $(1 + \partial \xi / \partial a)^2$ (Truesdell, 1954). Euler published the correct equation in 1765 (Euler, 1765), in somewhat more general form (allowing for a nonuniform ambient density ρ_0), and a year later gave the corresponding version in spatial coordinates (Euler, 1766).

Although out of chronological order, the version of Eq. (1.13) that holds for an adiabatic gas is given here for reference:

$$c_0^2 \frac{\partial^2 \xi}{\partial a^2} - \left(1 + \frac{\partial \xi}{\partial a}\right)^{\gamma+1} \frac{\partial^2 \xi}{\partial t^2} = 0. \tag{1.14}$$

This equation was obtained by Earnshaw (1860), but Truesdell (1956) points out that the form of the equation was also known to Brandes (1805).

Euler did nothing further with the exact wave equation except to propose tentatively that if the nonlinear terms were taken into account, the predicted propagation speed would be higher than the Newtonian value b, that is, closer to the experimentally measured speed. This kind of argument was to be advanced time and again for almost a hundred years as a means of solving the sound speed dilemma.

Lagrange (1760–1761), after deriving anew another incorrect equation [$(1 + \partial \xi / \partial a)$ in place of $(1 + \partial \xi / \partial a)^2$ in Eq. (1.13)], nevertheless obtained a very interesting solution valid to second order, $\xi = \psi(a - bt) + \frac{1}{4}bt[\psi'(a - bt)]^2$, where ψ is an arbitrary function and ψ' is its derivative. In a clever step, Lagrange saw that this is a two-term expansion of

$$\xi = \psi\left[a - (b + \tfrac{1}{4}u)t\right], \tag{1.15}$$

where u is the particle velocity. Lagrange correctly interpreted the factor $b + \frac{1}{4}u$ as the speed of propagation. Having grasped this important fruit, he immediately threw it away because "the new formula would destroy the uniformity of the speed of sound and would make it depend in some way on the nature of the original disturbances, that which is contrary to all experiments." He concluded that "the hypothesis of infinitely small disturbances is the only one acceptable in the theory of the propagation of sound."[4]

---

atmosphere, and P_0 and ρ_0 are sea-level values of the pressure and density, respectively. Equation (1.13) may be derived as follows: In material coordinates, the equation of continuity is $\rho_0 = \rho(1 + \partial \xi / \partial a)$ and the momentum equation is $\rho_0 \partial^2 \xi / \partial t^2 + \partial P / \partial a = 0$, or, for a Boyle's law gas, $\rho_0 \partial^2 \xi / \partial t^2 + b^2 \partial \rho / \partial a = 0$. Elimination of ρ between the continuity and momentum equations leads to Eq. (1.13).

[4] If Lagrange had started with the correct equation, Eq. (1.13), and if he had put his solution in terms of u instead of ξ, he would have obtained $u = g[a - (b + u)t]$, which closely resembles Poisson's correct result, Eq. (1.17). The two expressions are not quite the same, however, because Lagrange used material coordinates while Poisson used spatial coordinates. The *exact* result for material coordinates is $u = g[a - b(\rho/\rho_0)t]$ (Blackstock, 1962). On the other hand, within the accuracy of Lagrange's method, the exact result reduces to $u = g[a - (b + u)t]$.

Poisson's (1808) very important contribution was to find an exact solution for progressive waves of finite amplitude. He assumed Boyle's law and used spatial coordinates x, t (today commonly called "Eulerian" coordinates). Although Poisson used velocity potential ϕ, defined by $u = \partial\phi/\partial x$, his work is presented here in terms of particle velocity. The wave equation for this case is (see Euler, 1766)[5]

$$b^2\frac{\partial^2 u}{\partial x^2} - \frac{\partial^2 u}{\partial t^2} = \frac{\partial}{\partial x}\left(\frac{\partial u^2}{\partial t} + u^2\frac{\partial u}{\partial x}\right). \tag{1.16}$$

Poisson's *exact* solution, the slightly incorrect form of which Lagrange had thrown away, is

$$u = g[x - (u + b)t], \tag{1.17}$$

where g is an arbitrary function. This solution is valid for outgoing waves (waves traveling in the direction of increasing x). Poisson regarded it necessary to couple this equation with the auxiliary relation

$$\frac{\partial u}{\partial t} + b\frac{\partial u}{\partial x} + u\frac{\partial u}{\partial x} = 0 \tag{1.18}$$

in order to complete the solution. Today we look upon Eq. (1.18) as a reduced wave equation valid for outgoing disturbances. Poisson also found the solution for incoming waves (waves traveling in the direction of decreasing x), $u = G[x - (u - b)t]$.

Ironically, Poisson failed to comprehend the far-reaching implication of his solutions. He focused his attention on a pulse that begins and ends with zero particle velocity. Since propagation would apparently not change the length of the pulse, he concluded that all "sound, loud or faint, is transmitted with the same speed."

Actually, Poisson's solutions should be interpreted as providing a special form of the following law of propagation of plane progressive waves, which we shall call the *first law*:

$$dx/dt|_u = u \pm b, \tag{1.19}$$

where the $+$ sign is for outgoing waves and the $-$ sign is for incoming waves.[6] Thus, while it is true that the beginning and end of Poisson's pulse propagate with the small-signal speed b, within the pulse the propagation speed varies.[7] This fact

---

[5] To obtain Eq. (1.16), start with Eqs. (1.11) and (1.12), note that $c = b$ for a Boyle's law gas, and then eliminate λ between the two equations.

[6] Hereinafter, when multiple signs are used, the upper sign pertains to outgoing waves.

[7] Thus "speed of sound" is an appropriate name only for small signals. For finite-amplitude waves, sound speed is not the same as propagation speed.

and its far-reaching consequence were not to be recognized until more than 40 years later (Stokes, 1848).

For future reference, the first law is restated here in a more general form appropriate for arbitrary lossless fluids (Earnshaw, 1860):

$$dx/dt|_u = u \pm c, \tag{1.20}$$

where c is defined by Eq. (1.5).

1.2.3 Lossless Theory at Its Zenith

Although nineteenth century acoustics generally belongs to European scientists, in midcentury an American named Eli W. Blake (1848) published a paper that contains an important law concerning finite-amplitude sound. Yet another who was still not convinced by Laplace's calculation of the speed of sound, Blake reasoned that the propagation speed of a pulse depends on the strength of the pulse. As the pulse passes over a group of particles, it compresses them more or less in accordance with its amplitude. The more compressed the particles, the more particles are passed over by the pulse in a given time. Hence, the propagation speed depends on the density ρ. In our notation, Blake's formula is $da/dt|_u = b(\rho/\rho_0)$. The quantity $da/dt|_u$ is the propagation speed in material coordinates, i.e., the number of particles traversed per unit time. Indeed, it is the correct material-coordinate version of the first law for an isothermal gas, Eq. (1.19) (outgoing waves). To Blake, the significance of his result was the opportunity to bring the theoretical value of the speed of sound into agreement with the measured value. All one needs to do is assume a large enough value of ρ. Conversely, disturbances associated with smaller values of ρ are not audible to the ear. By modern calculation this implies a threshold of hearing in the neighborhood of 180 dB (re $20\,\mu$Pa)! Blake's work went unappreciated and almost unnoticed (Helmholtz, 1852).

The lengthy period of dormancy following Poisson's definitive, but still uninterpreted, solution was ended with a burst of activity in 1848. In a squabble with Airy over the existence of plane waves of sound, Challis (1848) invoked Poisson's exact solution for a plane wave of initially sinusoidal shape $u = u_0 \sin k[x - (b + u)t]$, where $k = \omega/b$ is wave number, ω is angular frequency, and u_0 is amplitude. He showed that at time $t = \pi/2ku_0$ the wave peak is predicted to be at the same point in space as a zero. In Challis's words, "the points of no velocity are also points of maximum velocity. This is manifest absurdity." He concluded, "Plane waves are thus shown to be physically impossible."

This set the stage for an extraordinary contribution by Stokes (1848), who immediately picked up the "very remarkable difficulty" Challis had pointed out. He gave for the first time a clear description of the waveform distortion implied by Poisson's solution. Points on the waveform for which u is positive travel faster than points for which u is negative. Stokes's sketches, reproduced in Fig. 1.1,

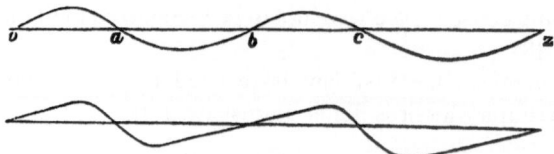

Fig. 1.1 The first sketches showing waveform distortion caused by the dependence of propagation speed on particle velocity (Stokes, 1848). The upper trace is the original waveform; the lower after some time has elapsed.

demonstrate the distortion. They are the first illustrations of "waveform steepening" now commonly found in textbooks.

Recognition of the distortion implied by Poisson's solution opened up a whole new vista. First, Stokes calculated the minimum time $\bar{t}$ required for a continuous wave to develop a vertical slope.[8] Thereafter, the wave motion would have to be qualitatively different. In a masterpiece of understatement, he muses

> Of course, after the instant at which the [slope] becomes infinite, some motion or other will go on, and we might wish to know ... the nature of that motion.

For the first time the concept of a shock wave appears, called by Stokes a "surface of discontinuity." He derived two conservation laws, conservation of mass and conservation of momentum, that must hold across the discontinuity. Later these would come to be known as two of the three Rankine–Hugoniot shock relations [see Eqs. (1.34) and (1.35)], but Stokes received no credit for being the first to derive them. His analysis suffers from the lack of the necessary tools of thermodynamics, which at the time were not yet well developed. In particular, he did not realize that shock propagation is accompanied by energy dissipation and that expansion shocks are impossible (Rayleigh, 1910).[9] Even so, Stokes's remarks are prescient. He concluded that shock formation destroys the progressive wave nature of the wave motion: "Apparently, something like reflexion must take place." He also saw that viscosity would limit the formation of shocks and keep true discontinuities from forming or "render the motion continuous again if it were for an instant discontinuous."

Stokes's paper touched off a torrent of controversy; a total of twelve papers by Challis, Stokes, and Airy followed during the next twelve months. Probably the most significant of these is one by Airy (1849), who argued by analogy with

---

[8] For the periodic wave considered by Challis, Stokes found that $\bar{t} = 1/ku_0$, which is the time needed for a point just behind a zero to catch up with the zero. The time $t = \pi/2ku_0$ picked by Challis is that needed for the peak behind to catch up with the zero.

[9] The error apparently weighed heavily on Stokes. In the version of his 1848 paper contained in his collected works (Stokes, 1883), Stokes omitted the last half of the article and replaced it with a paragraph explaining his error. From the modern vantage point, however, Stokes had nothing to be ashamed of. Not only was his 1848 paper a turning point in the development of nonlinear acoustics, it contains observations that were as much as a century ahead of their time.

water waves that something like a tidal bore must be formed as a result of the cumulative distortion suffered by a sound wave. He scolded Challis, remarking that "the expression 'a plane wave of air is impossible' stands on precisely the same footing as the expression 'a tide in the Severn is impossible.'" Using successive approximations, Airy also found that when the wave motion is due to a sinusoidally vibrating source, a second-harmonic component develops that grows with distance. Indeed, his expression, converted to pressure, shows the now familiar result that the relative pressure amplitude of the second harmonic increases linearly with distance, source frequency, and source amplitude.

Perhaps the most important part of Airy's paper, however, concerns a new partial solution of the exact differential equations. Airy credits De Morgan with first discovering this relation. For sound waves in an isothermal gas, the relation is

$$u = b \ln(\rho/\rho_0) \tag{1.21}$$

for outgoing waves. The more general form, valid for arbitrary (isentropic) fluids, was obtained by Earnshaw (1860) and independently by Riemann (1860):

$$\lambda = \pm u. \tag{1.22}$$

We shall call Eq. (1.22) the *second law* for plane progressive waves. For an isothermal gas [λ given by Eq. (1.9)], Eq. (1.22) reduces to Airy's result, Eq. (1.21). For an adiabatic gas [λ given by Eq. (1.10)], one obtains

$$c = c_0 \pm \tfrac{1}{2}(\gamma - 1)u. \tag{1.23}$$

Perhaps the simplest way to describe this law is to say that it is an antecedent form of the characteristic impedance relation. In the case of an adiabatic gas, for example, Eqs. (1.6), (1.7), and (1.22) may be combined to yield

$$P/P_0 = \left[1 + \tfrac{1}{2}(\gamma - 1)u/c_0\right]^{2\gamma/(\gamma-1)}. \tag{1.24}$$

This is the exact pressure–particle velocity relation. It shows that in general the characteristic impedance, defined as the ratio p/u for an outgoing wave, where $p = P - P_0$ is the acoustic pressure, varies from point to point along the wave. Only when the amplitude is low enough for a first-order expansion of Eq. (1.24) to be accurate does the characteristic impedance turn out to be the constant $\rho_0 c_0$.

Earnshaw's work (1860) represents the high-water mark of the era on the subject of progressive waves. It is a fitting valedictory. After rederiving the two laws by a new method, Earnshaw applied them to the now-classic problem of the wave motion generated by arbitrary movement of a piston in a lossless tube. His treatment was exhaustive. He began by considering waves in a gas obeying Boyle's law, which was the equation of state most often used by previous investigators, moved on to the case of an adiabatic gas, and finally generalized his analysis to cover an arbitrary pressure–density relation $P = P(\rho)$. We give here his solution of the piston problem for an adiabatic gas. If the piston, located initially at $x = 0$, has displacement $X(t)$ and velocity $U(t) = dX/dt$, the solution is

$$u = U(\phi), \qquad t > \pm x/c_0, \tag{1.25}$$

$$\phi = t - \frac{x - X(\phi)}{\beta U(\phi) \pm c_0}, \tag{1.26}$$

where $\beta = \frac{1}{2}(\gamma + 1)$ is today called the coefficient of nonlinearity [see Eqs. (3.16)–(3.18)]. It is assumed that the piston starts from rest and that the gas is initially quiet. The parameter ϕ in Earnshaw's solution represents the time a given point on a waveform—e.g., peak, trough, or zero crossing—left the piston. Once the particle velocity is found, Eq. (1.24) is used to obtain the pressure.

From Eq. (1.26), or from Eqs. (1.20) and (1.23), it can be seen that for an adiabatic gas, the propagation speed is

$$dx/dt|_u = \beta u \pm c_0. \tag{1.27}$$

For this case, in place of Eq. (1.18), the reduced wave equation is

$$\frac{\partial u}{\partial t} \pm c_0 \frac{\partial u}{\partial x} + \beta u \frac{\partial u}{\partial x} = 0, \tag{1.28}$$

the "Poisson" solution of which is, for outgoing waves,

$$u = g[x - (\beta u + c_0)t]. \tag{1.29}$$

Although the Poisson and Earnshaw solutions appear very different, they are in fact equivalent. Observe that Eq. (1.29) applies to an *initial-value* problem, in which $g(x)$ represents the particle velocity everywhere in space at time $t = 0$. Earnshaw, on the other hand, found the solution of a *source* problem, in which the velocity of the piston is specified for all time at the piston face. To adapt the Poisson solution to the piston problem, consider its equivalent form

$$u = f\left(t - \frac{x}{\beta u + c_0}\right). \tag{1.30}$$

Now let ϕ stand for the argument of f, i.e., $u = f(\phi)$, where $\phi = t - x/(\beta u + c_0)$. In this form the Poisson solution closely resembles the Earnshaw solution. The only difference is that f represents the source velocity at the origin $x = 0$, whereas in the Earnshaw solution $U(t)$ represents the velocity at the moving point $X(t)$. The two solutions may be made to coincide by specifying identical source conditions, as shown in Sect. 4.2.1.

Why physically does the propagation speed deviate from c_0? Earnshaw was the first to attempt an explanation. For a gas obeying Boyle's law, the excess speed

$(dx/dt|_u - b)$ is simply u. Earnshaw saw that convection is the explanation. In his words, "we may consider the velocity u to be a wind velocity in that part of the medium ... with the velocity $[b]$ imposed upon the wind." For an adiabatic gas, however, he noted that the simple explanation is not sufficient. In this case, the excess propagation speed is $\beta u = \frac{1}{2}(\gamma + 1)u$ [see Eqs. (1.20) and (1.23), outgoing waves], which result he felt "renders the property of [wind superposition] inapplicable here." Actually the excess speed can be expressed as the sum of two terms, $\beta u = u + \frac{1}{2}(\gamma - 1)u$, to show that the wind is still at work; the second term is due to nonlinearity of the pressure–density relation.[10] In this connection, the often-used phrase "nonlinearity of the medium" is appropriate.

Among Earnshaw's other contributions in this long work are the following:

1. A discussion of the formation of shocks (he called them bores), including a calculation of the time and place at which a shock first occurs. But Earnshaw thought rarefaction shocks possible.
2. A rather fuzzy argument that the shock speed would exceed c_0.
3. A calculation of the speed at which a piston must be withdrawn to create a vacuum, namely $2c_0/(\gamma - 1)$, today called the *escape speed*.
4. A deduction that waves of permanent shape are possible only if the pressure–density relation has the form $P = A - B/\rho$. This was to become known as "Earnshaw's law" (Rayleigh, 1910). That no distortion occurs in such a fluid may be shown by using Eqs. (1.5), (1.8), and (1.22) (outgoing waves) to obtain $c = c_0 - u$, in which case the first law reduces to $dx/dt|_u = c_0$.

A rather amusing sequence of events is connected with item 2. Earnshaw communicated the essential features of his treatise at the 28th Meeting of the British Association for the Advancement of Science in 1858. We now quote from an account apparently written by someone in the audience (Earnshaw, 1858):

> The velocity with which a sound is transmitted through the atmosphere depends on the degree of violence with which it was produced ... so that sound of every pitch will travel at the same rate, if their genesis do not differ much in violence; but a violent sound, as the report of the firearms, will travel sensibly faster than a gentle sound, such as the human voice. This last property the author stated to have caused him much trouble, in consequence of its being directly opposed to the testimony of almost every experimenter. For many affirmed, as the direct result of their observations, and others assumed, that all sounds travel at the same rate. Fortunately, it transpired at the Meeting, that in Capt. Parry's Expedition to the North [1819–20], whilst making experiments on sound, during which it was necessary to fire a cannon at the word of command given by an officer, it was found that the persons stationed at the distance of three miles to mark the arrival of the report of the gun, always heard the report of the gun before they heard the command to fire, thus proving that the sound of the gun's report had outstripped the sound of the officer's voice; and confirming in a remarkable manner the result of the author's mathematical investigation, that the velocity of sound depends in some degree on its intensity.

---

[10] To see why the second cause of distortion is nonlinearity of the P–ρ relation, observe that if the P–ρ relation is linear, as for Boyle's law or for an adiabatic gas for which $\gamma = 1$, the second term disappears.

So Earnshaw had the courage of his convictions, whereas Lagrange had not.[11]

Up to this point, all the results presented are limited to progressive waves, or, as they are sometimes called, *simple waves*. Riemann (1860) found a way to deal with *compound* wave fields, that is, fields in which waves traveling in both directions are present, as for example in reflection. He obtained an important exact generalization of the first propagation law. In compound wave fields, propagation speeds $u \pm c$ are still appropriate, but in a more general sense. Form the following two linear combinations of u and λ:

$$J_+ = \tfrac{1}{2}(\lambda + u), \qquad J_- = \tfrac{1}{2}(\lambda - u), \tag{1.31}$$

which are now called Riemann invariants (Riemann used the symbols r and s instead of J_+ and J_-, respectively). The generalization of the first law, Eq. (1.20), is

$$dx/dt|_{J_+} = u + c, \qquad dx/dt|_{J_-} = u - c, \tag{1.32}$$

which means that the Riemann invariants J_+ and J_-, not u or λ by themselves, are propagated with speeds $u + c$ and $u - c$, respectively. Of course, progressive wave motion is included in Eq. (1.32) as a special case.

For compound wave fields, no general counterpart of the second law exists; instead, u and λ are related in a way that depends on the wave motion. Superposition does not work; the outgoing and incoming waves interact nonlinearly with each other. Nor is any easy solution analogous to that of Poisson or Earnshaw available. In the abstract, solutions are possible. The method, given by Riemann, is to invert the differential equations so that x and t become the dependent variables and J_+ and J_- the independent variables. A single, second-order linear(!) partial differential equation is obtained, but in this system the source or initial conditions are so tangled up that the solutions are extremely complicated. Few analytical examples have been worked out.

The great unanswered question at the end of the classical era was, "What happens after shocks form?" See Stokes's "masterpiece of understatement," quoted previously. A theory had been developed that was quite successful. But its very success also proved its undoing. Inherent in the lossless equations of motion, which form the sole basis of the theory, is the assumption of continuous functions. Yet, except

---

[11] The account above is somewhat misleading. Parry's journal (Anon., 1825) indeed contains a description of experiments with a cannon to measure the speed of sound. However, only on *one* particular evening (9 February 1822, propagation distance 5645 ft, not 3 mi) did listeners several times hear the command "Fire" after the gunshot. The delay was about 3/8 s. Making use of modern knowledge of shock propagation speed and the interaction of weak signals with shocks, the present author analyzed the experiments and concluded that it was very unlikely that the reverse order of reception of the two sounds was due to finite-amplitude effects (Blackstock, 1983). Even though the 1858 interpretation of the 1822 experiment was probably spurious, however, the support it gave to Earnshaw was significant. No other experimental evidence was available at the time to corroborate the dependence of propagation speed on wave strength.

in the case of a wave of pure expansion, the prediction is that discontinuities always form. Furthermore, although it might be possible to modify the theory to account for the discontinuities, an additional complication loomed. When discontinuities form, reflected waves are generated (Stokes, 1848). Therefore, even if the wave field is progressive to begin with, so that the relatively simple theory of Poisson or Earnshaw may be used, once shocks form, it is necessary, in principle, to treat the flow as compound. The challenge after 1860 was to find ways to deal with shocks that develop in the waveform.

At bottom, the trouble lay in the neglect of dissipation (Stokes, 1848). Dissipation prevents the formation of true discontinuities. Put another way, shock propagation is always accompanied by energy loss. From here on, success generally depended on an investigator's ability to account in some way for dissipation.

1.3   1870–1910: Shock Waves

During the half century after Earnshaw and Riemann, the only substantial progress made was on shock waves. The theory of shock waves was given a firm foundation by the establishment of what are now called the Rankine–Hugoniot shock relations (Rankine, 1870; Hugoniot, 1887, 1889). At the same time, the problem of the profile of the steady shock wave in a viscous, heat-conducting gas was found to be soluble (Rankine, 1870; Rayleigh, 1910; Taylor, 1910).

Rankine and Hugoniot arrived at the shock relations independently and from quite different starting points. Rankine wanted to find the heat transfer within the gas necessary for a waveform not to change—that is, for Earnshaw's law to be fulfilled. He ended up with not only the shock relations but also the profile of a steady shock in a heat-conducting, but inviscid, gas. Hugoniot, on the other hand, simply sought the relations necessary for steady discontinuities to exist. Not realizing that dissipation is essential to shock propagation, Hugoniot assumed an inviscid, thermally nonconducting gas at the outset.

We pause here to report a side issue because of its relation to work done during the 1930s. Besides his work on the shock relations, Hugoniot (1889) also considered the problem of lossless propagation of the plane wave generated by a piston vibrating in a tube with velocity $u_0 \sin \omega t$, where u_0 is the velocity amplitude and ω is angular frequency. By using a solution similar to Eq. (1.29) (but in material coordinates), Hugoniot constructed Fig. 1.2, which is the spatial waveform of the field in front of the piston (Hugoniot's v is particle velocity). The figure illustrates the cumulative nature of the distortion, which is least near the piston and greatest at the head of the wave. He also calculated the distance to the point at which a shock first forms, or shock formation distance:

$$\bar{x} = \frac{c_0^2}{\beta u_0 \omega} = \frac{1}{\beta \varepsilon k},$$

(1.33)

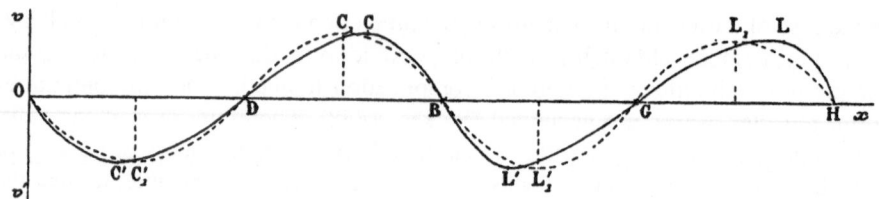

Fig. 1.2 Spatial waveform of particle velocity in front of a sinusoidally vibrating piston (Hugoniot, 1889). The dotted waveform represents linear theory.

which corresponds to Stokes's shock formation time $\bar{t}$ (recall that Stokes's calculation was for an isothermal gas, for which $\beta = 1$). Here $\varepsilon = u_0/c_0$ is the peak particle velocity Mach number and $k = \omega/c_0$ is the wave number. Now back to the main course.

The Rankine–Hugoniot relations are conservation equations that connect the flow field behind a shock (P_b, ρ_b, T_b, u_b, where T is absolute temperature) with that ahead of it (P_a, ρ_a, T_a, u_a). Let the shock be moving with constant velocity U_{sh}. The conservation equations are most simply expressed in a reference frame in which the shock is at rest, since then the flow is steady. In this frame the particle velocity behind the shock is $v_b = u_b - U_{sh}$ and that ahead is $v_a = u_a - U_{sh}$. Conservation of mass and momentum for this case are

$$\text{Mass:} \qquad \rho_a v_a = \rho_b v_b = m, \qquad (1.34)$$

$$\text{Momentum:} \quad P_a + \rho_a v_a^2 = P_b + \rho_b v_b^2, \qquad (1.35)$$

where m is the mass flow (per unit area) through the shock. As already noted, Stokes (1848) was the first to obtain a form of these equations. The trap into which he fell, as did Riemann (1860) after him, was to close the system by adding a lossless pressure–density relation, such as Boyle's law. Kelvin and later Rayleigh (Truesdell, 1966) pointed out to Stokes that the ensuing result violates conservation of energy. If a *lossless* energy equation is used, $\int dP/\rho = (v_a^2 - v_b^2)/2$ (Bernoulli's equation for a compressible fluid), and Boyle's law [Eq. (1.3)] is adopted as the equation of state, the combination of these expressions with Eqs. (1.34) and (1.35) has no solution other than the trivial one $\rho_b = \rho_a$ (Stokes, 1883). Therefore, no shock or waveform of permanent shape is possible if the flow is lossless.

The key, found by both Rankine (1870) (for perfect gases) and Hugoniot (1887, 1889), is to use an energy equation in which losses may be included. For the general case (not limited to perfect gases), the third Rankine–Hugoniot relation is

$$\text{Energy:} \quad \frac{v_a^2}{2} + e_a + \frac{P_a}{\rho_a} = \frac{v_b^2}{2} + e_b + \frac{P_b}{\rho_b}, \qquad (1.36)$$

where e is the internal energy per unit mass. If the fluid is a perfect gas, the following relations hold: $P = R\rho T$ and $e_b - e_a = c_v(T_b - T_a) = (P_b/\rho_b - P_a/\rho_a)/(\gamma - 1)$, where the specific heat at constant volume c_v and the gas constant R are related by $R/c_v = \gamma - 1$. In this case, the energy equation becomes

$$\text{Energy:} \quad v_b^2 - v_a^2 = \frac{2\gamma}{\gamma - 1}\left(\frac{P_a}{\rho_a} - \frac{P_b}{\rho_b}\right). \tag{1.37}$$

Although presented here for steady shocks, the Rankine–Hugoniot relations hold for unsteady shocks as well.

A glimpse of how the Rankine–Hugoniot relations are used in modern nonlinear acoustics is worthwhile at this point. If Eqs. (1.34), (1.35), and (1.37) are combined to eliminate the pressure and density, and if for simplicity we take $u_a = 0$ (the shock propagates into a quiet fluid), the following relation for the shock speed may be obtained: $U_{sh}^2 - \beta u_b U_{sh} - c_0^2 = 0$. The solution, $U_{sh} = [c_0^2 + (\beta u_b/2)^2]^{1/2} + \beta u_b/2$, shows that shocks propagate faster than small-signal sound waves, a result Rankine (1870) found by a different but equivalent equation. An important special case is that of weak shocks ($u_b/c_0 \ll 1$), for which a two-term expansion of the expression for U_{sh} yields $U_{sh} = c_0 + \beta u_b/2$. The more general expression, valid when $u_a \neq 0$, is

$$U_{sh} = c_0 + \beta\frac{(u_a + u_b)}{2}, \tag{1.38}$$

where again terms of order $u_{a,b}^2$ have been dropped. Equation (1.38) shows that weak shocks travel with a speed that is the mean of the ordinary finite-amplitude speeds [see Eq. (1.27)] just ahead of and just behind the shock. This relation is one of the foundations of weak shock theory (Sects. 4.3.2 and 4.4.1).

Now for the profile of a plane steady shock wave. The key words are "plane" and "steady." Although the general conservation equations [see Eqs. (3.1)–(3.3)[12]] are very complicated, they simplify greatly when applied to a steady shock, for which the flow is planar and time-independent. A set of ordinary differential equations is obtained. These equations have been solved, in some cases analytically, for a variety of different cases. The first to tackle the shock profile problem was Rankine (1870), who considered a thermally conducting but inviscid gas. For this case, he obtained an exact expression for the profile. Rayleigh (1910) later pointed out, however, that Rankine's result is limited to shocks for which the pressure ratio P_b/P_a does not exceed $(\gamma + 1)/(3 - \gamma)$ ($= 1.5$ for air). Heat conduction is therefore incapable of providing enough dissipation to keep strong shocks single-valued. Rayleigh (1910) was the first to include viscosity.[13] Of the third-order differential equation he derived

---

[12] If desired, the relation $T Ds/Dt = De/Dt + P D(\rho^{-1})/Dt$ may be used to turn the entropy equation, Eq. (3.3), into an energy equation.

[13] Besides being an excellent review and critique of finite-amplitude sound, Rayleigh's 1910 article also contains many gems. One is Rayleigh's disposal of the erroneous notion, which had been held

for the shock profile in a thermoviscous gas, he remarked ruefully, "I suppose a complete analytical solution ... is not to be expected." However, by numerical means he did demonstrate that a single-valued solution does exist even for the strongest of shocks. For one example, $P_b/P_a = 5.67$ in air, he calculated the shock thickness to be of order $30\,\mu\mathrm{m}$, which he reckoned to be "well below the microscopic limit."

Less than 100 pages later in the 1910 volume of the *Proceedings of the Royal Society* is Taylor's article on the shock profile for a viscous, heat-conducting gas. By limiting consideration to weak shocks, Taylor was able to find an analytical solution of the profile equation and from it calculated the shock thickness. Ever since, the thickness of a weak shock in a thermoviscous gas has been called the "Taylor shock thickness." A modern expression for the profile, for a shock traveling into a quiet gas ($u_a = 0$), is given by

$$u = \frac{u_b}{2}\left(1 - \tanh\frac{\beta u_b \chi}{2\delta}\right). \tag{1.39}$$

Here $\chi = x - U_{\mathrm{sh}}t$ is the moving coordinate in which the shock appears at rest (the center of the shock, $u = \frac{1}{2}u_b$, is at the origin $\chi = 0$), $\delta = (\mu/\rho_0)[4/3 + \mu_B/\mu + (\gamma - 1)/\mathrm{Pr}]$ is the sound diffusivity (representing the thermoviscous dissipation of the fluid), μ is the shear viscosity, μ_B is the bulk viscosity, and Pr is the Prandtl number. If the shock thickness h is taken to be the distance for the particle velocity to rise from 10% to 90% of its final value u_b (Taylor's criterion), we find[14]

$$h = \frac{2\delta \ln 9}{\beta u_b}. \tag{1.40}$$

Thus very weak shocks (low amplitude or strong dissipation) are thick, whereas robust ones (high amplitude or low dissipation) are thin. As Lighthill (1956) pointed out, Eq. (1.40) is one of the clearest demonstrations of the battle between nonlinearity and dissipation. Nonlinearity, represented by the excess propagation speed βu_b, tends to steepen the compression, while dissipation, represented by δ, tends to spread it out. In a nutshell, nonlinearity makes a wave interesting, dissipation keeps it honest. Now more than ever we can appreciate the work of the earlier investigators like Poisson, Stokes, and Earnshaw. They made progress on finite-amplitude waves even though they had only half the deck (lossless theory) to work with.

---

by many, that expansion shocks are possible in air. Rayleigh showed that they are not because they violate the second law of thermodynamics. A simpler argument, made by Kelvin to Rankine (Rankine, 1870), is that expansion shocks are inherently unstable. If an expansion shock were ever to exist, it would immediately "unshock" because the top of the shock would run away from the bottom [see Eq. (1.27)].

[14] The corresponding shock rise time, the quantity of interest when the waveform is measured with a microphone, is $t_{\mathrm{rise}} = h/c_0$, since for weak shocks, $U_{\mathrm{sh}} \simeq c_0$.

1.4   1930s: Precursors of the Modern Era

In the words of Beyer (1984), "a long pause" followed the papers of Rayleigh and Taylor. Shock-wave theory had been launched, but still no one knew how to solve the conservation equations with dissipation included, except for the very special case of a step shock. Lossless theory was still stuck at about the place where Earnshaw and Riemann had left it. The long pause was ended in the 1930s by important contributions that set the stage for the groundswell of interest in nonlinear acoustics that developed after World War II. Reviewed here are the theoretical papers of Fay (1931) and Fubini (1935), the experimental work of Thuras, Jenkins, and O'Neil (1935), and the combined theoretical and experimental work of Langevin reported by Biquard (1936). Each paper is about the traditional acoustical problem of the plane wave generated by a sinusoidally vibrating source. Although Airy (1849) and later Hugoniot (1889) had made a start on this problem [see Eq. (1.33)], much more extensive work was done by the 1930s investigators. Their results may be expressed generically as a Fourier series composed of fundamental and higher harmonic components:

$$u(x, \tau) = u_0 \sum_{n=1}^{\infty} B_n(x) \sin n\omega\tau, \qquad (1.41)$$

where u_0 is a reference amplitude (often the amplitude of the fundamental at the source), B_n is the relative amplitude of the harmonic components, ω is the source angular frequency, and $\tau = t - x/c_0$ is the retarded time. The problem was to find an expression for the amplitudes B_n.

Before presenting the results of the three investigations, we note the difference between *cumulative* and *local* nonlinear effects (see Sects. 3.6 and 4.2.3). Cumulative effects are those due to variation of propagation speed over the waveform [see Eq. (1.27)], which causes distortion that accumulates with distance. Other effects leading to deviation from small-signal behavior, such as the difference between spatial and material coordinates, the finite displacement of a vibrating source, and nonlinearity of the pressure–particle velocity (impedance) relation, are termed local because the distortion they produce does not increase with propagation distance. Since it is small compared with cumulative distortion (except close to the source), local distortion may often be neglected (Blackstock, 1962). This fact was not appreciated in the 1930s, and solutions of that time frequently appear more complicated than is necessary in practice. It has therefore become traditional to report only the cumulative part of their solutions. For example, although material coordinates were used in all four investigations described here, we report solutions in spatial coordinates, which are appropriate for most laboratory experiments, because the difference is usually insignificant. As another example, Fubini (1935) criticized Fay (1931) for stating a momentum equation in material coordinates but using spatial coordinates to describe the viscous force. While Fubini was correct, we now recognize that the error committed by Fay was insignificant.

Fay's remarkable solution (1931) was developed without regard for the source other than that it be periodic. He sought the most nearly stable waveform in a viscous gas. Since distortion generally tends to enrich the higher harmonic components at the expense of the lower ones (energy transfer effect), while viscosity damps out the higher components more rapidly than the lower ones, he reasoned that a balance should be reached in which a given component loses as much energy by absorption as it gains from nonlinear distortion. He recognized, however, that since the conditions for stability depend on the amplitude of the wave, which slowly decreases with propagation distance, the wave can never be completely stable, only relatively so. Over the years the Fay solution has been generalized to apply to thermoviscous fluids of arbitrary equation of state, a correction to a numeric has been made, and an interpretation of an unspecified coefficient has been given (see, for example, Blackstock, 1964). The harmonic amplitudes for what is now called the Fay solution are as follows:

$$B_n = \frac{2}{\Gamma \sinh[n(1+\sigma)/\Gamma]}. \tag{1.42}$$

Here $\Gamma = \beta \varepsilon k/\alpha$ is the Gol'dberg number (a measure of the strength of nonlinearity relative to that of dissipation), $\alpha = \delta \omega^2/2c_0^3$ is the small-signal absorption coefficient for the fundamental, and $\sigma = x/\bar{x}$ is the distance relative to the shock formation distance [see Eq. (1.33)].

Of several interesting properties of the Fay solution, two are noted here. First, for strong waves at points not greatly distant [$\Gamma \gg n(1+\sigma)$], the hyperbolic sine function may be replaced by its argument. The resulting harmonic amplitudes $B_n = 2/n(1+\sigma)$ are those for a sawtooth waveform (see Sect. 4.4.3.2). Moreover, the amplitude of the sawtooth decays inversely with distance, not exponentially. Although Fay did not present the limiting expression shown here, he did recognize the near-sawtooth character of his solution, and also the fact that absorption is much increased by the energy transfer effect. Second, although Fay did not mention it, his solution shows that when strongly affected by nonlinearity, the wave forgets its origin. The *dimensional* amplitude of the fundamental component is $u_1 \equiv u_0 B_1 = 2\alpha c_0/\beta k \sinh[(1+\sigma)/\Gamma]$. At great distance ($\sigma \gg 1$), this reduces to $u_1 = (4\alpha c_0/\beta k)e^{-\alpha x}$, a value that is independent of the source amplitude u_0. Saturation, as it is now called, is discussed in Sect. 4.4.3.4.

The solution derived by Fubini for the vibrating piston problem is explicit, that is, it is of the form $u = u(x, t)$. By contrast, the solution given previously by Hugoniot (1889) is implicit, i.e., it is in the Poisson form $u = u(x, t, u)$ (but in material coordinates). Fubini's solution is one of those for which history has filtered out all but the most important part.[15] The harmonic amplitudes are

---

[15] Because Fubini's article was in Italian and published in a journal not widely known, it went unappreciated for many years. Indeed, several investigators in the 1950s unknowingly repeated Fubini's ingenious derivation. The parts of Fubini's solution that are not usually reported are associated with local effects.

$$B_n = \frac{2}{n\sigma} J_n(n\sigma),$$ (1.43)

where J_n is the ordinary Bessel function. Since the solution is for a lossless gas, it is valid only up to the shock formation point, that is, for values of σ up to unity. If the expression for the second-harmonic amplitude is expanded, the leading term is $B_2 = \sigma/2 = \beta\varepsilon kx/2$, which agrees with the successive approximations solution found by Airy (1849). Note that the Fay solution does not reduce to the Fubini solution in the limit of vanishing viscosity. Fubini found this puzzling. We now know that the reason the two solutions do not agree in the limit is that the Fubini solution is valid for the region near the source, $\sigma \leq 1$, whereas the Fay solution is valid in the sawtooth region, $\sigma \geq 3$, approximately (Blackstock, 1966).

The first known experiment on airborne finite-amplitude waves generated by a sinusoidally vibrating piston was done by Thuras, Jenkins, and O'Neil (1935). The measurements were done in a plane-wave tube, terminated at the far end to prevent reflections. The ratio of the second-harmonic amplitude (pressure) to the amplitude of the fundamental B_2/B_1 was measured for various (audio) frequencies, distances, and source amplitudes; in terms of σ, the range was $0.01 < \sigma < 0.4$. Their theoretical model was a second-order solution of Eq. (1.14), but modified to include the small effect of tube wall losses. Although the measurements confirmed the linear growth of B_2/B_1 with distance, frequency, and source amplitude, as predicted by Airy and Fubini, the measured values were consistently $\sim$3 dB lower than the predicted values. The puzzling discrepancy (Blackstock, 1962) was put to rest after World War II when experiments by several other investigators showed good agreement between theory and experiment. Also included in the paper by Thuras, Jenkins, and O'Neil are measurements of (1) sum and difference frequency components generated when the source emits a two-frequency wave, and (2) second-harmonic distortion in an exponential horn.

Finally we come to the work of Langevin, which has long been overlooked. Langevin's theoretical and experimental efforts are reported in Chap. 5 of a long article by Biquard (1936) on ultrasonic absorption. Biquard attributes Chap. 5 to Rayleigh (1945) and to lectures presented by Langevin in a course at the College of France in 1923. Since Rayleigh's work is well known, it is easy to identify the very remarkable results of Langevin, several of which were to be rediscovered in the 1940s, 1950s, and 1960s. For liquids, an expression for the coefficient of nonlinearity β was derived that is equivalent to the one currently used ($\beta = 1 + B/2A$, where A and B are coefficients in a series expansion of the isentropic pressure–density relation; see Sects. 2.2 and 2.3). Measurements were made of the values of β for water, benzene, chloroform, and ether.[16] An expression for the profile of a steady shock wave in a liquid was found. This profile was used to substitute for the discontinuities in an ideal sawtooth waveform to obtain a realistic sawtooth, which appears to be the same as the stable waveform of Fay. Moreover, it was

---

[16] For water, Langevin found $\beta = 3$, a value reasonably close to 3.5, the value accepted today.

observed that amplitude decay, accompanied by thickening of the shocks, would eventually restore the wave to sinusoidal form. Finally, the amplitude decay formula, equivalent to the asymptotic result pointed out above for the Fay solution, was found for the ideal sawtooth.

Langevin's work is a good note on which to end this historical review. It built on all the work of previous investigators. At the same time, it foreshadowed some of the most significant topics on which research would be done after World War II.

References

Airy, G. B. (1849). On a difficulty in the problem of sound. *Phil. Mag.* (Series 3) **34**, 401–405, or Beyer (1984), 37–41.

Anon. (1825). Abstract of experiments to determine the velocity of sound at low temperature. *Appendix to Captain Parry's Journal of a Second Voyage for the Discovery of a North-West Passage from the Atlantic to the Pacific; Performed in the Years 1821-22-23, in His Majesty's Ships Fury and Hecla* (John Murray, London), 237–239.

Beyer, R. T. (1984). *Nonlinear Acoustics in Fluids* (Van Nostrand Reinhold, New York), p. 63.

Biquard, P. (1936). Sur l'absorption des ondes ultra-sonores par les liquides. *Ann. Physique* (Ser. 11) **6**, 195–304.

Blackstock, D. T. (1962). Propagation of plane sound waves of finite amplitude in nondissipative fluids. *J. Acoust. Soc. Am.* **34**, 9–30.

Blackstock, D. T. (1964). Thermoviscous attenuation of plane, periodic, finite-amplitude sound waves. *J. Acoust. Soc. Am.* **36**, 534–542.

Blackstock, D. T. (1966). Connection between the Fay and Fubini solutions for plane sound waves of finite amplitude. *J. Acoust. Soc. Am.* **39**, 1019–1026.

Blackstock, D. T. (1983). Propagation of a weak shock followed by a tail of arbitrary waveform. *Proceedings, 11th Int. Cong. Acoustics, Paris, France,* **I**, 305–308.

Blake, E. W. (1848). A determination of the general law according to which pulses differing in intensity are propagated in elastic media, with remarks on the received theory of the velocity of sound. *Silliman's Amer. J.* (now *Amer. J. Sci.*) **5**, 372–377.

Brandes, H. W. (1805). *Die Gesetze des Gleichgewichts und der Bewegung flüssiger Körper. Dargestellt von Leonhard Euler,* German translation of Euler's *Treatise on Fluid Mechanics* (Leipzig).

Challis, J. (1848). On the velocity of sound, in reply to the remarks of the Astronomer Royal. *Phil. Mag.* (Series 3) **32**, 494–499.

Earnshaw, S. (1858). On the mathematical theory of sound. *Brit. Assn. Adv. Sci.*, Report of the 28th Meeting, Notices and Abstracts Sec., 34–35.

Earnshaw, S. (1860). On the mathematical theory of sound. *Trans. Roy. Soc.* (*London*) **150**, 133–148.

Euler, L. (1755). Principes généraux du mouvement des fluides. *Mém. acad. sci. Berlin* **11** (1755), 274–315 (1757),[17] or *Euleri Opera Omnia II,* **12**, 54–91. For English translation of, and helpful commentary on, pertinent parts, see Truesdell (1954), pp. LXXXIV–LXXXV.

Euler, L. (1759). De la propagation du son. *Mém. acad. sci. Berlin* **15** (1759), 185–209 (1766), or *Euleri Opera Omnia II, Series III,* **1**, 428–451. See Truesdell (1954), p. CXXI.

---

[17] In the convention followed here, the first date given (in this case 1755) refers to the first disclosure of the material, the second date (in this case 1757) to the printed publication.

Euler, L. (1765). Eclaircissements plus détaillés sur la génération et la propagation du son, et sur la formation de l'écho. *Mém. acad. sci. Berlin* **21** (1765), 335–363 (1767), or *Euleri Opera Omnia II, Series III*, **1**, 540–567. See Truesdell (1956), pp. LIX–LX.

Euler, L. (1766). Sectio quarta de motu aëris in tubis. *Novi comm. acad. sci. Petrop.* **16** (1771), 281–425 (1772) (Euler's *Treatise on Fluid Mechanics*, Fourth Section, Chap. I, Arts. 1–5). See Truesdell (1956), p. LXIII.

Fay, R. D. (1931). Plane sound waves of finite amplitude. *J. Acoust. Soc. Am.* **3**, 222–241.

Fubini, E. (1935). Anomalie nella propagazione di ande acustiche de grande ampiezza. *Alta Frequenza* **4**, 530–581. English translation: Beyer (1984), 118–177.

Helmholtz, H. (1852). 1. Theoretische Akustik. *Fortschritte der Physik im Jahre 1848* **4**, 101–118.

Hugoniot, H. (1887, 1889). Mémoire sur la propagation du mouvement dans les corps et spécialement dans les gaz parfaits. *J. l'école polytech.* (Paris) **57**, 3–97, and *J. l'école polytech.* (Paris) **58**, 1–125. English translation of Sections 90–92, 110–112, 123–127 of 1889 article: Beyer (1984), 77–89.

Hunt, F. V. (1978). *Origins in Acoustics* (Yale University Press, New Haven), 148–161.

Lagrange, J. L. (1760–1761). Sec. 42 in Nouvelles recherches sur la nature et la propagation du son. *Miscellanea Taurinensis* **II**, 11–172, or *Oeuvres de Lagrange* **I** (Gauthier-Villars, Paris, 1867), 151–316.

Laplace, P. S. (1816). On the velocity of sound through air and through water. *Ann. Chim. Phys.* (2) **3**, 338–241. English translation: Lindsay (1972), 180–182.

Lighthill, M. J. (1956). Viscosity effects in sound waves of finite amplitude. In *Surveys in Mechanics*, G. K. Batchelor and R. M. Davies, eds. (Cambridge University Press, Cambridge, England), 250–351.

Lindsay, R. B. (1972). *Acoustics: Historical and Philosophical Development* (Dowden, Hutchinson & Ross, Stroudsburg, Pa.).

Lindsay, R. B. (1974). *Physical Acoustics* (Hutchinson & Ross, Stroudsburg, Pa.).

Newton, I. (1686). *Principia Mathematica*, Sec. VIII. Of motion propagated through fluids. English translation by Andrew Motte (1729), first American edition (1848) (D. Adee, New York), 356–357. Reprint: Lindsay (1972), 75–86. See Pierce, A. D. (1989), *Acoustics* (Acoustical Society of America, New York), 4–5 for other references.

Poisson, S. D. (1808). Mémoire sur la théorie du son. *J. l'école polytech.* (Paris) **7**, 319–392. See Secs. 24–25, pp. 365–370. English translation of Sections 23–25: Beyer (1984), 23–28.

Rankine, W. J. M. (1870). On the thermodynamic theory of waves of finite longitudinal disturbance. *Phil. Trans. Roy. Soc.* **160**, 277–288, or Beyer (1984), 65–76.

Rayleigh, Lord (1910). Aerial plane waves of finite amplitude. *Proc. Roy. Soc.* **A 84**, 247–284, or *Scientific Papers of Lord Rayleigh* **V** (Dover, New York, 1964), or Lindsay (1974), 135–173.

Rayleigh, Lord (1945). *Theory of Sound* (Dover, New York).

Riemann, B. (1860). Ueber die Fortpflanzung ebener Luftwellen von endlicher Schwingungsweite. *Abhandl. Ges. Wiss. Göttingen, Math.-physik.* **8**, 43–65 (1858–1859), reprinted in *The Collected Works of Bernard Riemann* (Dover, New York, 1953), 156–175. English translation of Sections 1–10; Beyer (1984), 42–60.

Stokes, G. G. (1848). On a difficulty in the theory of sound. *Phil. Mag.* (Series 3) **33**, 349–356, or Beyer (1984), 29–36.

Stokes, G. G. (1883). *Mathematical and Physical Papers* **II** (Cambridge University Press, Cambridge, England), 51–55.

Taylor, G. I. (1910). The conditions necessary for discontinuous motion in gases. *Proc. Roy. Soc.* **A 84**, 371–377.

Thuras, A. L., Jenkins, R. T., and O'Neil, H. T. (1935). Extraneous frequencies generated in air carrying intense sound waves. *J. Acoust. Soc. Am.* **6**, 173–180.

Truesdell, C. (1954). *Rational Fluid Mechanics 1687–1765*, Editor's Introduction to Vol. II **12** of *Euler's Works* (Orell Füssli, Zürich).

Truesdell, C. (1956). *I. The First Three Sections of Euler's Treatise of Fluid Mechanics 1766. II. The Theory of Aerial Sound 1687–1788. III. Rational Fluid Mechanics 1765–1788*, Editor's Introduction to Vol. II **13** of *Euler's Works* (Orell Füssli, Zürich).

Truesdell, C. (1966). *Mathematical and Physical Papers*, Editor's Preface to Vol. I of Stokes's collected works (Johnson reprint, New York).

Chapter 2
The Parameter B/A

Robert T. Beyer

Contents

2.1 Introduction

The ratio B/A, which has become a common term in the field of nonlinear acoustics, has its origin in the Taylor series expansion of the variations of the pressure in a medium in terms of variations of the density. The changes are carried out reversibly, adiabatically, and at constant chemical composition. In nonlinear acoustics, the current notation was perhaps first employed by Fox and Wallace (1954).[1] These

---

[1] Earlier, Biquard (1936) reported an equivalent ratio, attributed to lectures given in 1923 by Langevin, involving coefficients in the Taylor series expansion of density in terms of pressure, and alternatively coefficients in the expansion of pressure as a function of particle velocity. Using these coefficients, Langevin obtained an expression for the coefficient of nonlinearity β that is the same as the one used today (see the end of Sect. 1.4). Prior to this period, the emphasis in nonlinear acoustics was on propagation in perfect gases, for which explicit equations of state were known and therefore Taylor series expansions were unnecessary.

R. T. Beyer
Department of Physics, Brown University, Providence, RI, USA

© The Author(s) 2024
M. F. Hamilton, D. T. Blackstock (eds.), *Nonlinear Acoustics*,
https://doi.org/10.1007/978-3-031-58963-8_2

authors also pointed out the fact, not always noted, that the expansion should be carried out under adiabatic conditions. The quantity B/A is proportional to the ratio of coefficients of the quadratic and linear terms in the Taylor series. Consequently, it characterizes the dominant finite-amplitude contribution to the sound speed for an arbitrary fluid. Its effect on the propagation speed of a progressive plane wave is manifest through the coefficient of nonlinearity $\beta = 1 + B/2A$.

2.2 Definitions

Taylor series expansion of the equation of state $P = P(\rho, s)$ along the isentrope $s = s_0$ yields

$$P - P_0 = \left(\frac{\partial P}{\partial \rho}\right)_{s,0} (\rho - \rho_0) + \frac{1}{2!}\left(\frac{\partial^2 P}{\partial \rho^2}\right)_{s,0} (\rho - \rho_0)^2 + \cdots, \tag{2.1}$$

where P and ρ are pressure and density, respectively, P_0 and ρ_0 are their unperturbed or ambient values, and s is specific entropy. The partial derivatives $(\partial P/\partial \rho)_s$, $(\partial^2 P/\partial \rho^2)_s$, etc., in Eq. (2.1) are all evaluated at the unperturbed state (ρ_0, s_0); this is indicated by the subscript 0, and the same notation is used throughout this chapter for ambient values of partial derivatives. Equation (2.1) can be expressed more succinctly in the form

$$p = A\left(\frac{\rho'}{\rho_0}\right) + \frac{B}{2!}\left(\frac{\rho'}{\rho_0}\right)^2 + \frac{C}{3!}\left(\frac{\rho'}{\rho_0}\right)^3 + \cdots, \tag{2.2}$$

where $p = P - P_0$ is the sound pressure, $\rho' = \rho - \rho_0$ is the excess density, and

$$A = \rho_0\left(\frac{\partial P}{\partial \rho}\right)_{s,0} \equiv \rho_0 c_0^2, \tag{2.3}$$

$$B = \rho_0^2\left(\frac{\partial^2 P}{\partial \rho^2}\right)_{s,0}, \tag{2.4}$$

$$C = \rho_0^3\left(\frac{\partial^3 P}{\partial \rho^3}\right)_{s,0}. \tag{2.5}$$

The isentropic small-signal sound speed c_0 is defined via Eq. (2.3). Entropy variations due to heat conduction are taken into account in Eq. (3.3).

Measurement of the ratio B/A according to the definitions in Eqs. (2.3) and (2.4),

$$\frac{B}{A} = \frac{\rho_0}{c_0^2}\left(\frac{\partial^2 P}{\partial \rho^2}\right)_{s,0}, \tag{2.6}$$

requires varying the density adiabatically in order to measure the change in pressure, a difficult task with liquids owing to their low compressibilities. Alternative definitions of B/A are therefore desirable (Beyer, 1960, 1997). The following are expressed in terms of variations in the sound speed c, defined by $c^2 = (\partial P/\partial \rho)_s$. One alternative is obtained by writing $(\partial^2 P/\partial \rho^2)_s = (\partial c^2/\partial \rho)_s = 2c^3(\partial c/\partial P)_s$, which after substitution in Eq. (2.6) yields

$$\frac{B}{A} = 2\rho_0 c_0 \left(\frac{\partial c}{\partial P}\right)_{s,0}. \tag{2.7}$$

The pressure must be varied sufficiently smoothly and rapidly that isentropic conditions are maintained, which may be accomplished with sound waves. Another alternative follows from expanding the derivative in Eq. (2.7) as $(\partial c/\partial P)_s = (\partial c/\partial P)_T + (\partial T/\partial P)_s(\partial c/\partial T)_P$, where T is temperature, writing $(\partial T/\partial P)_s = (\partial \rho^{-1}/\partial s)_P = (\partial \rho^{-1}/\partial T)_P/(\partial s/\partial T)_P$, where $1/\rho$ is specific volume, and introducing standard definitions of thermodynamic quantities to obtain

$$\frac{B}{A} = 2\rho_0 c_0 \left(\frac{\partial c}{\partial P}\right)_{T,0} + \frac{2\alpha_T c_0 T_0}{c_p} \left(\frac{\partial c}{\partial T}\right)_{P,0}$$

$$\equiv (B/A)_1 + (B/A)_2, \tag{2.8}$$

where $\alpha_T = \rho_0(\partial \rho^{-1}/\partial T)_{P,0}$ is the volume coefficient of thermal expansion and c_p is the specific heat at constant pressure. An alternative form of the expression for C/A is also available (Coppens et al., 1965):

$$\frac{C}{A} = \frac{3}{2}\left(\frac{B}{A}\right)^2 + 2\rho_0^2 c_0^3 \left(\frac{\partial^2 c}{\partial P^2}\right)_{s,0}. \tag{2.9}$$

Values of C/A are seldom needed in nonlinear acoustics, and very few have been reported.

It is instructive to compare the above results with the expansion of the equation of state along an isentrope for a perfect gas, for which each coefficient is known explicitly:

$$P/P_0 = (\rho/\rho_0)^\gamma = (1 + \rho'/\rho_0)^\gamma$$

$$= 1 + \gamma\left(\frac{\rho'}{\rho_0}\right) + \frac{\gamma}{2!}(\gamma - 1)\left(\frac{\rho'}{\rho_0}\right)^2$$

$$+ \frac{\gamma}{3!}(\gamma - 1)(\gamma - 2)\left(\frac{\rho'}{\rho_0}\right)^3 + \cdots, \tag{2.10}$$

where γ is the ratio of specific heats. Term-by-term comparison of Eqs. (2.2) and (2.10) yields $c_0^2 = \gamma P_0/\rho_0$,

$$B/A = \gamma - 1, \tag{2.11}$$

$C/A = (\gamma - 1)(\gamma - 2)$, and so on. Whenever the lossless fluid equations are used to describe finite-amplitude sound propagation (as in Sect. 3.3) and higher-order nonlinear terms are discarded throughout (leaving only quadratic terms to account for finite-amplitude phenomena), a simple rule can be used to convert between perfect-gas and arbitrary-fluid results: Everywhere it appears, γ is replaced with $(1 + B/A)$, in accordance with Eq. (2.11). However, caution is advised when following this procedure, because Eq. (2.10) itself cannot be thus transformed as it stands without first eliminating the ambient pressure P_0.

The perfect gas is in fact a special case of a more general class of fluids whose isentropes are described by the Tait–Kirkwood equation (Sullivan, 1981):

$$\frac{P + \Pi}{P_0 + \Pi} = \left(\frac{\rho}{\rho_0}\right)^\nu, \tag{2.12}$$

where (Π, ν) are constants along any particular isentrope.[2] All such fluids have $B/A = \nu - 1$, $C/A = (\nu - 1)(\nu - 2)$, and so on. However, there is no connection in general between the index ν and the specific heat ratio of the fluid.

Insight into expected values of B/A for liquids may be gained from an analysis performed by Nomoto (1966) on the basis of Rao's (1940) rule for liquids that are not highly associated, $Mc^{1/3}/\rho = \text{const}$, where M is molecular weight. Nomoto thus derived the relations $(1/c_0)(\partial c/\partial T)_{P,0} = -3\alpha_T$ and $(1/c_0)(\partial c/\partial P)_{T,0} = 3\kappa_T$, where κ_T is the isothermal compressibility. These relations yield $(B/A)_1 = 6\gamma$, $(B/A)_2 = 6(1 - \gamma)$, and therefore $B/A = 6$, where γ is again the ratio of specific heats. Although B/A is not in fact a constant, the values for it for liquids and biological tissues reported in the tables below lie mainly in the range 5–10, in nominal agreement with Nomoto's prediction. Moreover, one obtains $(B/A)_2/(B/A)_1 = \gamma^{-1} - 1$, indicating that since γ is only slightly greater than unity for most liquids, $(B/A)_2$ (due to isobaric temperature changes) is a small negative correction to $(B/A)_1$ (due to isothermal pressure changes) in terms of its contribution to B/A, which is often the case. Finally, the value $\gamma = 1.4$ for a diatomic gas leads from Eq. (2.11) to $B/A = 0.4$, indicating that nonlinearity in the equation of state is considerably more important for liquids than for gases.

Ballou (1965) noticed a correlation between the value of B/A and the reciprocal of the sound velocity in many liquids. A best fit of data available at that time took the form $B/A = 1.2 \times 10^4 c_0^{-1} - 0.5$, where the sound velocity c_0 has units of meters per second. Further studies cast doubt on Ballou's rule (Sehgal et al., 1990), even though it was given some theoretical justification in the work by Hartmann (1979).

---

[2] An equivalent equation appears in Rayleigh's (1910) survey of finite-amplitude acoustic wave solutions, as his Eq. (19).

2.3   Physical Interpretation of B/A

The significance of B/A in acoustics is its effect on the sound speed. Using the relation $c^2 = (\partial P/\partial \rho)_s$, one obtains from Eq. (2.2)

$$\frac{c^2}{c_0^2} = 1 + \frac{B}{A}\left(\frac{\rho'}{\rho_0}\right) + \frac{C}{2A}\left(\frac{\rho'}{\rho_0}\right)^2 + \cdots . \tag{2.13}$$

Taking the square root and performing a binomial expansion yields

$$\frac{c}{c_0} = 1 + \frac{B}{2A}\left(\frac{\rho'}{\rho_0}\right) + \frac{1}{4}\left[\frac{C}{A} - \frac{1}{2}\left(\frac{B}{A}\right)^2\right]\left(\frac{\rho'}{\rho_0}\right)^2 + \cdots . \tag{2.14}$$

The parameter B/A thus determines the relative importance of the leading order finite-amplitude correction to the small-signal sound speed c_0. For a progressive plane wave the linear relation $\rho'/\rho_0 = u/c_0$, where u is the particle velocity, may be substituted into the second term on the right-hand side of Eq. (2.14) to obtain, with higher-order terms discarded,

$$c = c_0 + (B/2A)u, \tag{2.15}$$

which agrees with Eq. (1.23) when the relation in Eq. (2.11) is used. Furthermore, the propagation speed of a progressive plane wave becomes, from Eq. (1.27),

$$dx/dt|_u = c_0 + \beta u, \qquad \beta = 1 + B/2A, \tag{2.16}$$

where β is referred to as the coefficient of nonlinearity. The shock formation distance for a plane wave that is sinusoidal at the source is $\bar{x} = 1/\beta\varepsilon k$, where $\varepsilon = u_0/c_0$, u_0 is the peak particle velocity at the source, and k is the corresponding wave number [Eq. (4.22)]. Thus, β is the significant measure of the acoustic nonlinearity of the medium [see also Eqs. (3.15)–(3.17)].

2.4   Determination of B/A

Here we review briefly several methods that have been used to determine B/A. In all measurements that have been reported by the various methods, accuracies are rarely better than 5%. This should be kept in mind when using data from the tables in Sect. 2.6.

2.4.1 Thermodynamic Method

The commonest way of measuring B/A, and probably the most accurate, has been via measurement of the sound velocity as a function of the hydrostatic pressure and of the temperature, and the use of Eq. (2.8). The quantities in Eq. (2.8) (other than the two partial derivatives) are available in the thermodynamics literature for many liquids. The problem therefore reduces to measuring the dependence of the sound velocity on temperature and pressure. There is a large amount of data on the temperature variation of the sound velocity at constant pressure, but values of the pressure dependence of the sound velocity at constant temperature are hard to come by.[3] Nevertheless, values of B/A based on the thermodynamic method appear to be the most reliable of all data on the parameter and are often used as the standard for comparison with values obtained by other methods.

2.4.2 Finite-Amplitude Method

When a plane wave of finite amplitude travels through a medium, it becomes distorted, leading to generation of higher harmonics. Sufficiently close to a source of a sinusoidal plane wave at $x = 0$, in the region where the second harmonic pressure amplitude p_2 is still considerably larger than the pressure amplitude of all higher-order spectral components, the solution is [Thuras et al., 1935; see also Eq. (4.270), where $\alpha_2 = 4\alpha_1$]

$$p_2 = \left(1 + \frac{B}{2A}\right)\frac{p_0^2\omega}{2\rho_0 c_0^3}f(x), \qquad f(x) = \frac{e^{-2\alpha_1 x} - e^{-\alpha_2 x}}{\alpha_2 - 2\alpha_1}, \qquad (2.17)$$

where p_0 is the peak sound pressure, ω is the angular frequency at the source, and α_1 and α_2 are the attenuation coefficients at the fundamental and second-harmonic frequencies, respectively. Inversion of the first of Eqs. (2.17) yields

$$\frac{B}{A} = \left(\frac{p_2}{p_0}\right)\frac{4\rho_0 c_0^3}{p_0\omega f(x)} - 2. \qquad (2.18)$$

The difficulty here is that an independent measurement of p_0, the source amplitude, is required. This method was used frequently by Hiedemann and his students (Adler and Hiedemann, 1962), and has later been used, with greater accuracy, by Dunn and his group (Law et al., 1981). In both of these works, attenuation was ignored, in which case $f(x) = x$. The effect of absorption on this measurement

---

[3] Articles in which measurements of the isothermal variation of sound speed with pressure are reported for several organic liquids, not listed in Table 2.1 below, include those by Richardson and Tait (1957), Eden and Richardson (1960), Belinskii and Ergopulo (1968), and Hawley et al. (1970).

was taken into account by Dunn et al. (1981), who introduced the approximation $f(x) \simeq x \exp[-(\alpha_1 + \alpha_2/2)x]$ on the basis of the assumption $|\alpha_1 - \alpha_2/2|x \ll 1$. This approximation is normally appropriate for tissue, because the attenuation coefficients are often nearly proportional to frequency, and therefore $\alpha_2 \simeq 2\alpha_1$. A good review of related developments is contained in the paper by Bjørnø (1986).

2.4.3 Phase Comparison Method

Zhu et al. (1983) developed a phase comparison method for determination of B/A on the basis of Eq. (2.7). Measurements were made of the travel time of ultrasonic tone bursts under adiabatic variation of the hydrostatic pressure. These times were measured by comparison of the phase ϕ, in radians, of the tone burst and the phase of a reference signal. The resultant formula is

$$\frac{B}{A} = -\frac{2\rho_0 c_0^3}{\omega x}\left(\frac{\partial \phi}{\partial P}\right)_{s,0}, \tag{2.19}$$

where x is the travel distance of the tone burst.

2.4.4 Method for Aqueous Solutions

Sarvazyan et al. (1990) developed a differential method for measuring the effect of small concentrations on the value of B/A for aqueous solutions of proteins and amino acids [see also Chalikian et al. (1992)]. By differentiating Eq. (2.8) they obtained

$$\frac{\Delta(B/A)}{2\rho_0 c_0 \chi} = \frac{1}{\chi}\Delta\left(\frac{\partial c}{\partial P}\right)_{T,0} + ([c] + [\rho])\left(\frac{\partial c}{\partial P}\right)_{T,0}$$
$$+ \frac{\alpha_{T0} T_0}{\rho_0 c_{p0}}\left[\frac{1}{\chi}\Delta\left(\frac{\partial c}{\partial T}\right)_{P,0} + ([c] + [\alpha_T] - [c_p])\left(\frac{\partial c}{\partial T}\right)_{P,0}\right], \tag{2.20}$$

where χ is the concentration of the solution. Coefficients with subscript 0 correspond to ambient values for the solvent, and Δ indicates that the following quantity represents the difference between the value for the solution and that for the solvent. The bracketed quantities are the relative specific increments of the sound velocity, density, thermal expansion coefficient, and heat capacity at constant pressure, defined as

$$[c] = \frac{\Delta c}{\chi c_0}, \qquad [\rho] = \frac{\Delta \rho}{\chi \rho_0}, \qquad [\alpha_T] = \frac{\Delta \alpha_T}{\chi \alpha_{T0}}, \qquad [c_p] = \frac{\Delta c_p}{\chi c_{p0}}.$$

This group also succeeded in working with a much smaller chamber in making sound velocity measurements, so that less than 1 cc of the liquid was necessary.

2.4.5 Method for Immiscible Mixtures

Everbach et al. (1991) described a method for calculating an effective value of B/A for a mixture consisting of an arbitrary number of immiscible fluids. Here we let ρ be the density of the mixture and $v = 1/\rho$ be the corresponding specific volume. The ith component of the mixture has mass fraction $M_i = m_i/m$, where m_i is the mass of that component and $m = \sum_{i=1}^n m_i$ is the total mass of an n-component mixture. If the components are neither interactive nor mutually soluble the volumes must sum as follows:

$$v = \sum_{i=1}^n M_i v_i, \tag{2.21}$$

where v_i is the specific volume of the ith component. Since the mass fractions are independent of pressure we have

$$\left(\frac{\partial v}{\partial P}\right)_s = \sum_{i=1}^n M_i \left(\frac{\partial v_i}{\partial P}\right)_s, \qquad \left(\frac{\partial^2 v}{\partial P^2}\right)_s = \sum_{i=1}^n M_i \left(\frac{\partial^2 v_i}{\partial P^2}\right)_s. \tag{2.22}$$

Now employ the fundamental relations

$$\left(\frac{\partial v}{\partial P}\right)_s = -\frac{1}{\rho^2 c^2}, \qquad \left(\frac{\partial^2 v}{\partial P^2}\right)_s = \frac{2}{\rho^3 c^4}\left[1 + \frac{\rho}{2c^2}\left(\frac{\partial^2 P}{\partial \rho^2}\right)_s\right], \tag{2.23}$$

which apply to the mixture as a whole and to each of the n components individually [see Eq. (3.18) for interpretation of the second of Eqs. (2.23)]. Substitution of Eqs. (2.23) into Eqs. (2.22), combining the latter at equilibrium conditions, and using the definition of B/A in Eq. (2.6) yields

$$\beta = \left(\sum_{i=1}^n \frac{V_i}{\rho_i c_i^2}\right)^{-2} \sum_{i=1}^n \frac{V_i \beta_i}{\rho_i^2 c_i^4}, \tag{2.24}$$

where $\beta = 1 + B/2A$ is the effective coefficient of nonlinearity of the mixture and $\beta_i = 1 + (B/2A)_i$ the coefficient of the ith component, for which $V_i = (\rho/\rho_i)M_i$ is the corresponding volume fraction, ρ_i and c_i the ambient density and small-signal sound speed, respectively. Equation (2.24) has been applied to biological materials and to bubbly liquids (Everbach, 1989). In the latter case, the presence of gas bubbles greatly increases nonlinearity and hence the value of B/A [see Eq. (5.66) and Fig. 5.8].

2.5 Nonlinearity in Isotropic Solids

A coefficient of nonlinearity can also be derived for longitudinal waves in isotropic solids. When coupling with shear wave modes of propagation can be ignored (i.e., for compressional waves that are nearly planar, such as in directional beams), model equations can be derived having the same form as those for sound waves in fluids, but with the coefficient of nonlinearity replaced by [Gol'dberg (1961); see also Eq. (9.23)]

$$\beta = -\left(\frac{3}{2} + \frac{\mathcal{A} + 3\mathcal{B} + \mathcal{C}}{\rho_0 c_l^2}\right), \tag{2.25}$$

where c_l is the small-signal longitudinal wave speed, and $\mathcal{A}$, $\mathcal{B}$, and $\mathcal{C}$ are the third-order elastic (TOE) constants defined by Landau and Lifshitz (1986). For a fluid, these coefficients become $\mathcal{A} = 0$, $\mathcal{B} = -A$, and $\mathcal{C} = (A - B)/2$, where $A = \rho_0 c_l^2$, and Eq. (2.25) reduces to $\beta = 1 + B/2A$, as required (Kostek et al., 1993). Equation (2.25) may be rewritten in terms of a variety of other notations for the TOE constants using the relations provided in Table 9.1. Experimental methods used to measure the TOE constants are reviewed by Breazeale and Philip (1984).

2.6 Tables

In summary, the coefficient of nonlinearity defined by

$$\beta = \begin{cases} (\gamma + 1)/2 & \text{gas} \\ 1 + B/2A & \text{liquid} \\ -\left(\dfrac{3}{2} + \dfrac{\mathcal{A} + 3\mathcal{B} + \mathcal{C}}{\rho_0 c_l^2}\right) & \text{solid} \end{cases} \tag{2.26}$$

is the principal measure of finite-amplitude effects associated with the propagation of progressive sound waves, and the value of this coefficient is determined in part by the nonlinearity in the equation of state for a given medium. For diatomic gases such as air one has $\gamma = 1.4$ and therefore $\beta = 1.2$, and for distilled water one has $B/A = 5.0$ and therefore $\beta = 3.5$ at temperature 20 °C. The tables given here list values of B/A for selected fluids (Table 2.1), liquefied gases (Table 2.2), biological tissues (Table 2.3), and aqueous solutions of proteins and amino acids (Table 2.4).

Table 2.1 Values of B/A for fluids. Unless otherwise indicated, values are for atmospheric pressure. The data for water–glycol mixtures are from Table II by Swamy et al. (1975b), and all other data are from Table 3-1 by Beyer (1997).

Substance	$T, °C$	B/A
Distilled water	0	4.2
	20	5.0
	40	5.4
	60	5.7
	80	6.1
	100	6.1
Pressure dependence		
1 atm.	30	5.2
200 kg/cm^2	30	6.2
4000 kg/cm^2	30	6.2
8000 kg/cm^2	30	5.9
Sea water (salinity 3.5%)	20	5.25
Alcohols		
Methanol	20	9.6
Ethanol	0	10.4
	20	10.5
	40	10.6
n-propanol	20	10.7
n-butanol	20	10.7
Organic liquids		
Acetone	20	9.2
Benzene	20	9.0
Benzyl alcohol	30	10.2
Chlorobenzene	30	9.3
Cyclohexane	30	10.1
1,2-dichlorohexafluorocyclopentene (DHCP)	30	11.8
Diethylamine	30	10.3
Ethylene glycol	30	9.7
Ethyl formate	30	9.8
Glycerol (4% H_2O)	30	9.0
Heptane	30	10.0
Hexane	30	9.9
Methyl acetate	30	9.7
Methyl iodide	30	8.2
Nitrobenzene	30	9.9
Liquid metals		
Bismuth	318	7.1
Indium	160	4.6
Mercury	30	7.8
Potassium	100	2.9
Sodium	110	2.7
Tin	240	4.4

(continued)

Table 2.1 (continued)

Substance		T, °C	B/A
Other substances			
Sulfur		121	9.5
Monatomic gas		20	0.67
Diatomic gas		20	0.40
Mixtures			
Water–glycol	% glycol		
	0	30	5.2
	20	30	5.2
	33	30	5.1

Table 2.2 B/A for liquefied gases.

Argon		P, atm				
T, K		1	10	60	200	500
86	$(B/A)_1$	11.71	11.20	9.89		
	$(B/A)_2$	−6.70	−4.76	−3.98		
	B/A	5.01	6.44	5.91		
90	$(B/A)_1$	12.43	12.20	11.50		
	$(B/A)_2$	−6.76	−5.00	−4.40		
	B/A	5.67	7.20	7.10		
120	$(B/A)_1$			15.28	11.71	9.68
	$(B/A)_2$			−5.60	−4.00	−2.56
	B/A			9.68	7.71	7.12
150	$(B/A)_1$			33.67	12.60	8.82
	$(B/A)_2$			−11.13	−5.03	−3.15
	B/A			22.54	7.57	5.67

At atmospheric pressure:

Nitrogen		Hydrogen		Methane	
T, K	B/A	T, K	B/A	T, K	B/A
70	7.70	14	5.59	110	17.95
80	8.03	16	6.87	120	10.31
90	9.00	18	7.64	130	6.54
		20	7.79	135	5.41

After Tables I–IV by Swamy et al. (1975a)

Table 2.3 B/A for various biological tissues.

	Biological material (and state)	Method[a]	B/A (and uncertainty)	Ref.
1.	Bovine serum albumin (BSA) (20 g/100 cm³, 25 °C)	Therm.	6.23 (±0.25)	1
	BSA (22 g/100 cm³, 30 °C)	F.A.	6.45 (±0.30)	2
	BSA (38.9 g/100 cm³, 30 °C)	F.A.	6.64	3
	BSA (38.9 g/100 cm³, 30 °C)	Therm.	6.68 (±0.2)	3
2.	Haemoglobin (50%, 30 °C)	F.A.	7.6	4
3.	Whole porcine blood (12% haemoglobin, 7% plasma proteins, 30 °C)	F.A.	6.2 (±0.25)	4
4.	Beef liver (Whole, 23 °C)	F.A.	7.75 (±0.4)	4
	Beef liver (Homogenized, 23 °C)	F.A.	6.8 (±0.4)	4
	Beef liver (Whole, 30 °C)	F.A.	6.42	3
	Beef liver (Whole, 30 °C)	Therm.	6.88	3
	Beef liver (Whole, 30 °C)	Therm.	6.54 (±0.2)	5
	Dog liver (30 °C)	F.A.	7.6–7.9 (±0.8)	6
	Pig liver (25 °C)	F.A.	6.7 (±1.5)	7
	Human liver (Normal, 30 °C)	F.A.	7.6 (±0.8)	6
	Human liver (Congested, 30 °C)	F.A.	7.2 (±0.7)	6
5.	Pig fat	Therm.	10.9	8
	Pig fat	F.A.	11.0–11.3	8
	Human breast fat (22 °C)	Therm.	9.21	5
	Human breast fat (30 °C)	Therm.	9.91	5
	Human breast fat (37 °C)	Therm.	9.63	5
6.	Canine spleen	F.A.	6.8	9
	Dog spleen	F.A.	6.8 (±0.7)	6
	Human spleen (Congested)	F.A.	7.8	9
	Human spleen (Normal, 30 °C)	F.A.	7.8 (±0.8)	6
7.	Beef brain (30 °C)	F.A.	7.6	8
8.	Beef heart (30 °C)	F.A.	6.8–7.4	8
9.	Pig muscle (30 °C)	F.A.	7.5–8.1	8
	Pig muscle (25 °C)	F.A.	6.5 (±1.5)	7
10.	Dog kidney (Normal, 30 °C)	F.A.	7.2 (±0.7)	6
	Canine kidney (30 °C)	F.A.	7.2	9
11.	Human multiple myeloma (22 °C)	F.A.	5.6	5
	Human multiple myeloma (30 °C)	F.A.	5.8	5
	Human multiple myeloma (37 °C)	F.A.	6.2	5

[After Table 2.1 by Bjørnø (1986)]

[a] Therm. = thermodynamic method; F.A. = finite-amplitude method

Reference key: 1. Zhu et al. (1983); 2. Law et al. (1981); 3. Law et al. (1983); 4. Dunn et al. (1981); 5. Sehgal et al. (1984); 6. Cobb (1982); 7. Lewin and Bjørnø (1983); 8. Law et al. (1985); 9. Dunn et al. (1984)

Table 2.4 Specific increments of the nonlinearity parameter B/A of proteins (a) and amino acids (b) in aqueous solution at 25 °C.

(a)

Proteins	$\Delta(B/A)/\chi$ cm^3/g	Specific concentration mg/cm^3	Molecular weight kD
Collagen	4.3 ± 1.0	4	130
Rabbit serum albumin	4.1 ± 0.3	25–70	64[a]
Ribonuclease	3.7 ± 0.3	30	14
Hemoglobin	3.6 ± 0.3	25–50	64[b]
Lysozyme	3.3 ± 0.3	35	14
Myoglobin	2.8 ± 0.3	35	17

[a]Human. [b]16/subunit

(b)

Amino acids	$\Delta(B/A)/\chi$ cm^3/g	Specific concentration mg/cm^3
Glycine	6.4 ± 0.3	10–50
Glutamine	5.0 ± 0.4	25
Histidine	4.6 ± 0.4	10–25
Alanine	3.0 ± 0.3	50
Phenylalanine	2.4 ± 0.6	10–15
Valine	2.0 ± 0.2	50
Proline	1.6 ± 0.2	50
Norvaline	1.5 ± 0.2	50
Isoleucine	1.4 ± 0.3	15–35
Leucine	0.4 ± 0.2	20

After Tables I and II by Sarvazyan et al. (1990)

Measured values of the TOE constants for a variety of isotropic solids have been compiled by Hearmon (1979) and, particularly for rocks and synthetic materials, by Winkler and Liu (1996). The resulting values of β for homogeneous isotropic solids have magnitudes similar to those for liquids. However, whereas β is positive for fluids except under extraordinary conditions, e.g., in the neighborhood of a critical point (Thompson, 1984), it is negative for some solids, such as fused quartz and Pyrex glass [in which materials, in accordance with the first of Eqs. (2.16), the direction of waveform steepening is opposite that in fluids]. The microinhomogeneous features of rock can increase the magnitude of β by several orders.

References

Adler, L., and Hiedemann, E. A. (1962). Determination of the nonlinearity parameter B/A for water and m-Xylene. *J. Acoust. Soc. Am.* **34**, 410–412.

Ballou, J. F. (1965). An experimental determination of the parameter of nonlinearity for four indium–bismuth molten alloys. Sc.M. thesis, Brown University.

Belinskii, B.A., and Ergopulo, E. V. (1968). Investigation of the ultrasonic parameters of m-cresol and ethylene glycol as a function of the state parameters of the medium. *Sov. Phys. Acoust.* **14**, 94–96.

Beyer, R. T. (1960). Parameter of nonlinearity in fluids. *J. Acoust. Soc. Am.* **32**, 719–721.

Beyer, R. T. (1997). *Nonlinear Acoustics.* (Acoustical Society of America, New York), Chap. 3.

Biquard, P. (1936). Sur l'absorption des ondes ultra-sonores par les liquides. *Ann. Physique* (Ser. 11) **6**, 195–304.

Bjørnø, L. (1986). Characterization of biological media by means of their nonlinearity. *Ultrasonics* **24**, 254–259.

Breazeale, M. A., and Philip, J. (1984). Determination of third-order elastic constants from ultrasonic harmonic generation measurements. In *Physical Acoustics*, Vol. 17, W. P. Mason, ed. (Academic Press, New York).

Chalikian, T. V., Sarvazyan, A. P., Funck, Th., Belonenko, V. N., and Dunn, F. (1992). Temperature dependences of the acoustic nonlinearity parameter B/A of aqueous solutions of amino acids. *J. Acoust. Soc. Am.* **91**, 52–58.

Cobb, W. N. (1982). Measurement of the acoustic nonlinearity parameter for biological media. Ph.D. dissertation, Yale University.

Coppens, A. B., Beyer, R. T., Seiden, M. B., Donohue, J., Guepin, F., Hodson, R. H., and Townsend, C. (1965). Parameter of nonlinearity in fluids. II. *J. Acoust. Soc. Am.* **38**, 797–804.

Dunn, F., Law, W. K., and Frizzell, L. A. (1981). Nonlinear ultrasonic wave propagation in biological materials. *Proceedings of IEEE 1981 Ultrasonics Symposium*, Vol. 1, B. R. McAvoy, ed. (IEEE, New York), pp. 527–532.

Dunn, F., Law, W., and Frizzell, L. (1984). The nonlinearity parameter B/A of biological media. *Proceedings of the 10th International Symposium on Nonlinear Acoustics*, A. Nakamura, ed. (Teikohsha, Kadoma, Japan), pp. 221–225.

Eden, H. F., and Richardson, E. G. (1960). The propagation of ultrasonics in organic liquids under pressure. Variation of specific heat ratio and viscosity with pressure. *Acustica* **10**, 309–315.

Everbach, E. C. (1989). Tissue composition determination via measurement of the acoustic nonlinearity parameter. Ph.D. thesis, Yale University.

Everbach, E. C., Zhu, Z., Jiang, P., Chu, B. T., and Apfel, R. E. (1991). A corrected mixture law for B/A. *J. Acoust. Soc. Am.* **89**, 446–447.

Fox, F. E., and Wallace, W. A. (1954). Absorption of finite amplitude sound waves. *J. Acoust. Soc. Am.* **26**, 994–1006.

Gol'dberg, Z. A. (1961). Interaction of plane longitudinal and transverse elastic waves. *Sov. Phys. Acoust.* **6**, 306–310.

Hartmann, B. (1979). Potential energy effects on the sound speed in liquids. *J. Acoust. Soc. Am.* **65**, 1392–1396.

Hawley, S., Allegra, J., and Holton, G. (1970). Ultrasonic-absorption and sound-speed data for nine liquids at high pressures. *J. Acoust. Soc. Am.* **47**, 137–143.

Hearmon, R. F. S. (1979). The elastic constants of crystals and other anisotropic materials; The third- and higher-order elastic constants. In *Landolt–Börnstein: Numerical Data and Functional Relationships in Science and Technology*, Group III, Vol. 11, K.-H. Hellwege and A. M. Hellwege, eds. (Springer, New York), pp. 245–286.

Kostek, S., Sinha, B. K., and Norris, A. N. (1993). Third-order elastic constants for an inviscid fluid. *J. Acoust. Soc. Am.* **94**, 3014–3017.

Landau, L. D., and Lifshitz, E. M. (1986). *Theory of Elasticity*, 3rd ed. (Pergamon, New York), pp. 106–107.

Law, W. K., Frizzell, L. A., and Dunn, F. (1981). Ultrasonic determination of the nonlinearity parameter B/A for biological media. *J. Acoust. Soc. Am.* **69**, 1210–1212.

Law, W. K., Frizzell, L. A., and Dunn, F. (1983). Comparison of thermodynamic and finite amplitude methods of B/A measurement in biological materials. *J. Acoust. Soc. Am.* **74**, 1295–1297.

Law, W. K., Frizzell, L. A., and Dunn, F. (1985). Determination of the nonlinearity parameter B/A of biological media. *Ultrasound Med. Biol.* **11**, 307–318.

Lewin, P. A., and Bjørnø, L. (1983). Acoustic non-linearity parameters in tissue characterization. *Ultrasonic Imaging* **5**, 170(A).

Nomoto, O. (1966). Nonlinearity parameter of the "Rao liquid." *J. Phys. Soc. Japan* **21**, 569–571.

Rao, M. R. (1940). A relation between velocity of sound in liquid and molecular volumes. *Ind. J. Phys.* **14**, 109–116.

Rayleigh, Lord (1910). Aerial plane waves of finite amplitude. *Proc. Roy. Soc.* **A 84**, 247–284.

Richardson, E. G., and Tait, R. I. (1957). Ratios of specific heat and high-frequency viscosities in organic liquids under pressure, derived from ultrasonic propagation. *Phil. Mag.* **2**, 441–454.

Sarvazyan, A. P., Chalikian, T. V., and Dunn, F. (1990). Acoustic nonlinearity parameter B/A of aqueous solutions of some amino acids and proteins. *J. Acoust. Soc. Am.* **88**, 1555–1561.

Sehgal, C. M., Bahn, R. C., and Greenleaf, J. F. (1984). Measurement of the acoustic nonlinearity parameter B/A in human tissues by a thermodynamic method. *J. Acoust. Soc. Am.* **76**, 1023–1029.

Sehgal, C. M., Errabolu, R. L., and Greenleaf, J. F. (1990). Interrelationship between acoustic nonlinearity and sound speed in biological tissues. In *Frontiers of Nonlinear Acoustics: 12th ISNA*, M. F. Hamilton and D. T. Blackstock, eds. (Elsevier, London), pp. 402–407.

Sullivan, D. A. (1981). Historical review of real-fluid isentropic flow models. *J. Fluids Eng.* **103**, 258–267.

Swamy, K. M., Narayana, K. L., and Swamy, P. S. (1975a). A study of (B/A) in liquified gases as a function of temperature and pressure from ultrasonic velocity measurements. *Acustica* **32**, 339–341.

Swamy, K. M., Narayana, K. L., and Swamy, P. S. (1975b). Non-linear acoustical properties of liquids and liquid mixtures. *Acustica* **33**, 52–54.

Thompson, P. A. (1984). *Compressible-Fluid Dynamics* (Maple Press, New York), Sec. 5.2 [in which Eq. (5.8) is equivalent to Eq. (3.18) in the present text].

Thuras, A. L., Jenkins, R. T., and O'Neil, H. T. (1935). Extraneous frequencies generated in air carrying intense sound waves. *J. Acoust. Soc. Am.* **6**, 173–180.

Winkler, K. W., and Liu, X. (1996). Measurements of third-order elastic constants in rocks. *J. Acoust. Soc. Am.* **100**, 1392–1398.

Zhu, S., Roos, M. S., Cobb, W. N., and Jensen, K. (1983). Determination of the acoustic nonlinearity parameter B/A from phase measurements. *J. Acoust. Soc. Am.* **74**, 1518–1521.

Chapter 3
Model Equations

Mark F. Hamilton and Christopher L. Morfey

Contents

3.1 Introduction

To identify limitations and domains of applicability associated with model equations
of nonlinear acoustics, it is important to understand the assumptions and ordering
procedures on which the equations are based. In this chapter, the more widely used
model equations for homogeneous fluids are derived directly from the fundamental
equations of fluid mechanics given in Sect. 3.2. Exact relations for lossless fluids are
developed in Sect. 3.3, including characteristic equations and implicit solutions for
plane waves, as well as the nonplanar-wave equation for a perfect gas. Section 3.4
describes the ordering procedure and approximations used in Sects. 3.5–3.9 to

M. F. Hamilton
Department of Mechanical Engineering, The University of Texas at Austin, Austin, TX, USA

C. L. Morfey
Institute of Sound and Vibration Research, University of Southampton, Southampton, UK

© The Author(s) 2024
M. F. Hamilton, D. T. Blackstock (eds.), *Nonlinear Acoustics*,
https://doi.org/10.1007/978-3-031-58963-8_3

derive model equations for sound fields in thermoviscous fluids, outside boundary-layer regions. In Sect. 3.5 we obtain the complete second-order equation for arbitrary three-dimensional waves. Model equations developed in all subsequent sections pertain to progressive waves: quasiplane waves in Sect. 3.6; pure plane waves in Sect. 3.7; general one-dimensional waves in Sect. 3.8 (plane, spherical, and cylindrical waves, and waves in ducts with variable cross section); directional sound beams in Sect. 3.9.

Elsewhere in this book, the reader will encounter several complementary discussions that amplify and extend the present chapter. Chapter 1 reviews the history of theoretical developments leading up to the Burgers equation. The Rankine–Hugoniot and weak shock relations are derived in Chap. 4. Chapter 5 reviews the combined effects of dispersion and nonlinearity on wave propagation, e.g., in relaxing fluids and bubbly liquids. Radiation pressure is covered in Chap. 6, and acoustic streaming in Chap. 7. Model equations for nonlinear elastic waves in solids are obtained in Chap. 9. Nonlinear ray theory for inhomogeneous media is developed in Chap. 12.

3.2 Basic Equations

Four equations are required to describe the general motion of a viscous, heat-conducting fluid: (1) mass conservation, (2) momentum conservation, (3) entropy balance, and (4) thermodynamic state. We assume that the fluid is homogeneous in composition, that its unperturbed density and pressure are uniform, and that the dependence of the viscosity and heat conduction coefficients on the disturbance due to the sound wave may be neglected. This last assumption anticipates the perturbation ordering scheme introduced in Sect. 3.4, where dissipation terms are retained to lowest order only; by adopting it now, we avoid introducing unnecessary complication into the basic equations. The reader is referred to the text by Landau and Lifshitz (1987) for detailed derivation and discussion of the equations which follow.

The mass conservation, or continuity, equation is

$$\frac{D\rho}{Dt} + \rho \nabla \cdot \mathbf{u} = 0, \tag{3.1}$$

where ρ is the mass density, $\mathbf{u}$ is the fluid velocity vector, and $D/Dt = \partial/\partial t + \mathbf{u} \cdot \nabla$ is the total, or material, time derivative. The momentum equation may be written as

$$\rho \frac{D\mathbf{u}}{Dt} + \nabla P = \mu \nabla^2 \mathbf{u} + (\mu_B + \tfrac{1}{3}\mu) \nabla (\nabla \cdot \mathbf{u}), \tag{3.2}$$

where P is the thermodynamic pressure appearing in Eq. (3.4), μ is the shear viscosity, and μ_B is the bulk viscosity. Shear viscosity accounts for diffusion of momentum between adjacent fluid elements having different velocities. Bulk viscosity provides an approximate description, valid at low frequency, of nonequilibrium deviations between the actual local pressure and the thermodynamic pressure; the latter is given by the fluid equation of state $P = P(\rho, T)$, where T is the absolute temperature.

A more general description, covering a wide frequency range, would need to recognize that nonequilibrium deviations involve relaxation. Examples are vibrational relaxation of diatomic molecules (as in air) and chemical relaxation in seawater. The former occurs whenever the energy associated with molecular vibration fails to keep in step with the molecular translational energy associated with the fluctuating temperature in the gas. In air, vibrational relaxation is the dominant attenuation mechanism at audible frequencies, while chemical relaxation is the main cause of sound attenuation in seawater below 500 kHz. A general discussion is given in Sect. 81 of the book by Landau and Lifshitz (1987); the use of μ_B in the present chapter to account for nonequilibrium departures of the actual pressure from the thermodynamic pressure is valid provided the time scale of the disturbance is long compared with the relaxation times involved (Pierce, 1989). The effect of relaxation on the propagation of finite-amplitude sound at higher frequencies is discussed in Sect. 5.2.1.

For the present chapter, we assume for simplicity that all relaxation times are much shorter than the time scale of the acoustic disturbance. Equation (3.2) is then a valid approximate statement of momentum conservation, and the appropriate form of the entropy equation is

$$\rho T \frac{Ds}{Dt} = \kappa \nabla^2 T + \mu_B (\nabla \cdot \mathbf{u})^2 + \frac{1}{2}\mu \left(\frac{\partial u_i}{\partial x_j} + \frac{\partial u_j}{\partial x_i} - \frac{2}{3}\delta_{ij} \frac{\partial u_k}{\partial x_k} \right)^2, \tag{3.3}$$

where s is the specific entropy (per unit mass) and κ is the thermal conductivity. The final term in Eq. (3.3) is written in Cartesian tensor notation: u_i denotes the component of $\mathbf{u}$ in direction x_i, and δ_{ij} is the Kronecker delta, equal to unity for $i = j$ and zero otherwise.

Finally, for convenience in the analysis that follows, we choose to write the state equation in (P, ρ, s) variables, rather than (P, ρ, T):

$$P = P(\rho, s). \tag{3.4}$$

A commonly used explicit form of Eq. (3.4) is that for a perfect gas, i.e., a gas for which both $P/\rho T$ and the specific-heat ratio are constants:

$$P/P_0 = (\rho/\rho_0)^\gamma \exp[(s - s_0)/c_v]. \tag{3.5}$$

Here P_0, ρ_0, and s_0 are reference (hereafter taken to be ambient) values of the pressure, density, and specific entropy, respectively, and $\gamma = c_p/c_v$ is the ratio of the specific heats at constant pressure (c_p) and constant volume (c_v). However, the assumption of perfect-gas behavior is unnecessarily restrictive, and in order to obtain model equations that are valid in an arbitrary fluid, in later sections we shall expand Eq. (3.4) in a Taylor series about (ρ_0, s_0).

3.3   Lossless Theory

In this section we consider lossless fluids. We thus set μ, μ_B, and κ to zero. Equation (3.3) leads to the trivial conclusion $s = s_0$ (because the fluid is initially uniform with $s = s_0$ everywhere), and Eq. (3.4) reduces to $P = P(\rho)$.

3.3.1   Plane Waves

Exact solutions of the equations in Sect. 3.2 are available for plane waves in a uniform lossless fluid (Riemann, 1860; see also Sect. 104 of Landau and Lifshitz, 1987; Blackstock, 1962; and Sects. 8.2–8.4 of Thompson, 1984). Starting from the equation of state, we define

$$c^2 = \left(\frac{\partial P}{\partial \rho} \right)_s, \qquad \lambda = \int_{\rho_0}^{\rho} \frac{c}{\rho} \, d\rho = \int_{P_0}^{P} \frac{dP}{\rho c}, \tag{3.6}$$

where c is the sound speed. Although the lower limits on the integrals are arbitrary, they are conveniently identified with the uniform undisturbed state (density ρ_0, pressure P_0). Equations (3.6) may be used to write $\rho_t = (\rho/c)\lambda_t$, $\rho_x = (\rho/c)\lambda_x$, and $P_x = \rho c \lambda_x$, where the subscripts indicate partial derivatives, and propagation along the x axis is assumed. The one-dimensional forms of Eqs. (3.1) and (3.2) reduce, respectively, to Eqs. (1.11) and (1.12):

$$\frac{\partial \lambda}{\partial t} + u \frac{\partial \lambda}{\partial x} + c \frac{\partial u}{\partial x} = 0, \qquad \frac{\partial u}{\partial t} + u \frac{\partial u}{\partial x} + c \frac{\partial \lambda}{\partial x} = 0. \tag{3.7}$$

Addition and subtraction of Eqs. (3.7) yield

$$\frac{\partial J_+}{\partial t} + (u + c) \frac{\partial J_+}{\partial x} = 0, \qquad \frac{\partial J_-}{\partial t} + (u - c) \frac{\partial J_-}{\partial x} = 0. \tag{3.8}$$

The quantities

$$J_+ = \tfrac{1}{2}(\lambda + u), \qquad J_- = \tfrac{1}{2}(\lambda - u) \tag{3.9}$$

are called Riemann invariants (Riemann, 1860). These wavelike quantities propagate unchanged in the x direction at speeds $u + c$, $u - c$, respectively [Eqs. (1.32)].

The forward invariant J_+ is propagated unchanged along the forward characteristic $dx/dt = u + c$ in the xt plane, and the backward invariant J_- along the backward characteristic $dx/dt = u - c$. However, because quantities such as u, c, and P generally depend on both J_+ and J_-, the wavelike property is limited in the general case to the Riemann invariants themselves.

The exception is the important case of progressive-wave propagation, where one of the Riemann invariants is a constant. For example, if the fluid in the region $x > x_0$ is free of disturbances for all times $t < t_0$, there can be no incoming waves in view of Eqs. (3.8); i.e., $J_- = 0$ everywhere. It follows that $J_+ = \lambda = u$, and therefore

$$u = \int_{\rho_0}^{\rho} \frac{c}{\rho}\, d\rho \qquad (3.10)$$

from Eqs. (3.6), which means that a one-to-one relation exists between the fluid velocity disturbance and the density (or pressure) disturbance. Such progressive waves in a compressible fluid are called *simple* waves, as distinct from the general case of *compound* waves, which involve both J_+ and J_-.

For simple, forward-propagating waves, Eqs. (3.8) reduce to

$$\frac{\partial q}{\partial t} + (c + u)\frac{\partial q}{\partial x} = 0, \quad \text{i.e.,} \quad \left.\frac{dx}{dt}\right|_q = c + u. \qquad (3.11)$$

Here q stands for any of the quantities u, $P - P_0$, $\rho - \rho_0$; it can also be any combination of these, since they are one-to-one-related. Like Eqs. (3.8), the result is exact under the conditions assumed, namely, progressive plane waves traveling into a uniform region of lossless fluid. Historically, Eqs. (3.11) were derived first for an isothermal gas, a special case for which c is a constant independent of u or $\rho - \rho_0$ [see Eqs. (1.18) and (1.19)].

For the sake of definiteness, we put $q = u$ in Eqs. (3.11). A general implicit solution—remembering that c is not actually a constant, but is related to u by Eq. (3.10)—is an extension of Poisson's (1808) solution for an isothermal gas, Eq. (1.17):

$$u = g[x - (c + u)t]. \qquad (3.12)$$

Equation (3.12) is the solution of the initial-value problem $u = g(x)$ at $t = 0$. An alternative version, appropriate for boundary-value problems [e.g., a source waveform $u = f(t)$ is prescribed at $x = 0$ and a solution is sought for $x > 0$], follows from rewriting Eqs. (3.11) as

$$\frac{\partial q}{\partial x} + \frac{1}{c + u}\frac{\partial q}{\partial t} = 0, \quad \text{i.e.,} \quad \left.\frac{dt}{dx}\right|_q = \frac{1}{c + u}. \qquad (3.13)$$

Here the roles of t and x are interchanged, and the general implicit solution—again with $q = u$ for definiteness—is

$$u = f\left(t - \frac{x}{c+u}\right). \tag{3.14}$$

For a further detailed discussion and history of these solutions, see Chap. 1.

Equations (3.11)–(3.14) are elegant in their simplicity, particularly when applied to a perfect gas. In this case, putting $s = s_0$ in Eq. (3.5) and combining with the first of Eqs. (3.6) yields

$$P/P_0 = (\rho/\rho_0)^\gamma = (c/c_0)^{2\gamma/(\gamma-1)}, \quad c^2 = \gamma P/\rho, \tag{3.15}$$

where c_0 is the small-signal sound speed (c evaluated at the equilibrium state). Substitution into the differential relation $du = (c/\rho)\,d\rho = c\,d(\ln\rho)$ from Eq. (3.10) produces $du = 2c\,d(\ln c)/(\gamma-1)$, and it follows that $dc = \frac{1}{2}(\gamma-1)\,du$, which can be integrated to give, without approximation,

$$c = c_0 + \tfrac{1}{2}(\gamma-1)u, \quad c+u = c_0 + \beta u, \quad \beta = \tfrac{1}{2}(\gamma+1), \tag{3.16}$$

where β is the coefficient of nonlinearity. A further consequence is $u = \lambda = 2(c - c_0)/(\gamma-1)$. Equations (1.27)–(1.29) thus follow from Eqs. (3.11), (3.12), and (3.16). For an arbitrary fluid, the second of Eqs. (3.16) may be written as the Taylor series expansion

$$c + u = c_0 + \tilde{\beta}(0)u + \tfrac{1}{2}\tilde{\beta}'(0)u^2 + \cdots, \tag{3.17}$$

where $\tilde{\beta}(u)$ denotes the derivative $d(c+u)/du$ for simple, forward-propagating waves. Since $u = \lambda$ for such waves, $\tilde{\beta}$ is actually a thermodynamic quantity, which happens to be a constant for perfect gases. We shall not require the third or subsequent terms in expansion (3.17). The constant equilibrium value $\tilde{\beta}_0 \equiv \tilde{\beta}(0)$ is identified as the coefficient of nonlinearity β even when arbitrary fluids are discussed.

The quantity $\tilde{\beta}(\lambda)$ corresponds to a fundamental thermodynamic property encountered in gas dynamics (Landau and Lifshitz, 1987; Thompson, 1984):

$$\tilde{\beta} = \frac{c^4}{2v^3}\left(\frac{\partial^2 v}{\partial P^2}\right)_s = 1 + \frac{\rho}{2c^2}\left(\frac{\partial^2 P}{\partial \rho^2}\right)_s, \tag{3.18}$$

where $v = 1/\rho$ is specific volume. The second of Eqs. (3.18) yields $\tilde{\beta}_0 = 1 + B/2A \equiv \beta$, where $B/A = (\rho_0/c_0^2)(\partial^2 P/\partial\rho^2)_{s,0}$, and the definition in Eqs. (2.16) is thus recovered. From the first of Eqs. (3.18) we see that if the isentrope $v(P)$ for $s = s_0$ is a straight line—the isentropes being the contours of s in the Pv plane—then plane-wave propagation is linear for all amplitudes, since $\tilde{\beta}$ is zero and no cumulative waveform distortion occurs (in the absence of losses).[1] For a perfect

---

[1] Rayleigh called this Earnshaw's law; see item 4 on page 11. As discussed by Thompson (1984), $\tilde{\beta}$ may be negative near critical points.

gas, evaluation of $\tilde{\beta}$ on the basis of Eqs. (3.15) and substitution into Eq. (3.17) yields Eqs. (3.16) exactly. Although the result $\tilde{\beta} = \beta$ is obtained for a perfect gas, this is a special case, and in general $\tilde{\beta}$ is not a constant.

Finally, in situations where quadratic nonlinearity is sufficient to describe the dominant finite-amplitude effects, we can substitute first-order approximations into the nonlinear terms (such a substitution procedure is used extensively in Sect. 3.5). In particular, let $q = u$ in Eqs. (3.13) and write $c + u \simeq c_0 + \beta u$. Then binomial expansion of the coefficient $(c_0 + \beta u)^{-1}$ to first order in u leads to the following differential equation for u, written here with its corresponding solution (Blackstock, 1962):

$$\frac{\partial u}{\partial x} + \frac{1}{c_0} \frac{\partial u}{\partial t} = \frac{\beta}{c_0^2} u \frac{\partial u}{\partial t}, \quad u = f(t - x/c_0 + \beta x u/c_0^2). \tag{3.19}$$

To assess the error introduced by these approximations, write the second of Eqs. (3.19) in the form $\hat{u} = f(\hat{t})$, noting that $\hat{t} = t - x/c_0 + \beta x u/c_0^2$ is the approximation of the exact argument in Eq. (3.14). Taylor series expansion of the exact solution (3.14), following binomial expansion of its argument, yields for a perfect gas $u = f(\hat{t}) - (\beta^2 x/c_0^3) f^2(\hat{t}) f'(\hat{t}) + \cdots$, where $f' = df/d\hat{t}$. For plane waves in air ($\beta = 1.2$) with source condition $f(t) = u_0 \sin \omega t$, the relative error $|\hat{u} - u|/u_0$ at the point of shock formation is proportional to u_0 and less than 0.5% for $u_0/c_0 = 10^{-2}$ (154 dB re 20 μPa). Within this order of approximation—that is, with terms of *relative* order $(u_0/c_0)^2$ neglected—it is consistent for arbitrary fluids to replace the quantity $\tilde{\beta}$ by its unperturbed value $\beta = 1 + B/2A$.

3.3.2 Nonplanar Waves

We end this discussion of nonlinear waves in a lossless fluid by considering three-dimensional disturbances in which the flow remains irrotational. This last restriction requires not only that the fluid be inviscid, but also that the pressure and density be one-to-one related. Both these conditions are met in a lossless fluid whose undisturbed state is uniform, since Eq. (3.3) yields $Ds/Dt = 0$ when κ, μ, and μ_B are all zero. The practical application of such a restrictive model depends on recognizing that although all fluids have finite viscosity, and therefore vorticity generated at solid boundaries will diffuse into the body of the fluid, the diffusion length scale can be quite small at acoustic frequencies. Accordingly, setting the vorticity equal to zero and neglecting viscous stresses may prove to be appropriate over most of the flow field, provided that regions close to any solid boundaries are excluded. The same applies to thermal diffusion, which is likewise neglected in setting the material derivative of s equal to zero. See Sect. 3.4 for further discussion of boundary-layer effects.

Consider a lossless, homentropic, irrotational flow that extends to infinity, with $\mathbf{u} = \nabla\phi$ and $Ds/Dt = 0$. Here ϕ is the velocity potential. Equations (3.1) and (3.2), written in terms of ϕ, are

$$\frac{\partial\rho}{\partial t} + \nabla\rho \cdot \nabla\phi + \rho\nabla^2\phi = 0, \tag{3.20}$$

$$\nabla\left(\frac{\partial\phi}{\partial t} + \tfrac{1}{2}|\nabla\phi|^2\right) + \frac{\nabla P}{\rho} = 0, \tag{3.21}$$

where $|\nabla\phi|^2 = \nabla\phi \cdot \nabla\phi$. Now introduce the function

$$q = \int_{P_0}^{P} \frac{dP}{\rho} = \int_{\rho_0}^{\rho} \frac{c^2}{\rho}\, d\rho, \tag{3.22}$$

where the integrals are evaluated along $s = s_0$. Then $\nabla P = \rho\nabla q$, and Eq. (3.21) can be integrated with respect to spatial coordinates to obtain

$$\frac{\partial\phi}{\partial t} + \tfrac{1}{2}|\nabla\phi|^2 + q = 0. \tag{3.23}$$

The integration constant (actually, function of time) is zero, since the flow field extends to infinity and all acoustical disturbances vanish in this limit. Since q is defined by Eq. (3.22) and the equation of state, Eqs. (3.20) and (3.23) can be solved as a system of two equations in the two unknowns ϕ and ρ.

For a perfect gas, a wave equation in ϕ alone can be derived on the basis of Eqs. (3.15), as follows. Equation (3.22) gives $q = (c^2 - c_0^2)/(\gamma - 1)$, the substitution of which into Eq. (3.23) yields

$$c^2 = c_0^2 - (\gamma - 1)\left(\frac{\partial\phi}{\partial t} + \tfrac{1}{2}|\nabla\phi|^2\right). \tag{3.24}$$

The differential relation $dc^2/d\rho = (\gamma - 1)c^2/\rho$ also follows from Eqs. (3.15), and permits Eq. (3.20) to be rewritten as

$$\frac{\partial c^2}{\partial t} + \nabla c^2 \cdot \nabla\phi + (\gamma - 1)c^2\nabla^2\phi = 0. \tag{3.25}$$

Substitution of Eq. (3.24) in Eq. (3.25) leads to the following wave equation in ϕ:

$$c_0^2\nabla^2\phi - \frac{\partial^2\phi}{\partial t^2} = \left(2\nabla\frac{\partial\phi}{\partial t} + \tfrac{1}{2}\nabla|\nabla\phi|^2\right) \cdot \nabla\phi$$

$$+ (\gamma - 1)\left(\frac{\partial\phi}{\partial t} + \tfrac{1}{2}|\nabla\phi|^2\right)\nabla^2\phi. \tag{3.26}$$

Equation (3.26) is exact for a lossless perfect gas, but no analytical solutions are available for unsteady flow apart from those in Sect. 3.3.1 for plane waves. Equation (3.26) is often used in aeroelasticity, and it frequently serves as a starting point for perturbation analyses in nonlinear acoustics (see Sect. 10.1). When dissipation must be taken into account, approximations are normally introduced to obtain simplified wave equations that are more amenable to analysis. These approximations, and the resulting model equations, are introduced next.

3.4 Approximations for Thermoviscous Fluids

In order to describe nonlinear sound propagation in dissipative fluids, we abandon the search for exact solutions and turn instead to an approximation scheme based on the original full equations (3.1)–(3.4). Our limited aim is a set of model equations that describe three-dimensional finite-amplitude sound fields to second order in the acoustic Mach number $\varepsilon = u_0/c_0$ (where u_0 is a typical acoustic velocity magnitude). The suitability of ε as a small ordering parameter is indicated by its value $\varepsilon = 10^{-2}$ for 154 dB (re 20 µPa) in air and for 264 dB (re 1 µPa) in water.

However, even this limited objective produces unwieldy results if all $O(\varepsilon^2)$ terms are retained. Instead, we shall exploit the existence—in most situations of practical interest—of a second small parameter $\eta = \mu\omega/\rho_0 c_0^2$, where ω is a characteristic angular frequency. Physically, η is a measure of the importance of viscous stresses in a plane progressive sound wave, relative to the fluctuating pressure. At standard conditions one obtains $\eta \sim 10^{-6}$ for both 1 kHz in air and 1 MHz in water. The corresponding small parameter associated with heat conduction is $\kappa\omega/\rho_0 c_0^2 c_p = \eta/\mathrm{Pr}$, in which the Prandtl number $\mathrm{Pr} = \mu c_p/\kappa$ is $O(1)$ in terms of ε and η. With the expansion of Eqs. (3.1)–(3.4) limited to first order in η, and with only those $O(\varepsilon^2)$ terms that are zero order in η retained, a consistent nonlinear wave equation is derived in Sect. 3.5.

3.4.1 Small-Signal Modes and Ordering Scheme

The following discussion is based on Sects. 10-3 and 10-4 of Pierce's book (1989), to which the reader is referred for supporting details. In a viscous heat-conducting fluid, the variables $p = P - P_0$, $\rho' = \rho - \rho_0$, $T' = T - T_0$, $s' = s - s_0$, and $\mathbf{u}$ describe small disturbances relative to a uniform state of rest. Based on the linearized versions of Eqs. (3.1)–(3.4), dispersion relations can be obtained for three independent "modes" of small-signal disturbances in an unbounded fluid, called the acoustic, vorticity, and thermal (or entropy) modes. In general, each of the field variables contains contributions from each of the three modes; for example,

$$\mathbf{u} = \mathbf{u}_{\mathrm{ac}} + \mathbf{u}_{\mathrm{vor}} + \mathbf{u}_{\mathrm{ent}}. \tag{3.27}$$

Coupling between the modes occurs through imposition of boundary conditions. For example, at a fixed solid boundary $\mathbf{u}_{\mathrm{ac}} + \mathbf{u}_{\mathrm{vor}} + \mathbf{u}_{\mathrm{ent}} = 0$, and an incident disturbance in one mode (e.g., acoustic) is scattered by the boundary to produce reflected fields in all three modes.

The significance of these linearized modes in the present context is that in the weakly thermoviscous limit $\eta \ll 1$ (Pr is assumed of order one), the nonacoustic modes are governed by diffusion equations, with length scales of order $\eta^{1/2}$ times the acoustic wavelength. Specifically, it can be shown that near a boundary on which acoustic waves are incident, the magnitudes of the terms on the right-hand side of Eq. (3.27) are related as

$$|\mathbf{u}_{\mathrm{vor}}/\mathbf{u}_{\mathrm{ac}}| \simeq e^{-x/l_{\mathrm{vor}}}, \quad |\mathbf{u}_{\mathrm{ent}}/\mathbf{u}_{\mathrm{ac}}| \simeq [(\gamma - 1)/\mathrm{Pr}^{1/2}]\eta^{1/2}e^{-x/l_{\mathrm{ent}}}, \tag{3.28}$$

where x denotes distance from the boundary, the boundary-layer thicknesses l_{vor} and l_{ent} are defined by

$$kl_{\mathrm{vor}} = (2\eta)^{1/2}, \quad kl_{\mathrm{ent}} = (2\eta/\mathrm{Pr})^{1/2}, \tag{3.29}$$

and $k = \omega/c_0$ is the acoustic wave number. For small η, the boundary-layer thicknesses are thus very small fractions of an acoustic wavelength. Outside the thermoviscous boundary layer, the vorticity and entropy modes are effectively absent because of the exponential decay factors in Eqs. (3.28), and the variables p, ρ', T', and $\mathbf{u}$ may be identified with their acoustic-mode values p_{ac}, etc.

Following Lighthill (1956), we choose to treat ε and η as of comparable smallness, and we shall discard $O(\eta^2\varepsilon)$ and $O(\eta\varepsilon^2)$ terms along with $O(\varepsilon^3)$ terms in the expansions of Eqs. (3.1)–(3.4). The resulting model, with terms of order ε, $\eta\varepsilon$, and ε^2 retained, describes small-signal sound to leading order in η, and, more important, it is expected to account for the combined effects of nonlinearity and dissipation on weakly nonlinear three-dimensional sound waves. To facilitate implementation and discussion of this ordering scheme, we introduce a generic small parameter $\tilde{\varepsilon}$ that characterizes the smallness of both ε and η. Our primary objective in the remainder of this chapter is to derive model equations valid at order $\tilde{\varepsilon}^2$ outside thermoviscous boundary layers.

3.4.2   Second-Order Approximations

Begin by substituting $\rho = \rho_0 + \rho'$ in Eq. (3.1) and collecting $O(\tilde{\varepsilon})$ terms on the left, $O(\tilde{\varepsilon}^2)$ terms on the right:

$$\frac{\partial \rho'}{\partial t} + \rho_0 \nabla \cdot \mathbf{u} = -\rho' \nabla \cdot \mathbf{u} - \mathbf{u} \cdot \nabla \rho'. \tag{3.30}$$

Equation (3.30) remains exact. In Eq. (3.2), introduce the vector identities

$$\nabla(\nabla \cdot \mathbf{u}) = \nabla^2\mathbf{u} + \nabla \times \nabla \times \mathbf{u}$$

and, in $D\mathbf{u}/Dt$,

$$(\mathbf{u} \cdot \nabla)\mathbf{u} = \tfrac{1}{2}\nabla u^2 - \mathbf{u} \times \nabla \times \mathbf{u},$$

where $u^2 = \mathbf{u} \cdot \mathbf{u}$. Let $P = P_0 + p$, discard $O(\tilde{\varepsilon}^3)$ terms, and again collect $O(\tilde{\varepsilon})$ and $O(\tilde{\varepsilon}^2)$ terms on the left and right, respectively, to obtain

$$\rho_0\frac{\partial \mathbf{u}}{\partial t} + \nabla p = (\mu_B + \tfrac{4}{3}\mu)\nabla^2\mathbf{u} - \tfrac{1}{2}\rho_0\nabla u^2 - \rho'\frac{\partial \mathbf{u}}{\partial t}$$
$$+ (\mu_B + \tfrac{1}{3}\mu)\nabla \times \nabla \times \mathbf{u} + \rho_0\mathbf{u} \times \nabla \times \mathbf{u}. \tag{3.31}$$

Consider now the fourth and fifth terms on the right-hand side. Use of linear theory to approximate the vorticity field in these two terms[2] leads, via Eq. (3.27), to $\nabla \times \mathbf{u} \simeq \nabla \times \mathbf{u}_{\mathrm{vor}}$. The first of Eqs. (3.28) then shows that the fourth and fifth terms on the right-hand side of Eq. (3.31) decay exponentially away from boundaries and eventually become small in comparison with the corresponding first and second terms, which are dominated by acoustic-mode contributions. To determine where this occurs, let l_{vor} and k^{-1} characterize the length scales of the vorticity and acoustic modes, respectively, with $kl_{\mathrm{vor}} \ll 1$, and define $E = e^{-x/l_{\mathrm{vor}}}$. Taking Eqs. (3.28) and (3.29) into account, one finds that the magnitude of the fourth term relative to the first is $|\nabla \times \nabla \times \mathbf{u}_{\mathrm{vor}}|/|\nabla^2\mathbf{u}_{\mathrm{ac}}| \sim E/\eta$, and the magnitude of the fifth relative to the second is $|\mathbf{u}_{\mathrm{ac}} \times \nabla \times \mathbf{u}_{\mathrm{vor}}|/|\nabla u_{\mathrm{ac}}^2| \sim E/\eta^{1/2}$. Both the fourth and fifth terms may therefore be discarded for $E/\eta \ll 1$—for example, for $x/l_{\mathrm{vor}} \gtrsim 20$ with $\eta \sim 10^{-6}$. We shall henceforth assume that $E/\eta \ll 1$, which allows us to write Eq. (3.31) as

$$\rho_0\frac{\partial \mathbf{u}}{\partial t} + \nabla p = (\mu_B + \tfrac{4}{3}\mu)\nabla^2\mathbf{u} - \tfrac{1}{2}\rho_0\nabla u^2 - \rho'\frac{\partial \mathbf{u}}{\partial t}. \tag{3.32}$$

Note, however, that the vorticity terms that were discarded in order to arrive at this version of the momentum equation are essential to the analysis of acoustic streaming, a second-order phenomenon not considered here. For further details, see Chap. 7 and the discussion by Naze Tjøtta and Tjøtta (1990).

We next consider approximations of the entropy equation, Eq. (3.3). The presence of the coefficients κ, μ_B, and μ on the right side of Eq. (3.3) suggests that entropy perturbations s' generated as a by-product of a sound field are $O(\tilde{\varepsilon}^2)$. Closer investigation reveals this to be true only well away from solid boundaries, because

---

[2] For acoustic streaming calculations, this would not be valid—see the comment following Eq. (3.32).

at such boundaries, linear theory gives $|T_{ent}| \simeq |T_{ac}|$ (Pierce, 1989), and thus s' is $O(\varepsilon)$. Following steps similar to those leading to Eq. (3.32), one can show that for $E/\eta \ll 1$ (where now $E = e^{-x/l}$, and $l \sim l_{vor} \sim l_{ent}$ characterizes the thermoviscous boundary-layer thickness), the right side of Eq. (3.3) is dominated by the linear acoustic-mode term $\kappa \nabla^2 T'_{ac}$. We therefore obtain

$$\rho_0 T_0 \frac{\partial s'}{\partial t} = \kappa \nabla^2 T' \tag{3.33}$$

as the appropriate entropy equation, up to terms of second order in $\tilde{\varepsilon}$, for sound fields away from solid boundaries. Finally, expanding the equation of state, Eq. (3.4), in a Taylor series about the equilibrium state (ρ_0, s_0) and neglecting $O(\tilde{\varepsilon}^3)$ terms yields

$$p = c_0^2 \rho' + \frac{c_0^2}{\rho_0} \frac{B}{2A} \rho'^2 + \left(\frac{\partial P}{\partial s} \right)_{\rho,0} s', \tag{3.34}$$

where $B/A = (\rho_0/c_0^2)(\partial^2 P/\partial \rho^2)_{s,0}$ is the parameter of nonlinearity [Eq. (2.6)].

Equations (3.30) and (3.32)–(3.34) are the main results of this section. They represent consistent $O(\tilde{\varepsilon}^2)$ approximations of the full equations of motion and equation of state, Eqs. (3.1)–(3.4). More specifically, they permit the development of consistent second-order corrections to the linear acoustic wave equation for a nonreacting, weakly thermoviscous fluid ($\eta \ll 1$). Their domain of validity, $E/\eta \ll 1$, excludes thermoviscous boundary layers near solid surfaces.

3.5 Second-Order Wave Equation

Further manipulation of Eqs. (3.30) and (3.32)–(3.34) in order to combine them into a single wave equation requires repeated application of the following corollary to our ordering scheme: $O(\tilde{\varepsilon})$ acoustic-mode relations may be substituted into any $O(\tilde{\varepsilon}^2)$ terms, since the resulting errors are $O(\tilde{\varepsilon}^3)$. For example, the first term on the right-hand side of Eq. (3.30) may be rewritten as follows:

$$-\rho' \nabla \cdot \mathbf{u} = -\left(\frac{p}{c_0^2} \right) \left(-\frac{1}{\rho_0} \frac{\partial \rho'}{\partial t} \right) = \frac{p}{\rho_0 c_0^4} \frac{\partial p}{\partial t}, \tag{3.35}$$

where the equal signs signify equality at $O(\tilde{\varepsilon}^2)$. For Eqs. (3.30) and (3.32), respectively, this procedure yields (Aanonsen et al., 1984)

$$\frac{\partial \rho'}{\partial t} + \rho_0 \nabla \cdot \mathbf{u} = \frac{1}{\rho_0 c_0^4} \frac{\partial p^2}{\partial t} + \frac{1}{c_0^2} \frac{\partial \mathcal{L}}{\partial t}, \tag{3.36}$$

$$\rho_0 \frac{\partial \mathbf{u}}{\partial t} + \nabla p = -\frac{1}{\rho_0 c_0^2} (\mu_B + \tfrac{4}{3}\mu) \nabla \frac{\partial p}{\partial t} - \nabla \mathcal{L}, \tag{3.37}$$

where

$$\mathcal{L} = \tfrac{1}{2}\rho_0 u^2 - \frac{p^2}{2\rho_0 c_0^2} \tag{3.38}$$

is the second-order Lagrangian density. It is interesting to note that $\mathcal{L} = 0$ for plane progressive waves (since at first order, $p = \rho_0 c_0 u$), in which case the momentum equation, Eq. (3.37), is linear and therefore does not contribute at $O(\tilde{\varepsilon}^2)$ to finite-amplitude effects in such fields (Hamilton and Blackstock, 1990).

Now combine Eqs. (3.33) and (3.34) as follows: Use the $O(\tilde{\varepsilon})$ relation $\nabla^2 T' = c_0^{-2}\partial^2 T'/\partial t^2$ (i.e., the lossless, linear wave equation) to permit integration of Eq. (3.33) with respect to time, and thus eliminate s' in favor of T' in Eq. (3.34). To eliminate temperature, write $T = T(\rho, s)$ and use the $O(\tilde{\varepsilon})$ expansion $T' = (\partial T/\partial \rho)_{s,0}\rho'$ to obtain, following further $O(\tilde{\varepsilon})$ substitutions,

$$\rho' = \frac{p}{c_0^2} - \frac{1}{\rho_0 c_0^4} \frac{B}{2A} p^2 - \frac{\kappa}{\rho_0 c_0^6 T_0} \left(\frac{\partial T}{\partial \rho}\right)_{s,0} \left(\frac{\partial P}{\partial s}\right)_{\rho,0} \frac{\partial p}{\partial t}. \tag{3.39}$$

The thermodynamic relations $(\partial P/\partial s)_\rho = \rho^2(\partial T/\partial \rho)_s$ and $(\partial T/\partial \rho)_s^2 = RTc^2/c_v c_p \rho^2$, where $R = c_p - c_v$, may be used to rewrite the third term in Eq. (3.39) (see, e.g., Fetter and Walecka, 1980) to obtain

$$\rho' = \frac{p}{c_0^2} - \frac{1}{\rho_0 c_0^4} \frac{B}{2A} p^2 - \frac{\kappa}{\rho_0 c_0^4} \left(\frac{1}{c_v} - \frac{1}{c_p}\right) \frac{\partial p}{\partial t}. \tag{3.40}$$

For a perfect gas, one may use Eq. (3.5), together with $P = R\rho T$, to show that $(\partial T/\partial \rho)_{s,0} = RT_0/c_v\rho_0$ and $(\partial P/\partial s)_{\rho,0} = \rho_0 c_0^2/c_p$, the substitution of which into Eq. (3.39) also yields Eq. (3.40).

To obtain the desired second-order wave equation, subtract the time derivative of Eq. (3.36) from the divergence of Eq. (3.37), use Eq. (3.40) to eliminate ρ', and substitute the $O(\tilde{\varepsilon})$ relation $\nabla^2 p = c_0^{-2}\partial^2 p/\partial t^2$ into the viscosity term to give (Aanonsen et al., 1984)

$$\Box^2 p + \frac{\delta}{c_0^4} \frac{\partial^3 p}{\partial t^3} = -\frac{\beta}{\rho_0 c_0^4} \frac{\partial^2 p^2}{\partial t^2} - \left(\nabla^2 + \frac{1}{c_0^2} \frac{\partial^2}{\partial t^2}\right)\mathcal{L}, \tag{3.41}$$

where $\Box^2 = \nabla^2 - c_0^{-2}(\partial^2/\partial t^2)$ is the d'Alembertian operator, $\beta = 1 + B/2A$ is the coefficient of nonlinearity, and δ is the diffusivity of sound (Lighthill, 1978):

$$\delta = \frac{1}{\rho_0}\left(\frac{4}{3}\mu + \mu_B\right) + \frac{\kappa}{\rho_0}\left(\frac{1}{c_v} - \frac{1}{c_p}\right) = \nu\left(\frac{4}{3} + \frac{\mu_B}{\mu} + \frac{\gamma - 1}{\mathrm{Pr}}\right),\qquad(3.42)$$

where $\nu = \mu/\rho_0$ is the kinematic viscosity. Typical values for μ_B/μ are 0.6 for air at ambient temperature (reflecting rotational relaxation), and 3.0–2.7 for water at 0–60 °C (Pierce, 1989). The heat conduction term $(\gamma - 1)/\mathrm{Pr}$ is generally more significant for gases than for liquids.[3]

Equation (3.41) is a consistent $O(\tilde{\varepsilon}^2)$ wave equation for the sound pressure in a weakly thermoviscous fluid ($\eta \ll 1$), valid outside any thin thermoviscous boundary layers that may be present. It accounts for the mutual interaction of nonlinear and dissipative processes in modifying the propagation of sound in three dimensions. The wave field need not be progressive. Alternative, but lossless, forms of Eq. (3.41) were derived by Eckart (1948) and Westervelt (1957). Note that the small-signal attenuation coefficient $\alpha(\omega) = -\mathrm{Im}\,\chi$ follows from Eq. (3.41) if plane-wave solutions of the form $\exp[j(\omega t - \chi x)]$ are postulated. The characteristic wave number χ is found to be $k(1 - j\delta\omega/c_0^2)^{1/2}$, where $k = \omega/c_0$, and binomial expansion yields $\alpha \simeq \delta\omega^2/2c_0^3$ for $\alpha \ll k$.

3.6 Westervelt Equation

The Westervelt (1963) equation is obtained from Eq. (3.41) by discarding the term containing $\mathcal{L}$. We noted above that $\mathcal{L} = 0$ at $O(\tilde{\varepsilon}^2)$ for progressive plane waves. In general, however, justification for omitting the Lagrangian density is based on the distinction between cumulative and local nonlinear effects (Aanonsen et al., 1984). Begin by noting that at $O(\tilde{\varepsilon}^2)$, Eq. (3.38) can be rewritten in terms of the velocity potential as $\mathcal{L} = \frac{1}{4}\rho_0\Box^2\phi^2$, where use has been made of the first-order relation $p = -\rho_0\partial\phi/\partial t$ from Eq. (3.21). Equation (3.41) can therefore be rewritten as

$$\Box^2\left[p + \frac{\rho_0}{4}\left(\nabla^2 + \frac{1}{c_0^2}\frac{\partial^2}{\partial t^2}\right)\phi^2\right] + \frac{\delta}{c_0^4}\frac{\partial^3 p}{\partial t^3} = -\frac{\beta}{\rho_0 c_0^4}\frac{\partial^2 p^2}{\partial t^2}.\qquad(3.43)$$

Now introduce an auxiliary variable $\tilde{p}$ defined by

$$\tilde{p} = p + \frac{\rho_0}{4}\left(\nabla^2 + \frac{1}{c_0^2}\frac{\partial^2}{\partial t^2}\right)\phi^2,\qquad(3.44)$$

and eliminate p from Eq. (3.43) to obtain, ignoring $O(\tilde{\varepsilon}^3)$ terms,

---

[3] For a mixture of different fluids, δ contains a contribution due to molecular diffusion of species, which may be absorbed into μ_B in Eq. (3.42). In air, diffusion accounts for about 0.5% of the total value of δ.

$$\Box^2 \tilde{p} + \frac{\delta}{c_0^4} \frac{\partial^3 \tilde{p}}{\partial t^3} = -\frac{\beta}{\rho_0 c_0^4} \frac{\partial^2 \tilde{p}^2}{\partial t^2}. \tag{3.45}$$

The system defined by Eqs. (3.44) and (3.45) is consistent, at $O(\tilde{\varepsilon}^2)$, with Eq. (3.41). The Westervelt equation thus corresponds to assuming that $\tilde{p} \simeq p$:

$$\Box^2 p + \frac{\delta}{c_0^4} \frac{\partial^3 p}{\partial t^3} = -\frac{\beta}{\rho_0 c_0^4} \frac{\partial^2 p^2}{\partial t^2}. \tag{3.46}$$

The validity of the Westervelt equation, Eq. (3.46), depends on the extent to which the approximation $\tilde{p} \simeq p$ is valid.[4]

To evaluate this approximation, note that at $O(\tilde{\varepsilon}^2)$, the solution of Eq. (3.41) for the pressure p can be obtained by solving Eq. (3.45) for $\tilde{p}$ and then recovering the solution for p from Eq. (3.44) (Naze Tjøtta and Tjøtta, 1987). The desired solution p thus differs from the auxiliary solution $\tilde{p}$ (i.e., the solution of the Westervelt equation) by a function of ϕ^2 that depends only on the local properties of the sound field at the point of interest. Any and all cumulative nonlinear effects must therefore be associated with the term on the right-hand side of Eq. (3.45). The right-hand side may be regarded as a forcing function corresponding to a spatial distribution of virtual sources created by the sound wave itself. The integrated effect of these virtual sources on the sound wave accumulates with distance in the direction of propagation. Cumulative nonlinear effects produce the waveform steepening illustrated in Fig. 4.1, and the resonant harmonic interactions described in Sect. 4.2.4.

The Westervelt equation is thus an appropriate approximation of the full second-order wave equation, Eq. (3.41), when cumulative nonlinear effects dominate local nonlinear effects. The difficulty is knowing when local effects can be ignored. Errors due to ignoring the distinction between Lagrangian and Eulerian coordinates (e.g., linearizing a vibrating source condition by prescribing the velocity on a stationary surface) or due to using linear theory to transform the solution or model equation from one acoustical field variable to another (e.g., using the impedance relation $p = \rho_0 c_0 u$ for progressive plane waves) are noncumulative (Blackstock, 1962; see also Sect. 4.2.3). They become insignificant in comparison with the cumulative effects of waveform distortion once the propagation distance becomes much greater than a wavelength. The same applies to ignoring the distinction between the boundary conditions on p and $\tilde{p}$. As a general rule, except within one wavelength away from the source, local effects can be ignored for problems involving progressive quasi-plane waves, such as directional sound beams. Where local effects can be important are in compound wave fields, such as standing waves, guided waves in nonplanar

---

[4] The equation obtained by Westervelt (1963) is actually the lossless form of Eq. (3.46). We also note that a lossless form of Eqs. (3.44) and (3.45) was obtained independently by Morse and Ingard (1968), who used the equivalent $O(\varepsilon^2)$ relation $\tilde{p} = p + \frac{1}{2}\rho_0 u^2 + p^2/2\rho_0 c_0^2 + (\rho_0 c_0^2)^{-1}(\partial p/\partial t) \int p \, dt$ in place of Eq. (3.44).

modes (Sect. 5.3.1), and scattering of sound by sound from intersecting sound beams (Naze Tjøtta and Tjøtta, 1988). Radiation pressure, the subject of Chap. 6, is another example of local effects.

3.7 Burgers Equation

The Burgers equation is the simplest model that describes the combined effects of nonlinearity and losses on the propagation of plane progressive waves. Our derivation begins with the one-dimensional form of the Westervelt equation, Eq. (3.46):

$$
\left(\frac{\partial^2}{\partial x^2} - \frac{1}{c_0^2} \frac{\partial^2}{\partial t^2} \right) p + \frac{\delta}{c_0^4} \frac{\partial^3 p}{\partial t^3} = -\frac{\beta}{\rho_0 c_0^4} \frac{\partial^2 p^2}{\partial t^2}. \tag{3.47}
$$

A method for simplifying Eq. (3.47) can be deduced from examination of approximate solutions for plane progressive waves in two limiting cases, the first without dissipation and the second without nonlinearity. For the first case, we use the boundary-value solution for u presented in Sect. 3.3.1. An expression in terms of the sound pressure follows from substituting $u = p/\rho_0 c_0$ into the second of Eqs. (3.19):

$$
p = f[\tau + (\beta p/\rho_0 c_0^3)x], \tag{3.48}
$$

where the retarded time $\tau = t - x/c_0$ has been introduced. For the second case, we repeat here the linear solution obtained at the end of Sect. 3.5:

$$
p = p_0 \exp[j\omega\tau - (\delta\omega^2/2c_0^3)x]. \tag{3.49}
$$

In both Eqs. (3.48) and (3.49) the coefficient of x is $O(\tilde{\varepsilon})$. Each solution thus exhibits the functional form

$$
p = p(x_1, \tau), \quad x_1 = \tilde{\varepsilon}x, \quad \tau = t - x/c_0. \tag{3.50}
$$

The meaning of Eqs. (3.50) is that in the retarded time frame (i.e., for an observer in a reference frame that moves at speed c_0), nonlinearity and absorption separately produce only slow variations as functions of distance. Moreover, the relative order of the variations due to each effect is the same, i.e., it is $O(\tilde{\varepsilon})$. We thus anticipate that the combined effects of nonlinearity and absorption will introduce variations of the same order. The coordinate x_1 is referred to as the *slow scale* corresponding to the retarded time frame τ.

To derive a simplified progressive-wave equation that accounts for both absorption and nonlinearity, we first rewrite Eq. (3.47) in the new coordinate system (x_1, τ). Transformation of the partial derivatives yields

$$\left(\frac{\partial}{\partial x}\right)_t = \tilde{\varepsilon}\left(\frac{\partial}{\partial x_1}\right)_\tau - \frac{1}{c_0}\left(\frac{\partial}{\partial \tau}\right)_{x_1}, \quad \left(\frac{\partial}{\partial t}\right)_x = \left(\frac{\partial}{\partial \tau}\right)_{x_1}, \tag{3.51}$$

substitution of which into Eq. (3.47) gives

$$\tilde{\varepsilon}^2\frac{\partial^2 p}{\partial x_1^2} - \tilde{\varepsilon}\frac{2}{c_0}\frac{\partial^2 p}{\partial x_1 \partial \tau} + \frac{\delta}{c_0^4}\frac{\partial^3 p}{\partial \tau^3} = -\frac{\beta}{\rho_0 c_0^4}\frac{\partial^2 p^2}{\partial \tau^2}. \tag{3.52}$$

The first term in Eq. (3.52) is $O(\tilde{\varepsilon}^3)$ and is therefore discarded. Integration of the remaining terms with respect to τ (recognizing that the function of position that results from integration must vanish in the absence of sound, and must therefore be zero everywhere) and multiplication of the resulting equation by $-\frac{1}{2}c_0$ leads to

$$\tilde{\varepsilon}\frac{\partial p}{\partial x_1} - \frac{\delta}{2c_0^3}\frac{\partial^2 p}{\partial \tau^2} = \frac{\beta}{2\rho_0 c_0^3}\frac{\partial p^2}{\partial \tau}. \tag{3.53}$$

All three terms in Eq. (3.53)—the loss term, the nonlinear term, and the resulting rate of change with distance—are of the same order. Equation (3.53) is the desired $O(\tilde{\varepsilon}^2)$ equation in the coordinate system (x_1, τ).

Note that $\tilde{\varepsilon}$ is simply an ordering parameter, which we now remove from Eq. (3.53) by reinstating the physical coordinate x in place of x_1. This is accomplished by replacing $\tilde{\varepsilon}(\partial/\partial x_1)$ with $\partial/\partial x$ (but retaining τ):

$$\frac{\partial p}{\partial x} - \frac{\delta}{2c_0^3}\frac{\partial^2 p}{\partial \tau^2} = \frac{\beta p}{\rho_0 c_0^3}\frac{\partial p}{\partial \tau}. \tag{3.54}$$

Equation (3.54) is the Burgers equation,[5] and it is the most widely used model equation for studying the combined effects of dissipation and nonlinearity on progressive plane waves. Note that Eq. (3.48) is a solution of Eq. (3.54) for $\delta = 0$, and Eq. (3.49) is a solution for $\beta = 0$. The history of the Burgers equation in the field of acoustics is reviewed in Sect. 4.5.1. We call attention here only to landmark derivations of the Burgers equation by Mendousse (1953) for viscous fluids and by Lighthill (1956) for thermoviscous gases.

---

[5] The equation obtained by Burgers (1948) is $v_t + 2vv_y = \nu v_{yy}$ (subscripts denote partial derivatives), in which ν is a viscosity coefficient and v is a velocity associated with turbulent flow in a channel with cross-sectional coordinate y.

3.8 Generalized Burgers Equation

The generalized Burgers equation is an extension of Eq. (3.54) that takes into account the divergence (or convergence) of progressive spherical or cylindrical waves, and that reduces to Eq. (3.54) for plane waves. We begin the derivation by writing the Westervelt equation, Eq. (3.46), as follows:

$$
\left(\frac{\partial^2}{\partial r^2} + 2\frac{m}{r}\frac{\partial}{\partial r} - \frac{1}{c_0^2}\frac{\partial^2}{\partial t^2} \right) p + \frac{\delta}{c_0^4}\frac{\partial^3 p}{\partial t^3} = -\frac{\beta}{\rho_0 c_0^4}\frac{\partial^2 p^2}{\partial t^2}, \tag{3.55}
$$

where the Laplacian is expressed in terms of the radial coordinate r for the wave field, with $m = 0$, $\frac{1}{2}$, or 1 for plane, cylindrical, or spherical waves, respectively. The field is assumed to depend only on r and t. For plane waves, replace r by x and thus recover Eq. (3.47).

Recall that local nonlinear effects, which are taken into account by $\mathcal{L}$ in Eq. (3.41), are not included in Eq. (3.55). We also note again that at $O(\tilde{\varepsilon}^2)$, $\mathcal{L} = 0$ for progressive plane waves, for which $p = \pm\rho_0 c_0 u$ at $O(\tilde{\varepsilon})$ (the $\pm$ signs account for both outgoing and incoming waves). From linear theory, we know that to the same order the pressure and particle velocity in time-harmonic spherical and cylindrical waves are also related by $p = \pm\rho_0 c_0 u$ for $kr \gg 1$. Equation (3.55) is thus an appropriate model for lowest-order nonlinear effects in spherical and cylindrical waves with weak attenuation ($\alpha \ll k$) provided $kr_0 \gg 1$, where r_0 is the source (or starting) radius and k is the wave number corresponding to the lowest significant frequency component in the wave. Note that the restriction $kr_0 \gg 1$ on Eq. (3.55) can be replaced by $r_0 > \lambda$, that is, the source radius must exceed the maximum wavelength of the radiated sound. Such a restriction is normally weak in practice for diverging waves, because sources smaller than one wavelength are relatively inefficient and thus unlikely to radiate sound of finite amplitude. Conversely, Eq. (3.55) cannot be used closer than distances on the order of one wavelength away from the focus in a converging wave field.

On the basis of the preceding discussion, it is reasonable to seek a solution having a functional form similar to Eqs. (3.50) for plane waves:

$$
p = p(r_1, \tau), \quad r_1 = \tilde{\varepsilon}r, \quad \tau = t \mp (r - r_0)/c_0. \tag{3.56}
$$

Our sign convention when a choice exists, as in the retarded time, is to take the upper sign for diverging (outgoing) waves and the lower sign for converging (incoming) waves. The constant r_0/c_0 is introduced for convenience, in order to obtain $\tau = t$ at the source, $r = r_0$. The coordinate transformation from (r, t) to (r_1, τ) is similar to that for the plane-wave case, and Eq. (3.55) becomes, after $O(\tilde{\varepsilon}^3)$ terms are discarded and only $O(\tilde{\varepsilon}^2)$ terms remain,

$$\mp \tilde{\varepsilon} \frac{2}{c_0} \frac{\partial^2 p}{\partial r_1 \partial \tau} \mp \tilde{\varepsilon} \frac{2m}{c_0 r_1} \frac{\partial p}{\partial \tau} + \frac{\delta}{c_0^4} \frac{\partial^3 p}{\partial \tau^3} = -\frac{\beta}{\rho_0 c_0^4} \frac{\partial^2 p^2}{\partial \tau^2}. \tag{3.57}$$

Integration with respect to τ, multiplication by $\mp \frac{1}{2} c_0$, and transformation from the coordinates (r_1, τ) to (r, τ) yield

$$\frac{\partial p}{\partial r} + \frac{m}{r} p \mp \frac{\delta}{2c_0^3} \frac{\partial^2 p}{\partial \tau^2} = \pm \frac{\beta p}{\rho_0 c_0^3} \frac{\partial p}{\partial \tau}. \tag{3.58}$$

Equation (3.58) is the generalized Burgers equation, and it reduces to Eq. (3.54) for $m = 0$ (in which case one sets $r = x$ and $r_0 = 0$). Equation (3.58) was derived first by Khokhlov and coworkers (Naugol'nykh et al., 1963; Khokhlov et al., 1964), while a general form of Eq. (3.58) for lossless fluids was developed by Blackstock (1964). The relation $p = \pm \rho_0 c_0 u$ may be used to rewrite Eq. (3.58) in terms of particle velocity, because in the (r_1, τ) coordinate system one obtains $p = \pm \rho_0 c_0 u + O(\tilde{\varepsilon}^2)$ from Eq. (3.37), and the higher-order terms would appear at third order in Eq. (3.57).

We conclude by noting that Eq. (3.58) may be further generalized to accommodate progressive-wave propagation in narrow ducts with variable cross-sectional area $A(x)$, where x is distance along the duct. It is assumed that the duct diameter D is small enough ($kD < 1$), and that the area varies sufficiently slowly on the scale of a wavelength ($k^{-1} |A'/A| \ll 1$, where $A' = dA/dx$), to justify a one-dimensional propagation model. We also ignore dissipation. The continuity equation then becomes (Landau and Lifshitz, 1987)

$$\frac{\partial \rho}{\partial t} + \frac{\partial (\rho u)}{\partial x} + \frac{A'}{A} \rho u = 0, \tag{3.59}$$

whereas the momentum and state equations are unaltered. The assumption of lossless propagation requires l/D to be of smaller order than ε, where l represents the thermoviscous boundary-layer thicknesses defined in Eqs. (3.29). Taking now $k^{-1} A'/A = O(\varepsilon)$ and introducing a slow scale as above, with $\tau = t \mp (x - x_0)/c_0$, one may obtain the following progressive-wave equation:

$$\frac{\partial p}{\partial x} + \frac{A'}{2A} p = \pm \frac{\beta p}{\rho_0 c_0^3} \frac{\partial p}{\partial \tau}. \tag{3.60}$$

Here $A'/2A$ generalizes the factor m/r in Eq. (3.58), with x replacing the radial coordinate r ($A'/2A = m/x$ for $A \propto x^{2m}$). Equation (3.60) exhibits similarities with Eq. (12.58), which describes nonlinear propagation along rays. Moreover, it may be rewritten as (Blackstock, 1972)

$$\frac{\partial q}{\partial z} = \frac{\beta q}{\rho_0 c_0^3} \frac{\partial q}{\partial \tau}, \quad q = (A/A_0)^{1/2} p, \quad z = \pm \int_{x_0}^{x} (A_0/A)^{1/2} \, dx, \tag{3.61}$$

and $A_0 = A(x_0)$. Comparing the first of Eqs. (3.61) with the original Burgers equation, Eq. (3.54), we see that solutions of the latter for lossless plane waves may be transformed for application to ducts with variable cross section (see Sect. 4.6.1).

3.9   KZK Equation

The KZK (Khokhlov–Zabolotskaya–Kuznetsov) equation is an augmentation of the Burgers equation that accounts for the combined effects of diffraction, absorption, and nonlinearity in directional sound beams. Insight into how to choose an appropriate slow scale is provided by the derivation of Eq. (3.58). Let z designate the nominal axis of the beam pointing in the propagation direction, and let (x, y) be the coordinates perpendicular to that axis. The following assumptions are made regarding the source: It is defined in the plane $z = 0$, it has a characteristic radius a, and it radiates at frequencies that satisfy the relation $ka \gg 1$. The last assumption ensures that the beam is reasonably directional. Because of the assumed directionality of the beam, the sound is localized in the vicinity of the z axis, and the wavefronts are quasi-planar.

Linear theory for directional beams reveals the existence of near-field and far-field regions, with the latter beginning roughly at the Rayleigh distance $\frac{1}{2}ka^2$, measured from the source along the beam axis (see Sect. 8.3). The near field is characterized by wavefronts that are approximately planar, and the far field is characterized by wavefronts that are spherical [note that the relation $kz \gg 1$ is satisfied in the far field, where we have $z > \frac{1}{2}ka^2$ and therefore $kz > \frac{1}{2}(ka)^2 \gg 1$]. Consistent with the derivation of Eq. (3.58), a reasonable choice for the slow scale in the direction of propagation is therefore (z_1, τ), where $z_1 = \tilde{\varepsilon}z$ and $\tau = t - z/c_0$.

The appropriate scale for the coordinates perpendicular to the z axis is less apparent. We impose the stipulation that, like the effects of absorption and nonlinearity, the effect of diffraction must appear at $O(\tilde{\varepsilon}^2)$ so that all three effects contribute at the same order. We shall demonstrate by substitution that the following relations define the slow scale that accommodates this stipulation:

$$p = p(x_1, y_1, z_1, \tau), \quad (x_1, y_1, z_1) = (\tilde{\varepsilon}^{1/2}x, \tilde{\varepsilon}^{1/2}y, \tilde{\varepsilon}z), \quad \tau = t - z/c_0.$$
$$(3.62)$$

The postulated slow scale suggests that to an observer moving at speed c_0 in the z direction ($\tau = $ const), spatial variations are perceived to occur more slowly, by $O(\tilde{\varepsilon}^{1/2})$, along the axis of the beam than across the beam. Indeed, in the near field, the length scale for variations along the axis of the beam that are caused by diffraction is the Rayleigh distance $\frac{1}{2}ka^2 = \frac{1}{2}(ka)a$, which is greater than the length scale a across the beam (because $ka \gg 1$). An additional consequence of this reasoning is that ka is $O(\tilde{\varepsilon}^{-1/2})$.

We now write Eq. (3.46) at $O(\tilde{\varepsilon}^2)$ on the slow scale defined in Eqs. (3.62). Transformation of the Laplacian yields

$$\nabla^2 = \tilde{\varepsilon}\left(\frac{\partial^2}{\partial x_1^2} + \frac{\partial^2}{\partial y_1^2}\right) + \tilde{\varepsilon}^2 \frac{\partial^2}{\partial z_1^2} - \tilde{\varepsilon}\frac{2}{c_0}\frac{\partial^2}{\partial z_1 \partial \tau} + \frac{1}{c_0^2}\frac{\partial^2}{\partial \tau^2}, \tag{3.63}$$

and Eq. (3.46) becomes, after the $O(\tilde{\varepsilon}^3)$ term is discarded,

$$\tilde{\varepsilon}\left(\frac{\partial^2}{\partial x_1^2} + \frac{\partial^2}{\partial y_1^2}\right)p - \tilde{\varepsilon}\frac{2}{c_0}\frac{\partial^2 p}{\partial z_1 \partial \tau} + \frac{\delta}{c_0^4}\frac{\partial^3 p}{\partial \tau^3} = -\frac{\beta}{\rho_0 c_0^4}\frac{\partial^2 p^2}{\partial \tau^2}. \tag{3.64}$$

Equation (3.64) thus confirms that the slow scale introduced in Eqs. (3.62) accounts for diffraction at the same order as dissipation and nonlinearity, i.e., at $O(\tilde{\varepsilon}^2)$. Transformation of Eq. (3.64) from the slow scale (x_1, y_1, z_1) back to (x, y, z) yields

$$\frac{\partial^2 p}{\partial z \partial \tau} - \frac{c_0}{2}\nabla_\perp^2 p - \frac{\delta}{2c_0^3}\frac{\partial^3 p}{\partial \tau^3} = \frac{\beta}{2\rho_0 c_0^3}\frac{\partial^2 p^2}{\partial \tau^2}, \tag{3.65}$$

where $\nabla_\perp^2 = \partial^2/\partial x^2 + \partial^2/\partial y^2$ is a Laplacian that operates in the plane perpendicular to the axis of the beam. Equation (3.65) is the KZK equation, and it is the most widely used model equation for describing the combined effects of diffraction, nonlinearity, and absorption in directional sound beams (see Chap. 8). Equation (3.65) was derived first for $\delta = 0$ by Zabolotskaya and Khokhlov (1969), and the loss term was included subsequently by Kuznetsov (1971). A later derivation, based on the method of multiple scales, is provided by Naze Tjøtta and Tjøtta (1981). In the absence of diffraction ($\nabla_\perp^2 p = 0$), Eq. (3.65) reduces to the Burgers equation, Eq. (3.54).

The impedance relation that is consistent with use of the KZK equation is obtained from Eq. (3.37). First write the particle velocity vector in component form: $\mathbf{u} = (\mathbf{u}_\perp, u_z)$, where $\mathbf{u}_\perp = (u_x, u_y)$. On the slow scale in Eqs. (3.62), the z component of Eq. (3.37) yields the relation $p = \rho_0 c_0 u_z + O(\tilde{\varepsilon}^2)$. In analyses based on the KZK equation, it is thus consistent to use the linear plane-progressive-wave impedance relation $p = \rho_0 c_0 u_z$ to convert between pressure and the z component of the particle velocity. Equation (3.37) also reveals that $\mathbf{u}_\perp$ is $O(\tilde{\varepsilon}^{3/2})$ on the slow scale, which indicates that the particle motion is mainly in the z direction, because u_z is $O(\tilde{\varepsilon})$. These conclusions are consistent with the notion of directional beams composed of quasi-plane waves.

Finally, we note that ignoring the second term on the right-hand side of Eq. (3.63) is often referred to as the parabolic approximation, owing to the fact that the hyperbolic wave equation (two derivatives in the propagation direction z) is approximated by a parabolic equation (having one derivative in z). The parabolic approximation is a high-frequency limit, as may be shown by considering the $O(\varepsilon)$ equation $\Box^2 p = 0$. Assume radiation at angular frequency ω in predominantly the z direction and let $p = q(x, y, z)e^{j\omega\tau}$, where $\omega\tau = \omega t - kz$. The parabolic approximation corresponds to the assumption $|\partial^2 q/\partial z^2| \ll 2k|\partial q/\partial z|$, i.e., that q is a slowly varying function of z at high frequency. One then obtains $\Box^2 p \simeq [-j2k(\partial q/\partial z) + \nabla_\perp^2 q]e^{j\omega\tau}$, which following multiplication by $-\frac{1}{2}c_0$ is the frequency-domain form of the first two terms in Eq. (3.65).

References

Aanonsen, S. I., Barkve, T., Naze Tjøtta, J., and Tjøtta, S. (1984). Distortion and harmonic generation in the nearfield of a finite amplitude sound beam. *J. Acoust. Soc. Am.* **75**, 749–768.
Blackstock, D. T. (1962). Propagation of plane sound waves of finite amplitude in nondissipative fluids. *J. Acoust. Soc. Am.* **34**, 9–30.
Blackstock, D. T. (1964). On plane, spherical, and cylindrical sound waves of finite amplitude in lossless fluids. *J. Acoust. Soc. Am.* **36**, 217–219.
Blackstock, D. T. (1972). Nonlinear acoustics (theoretical). In *American Institute of Physics Handbook*, 3rd ed. (McGraw-Hill, New York), Chapter 3n.
Burgers, J. M. (1948). A mathematical model illustrating the theory of turbulence. In *Advances in Applied Mechanics*, Vol. 1, R. von Mises and T. von Kármán, eds. (Academic Press, New York), pp. 171–199.
Eckart, C. (1948). Vortices and streams caused by sound waves. *Phys. Rev.* **73**, 68–76.
Fetter, A. L., and Walecka, J. D. (1980). *Theoretical Mechanics of Particles and Continua* (McGraw-Hill, New York), Sec. 62.
Hamilton, M. F., and Blackstock, D. T. (1990). On the linearity of the momentum equation for progressive plane waves of finite amplitude. *J. Acoust. Soc. Am.* **88**, 2025–2026.
Khokhlov, R. V., Naugol'nykh, K. A., and Soluyan, S. I. (1964). Waves of moderate amplitudes in absorbing media. *Acustica* **14**, 248–253.
Kuznetsov, V. P. (1971). Equations of nonlinear acoustics. *Sov. Phys. Acoust.* **16**, 467–470.
Landau, L. D., and Lifshitz, E. M. (1987). *Fluid Mechanics*, 2nd ed. (Pergamon Press, New York).
Lighthill, M. J. (1956). Viscosity effects in sound waves of finite amplitude. In *Surveys in Mechanics*, G. K. Batchelor and R. M. Davies, eds. (Cambridge University Press, Cambridge, England), pp. 250–351.
Lighthill, M. J. (1978). *Waves in Fluids* (Cambridge University Press, Cambridge, England).
Mendousse, J. S. (1953). Nonlinear dissipative distortion of progressive sound waves at moderate amplitudes. *J. Acoust. Soc. Am.* **25**, 51–54.
Morse, P. M., and Ingard, K. U. (1968). *Theoretical Acoustics* (McGraw-Hill, New York), Eqs. (14.4.16) and (14.4.17).
Naugol'nykh, K. A., Soluyan, S. I., and Khokhlov, R. V. (1963). Spherical waves of finite amplitude in a viscous thermally conducting medium. *Sov. Phys. Acoust.* **9**, 42–46.
Naze Tjøtta, J., and Tjøtta, S. (1981). Nonlinear equations of acoustics, with application to parametric acoustic arrays. *J. Acoust. Soc. Am.* **69**, 1644–1652.
Naze Tjøtta, J., and Tjøtta, S. (1987). Interaction of sound waves. Part I: Basic equations and plane waves. *J. Acoust. Soc. Am.* **82**, 1425–1428.
Naze Tjøtta, J., and Tjøtta, S. (1988). Interaction of sound waves. Part III: Two real beams. *J. Acoust. Soc. Am*, **83**, 487–495.
Naze Tjøtta, J., and Tjøtta, S. (1990). Nonlinear equations of acoustics. In *Frontiers of Nonlinear Acoustics: 12th ISNA*, M. F. Hamilton and D. T. Blackstock, eds. (Elsevier, London), pp. 80–97.
Pierce, A. D. (1989). *Acoustics* (Acoustical Society of America, New York).
Poisson, S. D. (1808). Memoir on the theory of sound. *J. l'école polytech.* (Paris) **7**, 319–392. See Secs. 24–25, pp. 364–370. English translation of Sections 23–25: Beyer, R. T. (1984). *Nonlinear Acoustics in Fluids*, Benchmark Papers in Acoustics, Vol. 18 (Van Nostrand Reinhold, New York), pp. 23–28.
Riemann, B. (1860). The propagation of sound waves of finite amplitude. *Abhandl. Ges. Wiss. Göttingen, Math.-physik.* **8**, 43–65 (1858–1859), reprinted in *The Collected Works of Bernard*

Riemann (Dover, New York, 1953), pp. 156–175. English translation of Sections 1–10: Beyer, R. T. (1984). *Nonlinear Acoustics in Fluids*, Benchmark Papers in Acoustics, Vol. 18 (Van Nostrand Reinhold, New York), pp. 42–60.

Thompson, P. A. (1984). *Compressible-Fluid Dynamics* (Maple Press, New York).

Westervelt, P. J. (1957). Scattering of sound by sound. *J. Acoust. Soc. Am.* **29**, 199–203.

Westervelt, P. J. (1963). Parametric acoustic array. *J. Acoust. Soc. Am.* **35**, 535–537.

Zabolotskaya, E. A., and Khokhlov, R. V. (1969). Quasi-plane waves in the nonlinear acoustics of confined beams. *Sov. Phys. Acoust.* **15**, 35–40.

Chapter 4
Progressive Waves in Lossless and Lossy Fluids

David T. Blackstock, Mark F. Hamilton, and Allan D. Pierce

Contents

D. T. Blackstock · M. F. Hamilton
Department of Mechanical Engineering, The University of Texas at Austin, Austin, TX, USA

A. D. Pierce
Department of Aerospace and Mechanical Engineering, Boston University, Boston, MA, USA

© The Author(s) 2024
M. F. Hamilton, D. T. Blackstock (eds.), *Nonlinear Acoustics*,
https://doi.org/10.1007/978-3-031-58963-8_4

4.1 Introduction

The subject of this chapter is progressive-wave motion. For most of the chapter, for simplicity, plane waves are assumed. Both lossless and lossy fluids are considered. Section 4.2 is devoted to progressive plane-wave motion in lossless fluids. After exact, implicit solutions—those of Poisson and Earnshaw—are discussed, a second-order approximation of the theory is given. Explicit solutions for periodic waves are then obtained, including the Fubini solution for a monofrequency source and the Fenlon solution for a bifrequency source. Since the waveform steepening described by all these solutions inevitably leads to formation of shock waves, which invalidates the lossless fluid assumption, the next step, Sect. 4.3, is to discuss the Rankine–Hugoniot shock relations. Of particular interest for acoustics are the results for weak shocks. The next two Sects. 4.4 and 4.5, are about propagation with dissipation taken into account. In weak shock theory, the subject of Sect. 4.4, dissipation is assumed to be concentrated at whatever shocks are present in the waveform. Section 4.5 presents the history and solutions of the Burgers equation. This equation takes explicit account of dissipation everywhere in the wave, not just at the shocks. Section 4.6 takes up two special topics, generalization of the theory to cover nonplanar waves (spherical and cylindrical waves and waves in horns), and calculation of intensity and absorption of finite-amplitude waves.

As noted in Sects. 1.2.3 and 3.3.1, two viewpoints of progressive-wave motion are common. The wave may have been generated by an initial disturbance, known everywhere in space, at time $t = 0$. Or the wave may have been generated by a source, specified by the time variation of a disturbance at a known point, usually $x = 0$. The former is called an initial-value problem, the latter a source problem. Since most practical acoustics problems are of the source variety, in this chapter we concentrate on waves generated by sources.

4.2 Lossless Progressive Waves

For an introduction to the problem of plane progressive waves in a lossless fluid, the reader may wish to review Sects. 1.2.2, 1.2.3, and 3.3.1. Throughout the present chapter, propagation in the positive x direction (outgoing waves) is assumed. For this case, the reduced wave equation is [see Eq. (1.28), or Eqs. (3.11) and (3.16)]

$$\frac{\partial u}{\partial t} + (c_0 + \beta u)\frac{\partial u}{\partial x} = 0, \tag{4.1}$$

where u is particle velocity, c_0 is small-signal sound speed, β is the coefficient of nonlinearity [see Eq. (2.16) or (3.18)], x is the (Cartesian) spatial variable, and t is time. This equation is exact for perfect gases and a very good approximation for liquids.

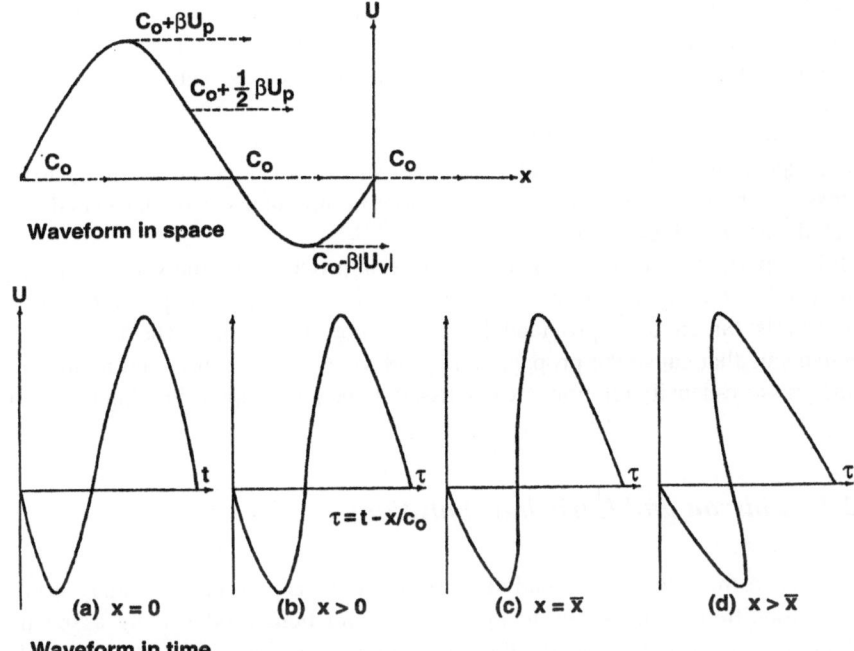

Fig. 4.1 Progressive distortion of a finite-amplitude wave (after Blackstock, 1972). The upper sketch is a spatial plot of the initial waveform ($t = 0$). The lower sketches are time waveforms of the traveling wave at various distances x.

Equation (4.1) may be used to calculate the propagation speed of a given phase point, identified by its value of u, on the waveform of a traveling disturbance. Start with the relation $du = (\partial u/\partial x)\,dx + (\partial u/\partial t)\,dt$. Because $du = 0$ for the phase point, we have $dx/dt|_u = -(\partial u/\partial t)/(\partial u/\partial x)$, or, in view of Eq. (4.1),

$$dx/dt|_u = c_0 + \beta u. \tag{4.2}$$

Equation (4.2) shows that the propagation speed varies from point to point on the waveform.

Figure 4.1 illustrates the effect of the varying propagation speed. The upper sketch is the spatial waveform of a given signal at time $t = 0$. Arrows indicate the propagation speed of various points on the waveform. At $t = 0$, the front end of the wave is at $x = 0$. A receiver at that point records the time waveform as the wave passes by; see sketch a in the lower set. Succeeding sketches in the lower set, b to d, show the time waveform recorded by receivers at progressively more distant locations. Because waveform peaks travel faster than troughs, compression sections of the waveform (negative $\partial u/\partial x$, or positive $\partial u/\partial t$; see sketch b) become more steep, while expansion sections (positive $\partial u/\partial x$, or negative $\partial u/\partial t$) become less steep. Unchecked, waveform steepening would produce a triple-valued waveform, as shown in sketch d. Of course, no waveform ever reaches the triple-valued stage. Instead, shocks form (at $x = \bar{x}$, shown in sketch c) and grow, as faster-traveling phase points behind the shock overtake it. Shock growth is the check that prevents the wave from folding over on itself. As shown in Sect. 4.3.5, shocks are accompanied by dissipation. Since dissipation is not accounted for in Eq. (4.1), strictly speaking the lossless model fails after shocks form. However, many aspects of lossless theory are incorporated in a more comprehensive model called weak shock theory; see Sect. 4.4.1.

It is important to recognize that the distortion produced by the varying propagation speed is cumulative. Even a relatively weak wave can steepen into a shock if it travels far enough (provided losses are not significant). The two physical mechanisms that cause the propagation speed to vary, convection and nonlinearity of the pressure-density relation, are discussed in the paragraph following Eq. (1.30).

4.2.1 Poisson and Earnshaw Solutions

Presented in this section are analytical solutions of Eq. (4.1) that satisfy a given source condition. One wishes to know the sound field produced by a specified excitation $u = f(t)$ at a source. Often the source condition is taken to be specified at a fixed point, usually the origin $x = 0$:

$$u(0, t) = f(t). \tag{4.3}$$

An implicit solution of Eq. (4.1) that satisfies Eq. (4.3) is [see Eq. (1.30), or Eqs. (3.14) and (3.16)]

$$u = f\left(t - \frac{x}{c_0 + \beta u}\right). \tag{4.4}$$

It is traditional to call this the "Poisson solution" even though Poisson (1808) derived it only for an isothermal gas ($\beta = 1$); see Sect. 1.2.2. Notice that the form of the solution is implicit in that u is given as $u(x, t, u)$, not $u(x, t)$.

For reference, we give here the "initial-value" version of the Poisson solution (this is the form Poisson himself presented):

$$u = g[x - (c_0 + \beta u)t]. \tag{4.5}$$

Another implicit solution is the one found by Earnshaw (1860). In this case, specific account is taken of what is often called the finite displacement of the source. A typical source, such as a moving piston, produces a given velocity signal not at a fixed point, such as the origin, but at a moving point, such as the face of the piston. If the piston displacement is designated $X(t)$, the exact source condition is

$$u = \dot{X}(t) \quad \text{at} \quad x = X(t), \tag{4.6}$$

where $\dot{X}(t)$ means dX/dt. At time ϕ the piston is at the place $x = X(\phi)$ and has the velocity $\dot{X}(\phi)$. This velocity is imparted to the fluid as a disturbance, namely the point on the waveform where the particle velocity is $u = \dot{X}(\phi)$. During time $\Delta t = t - \phi$, the disturbance travels a distance $\Delta x = x - X(\phi)$. Since by Eq. (4.2) the propagation speed for the travel has the constant value $c_0 + \beta \dot{X}(\phi)$, we have

$$\frac{x - X(\phi)}{t - \phi} = c_0 + \beta \dot{X}(\phi), \tag{4.7}$$

or, rearranged and combined with Eq. (4.6),

$$u = \dot{X}(\phi), \quad \phi = t - \frac{x - X(\phi)}{c_0 + \beta \dot{X}(\phi)}. \tag{4.8}$$

This is the Earnshaw solution. In order to obtain an explicit expression $u(x, t)$, one must eliminate the Earnshaw phase variable ϕ between the two parts of Eq. (4.8). The extra factor $X(\phi)$ in the expression for ϕ is due to finite displacement of the source.

Despite appearances, the Earnshaw and Poisson solutions are equivalent. To show this, apply the Poisson solution to the moving piston problem. First use Eq. (4.6) to evaluate Eq. (4.4) at the piston face:

$$\dot{X}(t) = f\left(t - \frac{X}{c_0 + \beta \dot{X}}\right). \tag{4.9}$$

In order to solve for $f(t)$, replace t by $t + X/(c_0 + \beta \dot{X})$:

$$f(t) = \dot{X}\left(t + \frac{X}{c_0 + \beta \dot{X}}\right). \tag{4.10}$$

Notice that f turns out to be the piston velocity $\dot{X}$, but with a peculiar argument. Now replace t by $t - x/(c_0 + \beta u)$ in Eq. (4.10). When the result is combined with Eq. (4.4), the Poisson solution becomes

$$u = \dot{X}\left(t - \frac{x - X}{c_0 + \beta \dot{X}}\right). \tag{4.11}$$

Letting ϕ stand for the argument of $\dot{X}$, we recapture the Earnshaw solution. That the argument is indeed ϕ is shown by the fact that $\phi = t$ when $x = X$. As advertised, the Earnshaw phase variable ϕ represents the time the signal leaves the piston.

As an example, let the source excitation be sinusoidal, that is, $u_{\text{source}} = u_0 \sin \omega t$, where u_0 is the amplitude and ω is the angular frequency. If a sinusoidal excitation is imposed at the origin,

$$u(0, t) = u_0 \sin \omega t, \tag{4.12}$$

the Poisson solution is

$$u = u_0 \sin \omega \left(t - \frac{x}{c_0 + \beta u} \right). \tag{4.13}$$

If, however, the source is a vibrating piston, for which the boundary condition is

$$u = u_0 \sin \omega t \quad \text{at} \quad x = \frac{u_0}{\omega}(1 - \cos \omega t), \tag{4.14}$$

the solution by the Earnshaw formula is

$$u = u_0 \sin \omega \phi, \quad \phi = t - \frac{x - (u_0/\omega)(1 - \cos \omega \phi)}{c_0 + \beta u_0 \sin \omega \phi}. \tag{4.15}$$

It is clear that including the finite displacement of the source greatly complicates the solution.

Once the particle velocity has been found, how does one calculate the acoustic pressure $p = P - P_0$, where P and P_0 are total and ambient pressure, respectively? Since the wave motion is progressive, a characteristic impedance relation may be anticipated. Start with Eq. (3.10). Given $d\rho = dP/c^2$, that equation may be rewritten as

$$\frac{dP}{du} = \frac{dp}{du} = \rho c. \tag{4.16}$$

This is the general characteristic impedance relation. For small-signal waves, the right-hand side has the constant value $\rho_0 c_0$, and the familiar expression $p = \rho_0 c_0 u$ is obtained. When the waves have finite amplitude, however, the relation is not so simple. For an adiabatic gas, Eq. (4.16) may be integrated to yield Eq. (1.24). A series expansion of that equation is

$$p = \rho_0 c_0^2 \left[\frac{u}{c_0} + \frac{\beta}{2}\left(\frac{u}{c_0}\right)^2 + \frac{\beta}{6}\left(\frac{u}{c_0}\right)^3 + \cdots \right]. \tag{4.17}$$

For other fluids ($\beta = 1 + B/2A$), one may use Eq. (2.14) to put ρ in terms of c, and then Eq. (2.15) to put c in terms of u, in order to express ρc in terms of u. Integration of Eq. (4.16) then leads to a pressure–particle velocity relation that is the same as Eq. (4.17) through second order (see also Blackstock, 1962). For both liquids and gases, therefore, pressure is linearly proportional to the particle velocity via the small-signal characteristic impedance relation, but the exact relation contains quadratic and higher-order terms.

4.2.2 Shock Formation

The tendency for waveform steepening to cause discontinuities or shocks to develop in the waveform has been noted and illustrated in Fig. 4.1. Shocks imply dissipation (see Sect. 4.3.5), which is not included in the lossless wave equation, Eq. (4.1). Moreover, the sudden change in fluid properties across a shock can cause reflections (Stokes, 1848; see also Sect. 4.3.3). The assumption of progressive-wave motion must then be dropped. On at least two accounts, therefore, shock formation limits the validity of the Poisson and Earnshaw solutions. In this section we develop formulas to estimate the spatial point $x = \bar{x}$ at which a shock first forms. Lossless theory is then known to be limited to the region $x \le \bar{x}$.

To calculate $\bar{x}$, we find the distance at which the waveform first develops a vertical tangent, that is, $\partial u/\partial t = \infty$. If the source condition is given by Eq. (4.3), $\partial u/\partial t$ is found from Eq. (4.4):

$$\frac{\partial u}{\partial t} = \left[1 + \frac{\beta x (\partial u/\partial t)}{(c_0 + \beta u)^2}\right] f', \tag{4.18}$$

where f' means the derivative of f with respect to its argument. Solve for $\partial u/\partial t$:

$$\frac{\partial u}{\partial t} = \frac{f'}{1 - \beta x f'/(c_0 + \beta u)^2}. \tag{4.19}$$

A vertical tangent $\partial u/\partial t = \infty$ occurs when the denominator vanishes. The distance $x = x_{vt}$ at which this happens is

$$x_{vt} = \frac{(c_0 + \beta u)^2}{\beta f'}, \tag{4.20}$$

which is seen to depend on the slope f' of the source function (the slope of the velocity waveform at the source) and also on the value of u [given by Eq. (4.4)] itself. The smallest value of x_{vt} is the distance $\bar{x}$ that we seek. A general solution of this problem has been given (Blackstock, 1962), but for most cases the following approximate treatment is sufficient. If $\beta|u| \ll c_0$ is assumed, a condition widely met in nonlinear acoustics, the explicit dependence on u in Eq. (4.20) may be ignored. The smallest value of x_{vt} is then associated with the maximum positive slope of the source time waveform, whence

$$\bar{x} = \frac{c_0^2}{\beta f'_{max}}. \tag{4.21}$$

As an example, consider sinusoidal source excitation. If Eq. (4.12) is the source condition, the maximum positive slope at the source occurs at the zero crossings $\omega t = 2n\pi$ $(n = 0, \pm 1, \pm 2, \ldots)$ and has the value $f'_{max} = u_0 \omega$. Equation (4.21) gives

$$\bar{x} = \frac{1}{\beta \varepsilon k},$$
(4.22)

where $\varepsilon = u_0/c_0$ is the acoustic Mach number at the source and $k = \omega/c_0$ is the wave number. The higher the amplitude and/or the frequency, the more quickly shocks form. The reason is that when the waveform at the zero crossings is steep to begin with, the wave does not have to travel very far before the slope becomes infinite.

For source signals that are periodic, the combination of quantities

$$\sigma = \beta \varepsilon k x$$
(4.23)

turns out to be a convenient dimensionless distance with which to gauge distortion. Here ω (in the wave number k) and u_0 (in the factor ε) are the characteristic angular frequency and amplitude, respectively, of the source signal. Even when the source signal is not strictly periodic but has an oscillatory character—for example, a tone burst—σ is often the natural choice for organizing the variables. In the special case of a pure sinusoid, Eqs. (4.22) and (4.23) show that σ is the convenient ratio $x/\bar{x}$. Thus $\sigma = 1$ signifies shock formation. In the general case, however, $\sigma = 1$ does not necessarily signify shock formation. See, for example, Sect. 4.2.4.

4.2.3   Second-Order Approximation Theory

A simplification of the theory is developed in this section. An approximation procedure is given in Sects. 3.4–3.9, by which the full-fledged conservation equations are simplified to obtain more manageable models for various problems. Here the problem of interest is lossless plane-progressive-wave motion. An expanded discussion is given of that part of the approximation procedure that applies to progressive waves. Throughout, the focus is on source problems, not initial-value problems. Moreover, in keeping with modern practice, we adopt pressure p as the main field variable, not particle velocity u. Although only plane waves in lossless fluids are treated here, the approach has broader application. See Sect. 4.6.1 for a generalization that covers other one-dimensional waves, namely spherical and cylindrical waves and waves in horns. Moreover, much of what is said here continues to apply when losses are included (Sects. 4.4 and 4.5, and Sects. 3.7 and 3.8), and even when the propagation is three-dimensional, as in the case of sound beams (Sect. 3.9).

As noted in Sects. 1.4 and 3.6, finite-amplitude sound is subject to two different kinds of nonlinear effects, *cumulative* and *local*. In the case of progressive waves, cumulative effects generally dominate. Convection and nonlinearity of the pressure–density relation are classified as cumulative nonlinear effects because they cause waveform steepening, which accumulates with propagation distance. Distortion occurring at one location builds on all previous distortion and extends it. Local nonlinear effects, on the other hand, produce distortion that does not grow with propagation distance. An example of the latter is nonlinearity of the characteristic impedance relation. If the particle velocity waveform has been calculated, Eq. (4.17) is used to find the pressure waveform. The linear term in that equation yields a pressure waveform that has the same cumulative distortion as the particle velocity waveform. Since the remaining terms in Eq. (4.17) are smaller than the linear term by at least $O(\varepsilon)$, the added distortion that they introduce is small. Furthermore, because the added distortion depends only on the local waveform (hence the term *local effect*), it tends to remain constant, not grow, with propagation distance. Thus the only place where local distortion is dominant is near the source, where cumulative distortion is still very small.

Local effects greatly complicate the analysis of progressive-wave propagation, and they contribute little to the solution because the distortion they cause is usually minor compared to that caused by cumulative effects. Much simplification, at small cost, may therefore be achieved by ignoring local effects.

Two assumptions underlie second-order approximation theory. First, the waves are not exceedingly strong; i.e., $|u| \ll c_0$ (or $\varepsilon \ll 1$)—a restriction rarely violated even in nonlinear acoustics. On this basis, although $O(\varepsilon^2)$ terms must be retained, $O(\varepsilon^3)$ terms may safely be dropped. Second, distortion is dominated by cumulative effects; i.e., the observation point is not close to the source.[1] The immediate consequences of these two assumptions are as follows:

1. The linear characteristic impedance relation (or the comparable relation between excess density $\rho' = \rho - \rho_0$ and u) may be used.
2. Finite displacement of a source from its rest position may be ignored—i.e., Eq. (4.6) need not be used; Eq. (4.3) is sufficient.
3. The difference between material (Lagrangian) and spatial (Eulerian) representations may be ignored (Blackstock, 1962).
4. In $O(\varepsilon^2)$ terms (relating to cumulative distortion), any $O(\varepsilon)$ factor may be replaced by its *progressive-wave* $O(\varepsilon)$ equivalent. The justification is that the error introduced is $O(\varepsilon^3)$. This principle is introduced at the end of Sect. 3.3.1 and beginning of Sect. 3.5.

Use of Item 4 leads to an approximate wave equation for progressive signals and to corresponding approximate Poisson and Earnshaw solutions. The latter are particularly convenient and easy to use for source problems. The approximate wave equation may be obtained from Eq. (4.1) by using the $O(\varepsilon)$ progressive-wave relation $\partial u/\partial x = -c_0^{-1}(\partial u/\partial t)$ in the nonlinear term $\beta u(\partial u/\partial x)$. The result is the wave equation shown as the first of Eqs. (3.19). The second of Eqs. (3.19) gives

---

[1] For periodic waves, "not close to the source" means $x \gg \lambda/2\pi\beta$, where λ is the wavelength of the fundamental frequency component (Blackstock, 1962). For most fluids, this means x greater than about a wavelength.

the exact solution, as may be verified by direct substitution or by using the same approximations in Eq. (4.4).[2] Immediately following Eqs. (3.19) is a numerical example to demonstrate how precise the approximate theory is even for a very intense wave. Similar steps applied to Eqs. (4.8) produce the approximate Earnshaw solution:

$$u = \dot{X}(\phi), \quad \phi = t - \frac{x}{c_0} + \frac{\beta x \dot{X}(\phi)}{c_0^2}. \tag{4.24}$$

Thus, in place of Eqs. (4.1), (4.4), and (4.8), one uses Eqs. (3.19) and (4.24) when the second-order approximation model is used. The advantage of the approximate expressions is that analytical solutions for specific source problems are more easily obtained from them than from the exact expressions.

Finally, pressure p often replaces particle velocity u as the field variable of choice in second-order approximation theory. One reason is that pressure is the variable measured in almost all acoustical experiments. Pressure is used for much of the rest of this chapter, and indeed for many succeeding chapters of the book. We therefore end this section by converting the main results to expressions in terms of p. Although several of these are given in Chap. 3, they are gathered together here for convenience.

To obtain the exact wave equation for pressure, start with Eq. (4.1), convert the derivatives of u to derivatives of p by the transformation $\partial u/\partial(\cdot) = (du/dp)\partial p/\partial(\cdot)$, and cancel the common factor du/dp. The exact wave equation in p has the same form as Eq. (4.1) [see Eqs. (3.11) and (3.16)]. Next, in accordance with Item 4, replace the nonlinear term $\beta u(\partial p/\partial x)$ with $-(\beta/\rho_0 c_0^2)p(\partial p/\partial t)$. The result is the approximate wave equation in p, equivalent to the first of Eqs. (3.19),

$$\frac{\partial p}{\partial t} + c_0 \frac{\partial p}{\partial x} = \frac{\beta p}{\rho_0 c_0^2} \frac{\partial p}{\partial t}, \tag{4.25}$$

and its exact solution (Earnshaw form), equivalent to the second of Eqs. (3.19),

$$p = f(\phi), \quad \phi = t - \frac{x}{c_0} + \frac{\beta x p}{\rho_0 c_0^3}. \tag{4.26}$$

The reader is warned that it is common practice to use the same symbol f in expressions for both pressure and particle velocity. Here, since f stands for pressure, it is $\rho_0 c_0$ times the same symbol used for particle velocity, for example, in the second of Eqs. (3.19). For reference, the shock formation distance in terms of pressure is

---

[2] The argument of f is expanded as $t - (x/c_0)[1 - (\beta u/c_0) + (\beta u/c_0)^2 + \cdots]$. The expansion is terminated after the second term because subsequent terms have an $O(\varepsilon^3)$ effect. To show this, expand f in a Taylor series: $f = f(t - x/c_0) - [(\beta u/c_0) - (\beta u/c_0)^2 + \cdots]f'(t - x/c_0) + \cdots$. Because f itself is $O(\varepsilon)$, retaining the $(\beta u/c_0)^2$ term would amount to keeping an $O(\varepsilon^3)$ correction.

$$\bar{x} = \frac{\rho_0 c_0^3}{\beta f'_{max}}. \tag{4.27}$$

If the retarded time $\tau = t - x/c_0$ is introduced, Eqs. (4.25) and (4.26) take on a simpler form, which is useful for further developments in this chapter and throughout the book. In the x, τ coordinate system, the first derivatives become $\partial p/\partial t = \partial p/\partial \tau$ and $\partial p/\partial x = -(1/c_0)\partial p/\partial \tau + \partial p/\partial x$, and Eq. (4.25) becomes

$$\frac{\partial p}{\partial x} = \frac{\beta p}{\rho_0 c_0^3} \frac{\partial p}{\partial \tau}. \tag{4.28}$$

This equation is the lossless form of the Burgers equation, Eqs. (3.54) and (4.189). The solution, Eq. (4.26), becomes

$$p = f(\phi), \quad \phi = \tau + \frac{\beta x p}{\rho_0 c_0^3}, \tag{4.29}$$

which is the same as Eq. (3.48).

Finally, consider the *slowness* of the wave, defined as the inverse of the propagation speed. If u on the right-hand side of Eq. (4.2) is replaced by $p/\rho_0 c_0$, the slowness becomes

$$\left.\frac{dt}{dx}\right|_p = \frac{1}{c_0 + \beta p/\rho_0 c_0} = \frac{1}{c_0} - \frac{\beta p}{\rho_0 c_0^3} \tag{4.30}$$

within the limits of second-order theory. Put this expression in terms of retarded time by noting that $dt/dx|_p - 1/c_0 = d\tau/dx|_p$:

$$\left.\frac{d\tau}{dx}\right|_p = -\frac{\beta p}{\rho_0 c_0^3}, \tag{4.31}$$

which shows how the retarded time decreases with distance for any phase point on the waveform. In particular, points of high pressure arrive in a shorter time than points of low pressure. We return to this formula in Sect. 4.4.1.

4.2.4   Periodic Waves

For periodic waves, the Poisson and Earnshaw solutions may be turned into explicit functions of x and t by expanding u or p as a Fourier series. Pressure is taken as the field variable here. The explicit solutions display clearly the growth of harmonics and intermodulation distortion that characterize propagation of periodic waves. The first to use this approach was Fubini (1935), who was interested in the monofrequency source problem. The approach followed here, which is a generalization for multiple-frequency sources, is due to Fenlon (1973).

We begin by considering the source condition

$$p(0, t) = p_0 F(\omega t), \tag{4.32}$$

where p_0 is a characteristic pressure amplitude, and the dimensionless function F is periodic in time with fundamental angular frequency ω and zero mean value. The corresponding Earnshaw form of the solution is Eq. (4.29):

$$p(\sigma, \tau) = p_0 F(\Phi), \quad \Phi = \omega\tau + \sigma F, \tag{4.33}$$

where $\sigma = \beta\varepsilon k x$, as defined in Eq. (4.23), is the nonlinear distortion scale with $\varepsilon = p_0/\rho_0 c_0^2$ and $k = \omega/c_0$. An expansion of Eq. (4.33) is sought in the form

$$p(\sigma, \tau) = p_0 \sum_{n=-\infty}^{\infty} C_n(\sigma) e^{jn\omega\tau} \tag{4.34}$$

$$= p_0 \sum_{n=1}^{\infty} [A_n(\sigma) \cos n\omega\tau + B_n(\sigma) \sin n\omega\tau], \tag{4.35}$$

where $A_n = C_n + C_n^*$ and $B_n = j(C_n - C_n^*)$. Note that dc pressure generation ($n = 0$), which is not taken into account by Eq. (4.33), is expressly ignored in Eq. (4.35) and in the following analysis.

The complex Fourier coefficients C_n are given by

$$C_n = \frac{1}{2\pi} \int_{-\pi}^{\pi} F(\Phi) e^{-jn\omega\tau} d(\omega\tau). \tag{4.36}$$

Let $u = F/2\pi$ and $dv = e^{-jn\omega\tau} d(\omega\tau)$, and integrate $u\, dv$ by parts to obtain

$$C_n = \frac{1}{j2\pi n} \left(-F e^{-jn\omega\tau} \Big|_{\omega\tau=-\pi}^{\omega\tau=\pi} + \int_{\omega\tau=-\pi}^{\omega\tau=\pi} e^{-jn\omega\tau} dF \right) \tag{4.37}$$

$$= \frac{1}{j2\pi n} \int_{\omega\tau=-\pi}^{\omega\tau=\pi} e^{-jn\omega\tau} dF, \tag{4.38}$$

noting that the first term in Eq. (4.37) vanishes because of periodicity. From Eq. (4.33) we have $dF = \sigma^{-1}[d\Phi - d(\omega\tau)]$ for σ fixed, and Eq. (4.38) becomes

$$C_n = \frac{1}{j2\pi n\sigma} \left(\int_{-\pi}^{\pi} e^{-jn(\Phi-\sigma F)} d\Phi - \int_{-\pi}^{\pi} e^{-jn\omega\tau} d(\omega\tau) \right), \tag{4.39}$$

where the integration limits pertain to the corresponding variable of integration. The second integral yields zero (except for the dc term $n = 0$, which is ignored), and the final result is

$$C_n(\sigma) = \frac{1}{j2\pi n\sigma} \int_{-\pi}^{\pi} e^{jn[\sigma F(\Phi)-\Phi]} d\Phi. \tag{4.40}$$

Thus, given the source function $F(\omega t)$, replace ωt by Φ and evaluate the integral in Eq. (4.40) to obtain expressions for the coefficients in Eqs. (4.34) and (4.35). The resulting solution is valid only in the preshock region, where Eq. (4.33) remains single-valued. This region is defined by $\sigma \leq \bar{\sigma}$, where

$$\bar{\sigma} = \beta \varepsilon k \bar{x} \tag{4.41}$$

is a dimensionless shock formation distance, and $\bar{x}$ is given by Eq. (4.27). In the notation used here, we have

$$\bar{\sigma} = \frac{1}{(dF/d\Phi)_{max}}. \tag{4.42}$$

For a monofrequency source with amplitude p_0, Eq. (4.42) yields $\bar{\sigma} = 1$, and therefore $\sigma = x/\bar{x}$, but in general $\sigma \neq x/\bar{x}$.

When F is an odd function of time, the coefficients A_n in Eq. (4.35) vanish, and only the sine terms remain:

$$p(\sigma, \tau) = p_0 \sum_{n=1}^{\infty} B_n(\sigma) \sin n\omega\tau, \tag{4.43}$$

where

$$B_n(\sigma) = \frac{2}{n\pi\sigma} \int_0^{\pi} \cos[n\Phi - n\sigma F(\Phi)] d\Phi. \tag{4.44}$$

In the special case where the source condition is such that none of the $B_n(0)$ is negative, Eq. (4.42) yields, in dimensionless and dimensional forms, respectively,

$$\bar{\sigma} = \left(\sum_{n=1}^{\infty} n B_n(0)\right)^{-1}, \quad \bar{x} = \left(\beta \sum_{n=1}^{\infty} \varepsilon_n k_n\right)^{-1}, \tag{4.45}$$

where $\varepsilon_n = p_{0n}/\rho_0 c_0^2$, p_{0n} is the pressure amplitude of the nth-harmonic component at the source, and $k_n = n\omega/c_0$.

In the following subsections we perform the calculations for two important specific sources, monofrequency and bifrequency.

4.2.4.1 Monofrequency Source (Fubini Solution)

The source condition considered here is

$$p(0, t) = p_0 \sin \omega t, \tag{4.46}$$

for which $F(\Phi) = \sin \Phi$ and $\bar{\sigma} = 1$. Since F is an odd function, Eq. (4.44) applies:

$$B_n = \frac{2}{n\pi\sigma} \int_0^\pi \cos(n\Phi - n\sigma \sin \Phi) \, d\Phi, \tag{4.47}$$

which is $2/n\sigma$ times Bessel's integral. Equation (4.47) may therefore be rewritten

$$B_n = \frac{2}{n\sigma} J_n(n\sigma), \tag{4.48}$$

where J_n is the Bessel function of order n. Substitution into Eq. (4.43) yields

$$p(\sigma, \tau) = p_0 \sum_{n=1}^\infty \frac{2}{n\sigma} J_n(n\sigma) \sin n\omega\tau, \tag{4.49}$$

which is the Fubini (1935) solution, valid only in the preshock region $\sigma \leq 1$.

Equation (4.49) predicts that the amplitude of the fundamental component ($n = 1$) will decrease continuously from $\sigma = 0$ to $\sigma = 1$ as energy is transferred to the nonlinearly generated higher-harmonic components. The latter grow monotonically in amplitude over the same region, with $B_{n+1} < B_n$ for all n. Further insight into the Fubini solution may be gained by investigating its asymptotic properties in the region near the source. The leading terms in expansions of the Bessel functions in Eq. (4.48) yield

$$B_1 = 1 - \tfrac{1}{8}\sigma^2 + O(\sigma^4), \tag{4.50}$$

$$B_2 = \tfrac{1}{2}\sigma + O(\sigma^3), \tag{4.51}$$

$$B_3 = \tfrac{3}{8}\sigma^2 + O(\sigma^4), \tag{4.52}$$

$$B_n = \frac{(n\sigma)^{n-1}}{2^{n-1}n!} + O(\sigma^{n+1}). \tag{4.53}$$

These asymptotic expressions for B_n are identical to those obtained by regular perturbation techniques (Sect. 10.2), and they are reasonably accurate over much of the preshock region. For example, at $\sigma = \tfrac{1}{2}$ the expressions for $n = 1, 2$, and 3 differ from Eq. (4.48), respectively, by 0.03%, 9%, and 15%. Substitution of Eq. (4.51) into Eq. (4.43) gives the following approximate solution for the second-harmonic signal p_2:

$$p_2(x, \tau) = \frac{\beta p_0^2 \omega}{2\rho_0 c_0^3} x \sin 2\omega\tau. \tag{4.54}$$

The experiments conducted by Thuras, Jenkins, and O'Neil (1935) in an air-filled plane-wave tube were the first to confirm the linear dependence of p_2 on distance and source frequency, and the quadratic dependence on source amplitude (see Sect. 1.4 for a discussion of their experiment). Graphical representation of $B_n(\sigma)$ is postponed to Sect. 4.4.3.3, where weak shock theory is used to extend the solution beyond $\sigma = 1$. See also Blackstock (1962) for further discussion.

4.2.4.2 Bifrequency Source (Fenlon Solution)

Here we consider the bifrequency source condition

$$p(0, t) = p_{0a} \sin \omega_a t + p_{0b} \sin \omega_b t. \tag{4.55}$$

To maintain periodicity, we take ω_a/ω_b to be the ratio of two integers n_a and n_b, i.e., $\omega_a = n_a\omega$ and $\omega_b = n_b\omega$, where ω remains the fundamental angular frequency of the signal. Now let $P_a = p_{0a}/p_0$ and $P_b = p_{0b}/p_0$ to obtain

$$F(\Phi) = P_a \sin n_a \Phi + P_b \sin n_b \Phi. \tag{4.56}$$

Although F is an odd function of time, it is more convenient in this case to begin with Eq. (4.40), not Eq. (4.44), to evaluate the Fourier coefficients:

$$C_n = \frac{1}{j2\pi n\sigma} \int_{-\pi}^{\pi} \exp[jn(\sigma P_a \sin n_a \Phi + \sigma P_b \sin n_b \Phi - \Phi)] \, d\Phi. \tag{4.57}$$

Use of the identity

$$e^{jz \sin\theta} = \sum_{n=-\infty}^{\infty} J_n(z)e^{jn\theta} \tag{4.58}$$

yields

$$C_n = \frac{1}{j2\pi n\sigma} \sum_{l=-\infty}^{\infty} \sum_{m=-\infty}^{\infty} J_l(n P_a\sigma) J_m(n P_b\sigma) \int_{-\pi}^{\pi} e^{j(ln_a + mn_b - n)\Phi} \, d\Phi. \tag{4.59}$$

The quantity $(ln_a + mn_b - n)$ is an integer, and therefore the integral equals 2π for $ln_a + mn_b = n$ but zero otherwise.

Noting that $A_n = 0$, we may express the solution in the form of Eq. (4.43), where

$$B_n = \frac{2}{n\sigma} \sum_{ln_a + mn_b = n} J_l(n P_a \sigma) J_m(n P_b \sigma). \tag{4.60}$$

The solution may be manipulated, with $\omega_a > \omega_b$ assumed, to obtain (Fenlon, 1972)

$$p(\sigma, \tau) = p_0 \sum_{l=1}^{\infty} \frac{2}{ln_a\sigma} J_l(ln_a P_a \sigma) J_0(ln_a P_b \sigma) \sin l\omega_a \tau$$

$$+ p_0 \sum_{m=1}^{\infty} \frac{2}{mn_b\sigma} J_0(mn_b P_a \sigma) J_m(mn_b P_b \sigma) \sin m\omega_b \tau$$

$$+ p_0 \sum_{l=1}^{\infty} \sum_{m=1}^{\infty} \frac{2}{N_{lm}^{+}\sigma} J_l(N_{lm}^{+} P_a \sigma) J_m(N_{lm}^{+} P_b \sigma) \sin N_{lm}^{+} \omega \tau$$

$$+ p_0 \sum_{l=1}^{\infty} \sum_{m=1}^{\infty} (-1)^m \frac{2}{N_{lm}^{-}\sigma} J_l(N_{lm}^{-} P_a \sigma) J_m(N_{lm}^{-} P_b \sigma) \sin N_{lm}^{-} \omega \tau,$$

$$\tag{4.61}$$

where $N_{lm}^{\pm} = ln_a \pm mn_b$. Although the first two summations are expressed explicitly in terms of harmonics of the primary waves, contributions to these same frequency components are also contained in the last two summations. With $n_a = 1$, $P_a = 1$, and $P_b = 0$, the source condition in Eq. (4.46) is recovered, the last three summations in Eq. (4.61) vanish, and the first summation reduces to Eq. (4.49). Equation (4.61) is valid for $\sigma \leq \bar{\sigma}$, where $\bar{\sigma}$ is given by Eq. (4.42). For P_a and P_b each positive, Eq. (4.45) yields $\bar{\sigma} = (n_a P_a + n_b P_b)^{-1}$.

Arbitrary phase shifts in the frequency components at the source are easily included in the solution (Fenlon, 1972), and a generalization to N-frequency sources is also available (Fenlon, 1973).

Difference-Frequency Generation Often of interest is the solution for the difference-frequency pressure alone. To examine this case, for convenience take $\omega_- = \omega_a - \omega_b \equiv \omega$, i.e., choose ω_- to be the fundamental frequency, $n = 1$. Thus $n_a = n_b + 1$, and the condition on the indices in Eq. (4.60) becomes $l(n_b + 1) + mn_b = 1$, which is satisfied for $l = 1 + qn_b$ and $m = -[1 + q(n_b + 1)]$, where q is any integer. Equation (4.60) then gives, with $B_- \equiv B_1$,

$$B_- = \frac{2}{\sigma} \sum_{q=-\infty}^{\infty} J_{1+qn_b}(P_a \sigma) J_{-(1+qn_a)}(P_b \sigma), \quad n_a = n_b + 1. \tag{4.62}$$

The degenerate case $\omega_b = \omega_-$ ($n_b = 1$) is included in this result. For $\omega_- \ll \omega_a, \omega_b$ ($n_a, n_b \gg 1$) and $\sigma \leq \bar{\sigma}$, all terms in the summation for which $q \neq 0$ contain products of high-order Bessel functions that are small in comparison with the term for which $q = 0$, and in this case we may write

$$B_- \simeq -\frac{2}{\sigma} J_1(P_a\sigma)J_1(P_b\sigma). \tag{4.63}$$

This expression corresponds to the term for which $(l, m) = (1, 1)$ in the fourth summation of Eq. (4.61). For $n_b > 1$, the Bessel functions in Eq. (4.62) may be expanded as in Eqs. (4.50)–(4.53) to obtain $B_- = -\frac{1}{2}P_aP_b\sigma + O(\sigma^3)$, substitution of which into Eq. (4.43) yields the following expression for the difference-frequency pressure near the source:

$$p_-(x, \tau) = -\frac{\beta p_{0a} p_{0b} \omega_-}{2\rho_0 c_0^3} x \sin \omega_- \tau. \tag{4.64}$$

The amplitude thus increases linearly with distance, just as the second harmonic does for the monofrequency source [recall Eq. (4.54)]. However, the growth rate for p_- is not as strong, particularly for primary frequencies close together. Thuras et al. (1935) provided experimental confirmation of the properties of Eq. (4.64), as well as those of Eq. (4.54).

Parametric Amplification   In the degenerate case $(n_b, n_a) = (1, 2)$, the lower primary wave is also the difference-frequency component. This choice offers the possibility that the difference signal can add to the lower primary and thus amplify the latter. Here ω_b is the fundamental frequency, and $\omega_a = 2\omega_b$ its second harmonic. Equation (4.62) yields, with $B_b \equiv B_-$,

$$B_b = -\frac{2}{\sigma} \sum_{q=-\infty}^{\infty} J_{1+q}(P_a\sigma)J_{1+2q}(P_b\sigma), \quad (n_b, n_a) = (1, 2). \tag{4.65}$$

For $P_a = 0$ (i.e., for a monofrequency source), all terms vanish except for $q = -1$, and the Fubini solution for the fundamental, $B_b = (2/\sigma)J_1(P_b\sigma)$, is recovered. We now assume $|P_b| \ll |P_a|$ and investigate the influence of a strong second-harmonic component on a weak fundamental. The shock formation distance is then $\bar{\sigma} \simeq |2P_a|^{-1}$, where the argument of the second Bessel function has magnitude $|P_b/2P_a| \ll 1$. In this case, the dominant terms are those for which the second Bessel function in Eq. (4.65) is of order ± 1 (i.e., $q = -1, 0$), and the second Bessel function may also be approximated by the first term in its Taylor series expansion to obtain

$$B_b \simeq P_b[J_0(P_a\sigma) - J_1(P_a\sigma)], \quad |P_b| \ll |P_a|. \tag{4.66}$$

We examine this solution as a function of P_a. In small-signal theory, the weak fundamental propagates unchanged in amplitude, and we have $B_b = P_b$, as obtained from Eq. (4.66) for $P_a = 0$ (the condition $|P_b| \ll |P_a|$ notwithstanding). For $P_a > 0$, the amplitude of the fundamental decreases continuously as its energy is pumped by the second harmonic into other spectral components, with $B_b = 0.70P_b$ at $\sigma = \bar{\sigma}$. However, for $P_a < 0$ (i.e., when the phase of the second harmonic is shifted by 180°), the fundamental receives energy from the second harmonic and experiences continuous growth out to the shock formation distance, where $B_b = 1.18P_b$. This last case is referred to as parametric amplification. For all other relative phase shifts between the fundamental and the second harmonic, the values of B_b lie between those given above (Blackstock, 1982). The condition $P_a < 0$ is thus optimal for parametric amplification of a weak signal by another with twice its frequency, although the gain in amplitude is at most 18% in the preshock region (Novikov and Rudenko, 1976).

The distinctly different behavior of B_b as a function of the sign of P_a is associated with changes in the source waveform. This principle may also be used to "predistort" the source waveform in such a way as to postpone shock formation. In particular, the introduction of a second-harmonic component in antiphase with the fundamental predistorts the source waveform in a direction opposite that associated with subsequent finite-amplitude effects. For example, with $(n_b, n_a) = (1, 2)$, $P_b = 1$, and $P_a = 0$, $\frac{1}{4}$, $-\frac{1}{4}$, Eqs. (4.42) and (4.56) yield the corresponding shock formation distances $\bar{\sigma} = 1$, $\frac{2}{3}$, $\frac{4}{3}$. For $P_a = \frac{1}{4}$, the shock formation distance is reduced by 33% (in comparison with $P_a = 0$) because the second harmonic is in phase with the fundamental and thus increases the maximum positive slope in the source waveform. The signal need not travel as far before waveform steepening causes a shock to form. With $P_a = -\frac{1}{4}$, the shock formation distance is increased by 33%, even though the signal has slightly greater energy than for $P_a = 0$, because the maximum positive slope in the source waveform has been reduced. Notice that for both $P_a = 0$ and $P_a = \frac{1}{4}$ (and for any $P_a > 0$), shock formation occurs at the zero crossing in the waveform, defined by $\Phi = 0$ in Eq. (4.56), whereas it occurs at $\Phi = \pi/3$ for $P_a = -\frac{1}{4}$.

Suppression of Sound by Sound Equations (4.60) and (4.61) can also be used to investigate the modulation of one primary wave by another of lower frequency. One application is known as suppression of sound by sound (Schaffer, 1975; Moffett et al., 1978). Consider the solution for the primary wave at frequency ω_a, which is designated by $n = n_a$. For convenience, take $\omega_b = \omega$, for which $n_b = 1$, and Eq. (4.60) yields

$$B_a = \frac{2}{n_a \sigma} \sum_{l=-\infty}^{\infty} J_l(n_a P_a \sigma) J_{(1-l)n_a}(n_a P_b \sigma), \quad n_b = 1. \qquad (4.67)$$

An interesting special case arises when the low-frequency wave is much stronger than the high-frequency wave, and much lower in frequency. We shall refer to these as the "strong" and "weak" waves, respectively. First, for $\omega_a \gg \omega_b$ ($n_a \gg 1$), the second Bessel function in Eq. (4.67) is of very high order, and therefore very small in value, for all $l \neq 1$. The series is then well approximated by the term for which $l = 1$:

$$B_a \simeq \frac{2}{n_a \sigma} J_1(n_a P_a \sigma) J_0(n_a P_b \sigma). \qquad (4.68)$$

Equation (4.68) is the spectral amplitude given by the term for which $l = 1$ in the first summation in Eq. (4.61). Now take $p_{0a} \ll p_{0b}$ (here each assumed positive), in which case J_1 is a more slowly oscillating function of σ than J_0. For $n_a P_a \sigma \ll 1$, we may approximate the value of J_1 by the leading term in its Taylor series expansion to obtain

$$B_a \simeq P_a J_0(n_a P_b \sigma). \tag{4.69}$$

Substitution into Eq. (4.43) yields the following approximate expression for the pressure p_a of the weak primary wave at the high frequency ω_a:

$$p_a(x, \tau) \simeq p_{0a} J_0 \left(\frac{\beta p_{0b} \omega_a x}{\rho_0 c_0^3} \right) \sin \omega_a \tau. \tag{4.70}$$

Note that the argument of J_0 depends on p_{0b}, the source amplitude of the strong primary wave, and ω_a, the frequency of the weak wave.

The form of Eq. (4.70) suggests a suppression phenomenon. That is, at distances for which $J_0 = 0$, the high-frequency primary wave vanishes. These distances x_n ($n = 1, 2, \cdots$) correspond to the zeros defined by $J_0(j_{0n}) = 0$, where $j_{01} = 2.4$, $j_{02} = 5.5$, etc., and therefore

$$x_n = \frac{j_{0n} \rho_0 c_0^3}{\beta p_{0b} \omega_a}. \tag{4.71}$$

An additional restriction on the use of Eq. (4.70) to predict suppression at location x_n thus arises, $x_n \leq \bar{x}$, where $\bar{x} = [\beta(\varepsilon_a k_a + \varepsilon_b k_b)]^{-1}$ is the shock formation distance ($\varepsilon_i = p_{0i}/\rho_0 c_0^2$, $k_i = n_i \omega/c_0$, $i = a, b$). This inequality may be rewritten as $(\omega_b/\omega_a + p_{0a}/p_{0b}) j_{0n} \leq 1$, which is satisfied for the first few values of x_n under the conditions assumed, namely, $\omega_b \ll \omega_a$ and $p_{0a} \ll p_{0b}$ [which also assure that the restriction $n_a P_a \sigma \ll 1$ leading to Eq. (4.69) is satisfied]. Subject to these restrictions, the amplitude of the weak, high-frequency signal p_a at a fixed location x decreases monotonically according to Eq. (4.70) as the source amplitude p_{0b} of the low-frequency signal is increased from zero, ultimately leading to the suppression of p_a. Further increase in p_{0b} causes the signal p_a to reappear and grow in amplitude before eventually being suppressed once again.

Examination of other terms in the Fenlon solution reveals that the energy pumped out of the high-frequency wave by the low-frequency wave ends up mainly in the sidebands at frequencies $\omega_q = \omega_a + q\omega_b$, where $q = \pm 1, \pm 2, \cdots$. The amplitudes of these sidebands, including that of the "center frequency" ω_a for $q = 0$, are obtained by setting $n = n_a + q$ in Eq. (4.60), with $n_b = 1$ as before. Making the same assumptions that lead to Eq. (4.69), here with the frequency restriction $\omega_q \gg \omega_b$ for all q of interest, one readily arrives at the result $B_{n_a+q} \simeq P_a J_q[(n_a + q) P_b \sigma]$, from which Eq. (4.69) is recovered for $q = 0$. Substitution into Eq. (4.43) and summation over the sidebands yields

$$p_w(\sigma, \tau) \simeq p_{0a} \sum_{q=-N}^{N} J_q[(n_a + q)P_b\sigma] \sin(\omega_a + q\omega_b)\tau, \qquad (4.72)$$

where N is chosen not too large, such that $n_a - N \gg 1$ is satisfied. Here p_w is taken to mean the weak primary wave together with its sidebands, with the term $q = 0$ corresponding to Eq. (4.70). If q did not appear in the argument of the Bessel function, Eq. (4.72) would resemble the solution for frequency modulation of a carrier wave of frequency ω_a by a signal of frequency ω_b, with modulation index $n_a P_b\sigma$. In contrast with FM theory, the presence of q in the argument produces a nonsymmetric sideband structure; i.e., $|B_{n_a+q}| \neq |B_{n_a-q}|$. Nevertheless, the actual nonlinear distortion process described by Eq. (4.72) does indeed correspond to frequency modulation. The weak, high-frequency signal rides atop the strong, low-frequency signal. As the latter distorts, the high-frequency cycles are compressed together where waveform steepening occurs, i.e., in regions of positive slope on the low-frequency waveform, and they are stretched apart in regions of negative slope.

4.3   Shock Waves

In Sect. 1.3, the theory of shock waves is mentioned, and the role this theory played in the history of nonlinear acoustics is discussed. Basic equations in this theory include the Rankine–Hugoniot equations, which appear as Eqs. (1.34)–(1.36). In the present section a careful derivation of these equations for circumstances of general interest in nonlinear acoustics is given.

In typical presentations of fluid-dynamic equations, such as in the books by Landau and Lifshitz (1987) and by Pierce (1989), the basic equations are first given in an integral form. Partial differential equations such as Eqs. (3.1)–(3.3) are then derived by invoking various seemingly plausible mathematical arguments (including the use of Gauss's theorem to convert surface integrals to volume integrals), along with the arbitrariness of the integration volume. Such arguments ideally require, however, that the resulting integrands be everywhere finite. If dissipation terms are explicitly taken into account in the original formulation, this requirement is invariably met.

In many situations, one can achieve a very good approximation, and even an excellent approximation, by neglecting the dissipation terms completely, both in the integral equation versions of the governing equations and in the partial differential equation versions. There are also other situations in which such neglect still gives a good approximation, with the exception of small spatial regions. The situations of interest here are where those regions are thin, generally moving, sheets of space known as shocks (see Fig. 4.2). The principal feature of a fluid-dynamic field within such a sheet is that the gradients of certain fluid-dynamic variables are inordinately high. In the limit when the parameters characterizing the strength of the dissipation become smaller and smaller, the sheets can be regarded as thinner and thinner,

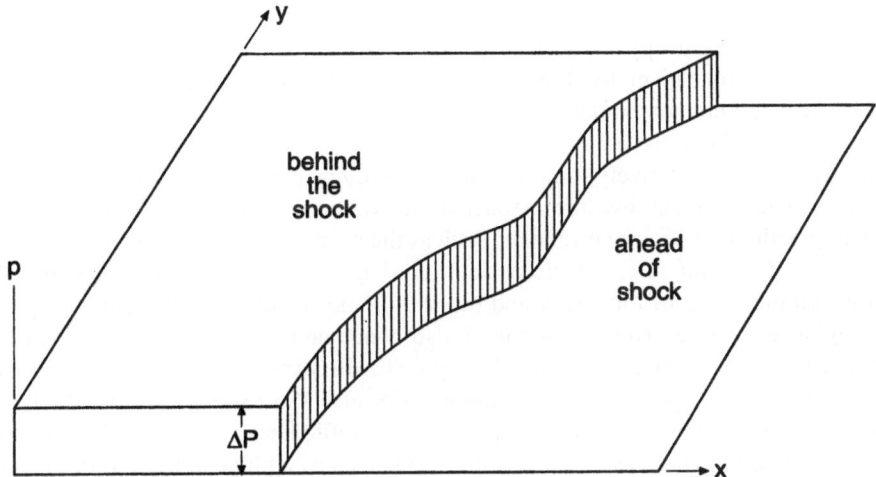

Fig. 4.2 Shock wave depicted as a discontinuity.

but the gradients within them become larger and larger. Thus, one can conceive of a mathematical model in which the idealized partial differential equations hold separately on each side of a sheet, but not within the sheet. With such a model, one has the possibility that fluid-dynamic variables may have values on one side of a sheet that are substantially different from those on the other side. The question then naturally arises as to what, if anything, should turn out to be continuous across the sheet in the limit when the dissipation parameters go to zero and the sheet becomes infinitesimally thin.

4.3.1 Rankine–Hugoniot Shock Relations

The Rankine–Hugoniot relations are a collective statement of what is continuous across a shock. Their formulation and understanding evolved slowly in the history of fluid mechanics, and several authors besides Rankine (1870) and Hugoniot (1887, 1889) contributed to this. Discussions of the difficulties that impeded the progress can be found in the articles by G. I. Taylor (1910) and by Rayleigh (1910). A typical, and common, pitfall is to adopt the notion that, for an ideal gas, the adiabatic relation $P = K\rho^{\gamma}$ should hold on both sides of a shock, with the same choice for the constant K. When one examines the old classic papers, the key word that is invariably absent from the writing is *entropy*. Even as late as 1910, the vocabulary of modern thermodynamics, which is now standard in virtually every undergraduate curriculum in physics and engineering, was used by only relatively few of the prominent researchers. What would be said today is that K has to depend on the entropy, and the entropy changes at a shock.[3]

The derivations given by Rankine (1870) and Taylor (1910), as well as in almost all fluid dynamics textbooks, assume that the flow is steady in a reference frame in which the shock appears to be nonmoving. With this steady flow assumption, the derivation is relatively simple, but even though the results turn out to be of more general applicability, the manner of derivation can be disquieting to anyone dealing with more general problems, such as the propagation of pulses, in nonlinear acoustics. Consequently, the discussion here begins at the outset with a situation in which the speed of the shock and the magnitudes of the discontinuities may be changing with time. The discussion is also concerned with a general fluid rather than with the special case of an ideal gas. To discover what is continuous, one backs up to the original integral equations. Because shocks are such that the fluid velocity tangential to a shock is continuous, it is sufficient to limit oneself to a one-dimensional model, where a shock can be idealized as a plane normal to the x axis and the flow is only in the x direction. In a given inertial frame, one considers a fixed control volume (Fig. 4.3) of unit cross section and with fixed endpoints x_1 and x_2.[4] The integral equation corresponding to conservation of mass or continuity can be written in this one-dimensional case and for the stated control volume as

$$\frac{d}{dt} \int_{x_1}^{x_2} \rho \, dx = (\rho u)_{x_1} - (\rho u)_{x_2}. \tag{4.73}$$

This is just the statement that the time rate of change of mass in the volume is the difference of the rate at which the mass is flowing in at x_1 minus that at which it is flowing out at x_2. (Here the subscript denotes the point at which the indicated quantity is evaluated.)

The analogous integral equation accounting for changes in momentum states that the time rate of change of momentum in the volume is equal to the rate (per unit area) $(\rho u)_{x_1} (u)_{x_1}$ at which momentum is flowing in minus the rate $(\rho u)_{x_2} (u)_{x_2}$ at which it is flowing out plus the net force (per unit area) exerted on the control volume. For our present purposes this force can be expressed in the limit of vanishing viscosity, so that it is simply $P_{x_1} - P_{x_2}$. Thus one has

---

[3] One can easily see how one might fall into the trap of thinking of K as continuous at a shock, if one begins with Eq. (1.6), which implies that $K = P_0 \rho_0^{-\gamma}$, where P_0 and ρ_0 are *ambient* pressure and density. An illusion that K can be taken as constant comes about because of an imprecise definition of these ambient variables. After the shock sweeps by, the gas is slightly heated, and the ambient variables do actually change. They may change only slightly, and in many acoustic relations it is an appropriate approximation to use the same values for P_0 and ρ_0 on both sides of the shock, but it is not appropriate to do so in any derivations that make use of the adherence of the pressure and the density to the adiabatic law. The escalation from linear acoustics to nonlinear acoustics requires that one take special care in dealing with seemingly small effects.

[4] An alternative approach (Courant and Friedrichs, 1948), leading to the same results, is to consider time-dependent endpoints moving with the fluid. It seems to be an even trade-off as to which approach is more easily grasped.

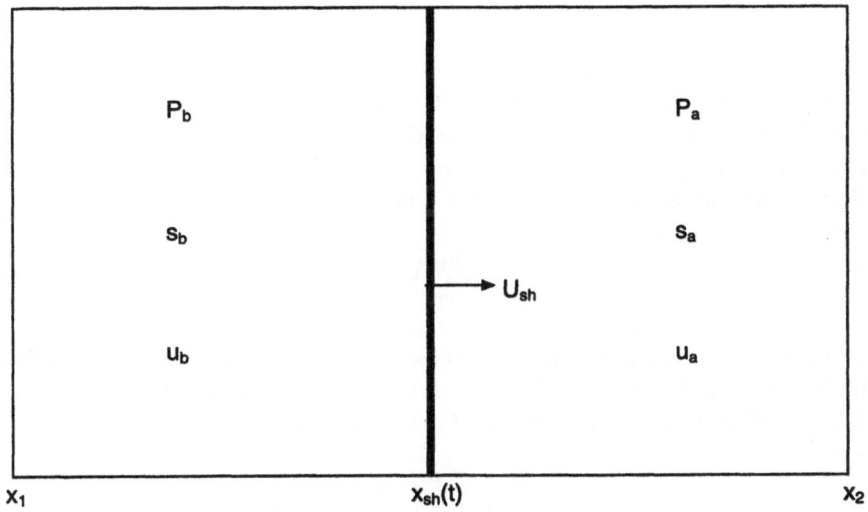

Fig. 4.3 Sketch of a fixed control volume containing a shock.

$$\frac{d}{dt}\int_{x_1}^{x_2} \rho u\, dx = (\rho u^2 + P)_{x_1} - (\rho u^2 + P)_{x_2}.$$ (4.74)

The third integral equation is one that accounts for changes of energy. The energy density within the fluid is equal to $\frac{1}{2}\rho u^2$ plus ρe, where $\frac{1}{2}\rho u^2$ represents kinetic energy per unit volume and e represents internal energy per unit mass. The time rate of change of energy within the control volume, when dissipation is neglected, should equal the rate $(\frac{1}{2}\rho u^2 + \rho e)_{x_1} u_{x_1}$ at which energy is being convected in by the flow minus the rate at which it is being convected out, plus the rate $(Pu)_{x_1} - (Pu)_{x_2}$ at which work is being done on the control volume by external pressures, or

$$\frac{d}{dt}\int_{x_1}^{x_2} \rho(\tfrac{1}{2}u^2 + e)\, dx = [(\tfrac{1}{2}\rho u^2 + \rho e + P)u]_{x_1} - [(\tfrac{1}{2}\rho u^2 + \rho e + P)u]_{x_2}.$$

(4.75)

Before the above integral equations are used to derive the Rankine–Hugoniot equations, the reader should be assured that these are indeed consistent with Eqs. (3.1)–(3.3). If one considers x_1 and x_2 as arbitrary and all quantities as continuous and differentiable, the above displayed equations are readily seen to lead to

$$\frac{\partial \rho}{\partial t} + \frac{\partial}{\partial x}(\rho u) = 0,$$ (4.76)

$$\frac{\partial}{\partial t}(\rho u) + \frac{\partial}{\partial x}(\rho u^2 + P) = 0,$$ (4.77)

$$\frac{\partial}{\partial t}(\tfrac{1}{2}\rho u^2 + \rho e) + \frac{\partial}{\partial x}(\tfrac{1}{2}\rho u^3 + \rho e u + P u) = 0. \tag{4.78}$$

The first of these is readily seen to be mathematically equivalent to the one-dimensional version of Eq. (3.1). The one-dimensional version (albeit in the absence of viscosity) of Eq. (3.2) is obtained if Eq. (4.76) is multiplied by u and the result subsequently subtracted from Eq. (4.77), which yields

$$\rho\left(\frac{\partial u}{\partial t} + u\frac{\partial u}{\partial x}\right) + \frac{\partial P}{\partial x} = 0. \tag{4.79}$$

To derive the one-dimensional counterpart of Eq. (3.3), one first multiplies Eq. (4.76) by $e + (P/\rho) - \tfrac{1}{2}u^2$ and Eq. (4.77) by u, adds the two resulting equations, then subtracts the sum from Eq. (4.78), to obtain

$$\rho\left(\frac{\partial e}{\partial t} + u\frac{\partial e}{\partial x}\right) - \frac{P}{\rho}\left(\frac{\partial \rho}{\partial t} + u\frac{\partial \rho}{\partial x}\right) = 0. \tag{4.80}$$

One subsequently uses the second law of thermodynamics in the form

$$T\,ds = de + P\,d\rho^{-1} \tag{4.81}$$

(with ρ^{-1} being recognized as the specific volume), which is readily interpreted to imply

$$T\frac{Ds}{Dt} = \frac{De}{Dt} + P\frac{D\rho^{-1}}{Dt}, \tag{4.82}$$

where D/Dt is the time derivative following a material point. Thus Eq. (4.80) reduces to

$$\rho T\frac{Ds}{Dt} = 0, \tag{4.83}$$

which is the same as Eq. (3.3) in the absence of viscosity and thermal conductivity. It is important to note, however, that the steps as described above, which led from Eqs. (4.73), (4.74), and (4.75) to Eqs. (4.76), (4.79), and (4.83), cannot be carried through should there be a surface of discontinuity in the control volume. In actuality, the so-derived differential equations can yield results that, even with a generalized interpretation of a derivative or even after integration over a finite volume, are incorrect. For example, Eq. (4.83) implies that the entropy of a fixed mass of fluid stays constant, even if that mass is swept over by a shock. This turns out not to be the case.

To see what results when a discontinuity is present, one postulates a moving point $x_{sh}(t)$ (eventually identified as the location of a shock) between x_1 and x_2 at which P, u, ρ, and e are discontinuous. Each of the integrals over x in Eqs. (4.73)–(4.75) can be split into integrals from x_1 to x_{sh} and from x_{sh} to x_2. Then, standard rules for differentiation (especially those that allow for an upper or lower limit that depends on the parameter with respect to which one is taking the derivative) yield, for example,

$$\frac{d}{dt}\int_{x_1}^{x_2} \rho\, dx = (\rho_b - \rho_a)U_{sh} + \int_{x_1}^{x_{sh-}} \frac{\partial \rho}{\partial t}\, dx + \int_{x_{sh+}}^{x_2} \frac{\partial \rho}{\partial t}\, dx, \qquad (4.84)$$

where ρ_b and ρ_a represent the values of ρ on the $-x$ and $+x$ sides of the discontinuity, and $U_{sh} = dx_{sh}/dt$ is the velocity of the discontinuity surface. (The letters a and b can here be regarded as abbreviations for *ahead of* and *behind*, as much of the ensuing discussion is concerned with a shock advancing in the direction of positive x.) In the limit in which x_1 and x_2 are arbitrarily close to x_{sh}, the integrals on the right become negligible and $(\rho u)_{x_1} \to (\rho u)_b$, $(\rho u)_{x_2} \to (\rho u)_a$, so Eq. (4.73) yields

$$[\rho(u - U_{sh})]_a = [\rho(u - U_{sh})]_b. \qquad (4.85)$$

In a similar manner, Eqs. (4.74) and (4.75) imply

$$[\rho u(u - U_{sh}) + P]_a = [\rho u(u - U_{sh}) + P]_b, \qquad (4.86)$$

$$[\rho(\tfrac{1}{2}u^2 + e)(u - U_{sh}) + Pu]_a = [\rho(\tfrac{1}{2}u^2 + e)(u - U_{sh}) + Pu]_b. \qquad (4.87)$$

Equations (4.85)–(4.87) are the Rankine–Hugoniot relations in their most primitive form. The three-dimensional version is the same, except that the velocities u and U_{sh} should be regarded as components that are normal to the shock surface. One should note that the three quantities that emerge as being continuous across a shock actually involve the speed of the shock. Since there are only three equations, but four quantities (ρ, u, P, and e) are suspected of possibly being discontinuous, one has little hope that these equations can lead to a rigorous conclusion that, say, P is automatically continuous.

Since the formulation of fluid dynamics should be insensitive to shifts from one inertial frame to another (Galilean invariance), one expects the *total* information content of the above three equations to be unchanged if u and U_{sh} are simultaneously everywhere replaced by $u - K$ and $U_{sh} - K$, where K is any constant. In retrospect, one can also argue that K can be a quantity not necessarily independent of time, since the relations as stated do not involve time differentiation. In particular, one can take $K = U_{sh}$. Such a substitution would then yield the equivalent set of equations

$$[\rho v]_a = [\rho v]_b, \qquad (4.88)$$

$$[\rho v^2 + P]_a = [\rho v^2 + P]_b, \qquad (4.89)$$

$$[\tfrac{1}{2}\rho v^3 + \rho ev + Pv]_a = [\tfrac{1}{2}\rho v^3 + \rho ev + Pv]_b, \qquad (4.90)$$

with the abbreviation

$$v = u - U_{sh}. \tag{4.91}$$

To confirm that Galilean invariance really does apply and that this set of equations is indeed equivalent in total information content to Eqs. (4.85)–(4.87), one can obtain Eq. (4.89) by subtracting U_{sh} times Eq. (4.85) from Eq. (4.86). Similarly, one can obtain Eq. (4.90) by subtracting U_{sh} times Eq. (4.86) and adding $\frac{1}{2}U_{sh}^2$ times Eq. (4.85) to Eq. (4.87).

One further simplification results from factoring out ρv on each side of Eq. (4.90). Since ρv, according to Eq. (4.88), is continuous across a shock, one can divide both sides by this common factor, with the result

$$[\tfrac{1}{2}v^2 + h]_a = [\tfrac{1}{2}v^2 + h]_b, \tag{4.92}$$

where

$$h = e + \rho^{-1}P \tag{4.93}$$

is identified as the enthalpy per unit mass. Carrying out the division of course requires that ρv not be zero, so one automatically rules out consideration of contact discontinuities, at which the density is discontinuous, but the discontinuity surface moves with the flow.

One may note that, apart from slight changes in notation, Eqs. (4.88), (4.89), and (4.92) are identical with Eqs. (1.34)–(1.36). [This is not, strictly speaking, the full set of conditions that apply at a shock. There is also an inequality condition, given further below in Eq. (4.105).]

In the applications of the Rankine–Hugoniot relations, it is generally useful to reexpress them in terms of the values of the discontinuities that occur at a shock, as is done, for example, by Hayes (1958). With the abbreviations $\Delta v = v_b - v_a$, $\Delta h = h_b - h_a$, $v_{av} = \frac{1}{2}(v_a + v_b)$, etc., Eqs. (4.88), (4.89), and (4.92) can be rewritten

$$\rho_{av}\Delta v + v_{av}\Delta\rho = 0, \tag{4.94}$$

$$v_{av}^2\Delta\rho + 2v_{av}\rho_{av}\Delta v + \Delta P + \tfrac{1}{4}(\Delta v)^2\Delta\rho = 0, \tag{4.95}$$

$$v_{av}\Delta v + \Delta h = 0. \tag{4.96}$$

Alternatively, one can use instead $(1/\rho)_{av}$, whereby

$$\frac{1}{(1/\rho)_{av}\rho_{av}} = 1 - \frac{1}{4}\left(\frac{\Delta\rho}{\rho_{av}}\right)^2. \tag{4.97}$$

This enables Eq. (4.94) to yield the substitutions

$$\frac{1}{(1/\rho)_{\mathrm{av}}\rho_{\mathrm{av}}} = 1 - \frac{1}{4}\left(\frac{\Delta v}{v_{\mathrm{av}}}\right)^2,\tag{4.98}$$

$$v_{\mathrm{av}}^2\Delta\rho + 2v_{\mathrm{av}}\rho_{\mathrm{av}}\Delta v + \tfrac{1}{4}(\Delta v)^2\Delta\rho = v_{\mathrm{av}}[(1/\rho)_{\mathrm{av}}]^{-1}\Delta v,\tag{4.99}$$

and the latter of these allows Eq. (4.95) to be reexpressed as

$$v_{\mathrm{av}}\Delta v + (1/\rho)_{\mathrm{av}}\Delta P = 0.\tag{4.100}$$

With Eqs. (4.94), (4.96), and (4.100), one has three equations that are formally linear in the four discontinuities ΔP, $\Delta\rho$, Δh, and Δv. Among the algebraic rearrangements of these equations, the set that is of especial use in the discussion of nonlinear acoustics is

$$\Delta h - (1/\rho)_{\mathrm{av}}\Delta P = 0,\tag{4.101}$$

$$(U_{\mathrm{sh}} - u_{\mathrm{av}})\Delta u = (1/\rho)_{\mathrm{av}}\Delta P,\tag{4.102}$$

$$(U_{\mathrm{sh}} - u_{\mathrm{av}})^2 = (1/\rho)_{\mathrm{av}}\rho_{\mathrm{av}}\frac{\Delta P}{\Delta\rho}.\tag{4.103}$$

Here we have replaced the auxiliary variable v by its explicit definition $u - U_{\mathrm{sh}}$, as given in Eq. (4.91). It follows from the above equations that ΔP, $\Delta\rho$, Δh, and $\Delta u/(U_{\mathrm{sh}} - u_{\mathrm{av}})$ must all have the same sign.

Equation (4.101) yields what is frequently referred to as the *Hugoniot equation*. It is expected to apply even if the equation of state of the fluid is changed by passage of the shock. If the equation of state of the fluid ahead of the shock is used to express h_a in terms of P_a and $(1/\rho)_a$, and if the equation of state of the fluid behind the shock is used to express h_b in terms of P_b and $(1/\rho)_b$, then Eq. (4.101) can be regarded as a relation of the general form

$$F[P_a, (1/\rho)_a, P_b, (1/\rho)_b] = 0,\tag{4.104}$$

where ρ^{-1} is recognized as specific volume. The relation is independent of flow velocities and shock speeds and is consequently a purely thermodynamic relation. As discussed by Hayes (1958), this relation is of great importance in considerations of detonations and deflagrations. For fixed pressure P_a and specific volume $(1/\rho)_a$ ahead of the shock, the relation allows one to conceive of a plot of pressure P_b versus specific volume $(1/\rho)_b$. Such a plot is referred to as a *Hugoniot diagram*; lines corresponding to different choices of P_a and $(1/\rho)_a$ are referred to as *Hugoniots*. If the equation of state is the same on both sides of the shock, then the point $[P_a, (1/\rho)_a]$ lies on the corresponding Hugoniot.

One further restriction comes from the inequality version of the second law of thermodynamics. If a shock is advancing in the $+x$ direction relative to the fluid, and therefore $U_{\mathrm{sh}} - u_{\mathrm{av}} > 0$, then $s_b \geq s_a$; a fluid particle's entropy cannot be decreased by passage of the shock. This yields the inequality

$$\frac{\Delta s}{U_{sh} - u_{av}} \geq 0. \tag{4.105}$$

It should be evident from a comparison of Eq. (4.101) with the thermodynamic differential relation $dh = T\,ds + (1/\rho)\,dP$ that the entropy jump in a shock should be relatively small. It is, nevertheless, not identically zero. The quantification of the relation of Δs to ΔP is derived further below.

4.3.2   Weak Shocks

The consequences of the Rankine–Hugoniot equations for weak shocks, where $|\Delta\rho|/\rho_{av} \ll 1$, can be explored by expanding $h(P, s)$ in a Taylor series in $\delta P = P - P_{av}$ and $\delta s = s - s_{av}$, the various coefficients being denoted by h^0, h_P^0, h_s^0, h_{PP}^0, h_{Ps}^0, etc., such that, for example, h_{Ps}^0 is $\partial^2 h/\partial P\,\partial s$ evaluated at P_{av} and s_{av}. It follows that

$$\begin{aligned}
h = {} & h_0 + h_P^0\,\delta P + h_s^0\,\delta s + \tfrac{1}{2}[h_{PP}^0(\delta p)^2 + 2h_{Ps}^0(\delta p)(\delta s) + h_{ss}^0(\delta s)^2] \\
& + \tfrac{1}{6}[h_{PPP}^0(\delta P)^3 + 3h_{PPs}^0(\delta P)^2\delta s + 3h_{ssP}^0\delta P(\delta s)^2 + h_{sss}^0(\delta s)^3] \\
& + \tfrac{1}{24}[h_{PPPP}^0(\delta P)^4 + 4h_{PPPs}^0(\delta P)^3\delta s + 6h_{PPss}^0(\delta P)^2(\delta s)^2 \\
& + 4h_{Psss}^0(\delta P)(\delta s)^3 + h_{ssss}^0(\delta s)^4] + \cdots .
\end{aligned} \tag{4.106}$$

To obtain h_a, one sets $\delta P = -\tfrac{1}{2}\Delta P$, $\delta s = -\tfrac{1}{2}\Delta s$ in this expansion; to obtain h_b, one sets $\delta P = \tfrac{1}{2}\Delta P$, $\delta s = \tfrac{1}{2}\Delta s$. Consequently, one has

$$\begin{aligned}
\Delta h = {} & h_P^0\Delta P + h_s^0\Delta s + \tfrac{1}{24}[h_{PPP}^0(\Delta P)^3 + 3h_{PPs}^0(\Delta P)^2\Delta s \\
& + 3h_{ssP}^0\Delta P(\Delta s)^2 + h_{sss}^0(\Delta s)^3] + \cdots .
\end{aligned} \tag{4.107}$$

Expansions for the specific volume ρ^{-1} and the absolute temperature T follow from Eqs. (4.81) and (4.93):

$$dh = de + P\,d\rho^{-1} + (1/\rho)\,dP = T\,ds + (1/\rho)\,dP. \tag{4.108}$$

Thus, one has

$$\begin{aligned}
\frac{1}{\rho} = {} & \frac{\partial h}{\partial P} = h_P^0 + \tfrac{1}{2}[2h_{PP}^0(\delta p) + 2h_{Ps}^0(\delta s)] \\
& + \tfrac{1}{6}[3h_{PPP}^0(\delta P)^2 + 6h_{PPs}^0(\delta P)(\delta s) + 3h_{ssP}^0(\delta s)^2] \\
& + \tfrac{1}{24}[4h_{PPPP}^0(\delta P)^3 + 12h_{PPPs}^0(\delta P)^2\delta s + 12h_{PPss}^0(\delta P)(\delta s)^2
\end{aligned}$$

$$+ 4h^0_{Psss}(\delta s)^3] + \cdots , \tag{4.109}$$

$$(1/\rho)_{\mathrm{av}} = h^0_P + \tfrac{1}{24}[3h^0_{PPP}(\Delta P)^2 + 6h^0_{PPs}(\Delta P)(\Delta s) + 3h^0_{ssP}(\Delta s)^2] + \cdots , \tag{4.110}$$

$$\Delta(1/\rho) = \tfrac{1}{2}[2h^0_{PP}(\Delta P) + 2h^0_{Ps}(\Delta s)]$$
$$+ \tfrac{1}{96}[4h^0_{PPPP}(\Delta P)^3 + 12h^0_{PPPs}(\Delta P)^2\Delta s + 12h^0_{PPss}(\Delta P)(\Delta s)^2$$
$$+ 4h^0_{Psss}(\Delta s)^3] + \cdots . \tag{4.111}$$

The expansions derived above for Δh and $(1/\rho)_{\mathrm{av}}$ lead in turn to

$$\Delta h - (1/\rho)_{\mathrm{av}}\Delta P = h^0_s\Delta s - \tfrac{1}{12}h^0_{PPP}(\Delta P)^3 - \tfrac{1}{8}h^0_{PPs}(\Delta P)^2\Delta s - \cdots . \tag{4.112}$$

The left side of Eq. (4.112) is zero, according to the Rankine–Hugoniot relation Eq. (4.101); the resulting equation, when solved by iteration for Δs in terms of ΔP yields, to lowest nonvanishing order,

$$\Delta s = \frac{h^0_{PPP}}{12h^0_s}(\Delta P)^3. \tag{4.113}$$

The entropy change, being of third order in the pressure change, is consequently very small for a weak shock, but nevertheless not identically zero, in contrast to what would be inferred from the partial differential equation appearing in Eq. (4.83).

The derivation here of Eq. (4.113) is essentially the same as that given in the text by Landau and Lifshitz (1987). It is also along lines similar to the derivation of Hayes (1958), who in turn states that his "derivation is essentially the same as that of Bethe," the citation being to a report written during World War II by Bethe (1942). The result may have been known to other researchers before 1942, as Landau (1945) has the parenthetical comment, "in carrying out the calculation it must be kept in mind that the change of entropy in the discontinuity is a quantity of the third order, while in the Riemann solution the entropy is constant," without any citing of references. [Bethe's report would not ordinarily have been available to Landau before 1945, although there is a possibility that it had been conveyed to the (then) Soviet Union via Fuchs.]

The most pertinent implication of Eq. (4.113), insofar as weak shock theory is concerned, is that the entropy discontinuity can be neglected in any approximate relation that is taken as being of second order or less in the pressure discontinuity ΔP. In this regard, one notes from Eq. (4.110) that

$$(1/\rho)_{\mathrm{av}} = h^0_P + O\{(\Delta P)^2\}, \tag{4.114}$$

while Eq. (4.111) yields

$$\Delta(1/\rho) = h^0_{PP}\Delta P + O\{(\Delta P)^3\}. \tag{4.115}$$

Consequently, it follows that

$$\rho_{av} = \frac{1}{h^0_P} + O\{(\Delta P)^2\}, \tag{4.116}$$

$$\Delta\rho = -\frac{h^0_{PP}}{(h^0_P)^2}\Delta P + O\{(\Delta P)^3\}. \tag{4.117}$$

Thus, the Rankine–Hugoniot relation of Eq. (4.103) leads to

$$(U_{sh} - u_{av})^2 = -\frac{(h^0_P)^2}{h^0_{PP}} + O\{(\Delta P)^2\}. \tag{4.118}$$

Alternatively, if one returns to the differential expression of Eq. (4.108) for the enthalpy, it is seen that Eq. (4.118) can be rewritten

$$(U_{sh} - u_{av})^2 = \left(\frac{\partial P}{\partial \rho}\right)_{\text{eval at av}} + O\{(\Delta P)^2\}, \tag{4.119}$$

where the indicated derivative is intended to be carried out at constant entropy and at the state that corresponds to entropy s_{av} and pressure P_{av}, the averages of values ahead of and behind the shock. This derivative is the square of a sound speed, but it differs in first order from the square of the sound speed ahead of the shock and in first order from the square of the sound speed behind the shock. However, it differs by only second order from the average of these two quantities, or by second order from the square of the average of the two sound speeds. This convenient fact, that averaging consistently accounts correctly for first-order corrections, follows in general from the observation that, for any function $f(\delta P)$, one has

$$f(0) = \tfrac{1}{2}[f(\tfrac{1}{2}\Delta P) + f(-\tfrac{1}{2}\Delta P)] + O\{(\Delta P)^2\}. \tag{4.120}$$

This holds, in addition, for any function of f, including its square, square root, and reciprocal. Thus, Eq. (4.119) can be rewritten

$$U_{sh} - u_{av} = \pm c_{av} + O\{(\Delta P)^2\}, \tag{4.121}$$

while Eq. (4.102) can be rewritten

$$\Delta u = \pm\frac{\Delta P}{\rho_{av}c_{av}} + O\{(\Delta P)^3\}, \tag{4.122}$$

where the $+$ signs correspond to a shock advancing in the $+x$ direction relative to the fluid.

Alternative versions of Eq. (4.121) that are consistent with Eq. (4.120) are

$$U_{\text{sh}} = \frac{dx_{\text{sh}}}{dt} = (\pm c + u)_{\text{av}} + O\{(\Delta P)^2\}, \tag{4.123}$$

$$\frac{1}{U_{\text{sh}}} = \frac{dt_{\text{sh}}}{dx} = \left(\frac{1}{\pm c + u}\right)_{\text{av}} + O\{(\Delta P)^2\}, \tag{4.124}$$

where $t_{\text{sh}}(x)$ is the time the shock arrives at point x. Use of the latter equation requires, of course, that u_a and u_b be regarded as first order in ΔP, while c_a and c_b are regarded as zeroth order.

Similar reasoning, with the aid of the differential thermodynamic relation of Eq. (4.108), allows one to reexpress Eq. (4.113) as

$$\Delta s = \left(\frac{\partial^2 \rho^{-1}/\partial P^2}{12T}\right)_{\text{av}} (\Delta P)^3, \tag{4.125}$$

in which the indicated derivative is

$$\frac{\partial^2 \rho^{-1}}{\partial P^2} = \frac{2}{\rho^3 c^4}\left(1 + \rho c \frac{\partial c}{\partial P}\right). \tag{4.126}$$

Now define

$$\tilde{\beta} = 1 + \rho c \frac{\partial c}{\partial P}, \tag{4.127}$$

where $\tilde{\beta}$ is the coefficient of nonlinearity evaluated at the considered thermodynamic state [see Eq. (3.18)]. Thus Eq. (4.125) can be rewritten in terms of familiar symbols as

$$\Delta s = \left(\frac{\tilde{\beta}}{6\rho^3 c^4 T}\right)_{\text{av}} (\Delta P)^3 = \frac{\beta}{6\rho_0^3 c_0^4 T_0}(\Delta P)^3. \tag{4.128}$$

To leading order in ΔP, it is consistent to evaluate the coefficient here at any convenient thermodynamic state—in particular, the ambient state; the zero subscript is therefore used in the second version of this relation.

For most occasions of interest, the parameter β is greater than 0, and so the discontinuities Δs and ΔP must have the same sign. Thus, the second law of thermodynamics would ordinarily require $p_b > p_a$ (or $\Delta p > 0$ for a shock advancing in the $+x$ direction). The pressure behind the shock front is higher because the specific entropy must be higher. For exceptions to this (*rarefaction shocks*), see the discussions and experimental results of Borisov et al. (1983), Cramer (1989), Lambrakis and Thompson (1972), Thompson (1971), Thompson and Lambrakis (1973), and Thompson et al. (1986).

4.3.3   Reflection of Waves by Shocks

A principle tenet on which the weak shock propagation theory of nonlinear acoustics is based is that the formation of a shock will not change the unidirectional character of a wave that is originally propagating in one direction. The Rankine–Hugoniot relation $U_{sh} = (c + u)_{av}$ predicts that, given $(c + u)_b > (c + u)_a$, the shock will be moving more slowly than the waveform behind the shock, so any fragment of the waveform immediately behind the shock must eventually overtake the shock. The question naturally arises as to what happens to such fragments. Do they simply "disappear" into the shock, or do they cause backward-propagating waves to be generated? As might be expected, neither answer is totally correct, but it turns out that assuming the former yields an extremely good approximation.

To confirm that such is the case, we consider the simple example shown in Fig. 4.4. Before time $t = 0$, two shocks are propagating in the $+x$ direction through an unbounded fluid. Ahead of the shock that is farthest to the right, the fluid-dynamic variables are constant, and have values labeled by the subscript a; in the middle region, between the two shocks, they are also constant, and have values labeled by the subscript m; then, in the region behind the shock farthest to the left, they are also constant, and have values labeled by the subscript b. According to the Rankine–Hugoniot relations and the fluid-dynamic equations, one expects the values of the variables in each of the regions to remain the same up until such time as the second shock catches up to the first. The instant when this happens is truly a defining moment, and is consequently the instant we here choose to take as $t = 0$. Also, the coordinate origin ($x = 0$) is chosen to be the point where the catch-up occurs.

Just before $t = 0$, the Rankine–Hugoniot relation Eq. (4.101), applied to each of the two shocks individually, requires

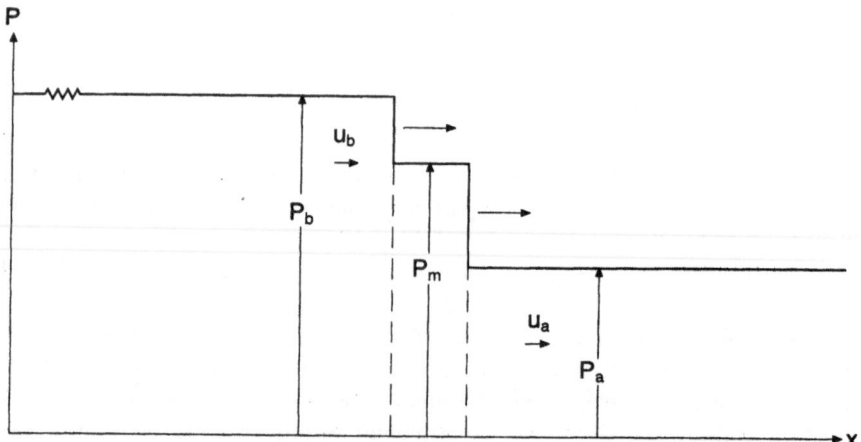

Fig. 4.4 Sketch of two shocks, one of which will overtake the other.

$$h_m - h_a = \tfrac{1}{2}(\rho_a^{-1} + \rho_m^{-1})(P_m - P_a), \tag{4.129}$$

$$h_b - h_m = \tfrac{1}{2}(\rho_b^{-1} + \rho_m^{-1})(P_b - P_m). \tag{4.130}$$

It is important to notice, however, that

$$h_b - h_a \neq \tfrac{1}{2}(\rho_b^{-1} + \rho_a^{-1})(P_b - P_a) \tag{4.131}$$

for $t > 0$; i.e., the two shocks do not lock together and propagate on indefinitely without some other secondary disturbance being generated.

Just at $t = 0$, the history of the disturbance before $t = 0$, and even the prior existence of the two shocks, is for the most part irrelevant. The fluid dynamics presents itself instead as an initial-value problem: Given the thermodynamic states and the fluid velocities for $x < 0$ and $x > 0$ at $t = 0$, one seeks to determine the disturbance for $t > 0$.

Since the initial-value problem just posed has no characteristic length scale and no characteristic time scale, whatever disturbances are to be generated must depend on x and t only in terms of their ratio, and the edges of any disturbances must propagate out from the point $x = 0$ with a constant speed. There are only two possibilities that can be considered (Landau and Lifshiftz, 1987). One is that the pressure in the intermediate region is everywhere larger than both P_b and P_a, and therefore the two speeds at the left and right edges correspond to shock velocities. The other possibility (Fig. 4.5) takes the pressure to be bigger than P_a but less than P_b, so the moving right edge corresponds to a shock, but the moving left edge does not. (Here, for simplicity, we consider the coefficient of nonlinearity to be positive, and we also consider P_b to be larger than P_a.) It turns out that the correct choice, at

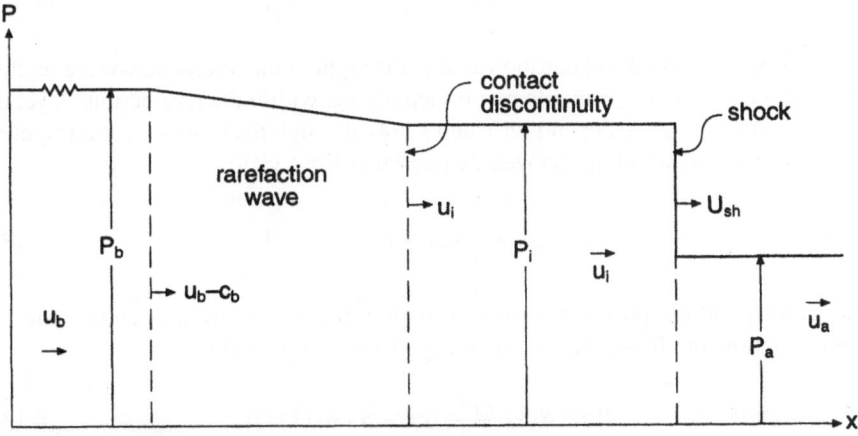

Fig. 4.5 Waveform showing features that occur shortly after one shock overtakes another.

least in the weak shock limit and for fluids of common interest, is the second case, and therefore the analysis here is restricted to that case.

The regime just behind the shock is at some constant pressure P_i and constant entropy s_i, and the fluid velocity has constant value u_i, but one allows for the possibility of a moving point at the left edge of this regime at which the density and entropy are discontinuous. The similitude principle allows one to pinpoint the constant-flow regime as consisting of points and times for which $u_i < (x/t) < U_{sh}$.

Although one does not know the fluid-dynamic variables in this intermediate constant-flow regime at the outset of the analysis, the Rankine–Hugoniot relations derived in Sect. 4.3.1 are applicable, and one can write

$$h_i - h_a = \tfrac{1}{2}(\rho_a^{-1} + \rho_i^{-1})(P_i - P_a), \tag{4.132}$$

$$u_i - u_a = (\rho_a^{-1} - \rho_i^{-1})^{1/2}(P_i - P_a)^{1/2}, \tag{4.133}$$

where the second equation is derived from Eqs. (4.102) and (4.103).

The flow regime farthest to the left is a *rarefaction wave*, the leftmost edge of which moves in the $-x$ direction with the sound speed relative to the fluid. The regime is thus defined by the inequality $-c_b + u_b < (x/t) < u_i$, where c_b is the sound speed in the region behind the overall disturbance. This rarefaction wave is the same as one would predict were a wall, originally at $x = 0$ and facing an unbounded fluid on its left ($x < 0$), suddenly at $t = 0$ to be moved to the right with constant speed u_i. The entropy in this wave is constant and equal to s_b, and the actual wave disturbance is governed by the nonlinear equation for a unidirectional wave propagating to the left, whereby points of given amplitude move with speed $-c(u) + u$. Also, u is related to the thermodynamic state by

$$\frac{dP}{du} = -\rho c, \quad \frac{dc}{du} = -\rho c \frac{dc}{dP}. \tag{4.134}$$

The entropy is constant and has the value s_b throughout the rarefaction-wave regime. Also, by the similitude principle mentioned above, within the rarefaction wave, the fluid-dynamic variables depend on x and t only through the ratio x/t, and therefore x/t is $-c(u) + u$, which in the weakly nonlinear limit yields

$$u = u_b + \frac{1}{\beta}[c_b - u_b + (x/t)]. \tag{4.135}$$

The corresponding pressure variation within the wave, to the same order of approximation and in accordance with Eq. (4.134), is given by

$$P = P_b - \frac{\rho_b c_b}{\beta}[c_b - u_b + (x/t)]. \tag{4.136}$$

Note that both of these expressions have a linear dependence on x for fixed t.

The claim can now be made that the equations as described above, given a knowledge of u_b and u_a along with the thermodynamic variables in states a and b, should be sufficient to determine the thermodynamic variables and fluid velocities in each of the two intermediate regions. To see that such is indeed the case, and also to reduce our consideration to weak shocks, we once again use the enthalpy expansion given by Eq. (4.106), with the reference state taken as that where the entropy and pressure are the averages of what exists in states a and b. Thus, in the previously derived expressions, Eqs. (4.106) and (4.109), one sets $\delta P = P_a - \frac{1}{2}(P_a + P_b)$ and $\delta s = s_a - \frac{1}{2}(s_a + s_b)$ to determine h_a and $1/\rho_a$, etc. For simplicity, we denote these differences as $(\delta P)_a$, $(\delta s)_a$, etc. The Rankine–Hugoniot relation Eq. (4.132) subsequently yields

$$(\delta s)_i - (\delta s)_a \simeq \frac{h^0_{PPP}}{12 h^0_s}[(\delta P)_i - (\delta P)_a]^3, \tag{4.137}$$

and similar equations hold for the entropy discontinuities $(\delta s)_m - (\delta s)_a$ and $(\delta s)_b - (\delta s)_m$. In addition, Eq. (4.133) yields

$$u_i - u_a \simeq [(\delta P)_i - (\delta P)_a]^{1/2}\{-h^0_{PP}[(\delta P)_i - (\delta P)_a] - h^0_{Ps}[(\delta s)_j - (\delta s)_a]$$
$$- \tfrac{1}{2}h^0_{PPP}[(\delta P)^2_i - (\delta P)^2_a] - \tfrac{1}{6}h^0_{PPPP}[(\delta P)^3_i - (\delta P)^3_a]\}^{1/2}, \tag{4.138}$$

and analogous equations hold for $u_m - u_a$ and $u_b - u_m$. Here the intent has been to consistently keep all terms of up to third order in the pressure differences, it being recognized that the entropy differences are of third order. (It is also recognized that h^0_{PP} is negative.) The final equation of the set comes from Eq. (4.134), which to the same order of approximation yields

$$u_b - u_i \simeq -\frac{1}{\rho_b c_b}(P_b - P_i) \simeq -(-h^0_{PP})^{1/2}[(\delta P)_b - (\delta P)_i], \tag{4.139}$$

where it is anticipated that $[(\delta P)_b - (\delta P)_i]$ is of third order.

The desire at this point is to "solve" the set of equations as described above for $(\delta P)_i$ in terms of the original pressure increments $(\delta P)_a$, $(\delta P)_b$, and $(\delta P)_m$. To do this, it is sufficient to use other equations for the fluid velocity differences in the combination

$$(u_b - u_i) + (u_i - u_a) = (u_b - u_m) + (u_m - u_a) \tag{4.140}$$

and to use Eq. (4.137) and its counterparts to make substitutions into the terms involving entropies.

To quantify the magnitude of the difference $(\delta P)_b - (\delta P)_i$ to lowest nonvanishing order, it is sufficient to replace $(\delta P)_i$ everywhere it occurs in the term $u_i - u_a$. This term can be subsequently expanded to third order in the pressure increments, with the result

$$(u_b - u_i) + (u_i - u_a) \simeq -(-h_{PP}^0)^{1/2}[(\delta P)_b - (\delta P)_i]$$

$$+ (-h_{PP}^0)^{1/2}\Big\{[(\delta P)_b - (\delta P)_a$$

$$+ \frac{1}{4}\frac{h_{PPP}^0}{h_{PP}^0}[(\delta P)_b^2 - (\delta P)_a^2]$$

$$+ \frac{1}{12}\frac{h_{PPPP}^0}{h_{PP}^0}[(\delta P)_b^3 - (\delta P)_a^3]$$

$$+ \frac{1}{24}\frac{h_{PPP}^0 h_{Ps}}{h_s^0 h_{PP}^0}[(\delta P)_b - (\delta P)_a]^3$$

$$- \frac{1}{32}\frac{(h_{PPP}^0)^2}{(h_{PP}^0)^2}[(\delta P)_b^2 - (\delta P)_a^2][(\delta P)_b + (\delta P)_a]\Big\}.$$

$$(4.141)$$

Similarly, an analogous expansion of the right side of Eq. (4.140) yields

$$(u_b - u_m) + (u_m - u_a)$$

$$\simeq (-h_{PP}^0)^{1/2}\Big\{[(\delta P)_b - (\delta P)_a] + \frac{1}{4}\frac{h_{PPP}^0}{h_{PP}^0}[(\delta P)_b^2 - (\delta P)_a^2]$$

$$+ \frac{1}{12}\frac{h_{PPPP}^0}{h_{PP}^0}[(\delta P)_b^3 - (\delta P)_a^3]$$

$$+ \frac{1}{24}\frac{h_{PPP}^0 h_{Ps}^0}{h_s^0 h_{PP}^0}([(\delta P)_b - (\delta P)_m]^3 + [(\delta P)_m - (\delta P)_a]^3)$$

$$- \frac{1}{32}\frac{(h_{PPP}^0)^2}{(h_{PP}^0)^2}([(\delta P)_b^2 - (\delta P)_m^2][(\delta P)_b + (\delta P)_m]$$

$$+ [(\delta P)_m^2 - (\delta P)_a^2][(\delta P)_m + (\delta P)_a])\Big\}. \qquad (4.142)$$

When these two expressions for $u_b - u_a$ are equated and solved to obtain $(\delta P)_b - (\delta P)_i$, the first- and second-order terms cancel identically. After some algebra, the result to lowest nonvanishing order is

$$(\delta P)_b - (\delta P)_i \simeq \frac{1}{8}\left(\frac{1}{4}\frac{(h_{PPP}^0)^2}{(h_{PP}^0)^2} + \frac{h_{PPP}^0 h_{Ps}^0}{h_s^0 h_{PP}^0}\right)$$

$$\times \ [(\delta P)_b - (\delta P)_a][(\delta P)_b - (\delta P)_m][(\delta P)_m - (\delta P)_a].$$
$$(4.143)$$

The coefficient here can be expressed in terms of familiar symbols by thermodynamic identities. One notes that $h_s = 1/T$, $h_{Ps} = \beta_{\text{vol}}T/c_p$, $h_{PP} = -1/\rho^2 c^2$, and $h_{PPP} = 2\tilde{\beta}/\rho^3 c^4$, where $\beta_{\text{vol}} = \rho(\partial \rho^{-1}/\partial T)_p$ is the coefficient of thermal expansion and c_p is the specific heat at constant pressure. To the same order of approximation, it is sufficient to evaluate these quantities at the ambient state (subscript 0). Thus the desired result takes the form

$$P_b - P_i \simeq \frac{\beta}{8\rho_0^2 c_0^4}[\beta - 2(\beta_{\text{vol}}c_0^2/c_p)](P_m - P_a)(P_b - P_m)(P_b - P_a). \qquad (4.144)$$

The factors $P_m - P_a$ and $P_b - P_m$ correspond to the pressure jumps in the first and second shocks, respectively; $P_b - P_a$ is the nominal pressure jump in the subsequent coalesced shock.

The thermodynamic coefficient is invariably positive. For an ideal gas, β is $\frac{1}{2}(\gamma + 1)$ and $\beta_{\text{vol}}c_0^2/c_p$ is $\gamma - 1$. The indicated difference is positive for air, where $\gamma = \frac{7}{5}$, but it does vanish for $\gamma = \frac{5}{3}$, which would correspond to a monatomic gas. For liquids, such as water, the quantity $\beta_{\text{vol}}c_0^2/c_p$ is typically much less than unity. The most important conclusion here, for the purposes of nonlinear acoustics, is that the "reflected wave" is extremely small, so one can proceed with an approximate theory of weak shock propagation in which such reflections are neglected at the outset. [For related discussions on this subject, the reader is referred to the books by Courant and Friedrichs (1948) and Landau and Lifshitz (1987), to Lighthill (1950), and to recent articles by Morfey and Sparrow (1993) and Morfey (1993).]

4.3.4 The Equal-Area Rule

A chief implication of the foregoing analysis is revealed by the approximate Rankine–Hugoniot relations of Eqs. (4.123) and (4.124). If a disturbance is moving in the $+x$ direction, then, once a discontinuity is formed, it moves with the average of the wave speeds behind and ahead of the shock. Suppose that both ahead of and behind the shock the disturbance is described by Eq. (4.24), which corresponds to the parametric equations

$$u = f(\phi), \qquad (4.145)$$

$$t(\phi, x) = \phi + \left(\frac{1}{c_0} - \frac{\beta u}{c_0^2}\right)x. \qquad (4.146)$$

Alternatively, one can use a description given by Eq. (4.5), which we here rewrite in the parametric form

$$u = g(\psi),$$ (4.147)

$$x(\psi, t) = \psi + [c_0 + \beta g(\psi)]t.$$ (4.148)

It is acknowledged that, for fixed x, the so-generated "waveform" plot in the (u, t) plane, as ϕ ranges over a continuum of values, can lead to multivalued waveforms, but that the waveform at any point and any instant corresponds to a definite value of ϕ. There may be some values of ϕ that do not correspond at all to the actual waveform at, say, a given value of x. Similarly, it is acknowledged that, for fixed t, the so-generated "waveform" plot in the (u, x) plane, as ψ ranges over a continuum of values, can lead to multivalued waveforms, but that the waveform at any point and any instant corresponds to a definite value of ψ. There may be some values of ψ that do not correspond at all to the actual waveform at, say, a given value of t.

If $g(\psi_b) = f(\phi_b)$ is the flow velocity behind the shock (or that just after the shock has passed by) and $g(\psi_a) = f(\phi_a)$ that in front of the shock (or just before the shock has arrived), then the intrinsic pressure dependence of the sound speed requires

$$c_a = c_0 + (\beta - 1)g(\psi_a), \quad c_b = c_0 + (\beta - 1)g(\psi_b),$$ (4.149)

or equivalently,

$$\frac{1}{c_a + u_a} = \frac{1}{c_0} - \frac{\beta}{c_0^2} f(\phi_a), \quad \frac{1}{c_b + u_b} = \frac{1}{c_0} - \frac{\beta}{c_0^2} f(\phi_b).$$ (4.150)

The approximate Rankine–Hugoniot relation $U_{\text{sh}} = c_{\text{av}} + u_{\text{av}}$ yields a shock speed equal to

$$\frac{dx_{\text{sh}}}{dt} = U_{\text{sh}} = c_0 + \tfrac{1}{2}\beta[g(\psi_a) + g(\psi_b)],$$ (4.151)

and its counterpart in Eq. (4.124) yields a shock slowness equal to

$$\frac{dt_{\text{sh}}}{dx} = \frac{1}{c_0} - \frac{\beta}{2c_0^2}[f(\phi_a) + f(\phi_b)].$$ (4.152)

(The quantities ψ_b, ψ_a, etc., are here regarded as functions of time t.) The location x_{sh} of the shock is given by Eq. (4.148) with ψ set to either ψ_a or ψ_b:

$$x_{\text{sh}} = \psi_a + [c_0 + \beta g(\psi_a)]t = \psi_b + [c_0 + \beta g(\psi_b)]t.$$ (4.153)

Similarly, the time of arrival t_{sh} is given by Eq. (4.146) with ϕ set to either ϕ_a or ϕ_b:

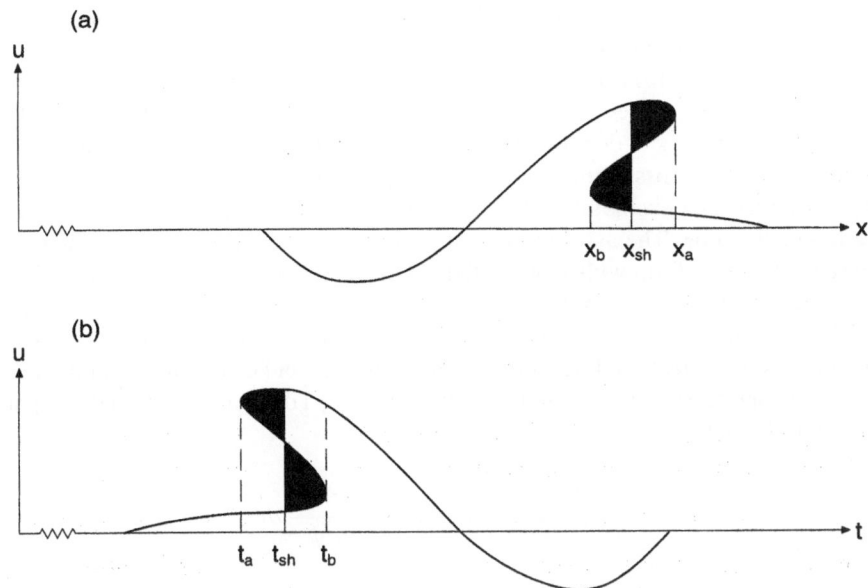

Fig. 4.6 Triple-valued wave: (**a**) Spatial waveform and (**b**) time waveform.

$$t_{sh} = \phi_a + \left[\frac{1}{c_0} - \frac{\beta}{c_0^2} f(\phi_a) \right] x = \phi_b + \left[\frac{1}{c_0} - \frac{\beta}{c_0^2} f(\phi_b) \right] x. \qquad (4.154)$$

As the shock moves, ψ_b decreases and ψ_a increases; the portion $g(\psi)$ for $\psi_b(t) < \psi < \psi_a(t)$ of the initial waveform does not contribute to the actual waveform at time t. The waveform, u versus x, so constructed is single-valued although discontinuous. Similarly, ϕ_b increases and ϕ_a decreases; the portion $f(\phi)$ for $\phi_a(x) < \phi < \phi_b(x)$ of the initial waveform (original value of x) does not contribute to the actual waveform at the point x. The waveform, u versus t, so constructed is single-valued although discontinuous.

Determination of the location of a shock at any instant (or of its arrival time at any point) is facilitated by the following theorem, which was stated first by Landau (1945) and derived independently by Whitham (1952). Suppose that one constructs the curve of u versus x from Eqs. (4.148) and (4.149) and that over the interval x_b to x_a the function u is triple-valued, the plot resembling a backward S (see Fig. 4.6a). The shock location x_{sh} is denoted by a vertical line connecting the upper and lower portions of the S, crossing the curve at some point g_{int} and thereby delimiting two areas, a lower area extending to the left of the line $x = x_{sh}$ and an upper area to the right of this line. The assertion is that x_{sh} must be such that these two areas are the same; the waveform with shock is then as sketched in Fig. 4.6a, with the vertical line replacing the two arcs of the S.

An alternative version of the theorem, applicable for the (u, ϕ) description given by Eqs. (4.145) and (4.146), is stated similarly. Suppose that one constructs the curve of u versus t from Eqs. (4.145) and (4.146) and that over the interval t_a to t_b the function u is triple-valued, the plot resembling a letter S (see Fig. 4.6b). The shock arrival time t_{sh} is denoted by a vertical line connecting the upper and lower portions of the S, crossing the curve at some point f_{int} and thereby delimiting two areas, a lower area extending to the right of the line $t = t_{sh}$ and an upper area to the left of this line. The assertion is that t_{sh} must be such that these two areas are the same; the waveform with shock is then as sketched in Fig. 4.6b, with the vertical line replacing the two arcs of the S.

The proof here proceeds for the latter version of the theorem. [The proof for the former version, corresponding to Fig. 4.6a, is similar, and can be found in the texts by Landau and Lifshitz (1987) and by Pierce (1989).] The total area, with due regard to sign, is given by

$$A(x) = -\int_{\phi_a(x)}^{\phi_b(x)} [t(\phi, x) - t_{sh}(x)] \frac{df(\phi)}{d\phi} \, d\phi. \tag{4.155}$$

Since $t(\phi_b, x)$ and $t(\phi_a, x)$ are both $t_{sh}(x)$, the integrand vanishes at the upper and lower limits. The derivative $dA(x)/dx$ is consequently given by an analogous expression; note that $t(\phi, x)$ is replaced by $\partial t(\phi, x)/\partial x$, or by $c_0^{-1} - \beta c_0^{-2} f(\phi)$ from Eq. (4.146). The resulting integration is readily performed, yielding

$$\frac{dA(x)}{dx} = -[f(\phi_b) - f(\phi_a)] \left\{ \frac{1}{c_0} - \frac{dt_{sh}}{dx} - \frac{\beta}{2c_0^2}[f(\phi_a) + f(\phi_b)] \right\}. \tag{4.156}$$

The factor in braces here, however, is zero because of the approximate Rankine–Hugoniot relation for the shock slowness, as rewritten in Eq. (4.152), so one concludes that $dA(x)/dx$ is zero. But $A(x) = 0$ at the position where the shock was first formed, so $A(x)$ is always zero and the equal-area rule is verified.

4.3.5   Energy Dissipation at a Shock

In the absence of shocks, nonlinear effects do not change the net acoustic energy associated with a pulse; they merely rearrange the frequency distribution of the energy. The energy density for a traveling wave can be expressed as either $\rho_0 u^2$ or $p^2/\rho_0 c_0^2$ because $u = p/\rho_0 c_0$. The net energy per unit area transverse to propagation direction for a pulse of finite duration is then

$$E(t) = \rho_0 \int_{-\infty}^{\infty} u^2 \, dx. \tag{4.157}$$

If Eqs. (4.147) and (4.148) are valid for a single-valued description of the pulse, Eq. (4.157) can alternatively be written

$$E(t) = \rho_0 \int_{-\infty}^{\infty} g^2(\psi) \frac{\partial x}{\partial \psi}\, d\psi = \rho_0 \int_{-\infty}^{\infty} g^2(\psi)[1 + \beta g'(\psi)t]\, d\psi, \qquad (4.158)$$

where $g'(\psi) = dg/d\psi$. The quantity $g^2(\psi)g'(\psi)$, however, integrates to zero, since $g^3(\psi) \to 0$ as $\psi \to \pm\infty$. Consequently, $E(t)$ is independent of time.

On the other hand, if a shock is present, the integral must be broken into integrals from $-\infty$ to $\psi_b(t)$ and from $\psi_a(t)$ to ∞. The time derivative of $E(t)$ consequently yields, with some algebraic manipulation, the relation

$$\frac{1}{\rho_0} \frac{dE(t)}{dt} = g^2(\psi_b) \frac{d}{dt}[\psi_b + \beta g(\psi_b)t]$$

$$- g^2(\psi_a) \frac{d}{dt}[\psi_a + \beta g(\psi_a)t] - \tfrac{2}{3}\beta[g^3(\psi_b) - g^3(\psi_a)]. \qquad (4.159)$$

The first two quantities in brackets here [see Eq. (4.153)] are $x_{\text{sh}} - c_0 t$, so their time derivatives [see Eq. (4.151)] are both $\tfrac{1}{2}\beta[g(\psi_a) + g(\psi_b)]$. Then, with additional manipulations, Eq. (4.159) yields

$$\frac{dE}{dt} = -\tfrac{1}{6}\beta\rho_0[g(\psi_b) - g(\psi_a)]^3 = -\tfrac{1}{6}\beta\rho_0(\Delta u)^3, \qquad (4.160)$$

where Δu is the jump in particle velocity at the shock. Alternatively, this is related to the pressure discontinuity by Eq. (4.122), which in turn is related to the entropy discontinuity by Eq. (4.128), and one has

$$\frac{dE}{dt} = -\frac{\beta}{6\rho_0^2 c_0^3}(\Delta P)^3 = -\rho_0 c_0 T_0 \Delta s. \qquad (4.161)$$

The physical interpretation of the latter version is that, as a shock (moving with a speed close to the ambient sound speed) sweeps through the fluid, it passes over a mass $\rho_0 c_0$ per unit time and per unit area of wavefront. This mass receives a heat increment $T_0 \Delta s$ per unit mass, and therefore heat of the magnitude of $\rho_0 c_0 T_0 \Delta s$ is being generated per unit area of wavefront and per unit time. This heat is generated at the expense of the energy in the waveform, and conservation of energy implies that the rate of energy loss in the waveform per unit area of waveform is as given above. The presence of the shock causes the energy in the wave to decrease with time. Because this is third order in ΔP, the loss is seemingly small in any given short time instant. However, over large propagation distances, these small losses accumulate, and their effect should be apparent. [The result here derived was discovered in the context of sawtooth waves by Rudnick (1953).]

4.4   Weak Shock Theory

The aim of this section, and of Sect. 4.5, is to describe propagation over the entire history of the wave motion. Achieving this goal requires that dissipation be taken into consideration. Two quite different methods for including dissipation are well known. The method based on the Burgers equation, which takes explicit account of dissipation, is presented in Sect. 4.5. The present section is devoted to a less general but simpler method called weak shock theory. The basis for weak shock theory was laid in Sect. 4.3, particularly Sects. 4.3.2–4.3.5. Here we develop the specific method.

4.4.1   General Method

Weak shock theory is based on the following somewhat interrelated assumptions:

1. Shocks present in the waveform are weak.
2. Dissipation is concentrated at the shocks. Dissipation associated with the rest of the waveform—i.e., the continuous segments between shocks—may be neglected.
3. Shocks are discontinuities; i.e., shock rise time is zero. Because real shocks, particularly weak ones, are not actually discontinuous, a more practical statement is as follows: Shock rise time is negligible compared with the wave's larger time scale, for example, the period for periodic signals or the wave duration for transients.

These assumptions are first used to develop a calculation procedure, which is then applied to predict the propagation of an N wave and a periodic wave. Finally, a restriction on the applicability of weak shock theory is developed.

The three assumptions lead to the calculation procedure as follows: Once shocks form in a waveform, in general they cause reflections and thus end the progressive-wave nature of the propagation (Stokes, 1848). For weak shocks (assumption 1), however, the reflections are so tiny, $O(\varepsilon^3)$ [see Eq. (4.144)], that they may be neglected within the bounds of second-order approximation theory. This fact, coupled with assumption 2, allows one to continue using lossless progressive-wave theory (second-order approximation) to describe distortion of the continuous segments of the waveform. Each continuous segment ends at a shock, where jump conditions provide the connection to the continuous segment on the other side. Assumption 3 then makes it possible to sketch the wave or find its spectrum.

To keep track of the location of the shocks in the waveform, we make use of results developed in Sect. 4.3.4. Shocks propagate at a speed different from that of continuous waves. The speed of a continuous wave is given by Eq. (4.2). For weak shocks, the speed is the mean of the finite-amplitude speeds just ahead of the shock, $c_0 + \beta u_a$, and just behind it, $c_0 + \beta u_b$. See Eq. (4.151) combined with Eq. (4.147),

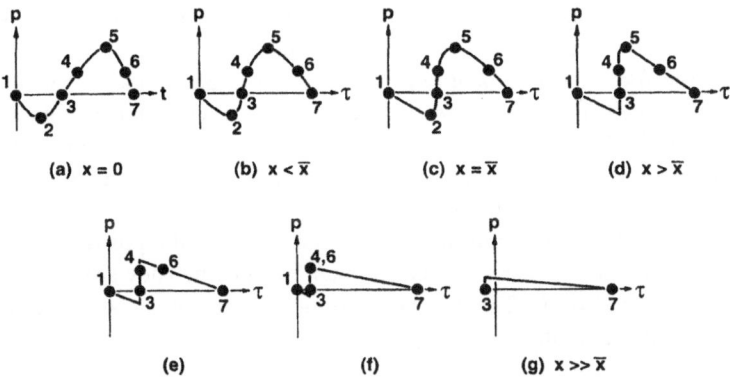

Fig. 4.7 Progressive distortion and shock-induced decay of a traveling wave (after Blackstock, 1966b). Numbers identify particular phase points on the waveform. Each phase point is known by its value of p or, the equivalent, the Earnshaw phase variable ϕ.

or Eq. (1.38). In terms of pressure, the shock speed is given by

$$U_{sh} = c_0 + \beta \frac{p_a + p_b}{2\rho_0 c_0}.$$ (4.162)

The corresponding slowness of the shock, $dt_{sh}/dx = U_{sh}^{-1}$, is given by Eq. (4.152). In terms of retarded time for the shock, $d\tau_{sh}/dx = dt_{sh}/dx - c_0^{-1}$, we have

$$\frac{d\tau_{sh}}{dx} = -\frac{\beta(p_a + p_b)}{2\rho_0 c_0^3}.$$ (4.163)

Compare this with Eq. (4.31). The two equations show that a point on the continuous segment of the waveform behind a shock tends to catch up with the shock, whereas a point ahead of the shock tends to be overtaken by the shock.

Figure 4.7 illustrates shock growth and decay in a waveform. Sketch a shows the source signal ($x = 0$), which was intentionally chosen for this example to be continuous but not symmetric. Typical points on the waveform, each tagged by its particular value of the Earnshaw phase variable ϕ, are identified by the numbers 1 through 7; note that points 1, 3, and 7 are zero crossings. Sketch b shows the early distortion of the wave, at a location $x < \bar{x}$. Since no shocks have formed yet, the entire waveform is well described by the approximate Earnshaw solution, Eqs. (4.29). The situation changes, however, when a shock forms, sketch c. The shock starts to grow as points behind, such as point 4, begin to catch up and join the shock, while at the same time the shock overtakes points ahead, such as point 2 [see Eq. (4.162)]. In sketch d, point 2 has disappeared because after it was overtaken, subsequent points overtaken were lower in "amplitude." By sketch e, a similar fate has befallen point 5 and its neighbors. Even more points on the initial waveform are missing in sketch f. In sketch g, the entire negative part of the wave has been "eaten up" by the shock. Of the original waveform shown in sketch a, all that is left are two small segments: (1) The left end of the section between points 3 and 4 survives as the shock, the foot of which is at point 3 in sketch g, and (2) a very small part of the tail end of the waveform between points 6 and 7 now stretches from the top of the shock to point 7. The attenuation of the wave is due to dissipation at the shock. The loss of most of the original waveform represents energy fed into the shock from behind or dropping into the shock from ahead.

Before Eq. (4.163) can be used to locate (and follow) the position of the shock in the waveform, the pressures p_a and p_b must be known. They are found from the Earnshaw solution, Eqs. (4.29). For the continuous segment ahead of the shock, the solution evaluated at point a is

$$p_a = f(\phi_a), \quad \phi_a = \tau + \frac{\beta x f(\phi_a)}{\rho_0 c_0^3}. \tag{4.164}$$

Similarly the solution for the segment behind the shock, evaluated at point b, yields

$$p_b = f(\phi_b), \quad \phi_b = \tau + \frac{\beta x f(\phi_b)}{\rho_0 c_0^3}. \tag{4.165}$$

Equations (4.163)–(4.165) are a coupled set that must be solved simultaneously. A method of solution is illustrated in the next two sections.

Another way to place the shock in the waveform is to apply the equal-area rule (see Sect. 4.3.4).

4.4.2 N Waves and Other Pulses

A particularly simple yet practical application of weak shock theory is to the propagation of N waves (see, for example, Blackstock, 1972). The N-shaped waveform is typical of disturbances produced by supersonic projectiles, supersonic aircraft (sonic booms, far away from the aircraft), bursting (spherical) balloons, and electric sparks in air. The pressure waveform is made up of a head shock, a tail shock, and a linear pressure decrease in between. At the source $x = 0$ (left-hand sketch in Fig. 4.8), we have

$$p(0, t) = -p_0 t / T_0, \qquad |t| < T_0, \tag{4.166}$$

$$= 0, \qquad |t| > T_0, \tag{4.167}$$

where p_0 and $2T_0$ are the head shock pressure and the duration, respectively, at the source.[5] To describe the pressure field between the two shocks, use the Earnshaw

---

[5] Note that T does not stand for temperature in this discussion.

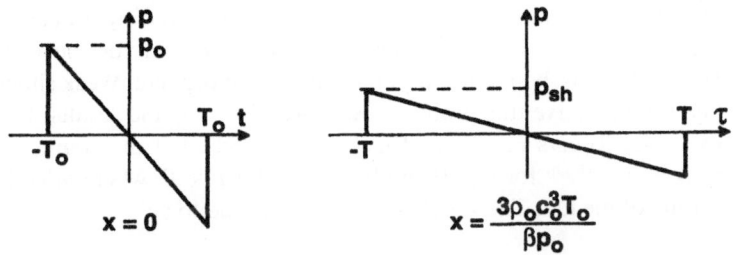

Fig. 4.8 N wave (after Blackstock, 1972).

solution, Eqs. (4.29): $p = f(\phi) = -p_0\phi/T_0$ and $\phi = \tau - \beta x p_0 \phi/\rho_0 c_0^3 T_0$. Solve the last relation for ϕ and substitute in $f(\phi)$ to give

$$p = -\frac{p_0\tau/T_0}{1+bx}, \qquad |\tau| < T, \tag{4.168}$$

$$= 0, \qquad |\tau| > T, \tag{4.169}$$

where $b = \beta p_0/\rho_0 c_0^3 T_0$ is a constant and T is a variable endpoint that must be determined from Eq. (4.163).

Next find the location $\tau_{\text{sh}} = -T$ of the head shock (and thus evaluate T). The pressure just ahead of the head shock is $p_a = 0$, and that just behind is $p_b = -(p_0\tau_{\text{sh}}/T_0)/(1+bx)$ [from Eq. (4.168) evaluated at $\tau = -T$]. Equation (4.163), converted into a differential equation in τ_{sh}, is thus

$$\frac{d\tau_{\text{sh}}}{dx} = \frac{b\tau_{\text{sh}}}{2(1+bx)}. \tag{4.170}$$

The solution satisfying the source condition $\tau_{\text{sh}} = -T_0$ at $x = 0$ is

$$\tau_{\text{sh}} = -T = -T_0\sqrt{1+bx}. \tag{4.171}$$

The head shock pressure p_{sh} $(= p_b)$ is found by evaluating Eq. (4.168) at $\tau = \tau_{\text{sh}}$:

$$p_{\text{sh}} = p_b = \frac{p_0}{\sqrt{1+bx}}. \tag{4.172}$$

By symmetry, the pressure at the tail shock has the same magnitude but is negative. Equations (4.172) and (4.171) show that besides decaying asymptotically as $x^{-1/2}$, the N wave spreads out as it travels (see Fig. 4.8, right-hand sketch). The spreading is due to the supersonic velocity of the head shock and the subsonic velocity of the tail shock [see Eq. (4.162)].

Another transient of practical interest is a shock followed by a decay tail that starts out as an exponential. This pulse is a simple model for the disturbance produced by explosions, both underwater and in the atmosphere. Weak shock theory may also be used to solve this problem. Here we give only the results; for details, see, for example, Rogers (1977) or Blackstock (1983). Let the source signal be defined by $p = p_0 e^{-t/t_0}$ for $t > 0$ and by $p = 0$ for $t < 0$, where t_0 is the initial e^{-1} decay time of the tail. The shock amplitude is found to be

$$p_{sh} = p_0 \frac{\sqrt{1 + 2bx} - 1}{bx}, \tag{4.173}$$

where in this case $b = \beta p_0 / \rho_0 c_0^3 t_0$. A measure of the increasing duration of the wave is the e^{-1} decay time t_{decay} of the waveform:

$$t_{decay} = t_0[1 + (1 - e^{-1})(\sqrt{1 + 2bx} - 1)]. \tag{4.174}$$

Comparison of Eqs. (4.172) and (4.173) shows that both the N wave and the exponential pulse decay asymptotically as $x^{-1/2}$. In the next section, we find that a sawtooth (infinite train of N waves) decays more rapidly.

4.4.3   Periodic Waves

Here we revisit the problem of radiation from a monofrequency source, previously solved only for the shock-free region (Fubini solution). See Sect. 4.2.4.1. Again the source condition is Eq. (4.46); the source waveform is shown as the $\sigma = 0$ sketch in the inset of Fig. 4.9. Shock formation occurs when the distortion range variable $\sigma = \beta \varepsilon kx$ attains the value $\sigma = 1$. See the sketch for that value of σ in the inset of Fig. 4.9. By using weak shock theory, we may extend the solution into the region $\sigma > 1$ (Blackstock, 1966b).

How is waveform distortion described when shocks are present? Consider first the continuous segments of the waveform. In dimensionless quantities, the Earnshaw solution is

$$P(\sigma, \omega\tau) = \sin \Phi, \quad \Phi = \omega\tau + \sigma \sin \Phi, \tag{4.175}$$

where $P = p/p_0$ and $\Phi = \omega\phi$ [ϕ is the Earnshaw phase variable; see Eq. (4.29)].[6] As for the shocks, they first appear in the waveform at the positive-slope zero crossings $\omega\tau = 2n\pi$, where $n = 0, \pm 1, \pm 2, \ldots$ (see Sect. 4.2.2). Because the wave is periodic, only a single cycle need be considered, say the one centered about

---

[6] Note that here and in Sects. 4.5.4 and 4.6.2, P is the acoustic pressure amplitude normalized to the source amplitude. It does not stand for total fluid pressure.

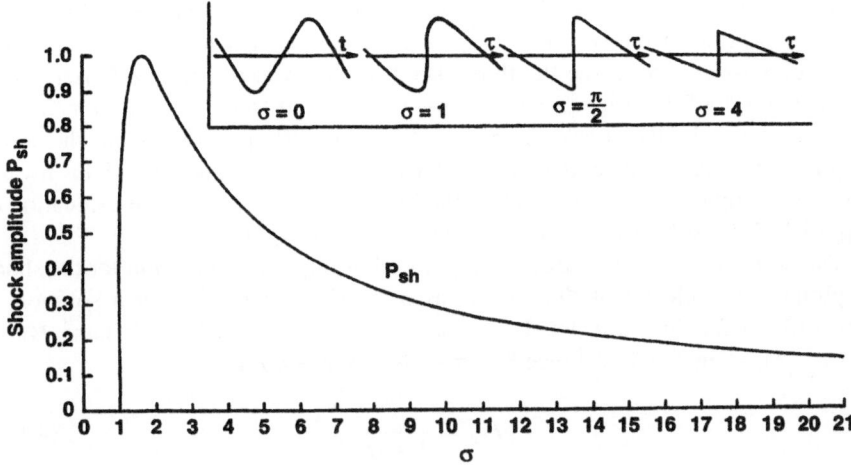

Fig. 4.9 Shock amplitude as a function of σ for a wave generated by a monofrequency source (after Blackstock, 1966b). Inset shows the time waveform at various distances from the source.

the origin, $-\pi < \omega\tau < \pi$. For this cycle, the shock forms at the midpoint, $\omega\tau = 0$. Its future track is determined by Eq. (4.163), restated here in dimensionless form:

$$d(\omega\tau_{sh})/d\sigma = -\tfrac{1}{2}(P_a + P_b). \tag{4.176}$$

Because of the symmetry of this particular wave, the pressure just ahead of the shock is equal to but opposite that just behind the shock; that is, $P_a = -P_b$. Equation (4.176) thus reduces to $d(\omega\tau_{sh})/d\sigma = 0$. The shock therefore remains at its birthplace $\omega\tau = 0$.

4.4.3.1 Shock Growth and Decay

Although the position of the shock in the cycle does not change, its amplitude $P_{sh} = P_b$ does.[7] Equations (4.175) evaluated just behind the shock, where $\Phi = \Phi_b$, $P = P_b$, and $\omega\tau = \omega\tau_{sh} = 0$, may be combined to yield

$$P_{sh} = \sin\sigma\, P_{sh} \quad \text{or} \quad \Phi_{sh} = \sigma \sin\Phi_{sh}. \tag{4.177}$$

The only solution for $\sigma \leq 1$ is $P_{sh} = 0$, but P_{sh} begins to take on finite values as σ increases beyond the value 1. Physically, what happens is that the shock overtakes points ahead of it ($\Phi < 0$) while points behind it ($\Phi > 0$) catch up. The shock thus begins to grow. The growth stage ends when the shock simultaneously overtakes

---

[7] Note that the total pressure jump at the shock is $2P_{sh}$.

the trough ahead of it ($\Phi_a = -\pi/2$) and is overtaken by the peak behind it ($\Phi_b = \pi/2$). Inspection of the second of Eqs. (4.177) shows that the distance at which this occurs is $\sigma = \pi/2$; see the third sketch in the inset of Fig. 4.9. Observe that at this point, half the original waveform—values of Φ from $-\pi/2$ to $+\pi/2$—has been consumed by the shock. As propagation continues past this point, the shock amplitude decreases, since successive values of P_b (and $-P_a$) now trend downward. See, for example, the $\sigma = 4$ sketch in the inset of Fig. 4.9 [numerical evaluation of Eqs. (4.177) for $\sigma = 4$ yields $\Phi_{sh} = 0.788\pi$ and $P_{sh} = 0.619$].

The description above leads to a very useful asymptotic expression for the shock amplitude. It is clear that $\Phi_{sh} \to \pi$ as σ becomes large. To get a solution of Eqs. (4.177) for this case, let $\Phi_{sh} = \pi - \delta$, where δ is small. We have $P_{sh} = \sin(\pi - \delta) = \sin\delta \simeq \delta$, or, since $\delta = \pi - \Phi_{sh} = \pi - \sigma P_{sh}$,

$$P_{sh} = \frac{\pi}{1 + \sigma}. \tag{4.178}$$

This expression combined with numerical solution of Eqs. (4.177) yields the curve shown in Fig. 4.9.

4.4.3.2 Sawtooth Wave

It is clear from the shape of the waveform in the $\sigma = 4$ inset of Fig. 4.9 that the wave has become a sawtooth. The close resemblance to a sawtooth actually sets in a little earlier. For $\sigma = 3$, the value of P_{sh} given by Eq. (4.178) differs by only 3.4% from the exact value found by numerical solution of Eqs. (4.177). It has become traditional to use $\sigma > 3$ to define the "sawtooth region." The sound field from a monofrequency source is thus characterized by three regions: the shock-free region $\sigma < 1$, the transition region $1 < \sigma < 3$, and the sawtooth region $\sigma > 3$. In the sawtooth region, where the amplitude is given by Eq. (4.178), the waveform may be represented by the following Fourier series:

$$p = \frac{2p_0}{1 + \sigma} \sum_{n=1}^{\infty} \frac{1}{n} \sin n\omega\tau, \quad \sigma > 3; \tag{4.179}$$

i.e., the harmonic amplitudes are $B_n = 2/n(1 + \sigma)$ (rigorous justification is presented in the next section). As shown in Sect. 4.5.4.2, the sawtooth solution is a limiting form of the Fay solution; see Eq. (4.275). The time-domain version of Eq. (4.179) is also of interest. If for ease of representation we change the limits of the cycle from $(-\pi, \pi)$ to $(0, 2\pi)$, the expression is

$$p = p_0 \frac{\pi - \omega\tau}{1 + \sigma}, \quad 0 < \omega\tau < 2\pi. \tag{4.180}$$

Both Eqs. (4.179) and (4.180) show the steady monotonic deterioration of the sawtooth wave. Notice that the sawtooth decays more rapidly than the N wave. The explanation is that the head shock of an N wave is free to move forward, the tail shock backward, whereas shock position in a sawtooth wave is fixed. Points behind the head shock of an N wave thus take longer to catch up to the shock than do equivalent points behind a sawtooth shock (similar remarks apply to the tail shock). As a result, the N-wave shocks are not "eaten away" so rapidly.

4.4.3.3 Transition Solution

How are two such disparate solutions, the Fubini series and the sawtooth function, to be patched together, and over such a short distance, $1 < \sigma < 3$? The key is to return to the analysis in Sect. 4.2.4 and modify the steps when shocks arise. The approach is like that leading from Eqs. (4.38) to (4.40) to find an integral expression for the Fourier coefficients. Since for the monofrequency source $F = \sin \omega t$, we begin with the Fourier sine series, Eq. (4.43). The coefficients B_n are given by

$$B_n = \frac{2}{\pi} \int_0^{\pi} P \sin n\omega\tau \, d(\omega\tau) = \frac{2}{\pi} \int_0^{\pi} \sin \Phi \sin n\omega\tau \, d(\omega\tau). \tag{4.181}$$

Integration by parts, in preparation for shifting from integration over $\omega\tau$ to integration over the (dimensionless) Earnshaw phase variable Φ, yields

$$B_n = -\frac{2}{n\pi} \sin \Phi \cos n\omega\tau \Big|_{\omega\tau=0}^{\omega\tau=\pi} + \frac{2}{n\pi} \int_{\omega\tau=0}^{\omega\tau=\pi} \cos n\omega\tau \cos \Phi \, d\Phi. \tag{4.182}$$

The second part of Eq. (4.175) shows that for the region $\sigma < 1$, the variables Φ and $\omega\tau$ have a one-to-one correspondence. In particular, $0 \le \omega\tau \le \pi$ has a one-to-one correspondence with $0 \le \Phi \le \pi$. However, as the discussion in Sect. 4.4.3.1 shows, after shocks form, $0 \le \omega\tau \le \pi$ corresponds to $\Phi_{sh} \le \Phi \le \pi$. The first term in Eq. (4.182) thus becomes $(2/n\pi)P_{sh}$, which vanishes in the shock-free region and takes on the value $2/n(1 + \sigma)$ in the sawtooth region [see Eq. (4.178)]. To evaluate the second term in Eq. (4.182), differentiate the second part of Eq. (4.175) to replace $\cos \Phi \, d\Phi$ with $\sigma^{-1}[d\Phi - d(\omega\tau)]$. Equation (4.182) becomes

$$B_n = \frac{2}{n\pi} P_{sh} + \frac{2}{n\pi\sigma} \int_{\Phi_{sh}}^{\pi} \cos n(\Phi - \sigma \sin \Phi) \, d\Phi \tag{4.183}$$

as the general expression for the harmonic amplitudes.

The interpretation is quite simple. As already argued, the first term is associated with the sawtooth solution. The second term reduces to the Fubini form $B_n = (2/n\sigma)J_n(n\sigma)$ [Eq. (4.48)] in the shock-free region ($\Phi_{sh} = 0$) and falls to zero in the sawtooth region as the lower limit of the integral approaches the upper limit ($\Phi_{sh} \to \pi$). For the fundamental amplitude B_1, Fig. 4.10 shows how the two terms

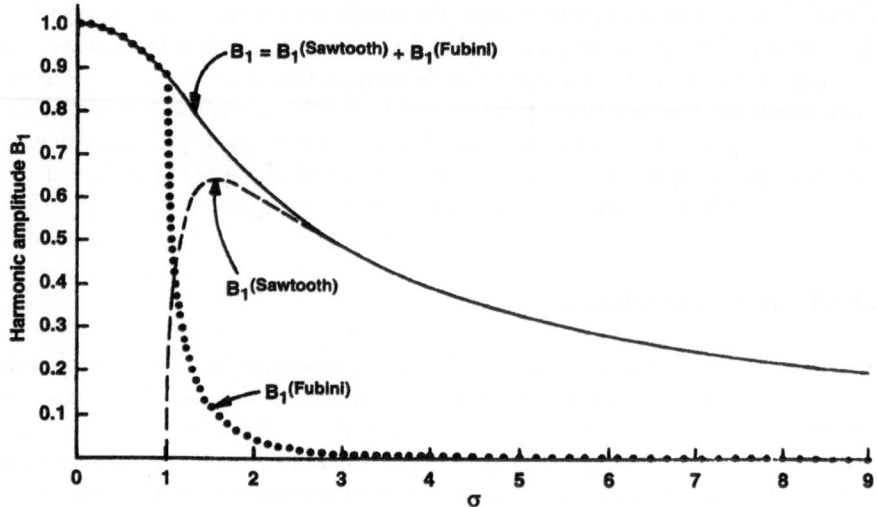

Fig. 4.10 Amplitude of the fundamental component B_1 as a function of σ for a wave generated by a monofrequency source (after Blackstock, 1966b).

contribute to the whole. The dotted curve represents the second term. It is identified on the figure as $B_1^{(\text{Fubini})}$ because it is the only contributor in the shock-free region. The first term, represented by the dashed curve, is labeled $B_1^{(\text{Sawtooth})}$ because it is the survivor in the sawtooth region. It is interesting to note how rapidly one curve decays while the other grows in the transition region. Figure 4.11 shows curves for B_1, B_2, and B_3, found by using the two limiting solutions for their respective regions and numerical evaluation of Eq. (4.183) for the transition region.

4.4.3.4   Acoustical Saturation

For σ large enough that $(1 + \sigma)$ may be replaced by σ in the denominator of Eq. (4.179), a remarkable result is obtained. Recalling that $P = p/p_0$ and $\sigma = \beta p_0 k x / \rho_0 c_0^2$, we obtain for Eq. (4.179), at distances well beyond the point of shock formation,

$$p = \frac{2\rho_0 c_0^2}{\beta k x} \sum_{n=1}^{\infty} \frac{1}{n} \sin n\omega\tau, \quad \sigma \gg 1. \tag{4.184}$$

The same limit applied to Eq. (4.180) yields

$$p = \frac{\rho_0 c_0^2 (\pi - \omega\tau)}{\beta k x}, \quad 0 < \omega\tau < 2\pi, \quad \sigma \gg 1. \tag{4.185}$$

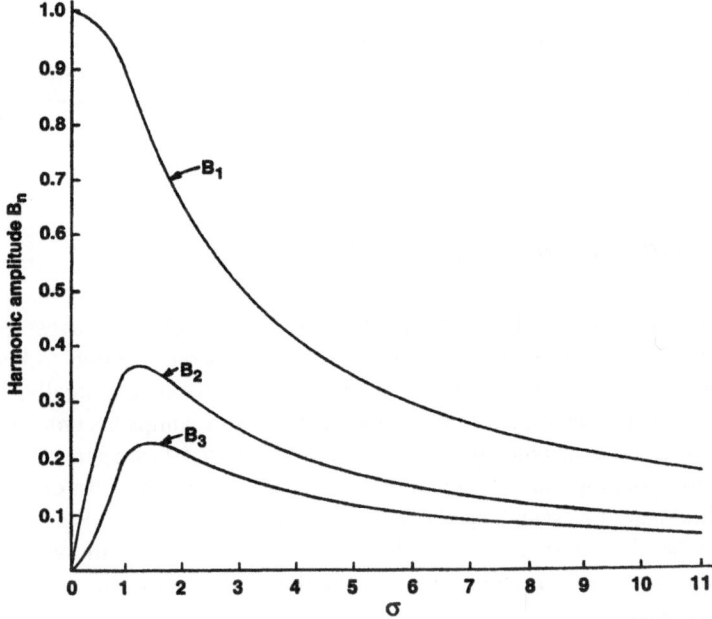

Fig. 4.11 Spectral amplitudes B_1, B_2, and B_3 as functions of σ for a wave generated by a monofrequency source (after Blackstock, 1966b).

Both results show the wave amplitude $|p| = \pi \rho_0 c_0^2 / \beta k x$ to be independent of the source amplitude p_0. The wave has forgotten its origins. The irreversible energy loss at the shock fronts thus imposes an upper bound on how much sound can be transmitted to a given distance. All additional acoustical energy pumped into the wave by the source is lost at the shock fronts, which form ever closer to the source. *Acoustical saturation* is said to have set in. The first to observe and correctly interpret acoustical saturation was Allen (1950). Webster and Blackstock (1977) report experiments that demonstrate the saturation of plane sound waves in air, and Shooter, Muir, and Blackstock (1974) report experiments on the saturation of directional sound beams in water.

Not all waves saturate. The N wave is a case in point. The amplitude of an N wave is given by Eq. (4.172), which at distances far from the source reduces to

$$p_{\text{sh}} = \sqrt{p_0 \rho_0 c_0^3 T_0 / \beta x}, \quad bx \gg 1, \tag{4.186}$$

where $b = \beta p_0 / \rho_0 c_0^3 T_0$. Although saturation (in the sense of no dependence on source amplitude) does not occur, the amplitude of the N wave increases only in proportion to the square root of the source amplitude. A 6-dB increase in source level therefore introduces only a 3-dB increase in the amplitude of the N wave.

One caveat must be noted before this discussion is closed. The saturation amplitude derived here is based on weak shock theory, which generally becomes inaccurate at very great distances (see Sect. 4.4.4). At such distances the saturation amplitude becomes a function of the absorption coefficient α. For plane waves in a thermoviscous fluid, see Sect. 4.5.4.2, in particular Eq. (4.277).

4.4.4 Limitations on Weak Shock Theory

Despite its name, weak shock theory does not work well for waves containing exceedingly weak shocks. Assume that the wave is strong enough to begin with for the assumptions of weak shock theory to be realized. Eventually, however, when the wave decays to the point at which the pressure jumps become very small, assumptions 2 and 3 on which the theory is based (Sect. 4.4.1) are generally violated. First consider assumption 3, which is that the shock is very thin. An expression for the rise time t_{rise} of a weak step shock is given in Sect. 4.5.3; see Eq. (4.250). If t_c stands for the characteristic time of the large-scale variation of the waveform, for instance the period of a periodic wave or the duration of a transient, assumption 3 may be stated as

$$t_{\text{rise}} \ll t_c. \tag{4.187}$$

Assumption 2, that dissipation is concentrated at the shocks, provides an alternative but equivalent test. Because energy dissipation at a shock varies as the cube of the pressure jump [see Eq. (4.161)], loss over the rest of the waveform cannot be neglected when the jump becomes tiny. For a periodic wave, the length scale associated with ordinary absorption is $\ell_a = 1/\alpha$ (see Sect. 4.5.4), where α is the small-signal absorption coefficient at a characteristic frequency, usually the fundamental, of the wave. When $x \sim \ell_a$, the rate of decay due to ordinary absorption equals that due to shock loss (Blackstock, 1966b). Weak shock theory may therefore be used for distances

$$x < \ell_a. \tag{4.188}$$

Equations (4.187) and (4.188) are obviously not independent. The latter may be more useful for periodic waves, the former for transients.

One related restriction should be noted. In order for weak shock theory to be applicable in the first place, nonlinearity must be strong enough, relative to absorption, that real shocks form. In quantitative terms, the requirement is $\bar{x} \ll \ell_a$. An analysis for thermoviscous fluids for the case $\bar{x} \sim \ell_a$ is given in Sect. 4.5.4.1.

4.5 The Burgers Equation

What has come to be known as the Burgers equation was derived earlier in this book as Eq. (3.54), and is repeated here for convenience:

$$\frac{\partial p}{\partial x} - \frac{\delta}{2c_0^3} \frac{\partial^2 p}{\partial \tau^2} = \frac{\beta p}{\rho_0 c_0^3} \frac{\partial p}{\partial \tau}, \tag{4.189}$$

where

$$\delta = \rho_0^{-1} [\tfrac{4}{3}\mu + \mu_B + \kappa(c_v^{-1} - c_p^{-1})] \tag{4.190}$$

is referred to as the diffusivity of sound (Lighthill, 1956), μ is shear viscosity, μ_B is bulk viscosity, κ is thermal conductivity, and c_v and c_p are the specific heats at constant volume and constant pressure, respectively. Equation (4.189) accounts explicitly for the effect of thermoviscous dissipation on the propagation of finite-amplitude sound. The present section discusses the properties of the Burgers equation and some of its solutions. Also, since the historical introduction in Chap. 1 stops with the beginning of World War II, the discussion here begins with a historical sketch of the origins of the Burgers equation, with an attempt to explain why this partial differential equation, which has played a central role in nonlinear acoustics over the second half of the twentieth century, happened to be called the Burgers equation.

4.5.1 History

Workers in applied mathematics and fields related to fluid mechanics loosely refer to any partial differential equation of the form

$$a\frac{\partial \theta}{\partial \xi_2} + b\frac{\partial \theta}{\partial \xi_1} + c\theta\frac{\partial \theta}{\partial \xi_1} + d\frac{\partial^2 \theta}{\partial \xi_1^2} = 0 \tag{4.191}$$

(generally with the coefficient b set to zero) as the Burgers equation. Here the dependent variable $\theta(\xi_1, \xi_2)$ is a function of two independent coordinates, one usually identified as a time and the other as a distance variable. The partial differential equation is quasilinear, because the nonlinear term involves only a first derivative, but the highest-order derivative, appearing in a linear term, is linear and second order. The earmarks of the Burgers equation are these two features, plus the feature that the equation has an intrinsic parabolic nature, analogous to the partial differential equation governing time-dependent heat conduction in one dimension. This parabolic nature is embodied in the fact that the second-derivative term involves differentiation with respect to only one of the two coordinates. The quantities a, b, c, and d are often taken as constants, but in some discussions, one or more are taken as functions of one or both of the independent coordinates.

The first appearance of such an equation in the archival literature, at least in a fluid-dynamic context, is generally considered to be in a 1915 paper published by Bateman in a meteorological journal not normally read by research workers in fluid dynamics. (The authors have been told that such an equation may have appeared in a purely mathematical context in one of Forsyth's books or papers, written sometime between 1895 and 1910, but they have not yet checked to confirm this.) Bateman's paper was largely review and cited an impressive list of fundamental research papers by his contemporaries and predecessors. The emphasis was on the question of the existence of discontinuous motion in fluids, primarily in the context of wakes behind moving bodies, for which one model is discontinuous potential flow (with the fluid taken as incompressible), and where the discontinuity was one of tangential fluid velocity, and not what we call a shock. It was understood that this discontinuity would not be abrupt were the viscosity taken into account, and to show a relatively simple example of an equation having some resemblance to the equations of fluid mechanics and where the limit of zero viscosity led to a discontinuity, Bateman wrote down the partial differential equation

$$\frac{\partial u}{\partial t} + u\frac{\partial u}{\partial x} = \nu\frac{\partial^2 u}{\partial x^2}. \tag{4.192}$$

This can be viewed as a one-dimensional version of the Navier–Stokes equation, which appears as Eq. (3.2), with the pressure term set to zero, with ρ taken as constant, and with $\nu = \rho^{-1}(\frac{4}{3}\mu + \mu_B)$. Bateman then proceeded to find a particular solution of this equation by a method similar to that discussed further in Sect. 4.5.3, and showed that his solution did indeed yield a discontinuity in the limit of vanishing ν. It appears, however, that no further attention was given to this partial differential equation in subsequent literature.

The contribution by Burgers to this history begins with a series of papers concerned with turbulence, which were published in the Netherlands during the early years of World War II, and of which the most frequently cited is Burgers (1940). Following the end of the war, von Kármán and von Mises established a set of review volumes to report advances in applied mechanics; for the inaugural volume of this series, Burgers was invited to write an expository paper (Burgers, 1948), intended to fully describe and extend the theoretical ideas that he had developed in the earlier papers. Burgers's 1948 article was apparently widely noticed and continues to have an influence on turbulence research to the present day. (The term "Burgerlence" to describe phenomena governed by Burgers's mathematical model has become part of the vocabulary of workers in turbulence.) At the outset of the paper, Burgers set down two coupled equations (without much of a motivational preamble), which he then proceeded to discuss in considerable detail with the objective of demonstrating that their solutions had many features analogous to what is commonly associated with turbulence. The second of these equations was

$$\frac{\partial v}{\partial t} = \frac{U}{b}v + \nu\frac{\partial^2 v}{\partial y^2} - 2v\frac{\partial v}{\partial y}, \tag{4.193}$$

where v and U were his two dependent variables, and b and ν were constants, it being understood that U depended only on t, while v depended on both t and y. Burgers's discussion is somewhat murky on the relation of the variables that appear here to actual physical variables; he states that U is the analog of "mean motion in the case of a liquid flowing through a channel," while v, when different from zero, corresponds to "turbulence in the channel," y corresponds to the coordinate "in the direction of the cross dimension of the channel," b corresponds to the width of the channel, and ν is a coefficient associated with "frictional effects." Farther on in the paper, he considers the case in which U is a constant and gives arguments to the effect that, even though U may be nonzero, there may be some domains in y where it may be an appropriate approximation to neglect the term $(U/b)v$ in the above equation, in which case the considered partial differential equation becomes

$$\frac{\partial v}{\partial t} + 2v\frac{\partial v}{\partial y} - \nu\frac{\partial^2 v}{\partial y^2} = 0. \tag{4.194}$$

Still farther on in the paper, Burgers develops an explicit solution of this partial differential equation that corresponds to a discontinuity in v in the limit as $\nu \to 0$. The solution is similar to that given by Bateman (1915) and to what is presented further below in Sect. 4.5.3. Burgers gives no reference to Bateman (1915), but in all fairness there is no reason to have expected him to have done so.

In the concluding section of his paper, Burgers (1948) states that "the group of terms

$$\frac{\partial v}{\partial t} + 2v\frac{\partial v}{\partial y} - \nu\frac{\partial^2 v}{\partial y^2} \tag{4.195}$$

will find its closest analogy (in fluid-dynamic contexts other than turbulence) in the terms

$$\left(\frac{\partial u}{\partial t} + u\frac{\partial u}{\partial x}\right) - \nu\frac{\partial^2 u}{\partial x^2} \tag{4.196}$$

which are decisive in determining the appearance of shock waves in the supersonic motion of a gas." This is the only place in the paper where there is any hint that there might be an analogous partial differential equation of the same general form that would govern nonlinear sound propagation.

The first appearance of an equation that is of the form of Eq. (4.191) and that also had a tangible connection with nonlinear acoustics can arguably be attributed to Julian Cole (1951). Lagerstrom, Cole, and Trilling (1949) (LCT) issued a report, which they stated was intended to be in the nature of a progress report rather than a report on finished research. The report was almost entirely concerned with

the linearized equations of viscous compressible flow, but Appendix B dealt with
"nonlinear longitudinal waves." A principal result, derived in the context of an ideal
gas, in that appendix was their Eq. (B24), which had the form

$$\phi_t + \tfrac{1}{4}(\gamma + 1)(\phi_x^2 - w_{-\infty}^2) = \tfrac{2}{3}v^*\phi_{xx}, \tag{4.197}$$

where the subscripts imply partial derivatives and ϕ is a velocity potential, v^* is a
kinematic viscosity, $\gamma = c_p/c_v$ is the specific heat ratio, and $w_{-\infty}$ is the value of
$u = \partial\phi/\partial x$ in the limit of $x \to -\infty$. The velocity u is the flow velocity relative
to a reference sound speed, here denoted by c^* (and denoted by c_0 in the following
discussion). If one takes the x derivative of the above equation and reexpresses it in
the notation of the present text, the result takes the form

$$u_t + \beta u u_x = (\delta/2)u_{xx}, \tag{4.198}$$

where $\beta = \tfrac{1}{2}(\gamma + 1)$ is the coefficient of nonlinearity for an ideal gas and δ is the
diffusivity of sound, given by Eq. (4.190), in the limit where the bulk viscosity μ_B
and the thermal conductivity κ are both set to zero. Although Eq. (4.198) does not
appear explicitly in the report, it was set down (without any derivation but with a
citation to the 1949 report) in a paper submitted in April 1950 by Cole (1951).

Because Eq. (4.198) is not invariant under a Galilean transformation, it is
incomplete unless it is accompanied by a specification of the coordinate system in
which it applies. The authors were concerned at that time with transonic flow, with
the flow velocity taken as $c_0+u(x, t)$. If one instead adopts a coordinate system with
its origin moving to the right with speed c_0 relative to the LCT coordinate origin,
and denotes the x, t, and u that appear in the above two equations by x_{LCT}, t_{LCT},
and u_{LCT}, then the transformed position and time coordinates are given by

$$x_{trans} = x_{LCT} - c_0 t_{LCT}, \quad t_{trans} = t_{LCT}, \tag{4.199}$$

and the total flow velocity in the transformed coordinate system, relative to a flow
at rest, is simply

$$u_{trans}(x_{trans}, t_{trans}) = u_{LCT}(x_{LCT}, t_{LCT}) \tag{4.200}$$

since the original u_{LCT} was flow velocity relative to the sound speed. It also follows
from the above relations that

$$\frac{\partial u_{LCT}}{\partial t_{LCT}} = \frac{\partial u_{trans}}{\partial t_{trans}} + c_0\frac{\partial u_{trans}}{\partial x_{trans}}, \quad \frac{\partial u_{LCT}}{\partial x_{LCT}} = \frac{\partial u_{trans}}{\partial x_{trans}}. \tag{4.201}$$

Consequently, Eq. (4.198) transforms to

$$u_t + (c_0 + \beta u)u_x = (\delta/2)u_{xx}. \tag{4.202}$$

Here, for brevity, the subscript "trans" has been omitted on x, t, and u, but it should be understood that these denote quantities that are different from what appears in Eq. (4.198).

Apart from the restriction of its derivation to ideal gases without thermal conductivity, Eq. (4.202) applies to the same circumstances as Eq. (4.189), which is what is referred to as the Burgers equation in this book. To show that the two equations are actually equivalent to the order of approximations with which either was derived, one first divides through by $c_0 + \beta u$ and, since $\beta|u| \ll c_0$, approximates the reciprocal by $1/c_0 - (\beta/c_0^2)u$, with the intermediate result

$$u_x + \frac{1}{c_0}u_t - \frac{\beta}{c_0^2}uu_t = \frac{1}{c_0}\frac{\delta}{2}u_{xx} - \frac{\delta\beta}{2c_0^2}uu_{xx}. \tag{4.203}$$

The two dominant terms are the first two on the left, although the others can have major accumulative effects over large propagation distances or large propagation times. Both δ and u are regarded as small in a relative sense, so the last term on the right can be discarded. Furthermore, in the first term on the right, it is consistent, given the approximate balancing out of the first two terms on the left, to make the substitution $\partial/\partial x \to -c_0^{-1}\partial/\partial t$. This then yields

$$u_x + \frac{1}{c_0}u_t - \frac{\beta}{c_0^2}uu_t = \frac{\delta}{2c_0^3}u_{tt}. \tag{4.204}$$

Finally, if one introduces the retarded time $\tau = t - x/c_0$, Eq. (4.204) reduces to

$$u_x - \frac{\beta}{c_0^2}uu_\tau = \frac{\delta}{2c_0^3}u_{\tau\tau}. \tag{4.205}$$

The substitution of the plane-wave relation $u = p/\rho_0 c_0$ then yields Eq. (4.189).

Although the above shows that the Cole equation, appearing here as Eq. (4.198), is trivially related to an equation governing nonlinear sound propagation through a medium nominally at rest, this was not explicitly pointed out in either the 1949 report or the closely associated journal article by Cole (1951). The first author to derive a partial differential equation specifically for plane-wave sound propagation was apparently Mendousse (1953). Mendousse refers to Cole (1951) only tangentially and does not refer to the Lagerstrom, Cole, and Trilling (1949) report at all; he apparently did not perceive the connection of Eq. (4.198) with sound propagation in the sense discussed above. His derivation begins with a Lagrangian description of one-dimensional flow in an ideal gas with viscosity. Rewritten in terms of the symbols used in Sect. 3.2, his starting point was a "Navier–Stokes equation," which would ordinarily have the correct form

$$\rho_0\frac{\partial^2\xi}{\partial t^2} - (4/3)\mu\frac{\partial}{\partial a}\left[\left(1 + \frac{\partial\xi}{\partial a}\right)\frac{\partial^2\xi}{\partial a\partial t}\right] + \frac{\partial P}{\partial a} = 0, \tag{4.206}$$

where ξ is particle displacement relative to a reference x position of a, in which case the actual x coordinate of the considered fluid particle is

$$x = a + \xi(a, t). \tag{4.207}$$

Mendousse omitted the factor $1 + (\partial\xi/\partial a)$ in the viscous term, but including it would have been of no consequence, given the further approximations discussed below. Also, with the neglect of the thermal conductivity and with the implicit neglect of second-order terms multiplied by the viscosity, it is consistent [see Eq. (3.3)] to take the pressure as a function of only density, and independent of entropy. Moreover, conservation of mass requires that

$$\rho_0 = \left(1 + \frac{\partial\xi}{\partial a}\right)\rho, \tag{4.208}$$

and therefore expansion of the pressure to second order in $\partial\xi/\partial a$ yields

$$P = P_0 - \rho_0 c_0^2 \left[\frac{\partial\xi}{\partial a} - \beta\left(\frac{\partial\xi}{\partial a}\right)^2\right], \tag{4.209}$$

where β is the coefficient of nonlinearity. Thus, with the approximations as described, Eq. (4.206) takes the form

$$\rho_0 \frac{\partial^2\xi}{\partial t^2} - \rho_0\delta \frac{\partial^3\xi}{\partial a^2 \partial t} - \rho_0 c_0^2 \frac{\partial^2\xi}{\partial a^2} + 2\rho_0 c_0^2\beta \frac{\partial\xi}{\partial a}\frac{\partial^2\xi}{\partial a^2} = 0, \tag{4.210}$$

which is comparable in form to, but not quite the same as, the one-dimensional version of the Westervelt equation, given by Eq. (3.47). [Mendousse does not number his equations, but the above equation appears, albeit in a different notation, at the end of page 53 in Mendousse (1953).]

Mendousse subsequently considered two transformations of Eq. (4.210) above. The first resulted from the change of coordinates to $\chi = a - c_0 t$ and t, in terms of which $\partial/\partial a \to \partial/\partial\xi$ and $\partial/\partial t \to \partial/\partial t - c_0\partial/\partial\chi$. In the accompanying discussion, Mendousse gives quantitative arguments for expecting that, for a wave propagating in the $+x$ direction, derivatives with respect to t should be much smaller than c_0 times derivatives with respect to χ, and he thus neglected the terms $\rho_0\partial^2\xi/\partial t^2$ and $-\rho_0\delta\partial^3\xi/\partial\chi^2\partial t$ in the resulting transformed expression. The remaining terms yielded the result

$$-2\rho_0 c_0 \frac{\partial\theta}{\partial t} + \rho_0 c_0\delta \frac{\partial^2\theta}{\partial\chi^2} + 2\rho_0 c_0^2\beta\theta \frac{\partial\theta}{\partial\chi} = 0, \tag{4.211}$$

where θ here abbreviates $\partial\xi/\partial\chi$. (Mendousse gives this in a dimensionless form in the middle of the first column of p. 54 as $\theta'' = \dot{\theta} - 2\theta\theta'$.)

The second considered transformation resulted from the change of coordinates to a and $\tau = t - a/c_0$, in terms of which $\partial/\partial a \to \partial/\partial a + c_0^{-1}\partial/\partial \tau$ and $\partial/\partial t \to \partial/\partial \tau$. Although Mendousse omits the details, it appears that his intent was that one neglect derivatives with respect to a when compared to c_0^{-1} times derivatives with respect to τ. Doing so would have yielded

$$-2\rho_0 c_0 \frac{\partial \phi}{\partial a} + \rho_0 c_0^{-2}\delta\frac{\partial^2 \phi}{\partial \tau^2} + 2\rho_0 c_0^{-1}\beta\phi\frac{\partial \phi}{\partial \tau} = 0, \tag{4.212}$$

where ϕ here abbreviates $\partial \xi/\partial \tau$. (Mendousse gives this in a dimensionless form in the middle of the first column of p. 54 as $\ddot{\theta} = \theta' + 2\theta\theta'$, his θ corresponding to a negative constant times the ϕ that appears above.)

In regard to the equivalence of Mendousse's two equations to other equations written in terms of Eulerian variables, one should realize that, in accordance with the statements following Eq. (3.53), all three terms in either of the two equations are of the same order. Consequently, the order of approximation of the two equations will be unchanged if a is replaced by x, so that τ can be regarded as $t - x/c_0$ and χ can be regarded as $x - c_0 t$. Also, the linear acoustic relations connecting field variables in a plane wave propagating in the $+x$ direction can be used in choosing replacements for either θ or ϕ without changing the extent of the validity of either equation. Thus, one can set

$$\frac{\partial \xi}{\partial \chi} = \frac{\partial \xi}{\partial a} \simeq \frac{\rho_0 - \rho}{\rho_0} \simeq -\frac{p}{\rho_0 c_0^2} \simeq -\frac{u}{c_0}, \tag{4.213}$$

$$\frac{\partial \xi}{\partial \tau} = \frac{\partial \xi}{\partial t} = u, \tag{4.214}$$

in which case Eqs. (4.211) and (4.212) reduce to

$$\frac{\partial u}{\partial t} + u\frac{\partial u}{\partial \chi} = \frac{\delta}{2}\frac{\partial^2 u}{\partial \chi^2}, \tag{4.215}$$

$$\frac{\partial u}{\partial x} - \frac{\beta}{c_0^2}u\frac{\partial u}{\partial \tau} = \frac{\delta}{2c_0^3}\frac{\partial^2 u}{\partial \tau^2}. \tag{4.216}$$

Equation (4.215) is the same as Cole's result, which appears above as Eq. (4.198), provided one interprets Cole's x as the χ that appears here. Equation (4.216) is the same as Eq. (4.205), which in turn is equivalent to Eq. (4.189).

The most influential article in this history was undoubtedly one published by Lighthill (1956) in a special volume commemorating the seventieth birthday of G. I. Taylor. This was a long article (152 pages), partly review, but conveying some new results and adding considerable fresh insight to the subject of dissipative propagation of nonlinear waves. Early in the article, Lighthill gives the disclaimer that although "the paper seeks to be a self-contained account of its subject, ... no attempt has been made to compile a bibliography, or even to give full references, especially when describing the classical parts of the paper." The cited references include Burgers (1948), Cole (1951), and Taylor (1910), but not Bateman (1915), Lagerstrom, Cole, and Trilling (1949), or Mendousse (1953). Section 7.1 of Lighthill's article has the title "Equation of progressive sound waves with convection and diffusion allowed for to first approximation ('Burgers's equation')." The principal result in that section is Lighthill's Eq. (121), which is the same as Eq. (4.215) above. Lighthill's δ, however, given by his Eq. (19) earlier in the article, includes bulk viscosity and thermal conduction [as in Eq. (4.190) above], which were not included in the considerations of either Cole [Lagerstrom, Cole, and Trilling (1949)] or Mendousse (1953). The derivation is in the context of an ideal gas. Lighthill appreciated that this equation might have been derived earlier, and made the subsequent statement, in relation to the papers giving a general solution of that equation by Hopf (1950) and Cole (1951), that "Cole and co-workers have followed up the implications of the solution for sound waves of finite amplitude, although they do not appear to have given an explicit derivation of [Eq. (4.215) above], and regard the equation more as an analogy than as an approximation of a definite order in a successive approximation scheme."

It is probably because of Lighthill's article that any one-dimensional partial differential equation governing nonlinear unidirectional sound propagation in a dissipative medium, such as those discussed in the present section, is now ubiquitously referred to as the Burgers equation. Lighthill may not have intended that the appellation be permanent, as he placed the term "Burgers's equation" within quotation marks. He did, however, perceive a stronger analogy between nonlinear sound propagation and the physical phenomena that Burgers (1948) was seeking to portray with his model than might have been apparent to a casual reader of Burgers's work, as he refers to Burgers's proffering of this equation as "the simplest equation embodying together the convective and diffusive effects whose conflict is the subject of [Lighthill's] article. [Burgers] has used it principally to throw light on turbulence, where also these two effects are fundamental, although the balance struck between them there is more complicated, being both three-dimensional and statistically random."

This short history concludes with the article, also of considerable subsequent influence and also a part of a survey volume, by Hayes (1958) in which the Burgers equation was derived, apparently for the first time, for a fluid with an arbitrary equation of state, rather than specifically for an ideal gas. His derivation also included thermal conductivity, as well as shear and bulk viscosity. Hayes cites Burgers (1948), Hopf (1950), and Lighthill (1956), but not Bateman (1915), Lagerstrom, Cole, and Trilling (1949), or Mendousse (1953). Apart from some minor differences in notation, Hayes's Eq. (5-42) is identical in form to Cole's result, which is written above, also with some minor changes in notation, as Eq. (4.198). Hayes (1958) uses the symbol Γ for the coefficient of nonlinearity β and ν'' for the diffusivity of sound δ. Like Cole's, Hayes's coordinate system is one in which the flow appears to be slightly transonic. His u is the total fluid velocity as seen in such a coordinate system, so the quantity $u - c_0$ is regarded as small. Alternatively, Hayes's result can be regarded as being the same as Eq. (4.215), with his x being the $\chi = x - c_0 t$ and his $u - c_0$ being the u that appear in Eq. (4.215). That the labeling of this partial differential equation in the context of nonlinear acoustics as the Burgers equation had taken hold is evident in Hayes's remarks: "It should be evident that [Hayes's] Eq. (5-42) is essentially the same as the equation extensively treated by Burgers [(1948)]. Burgers' equation was first obtained for the case of a perfect gas (with viscosity and heat conduction) by Lighthill [(1956)]."

4.5.2 General Solution

One of the most appealing features of the Burgers equation is that, when the coefficients are constants, it has an exact solution expressible in terms of definite integrals. An abbreviated form of the derivation of this solution appeared in Appendix B of the Lagerstrom, Cole, and Trilling (1949) report, and more comprehensive accounts were subsequently published by Hopf (1950) and Cole (1951). These two papers were apparently completely independent, as neither references any of the other's work. Also, Hopf (1950) states in a footnote that "[the solution] was known to me since the end of 1946. However, it was not until 1949 that [Hopf] became sufficiently acquainted with the recent development of fluid dynamics to be convinced that a theory of [the Burgers equation] could serve as an instructive introduction into some of the mathematical problems involved." (Hopf, incidentally, is the Hopf whose 1931 paper published with Norbert Wiener introduced the well-known Wiener–Hopf technique.)

The discussion here is in the context of the version of the Burgers equation given by Eq. (4.189). Let us suppose that one has an initial-value problem in which p is given at $x = 0$ as a function $F(t)$, this function being denoted by

$$p(0, t) = p_0 F(t), \qquad (4.217)$$

where p_0 is a pressure amplitude, and one seeks to determine p as a function of t for any value of x that is greater than or equal to zero. It is evident that p should vanish as $t \to -\infty$ for any positive value of x, given that $F(t)$ has this same property, and we expect the rate of vanishing to be sufficiently fast that it is meaningful to deal with the integral over τ of $p(x, \tau)$ from $-\infty$ up to any given value of τ. If this is so, then another function that should exist for any positive x is

$$\zeta(x, \tau) = \exp\left\{ \frac{1}{\rho_0 c_0 L} \int_{-\infty}^{\tau} p(x, \tau')\, d\tau' \right\}, \qquad (4.218)$$

where L is any positive constant having the units of length, this definition being such that

$$p(x, \tau) = \rho_0 c_0 L \frac{\partial}{\partial \tau} \ln \zeta. \qquad (4.219)$$

The method of solution is to determine some function $\zeta(x, \tau)$ and some choice of length L that will guarantee that the p determined subsequently from Eq. (4.219) does indeed satisfy the original partial differential equation (the Burgers equation) and the specified initial condition at $x = 0$. To discover what ζ and L should be, one inserts Eq. (4.219) into Eq. (4.189), with the result

$$
\zeta^2(\zeta_{\tau x} - \tfrac{1}{2}\delta c_0^{-3}\zeta_{\tau\tau\tau}) - \zeta\zeta_\tau[\zeta_x - (\tfrac{3}{2}\delta c_0^{-3} - \beta L c_0^{-2})\zeta\zeta_{\tau\tau}]
$$
$$
+ (\zeta_\tau)^3(\beta L c_0^{-2} - \delta c_0^{-3}) = 0. \tag{4.220}
$$

Given the latitude one has in choosing L, the most propitious choice is $L = \delta/\beta c_0$, which yields

$$
p = \frac{\rho_0\delta}{\beta}\frac{\partial}{\partial\tau}\ln\zeta = \frac{\rho_0\delta}{\beta}\frac{\zeta_\tau}{\zeta}. \tag{4.221}
$$

This choice of L has the effect of reducing Eq. (4.220) to

$$
\frac{\partial}{\partial\tau}\left\{\frac{1}{\zeta}\left(\zeta_x - \frac{\delta}{2c_0^3}\zeta_{\tau\tau}\right)\right\} = 0, \tag{4.222}
$$

which in turn will certainly be satisfied if one takes ζ to satisfy

$$
\frac{\partial\zeta}{\partial x} - \frac{\delta}{2c_0^3}\frac{\partial^2\zeta}{\partial\tau^2} = K(x)\zeta, \tag{4.223}
$$

where $K(x)$ is an arbitrary function of x.

The arbitrariness of ζ associated with the arbitrariness of $K(x)$ is illusory because one can always make the substitution

$$
\zeta(x, \tau) = \tilde{\zeta}(x, \tau)\exp\left[\int_0^x K(x')\,dx'\right] \tag{4.224}
$$

and find that $\tilde{\zeta}(x, \tau)$ satisfies the equation

$$
\frac{\partial\tilde{\zeta}}{\partial x} - \frac{\delta}{2c_0^3}\frac{\partial^2\tilde{\zeta}}{\partial\tau^2} = 0, \tag{4.225}
$$

which is of the same form as Eq. (4.223), only with $K(x)$ set to zero. Also, were Eq. (4.224) to be inserted into Eq. (4.221), the result would be independent of $K(x)$. Moreover, the initial ($x = 0$) values of ζ and $\tilde{\zeta}$ would be identical. Consequently, it is sufficient to proceed as if ζ satisfied Eq. (4.225) or, equivalently, Eq. (4.223) with $K(x)$ set identically to zero. With a relabeling of the variables,

Eq. (4.225) is recognized as the linear diffusion equation, which governs one-dimensional unsteady heat conduction and a variety of other phenomena that are studied in mathematical physics. The transformation Eq. (4.221) that leads from the (nonlinear) Burgers equation to the linear diffusion equation is often referred to as the *Hopf–Cole transformation*.

Solution of the initial-value problem for the linear diffusion equation, Eq. (4.225), proceeds with the writing of the solution as a convolution integral over the initial values in the form

$$\zeta(x, \tau) = \int_{-\infty}^{\infty} \zeta(0, \tau') G(x, \tau - \tau') \, d\tau', \tag{4.226}$$

where $G(x, \tau - \tau')$ is a "Green's function" that satisfies the homogeneous partial differential equation

$$\frac{\partial G}{\partial x} - \frac{\delta}{2c_0^3} \frac{\partial^2 G}{\partial \tau^2} = 0, \tag{4.227}$$

and that reduces to the Dirac delta function $\delta(\tau - \tau')$ in the limit as $x \to 0$, so that

$$\lim_{x \to 0} G(x, \tau - \tau') = 0, \quad \tau - \tau' \neq 0, \tag{4.228}$$

$$\lim_{x \to 0} \int_{-\infty}^{\infty} G(x, \tau - \tau') \, d\tau' = 1. \tag{4.229}$$

This Green's function is derived in many texts on mathematical physics and heat transfer, so the derivation is omitted here; the reader can independently verify that the equations above are satisfied by

$$G(x, \tau - \tau') = (c_0^3 / 2\pi x \delta)^{1/2} e^{-E_G}, \quad E_G = \frac{c_0^3 (\tau - \tau')^2}{2x\delta}. \tag{4.230}$$

As a consequence, Eq. (4.226) yields

$$\zeta(x, \tau) = (c_0^3 / 2\pi x \delta)^{1/2} \int_{-\infty}^{\infty} \zeta(0, \tau') e^{-E_G} d\tau' \tag{4.231}$$

as the general solution of the initial-value problem for the auxiliary function ζ.

The solution to the Burgers equation now results from substitution of Eq. (4.231) into Eq. (4.221) to obtain

$$p = \frac{\rho_0 c_0^3}{\beta x} \frac{\int_{-\infty}^{\infty} \zeta(0, \tau')(\tau - \tau') e^{-E_G} d\tau'}{\int_{-\infty}^{\infty} \zeta(0, \tau') e^{-E_G} d\tau'}, \tag{4.232}$$

where

$$\zeta(0, \tau') = e^{E_\zeta}, \qquad E_\zeta(\tau') = \frac{\beta p_0}{\rho_0 \delta} \int_{-\infty}^{\tau'} F(\tau'') \, d\tau''. \tag{4.233}$$

An alternative expression for the solution results after an integration by parts in the integral in the numerator:

$$p = p_0 \frac{\int_{-\infty}^{\infty} F(\tau') e^{E_\zeta} e^{-E_G} d\tau'}{\int_{-\infty}^{\infty} e^{E_\zeta} e^{-E_G} d\tau'}. \tag{4.234}$$

This form makes it manifestly evident that the solution so obtained reduces to the initial condition in the limit as $x \to 0$. Note also that the intrinsic nonlinear nature of the Burgers equation is exemplified by the appearance of the integral over the time history of F in the exponent E_ζ. If β is set to zero, but δ remains finite, the result reduces to the linear-acoustics prediction for the attenuation and distortion of a transient pulse, each of whose frequency components is being attenuated as $\exp(-\omega^2 x \delta / 2c_0^3)$ after propagation over a distance x. That such is so is confirmed by the computation

$$\frac{\int_{-\infty}^{\infty} \sin(\omega\tau') e^{-E_G} d\tau'}{\int_{-\infty}^{\infty} e^{-E_G} d\tau} = e^{-\omega^2 x \delta / 2c_0^3} \sin \omega\tau. \tag{4.235}$$

4.5.3 Rise Time and Thickness of Weak Shocks

The weak shock model discussed in Sect. 4.4 leads to abrupt discontinuities, but when the model incorporates dissipation processes, such discontinuities become instead transition regions over which the pressure and fluid velocity change rapidly. Insight into the nature of the transition results from consideration of the idealized model of a wave that moves without change of form in the x direction with speed U, and that is distinguished by the fact that the most rapid variations of the fluid-dynamic variables occur in the vicinity of points and times at which $x - Ut$ is relatively small. (The quantity U will subsequently be identified as the shock speed U_{sh}.) For $x \gg Ut$, p and u should be zero, while for $x \ll Ut$, p and $\rho_0 c_0 u$ approach the shock overpressure $\Delta P = \Delta p = \rho_0 c_0 \Delta u$.

The discussion here is based on the retarded-time version of the Burgers equation, as given by Eq. (4.189). In accordance with the notion of a frozen profile, we begin with the assumption of a profile moving with some speed U, but without change of form, i.e.,

$$p = p(t - x/U) = \rho_0 c_0 u(t'), \tag{4.236}$$

where

$$t' = t - x/U = \tau + (c_0^{-1} - U^{-1})x. \tag{4.237}$$

This, when inserted into Eq. (4.189), yields the ordinary differential equation

$$c_0^2(U - c_0)u_{t'} - (\delta/2)U u_{t't'} = \beta c_0 u U u_{t'}, \tag{4.238}$$

which integrates, with the (causality) boundary condition $u \to 0$ as $t' \to -\infty$, to

$$c_0^2(U - c_0)u - (\delta/2)U u_{t'} = \tfrac{1}{2}\beta c_0 u^2 U. \tag{4.239}$$

The second boundary condition, that $u \to \Delta u$ as $t' \to \infty$, requires that $u_{t'} \to 0$ as $u \to \Delta u$, which yields

$$c_0(c_0 - U) = -\tfrac{1}{2}\beta U \, \Delta u, \tag{4.240}$$

or

$$U = \frac{c_0^2}{c_0 - \tfrac{1}{2}\beta \Delta u}. \tag{4.241}$$

Since $\Delta u \ll c_0$, this can be replaced, to the same order of approximation, by

$$U = c_0 + \tfrac{1}{2}\beta \Delta u. \tag{4.242}$$

One should note that Eq. (4.242) is consistent with the approximate Rankine–Hugoniot relation

$$U_{\text{sh}} = c_{\text{av}} + u_{\text{av}} \tag{4.243}$$

in Eq. (4.123). Ahead of the shock, one has $c_a = c_0$ and $u_a = 0$, while far behind the shock, one has $c_b = c_0 + (\beta - 1)\Delta u$ and $u_b = \Delta u$. The identification of $U_{\text{sh}} = U$, $c_{\text{av}} = c_0 + \tfrac{1}{2}(\beta - 1)\Delta u$, and $u_{\text{av}} = \tfrac{1}{2}\Delta u$ results in Eq. (4.242).

In a spirit similar to that by which Eq. (4.241) is replaced by Eq. (4.242), one can further approximate Eq. (4.239) by replacing $U - c_0$ in the first term by $\tfrac{1}{2}\beta \Delta u$ and by replacing U by c_0 in the remaining terms, to obtain

$$\frac{\delta}{c_0}\frac{du}{dt'} = \beta u(\Delta u - u). \tag{4.244}$$

This first-order ordinary differential equation is readily integrated by rewriting it as

$$dt' = \frac{\delta/\beta c_0}{u(\Delta u - u)}\, du = \frac{\delta}{\beta c_0 \Delta u}\left(\frac{du}{u} + \frac{du}{\Delta u - u}\right) \tag{4.245}$$

and carrying out the indefinite integral, the result being

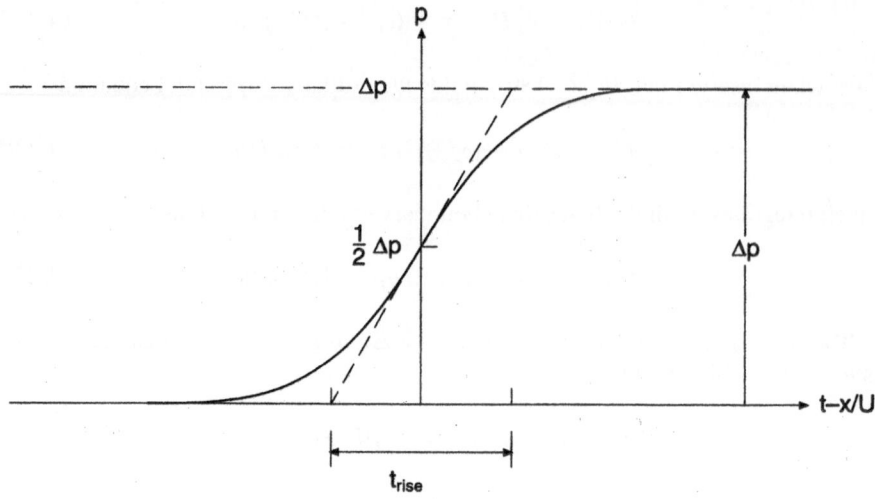

Fig. 4.12 Profile of a shock as a function of time for fixed position.

$$u = \tfrac{1}{2}\Delta u\{1 + \tanh[(\beta c_0 \Delta u/2\delta)(t' - t_0')]\}, \tag{4.246}$$

or, in terms of pressure

$$p = \tfrac{1}{2}\Delta p\{1 + \tanh[(\beta \Delta p/2\rho_0\delta)(t' - t_0')]\}, \tag{4.247}$$

where t_0' is a constant of integration, which is readily interpreted (Fig. 4.12) as corresponding to that value of $t - x/U$ at which the transition from $p = 0$ to $p = \Delta p$ passes through the halfway point, at which $p = \tfrac{1}{2}\Delta p$.

The nominal time of arrival t_{arr} of the shock is when u has reached its half-height, and this corresponds to $t' = t_0'$, whereby

$$t_{\mathrm{arr}} = \frac{x}{U_{\mathrm{sh}}} + t_0'. \tag{4.248}$$

Equation (4.247) can consequently be reexpressed as

$$p = \tfrac{1}{2}\Delta p\{1 + \tanh[(2/t_{\mathrm{rise}})(t - t_{\mathrm{arr}})]\}, \tag{4.249}$$

where the rise time is identified as

$$t_{\mathrm{rise}} = \frac{4\rho_0\delta}{\beta\Delta p} = \frac{4}{\beta\Delta p}[\tfrac{4}{3}\mu + \mu_B + (\gamma - 1)\kappa/c_p]. \tag{4.250}$$

In this last version, an explicit substitution has been made using Eq. (4.190) for the diffusivity δ of sound.

This particular definition of the rise time is based on the slope of the waveform at the half-peak point, i.e.,

$$t_{\text{rise}} = \frac{\Delta p}{(\partial p / \partial t)_{t=t_{\text{arr}}}}. \tag{4.251}$$

A straight line tangent (Fig. 4.12) to the waveform at its half-peak point reaches from the line $p = 0$ to the line $p = \Delta p$ over a time interval t_{rise}.

Alternatively, if one plots the solution in Eq. (4.249) as a function of x rather than of t, it takes the form

$$p = \tfrac{1}{2}\Delta p\{1 - \tanh[(2/\ell_{\text{sh}})(x - x_{\text{arr}})]\}, \tag{4.252}$$

where $x_{\text{arr}} = (t - t_0')U_{\text{sh}}$ is the nominal point at which the shock is located and where

$$\ell_{\text{sh}} = c_0 t_{\text{rise}} \tag{4.253}$$

is identified as the shock thickness.

The form of the shock profile expressed in terms of the hyperbolic tangent with the above identifications for shock thickness and shock rise time dates back to Taylor (1910), although the recognition of the above results from Taylor's paper is not trivial. That the Burgers equation allows solutions of the above form was recognized in the pioneering works of Bateman (1915), Burgers (1948), and Lagerstrom, Cole, and Trilling (1949).

4.5.4   Radiation from a Monofrequency Source

We shall investigate an explicit solution of the Burgers equation for a wave that is sinusoidal in time at $x = 0$, the source condition for which is Eq. (4.46) [or Eq. (4.217) with $F(t) = \sin \omega t$], repeated here for convenience:

$$p(0, t) = p_0 \sin \omega t. \tag{4.254}$$

As in Sect. 4.2.4, it is helpful to perform the analysis in dimensionless variables:

$$P = p/p_0, \quad \sigma = x/\bar{x}, \quad \theta = \omega \tau, \quad \Gamma = \ell_a/\bar{x}, \tag{4.255}$$

where $\bar{x} = 1/\beta \varepsilon k$ is the plane-wave shock formation distance in the absence of dissipation, $\varepsilon = p_0/\rho_0 c_0^2$ is the acoustic Mach number at the source, $k = \omega/c_0$ is the corresponding wave number, and $\ell_a = 1/\alpha$ is the absorption length associated with the thermoviscous attenuation coefficient $\alpha = \delta \omega^2/2c_0^3$ for small-signal propagation at the source frequency. In the present section, P is not to be confused with the total pressure. In terms of this notation, the Burgers equation, Eq. (4.189), becomes

$$\frac{\partial P}{\partial \sigma} - \frac{1}{\Gamma} \frac{\partial^2 P}{\partial \theta^2} = P \frac{\partial P}{\partial \theta}. \tag{4.256}$$

The parameter Γ, often referred to as the Gol'dberg (1956, 1957b) number, is thus anticipated to influence strongly the properties of the obtained solution. For $\Gamma \gg 1$ ($\bar{x} \ll \ell_a$, for which dissipation is negligible at distances on the order of the shock formation distance), we should expect to recover the solutions presented in Sect. 4.2.4.1. For $\Gamma \sim 1$ ($\bar{x} \sim \ell_a$), effects of nonlinearity and dissipation are expected to be of comparable order.

In dimensionless form, the general solution given by Eq. (4.221) is

$$P(\sigma, \theta) = \frac{2}{\Gamma} \frac{\zeta_\theta}{\zeta}, \tag{4.257}$$

where, from Eqs. (4.231) and (4.233), respectively,

$$\zeta(\sigma, \theta) = \left(\frac{\Gamma}{4\pi\sigma}\right)^{1/2} \int_{-\infty}^{\infty} \zeta(0, \theta') \exp\left[-\frac{\Gamma(\theta - \theta')^2}{4\sigma}\right] d\theta', \tag{4.258}$$

$$\zeta(0, \theta') = \exp\left[\frac{\Gamma}{2} \int_{-\infty}^{\theta'} P(0, \theta'') d\theta''\right]. \tag{4.259}$$

We now substitute $P(0, \theta) = \sin\theta$ from Eq. (4.254) into Eq. (4.259) to obtain $\zeta(0, \theta) = \exp(-\frac{1}{2}\Gamma \cos\theta)$, and Eq. (4.258) becomes

$$\zeta = \left(\frac{\Gamma}{4\pi\sigma}\right)^{1/2} \int_{-\infty}^{\infty} \exp\left[-\frac{1}{2}\Gamma \cos\theta' - \frac{\Gamma(\theta - \theta')^2}{4\sigma}\right] d\theta'. \tag{4.260}$$

By ignoring the contribution due to the lower integration limit in Eq. (4.259), $\theta'' = -\infty$, we assume that dissipation has caused any initial transient effects to vanish; see Blackstock (1964b) for the terms that describe the transients. Making use of the identity

$$\exp(-\tfrac{1}{2}\Gamma \cos\theta) = \sum_{n=-\infty}^{\infty} (-1)^n I_n(\tfrac{1}{2}\Gamma) e^{jn\theta}, \tag{4.261}$$

where I_n are modified Bessel functions, yields

$$\zeta = \left(\frac{\Gamma}{4\pi\sigma}\right)^{1/2} \sum_{n=-\infty}^{\infty} (-1)^n I_n(\tfrac{1}{2}\Gamma) \int_{-\infty}^{\infty} \exp\left[jn\theta' - \frac{\Gamma(\theta - \theta')^2}{4\sigma}\right] d\theta'. \tag{4.262}$$

The integral can be solved, and the summation rearranged on the basis of the relation $I_{-n} = I_n$, to obtain

$$\zeta = I_0(\tfrac{1}{2}\Gamma) + 2\sum_{n=1}^{\infty}(-1)^n I_n(\tfrac{1}{2}\Gamma)e^{-n^2\sigma/\Gamma}\cos n\theta. \tag{4.263}$$

Substitution into Eq. (4.257) gives the final form of the solution:

$$p = p_0 \frac{4\Gamma^{-1}\sum_{n=1}^{\infty}(-1)^{n+1}n I_n(\tfrac{1}{2}\Gamma)e^{-n^2\alpha x}\sin n\omega\tau}{I_0(\tfrac{1}{2}\Gamma) + 2\sum_{n=1}^{\infty}(-1)^n I_n(\tfrac{1}{2}\Gamma)e^{-n^2\alpha x}\cos n\omega\tau}, \tag{4.264}$$

where we have replaced σ/Γ by αx in the exponentials, and θ by $\omega\tau$ in the trigonometric functions. This solution was obtained first by Cole (1951) for the corresponding initial-value problem, while the above source-problem form of the solution was obtained by Mendousse (1953).

Because Eq. (4.264) is an odd function of time, it can be rewritten formally as

$$p(x,\tau) = p_0 \sum_{n=1}^{\infty} B_n(x)\sin n\omega\tau. \tag{4.265}$$

Exact expressions for the spectral amplitudes B_n are not available [although one should expect to recover the Fubini solution, Eq. (4.49), in the preshock region when there is no dissipation]. Here we use Eq. (4.264) to obtain approximate expressions for B_n in two limiting cases, weak waves ($\Gamma < 1$) and strong waves ($\Gamma \gg 1$). A time-domain solution that permits calculation of shock rise times in strong waves is also presented.

4.5.4.1 Weak Waves (Keck–Beyer Solution)

In this case ($\Gamma < 1$, or $\bar{x} > \ell_a$), Eq. (4.264) may be expanded directly in powers of the small quantity Γ, beginning with the following Taylor series expansion of the modified Bessel function:

$$I_n(\tfrac{1}{2}\Gamma) \sim \frac{\Gamma^n}{2^{2n}n!}\left(1 + \frac{\Gamma^2}{8(2n+1)} + \frac{\Gamma^4}{64(5n+1)} + \cdots\right). \tag{4.266}$$

Substitution into Eq. (4.264) and binomial expansion of the denominator [or, alternatively, using the formal algebraic procedure developed by Blackstock (1966a)] yields for the first three Fourier coefficients in Eq. (4.265)

$$B_1 = e^{-\alpha x} - \tfrac{1}{32}\Gamma^2 e^{-\alpha x}(1 - e^{-2\alpha x})^2 + O(\Gamma^4), \tag{4.267}$$

$$B_2 = \tfrac{1}{4}\Gamma(e^{-2\alpha x} - e^{-4\alpha x}) + O(\Gamma^3), \tag{4.268}$$

$$B_3 = \tfrac{1}{32}\Gamma^2(2e^{-3\alpha x} - 3e^{-5\alpha x} + e^{-9\alpha x}) + O(\Gamma^4). \tag{4.269}$$

Keck and Beyer (1960) derived these solutions using perturbation to solve the nonlinear wave equation directly. In the limit $\alpha x \to 0$, Eqs. (4.267)–(4.269) reduce to the asymptotic expressions obtained for the spectral coefficients in the Fubini solution, Eqs. (4.50)–(4.52).

Substitution of Eq. (4.268) into Eq. (4.265) yields for the dominant contribution to the second-harmonic component

$$p_2(x, \tau) = \frac{\beta p_0^2 \omega}{4\rho_0 c_0^3 \alpha}(e^{-2\alpha x} - e^{-4\alpha x}) \sin 2\omega\tau, \tag{4.270}$$

a result first obtained by Gol'dberg (1957a) by ordinary perturbation. The amplitude increases with x to a maximum value of $\beta p_0^2 \omega / 16\rho_0 c_0^3 \alpha = \tfrac{1}{16}\Gamma p_0$ at distance $x = \ell_a \ln\sqrt{2}$, beyond which it decays monotonically, ultimately as $e^{-2\alpha x}$. Equation (4.270) is a valid perturbation solution as long as $B_2/B_1 \ll 1$ is satisfied. From the leading terms in Eqs. (4.267) and (4.268), one finds that B_2/B_1 achieves a maximum value of $\Gamma/6\sqrt{3}$ at $x = \ell_a \ln\sqrt{3}$. Therefore, the amplitude of the second harmonic is everywhere less than 10% of that of the fundamental for $\Gamma < 1$, in which case Eq. (4.270) is a uniformly valid solution. Stated another way, for monofrequency radiation, an appropriate criterion for weak nonlinearity—i.e., one whereby linear theory accurately describes the primary wave—is $\Gamma < 1$.

Note that for $x \gg \ell_a$, the harmonic amplitudes B_n vary in proportion to $e^{-n\alpha x}$, not in proportion to the small-signal result for a thermoviscous fluid, $e^{-n^2\alpha x}$ (recall that small-signal attenuation increases in proportion to ω^2 in a thermoviscous fluid). This apparent contradiction may be explained by introducing the concept of virtual sources. Consider second-harmonic generation as an example. By virtual source in this case we mean the effective physical source that would correspond to the forcing function produced by substituting the linear solution p_1 for the primary wave in the right-hand side of Eq. (4.256). This interpretation was introduced by Berktay (1965) to explain the behavior of the parametric array on the basis of calculations corresponding to a Green's function approach (Sect. 8.3). We begin here by recognizing that the primary wave field (whose amplitude varies as $e^{-\alpha x}$) creates a distribution of virtual sources radiating second-harmonic sound with local source amplitudes $Ae^{-2\alpha x}$ (because the primary wave is squared). Consider now two distinct virtual sources, one at x_1 and a second a distance d farther away, at $x_2 = x_1 + d$, both of which radiate second-harmonic sound waves that propagate as small signals, decaying as $e^{-4\alpha x}$, out to location $x > x_2$. The signals received at x have amplitudes $A_i = Ae^{-2\alpha x_i}e^{-4\alpha(x-x_i)}$ ($i = 1, 2$). The ratio $A_2/A_1 = e^{2\alpha d}$ thus obtained reveals that for large values of αd (well-separated virtual sources), the received signal is dominated by the contribution from the nearer (second) virtual source. As a result, the decay law for the second harmonic matches asymptotically the decay law for the neighboring virtual sources themselves, $e^{-2\alpha x}$, because radiation from virtual sources far behind has been filtered out by the higher attenuation. Similar arguments apply to higher-harmonic signals. Their decay is determined asymptotically by the attenuation law for the most slowly decaying virtual sources that contribute to the given harmonic. As discussed in Chap. 8, the effect described above has a pronounced effect on harmonic directivity patterns in sound beams. A similar analysis has been applied to spherical waves by Webster and Blackstock (1978), and to sum- and difference-frequency generation by Darvennes et al. (1991).

4.5.4.2   Strong Waves (Fay Solution)

Here, for $\Gamma \gg 1$ and $\sigma > 3$ ($\bar{x} \ll \ell_a$ and $x > 3\bar{x}$), the following asymptotic relation is employed (Blackstock, 1964b):

$$I_n(\tfrac{1}{2}\Gamma) \sim I_0(\tfrac{1}{2}\Gamma)e^{-n^2/\Gamma}. \tag{4.271}$$

In this case, it is expedient to return to Eq. (4.263) and note that following substitution of Eq. (4.271), the former becomes

$$\zeta = I_0(\tfrac{1}{2}\Gamma)\theta_4(\tfrac{1}{2}\omega\tau,\, e^{-(1+\sigma)/\Gamma}), \tag{4.272}$$

where

$$\theta_4(z, q) = 1 + 2\sum_{n=1}^{\infty}(-1)^n q^{n^2} \cos 2nz \tag{4.273}$$

is the theta function of the fourth type (Abramowitz and Stegun, 1970), a property of which is

$$\frac{\theta_4'}{\theta_4} = 4\sum_{n=1}^{\infty}\frac{\sin 2nz}{q^{-n} - q^n}, \tag{4.274}$$

where $\theta_4' = d\theta_4/dz$. Substitution of Eq. (4.272) in Eq. (4.257) yields

$$p = p_0\frac{2}{\Gamma}\sum_{n=1}^{\infty}\frac{\sin n\omega\tau}{\sinh[n(1+\sigma)/\Gamma]}, \quad \sigma > 3. \tag{4.275}$$

This is known as the Fay (1931) solution, after the person who first derived it, in slightly different form, by entirely different means (see Sect. 1.4 for further discussion). The Fourier coefficients in Eq. (4.265) are thus identified as

$$B_n = \frac{2/\Gamma}{\sinh[n(1+\sigma)/\Gamma]}. \tag{4.276}$$

The domain of validity for Eq. (4.275) was established by Blackstock (1964b), who compared B_1 with numerical solution of the Burgers equation, which for $\Gamma = 50$ reveals an error less than 1% for $\sigma > 3.3$. In the limit $\Gamma \to \infty$, Eq. (4.276) reduces to $B_n = 2/n(1 + \sigma)$, and the sawtooth solution obtained from weak shock theory, Eq. (4.179), is recovered.

We consider now the asymptotic form of Eq. (4.275) for $\sigma \gg \Gamma \gg 1$ ($x \gg \ell_a \gg \bar{x}$), referred to as the "old age" region. In this case, we have $\sinh[n(1 + \sigma)/\Gamma] \simeq \frac{1}{2}(e^{n\sigma/\Gamma} - e^{-n\sigma/\Gamma}) \simeq \frac{1}{2}e^{n\alpha x}$, and Eq. (4.275) becomes

$$p = \frac{4\rho_0 c_0^3 \alpha}{\beta \omega} \sum_{n=1}^{\infty} e^{-n\alpha x} \sin n\omega \tau, \quad x \gg \ell_a \gg \bar{x}. \tag{4.277}$$

Two notable properties may be observed. First, acoustical saturation is evident— the solution is independent of the source pressure p_0. The harmonic amplitudes are different from those predicted for saturation in the sawtooth region by Eq. (4.184) because the observation point is well beyond the absorption length, where small-signal attenuation, not taken into account by weak shock theory, plays an important role. Second, note that, as in Eqs. (4.267)–(4.269), the harmonic amplitudes decay asymptotically as $e^{-n\alpha x}$, rather than according to the small-signal rate $e^{-n^2\alpha x}$. Thus, even though the wave seems to be a small signal, it carries with it forever telltale remnants from its nonlinear past.

We now have in hand a complete analytical model, in terms of explicit Fourier series solutions, for the entire history of a strong sound wave that starts out sinusoidal in a thermoviscous fluid, develops a sawtooth profile, and ultimately falls victim to effects of dissipation and reverts to a waveform resembling the signal at the source, although much reduced in amplitude. Just such a case history (i.e., with $\bar{x} \ll \ell_a$ assumed) is depicted schematically with time waveforms in Fig. 4.13, where the signal at the source ($x = 0$) is presented as sketch a. In the preshock region ($0 < x < \bar{x}$), characterized by sketch b, the waveform is described by the Fubini solution, Eq. (4.49). The signal at the shock formation distance ($x = \bar{x}$) is illustrated in sketch c, and that at the location of maximum shock amplitude in sketch d [$x = (\pi/2)\bar{x}$; see discussion of Eqs. (4.177)]. The transition region covered by sketches c to e, from shock formation to full sawtooth ($\bar{x} < x \lesssim 3\bar{x}$), is described by the harmonic amplitudes given by Eq. (4.183). Sketches e and f show propagation in the sawtooth region ($3\bar{x} < x \ll \ell_a$), where the wave is described by Eqs. (4.179) and (4.180) [as well as by the Fay solution, Eq. (4.275)]. This region is characterized by harmonic amplitudes that vary in relation to each other as $B_n \propto n^{-1}$, and by an overall decrease in wave amplitude due to dissipation at the shock that, for $\bar{x} \ll x \ll \ell_a$, is proportional to x^{-1}. Eventually, for $x \sim \ell_a$ (sketch g), dissipation reduces the wave amplitude to such an extent that finite-amplitude effects no longer sustain a shock with thickness negligible in comparison to the fundamental wavelength. The waveform in this region is described by Eq. (4.275) [or, alternatively, by the Khokhlov solution, Eq. (4.278) below]. Finally, as depicted in sketch h, for $x \gg \ell_a$, the old age region described by Eq. (4.277) is reached.

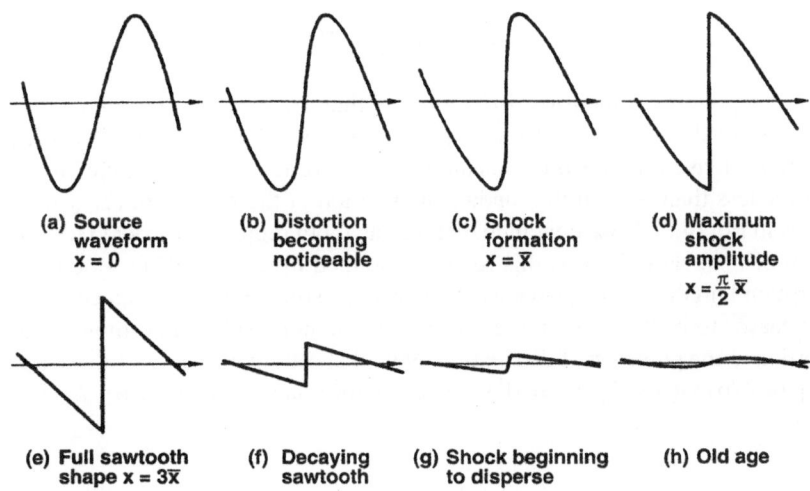

(a) Source   (b) Distortion   (c) Shock   (d) Maximum
    waveform   becoming   formation   shock
    x = 0   noticeable   x = X̄   amplitude
                                                                    x = $\frac{\pi}{2}$ X̄

(e) Full sawtooth   (f) Decaying   (g) Shock beginning   (h) Old age
    shape x = 3X̄   sawtooth   to disperse

Fig. 4.13 Case history of a strong wave (after Shooter, Muir, and Blackstock, 1974).

4.5.4.3   Shock Profile (Khokhlov Solution)

We conclude by presenting without formal derivation a time-domain solution that is
particularly useful for calculating shock rise times:

$$ p = \frac{p_0}{1+\sigma} \left[-\omega\tau + \pi \tanh \frac{\pi\Gamma\omega\tau}{2(1+\sigma)} \right], \quad 3 < \sigma < \Gamma, \tag{4.278} $$

which describes a full cycle of the waveform in the interval $-\pi \leq \omega\tau \leq \pi$.
Equation (4.278) was derived by Soluyan and Khokhlov (1961) [see also Rudenko
and Soluyan, 1977] by using the saddle-point method to approximate the integral in
Eq. (4.260). It can also be obtained as a limiting form of Eq. (4.275) (Blackstock,
1964b). Both derivations are somewhat too lengthy for our purposes here. We
note only the remarkable fact, as verified by direct substitution in Eq. (4.256),
that Eq. (4.278) is an exact solution of the Burgers equation, notwithstanding the
approximations introduced in the derivations. Equation (4.278) does not match
our monofrequency source condition, Eq. (4.254); it is nevertheless a very good
approximation of the desired solution for radiation from a monofrequency source
in the stated parameter range $3 < \sigma < \Gamma$, i.e., for strong waves beginning in the
sawtooth region up to distances on the order of the absorption length ℓ_a. Indeed, the
sawtooth solution obtained using weak shock theory, Eq. (4.180), is recovered from
Eq. (4.278) in the limit $\Gamma \to \infty$.

   To characterize the shock, we define a dimensionless rise time T_{rise} (based on the
fundamental periodicity 2π) to be the interval between the peak and the trough in
the waveform described by Eq. (4.278). From symmetry we have $\partial p / \partial(\omega\tau) = 0$ at
$\omega\tau = \frac{1}{2}T_{\text{rise}}$, which may be solved to give (Blackstock, 1964b)

$$T_{\text{rise}} = [4(1 + \sigma)/\pi \Gamma] \cosh^{-1} \sqrt{\pi^2 \Gamma / 2(1 + \sigma)}. \tag{4.279}$$

The corresponding dimensional rise time is thus $t_{\text{rise}} = T_{\text{rise}}/\omega$. As σ (distance) increases, so does T_{rise}, until the "sawtoothlike" wave profile disappears and the definition of rise time becomes meaningless. If we take "sawtoothlike" to mean a rise time less than 20% of the fundamental period of the waveform ($T_{\text{rise}} < 2\pi/5$), then from Eq. (4.279) we conclude that the sawtooth region is restricted to distances $\sigma < 0.6\Gamma - 1$. For $\Gamma \gg 1$, this region is defined by $\sigma < 0.6\Gamma$ ($x < 0.6\ell_a$). The absorption length ℓ_a thus provides a reasonable estimate of the distance at which a wave ceases to be "strong" in the sense of maintaining substantial finite-amplitude distortion in the presence of dissipation. Note that setting $\sigma = \Gamma \gg 1$ ($x = \ell_a \gg \bar{x}$) in Eq. (4.276) yields $B_2/B_1 = 0.32$, whereas for a sawtooth wave, $B_2/B_1 = 0.50$.

4.6 Special Topics

4.6.1 Nonplanar One-Dimensional Waves

Here we describe a method for transforming plane-wave solutions derived in previous sections to account for spreading (e.g., spherical or cylindrical) when small-signal absorption may be ignored. The transformations may be used for waves without shocks in ideal fluids, and beyond the shock formation distance when weak shock theory is valid, i.e., when energy loss may be attributed to dissipation just at the shocks (Blackstock, 1964a, 1972). Applications of the transformation when ordinary absorption is taken into account—for example, predictions of shock thicknesses in diverging and converging waves—are discussed by Naugol'nykh et al. (1963) and Khokhlov et al. (1964).

We begin with the generalized Burgers equation, Eq. (3.58), for an ideal fluid ($\delta = 0$):

$$\frac{\partial p}{\partial r} + \frac{m}{r} p = \pm \frac{\beta p}{\rho_0 c_0^3} \frac{\partial p}{\partial \tau}, \tag{4.280}$$

where $m = 1$ for spherical waves, $m = \frac{1}{2}$ for cylindrical waves, r is the radial coordinate (defined positive outward), and we let $\tau = t \mp (r - r_0)/c_0$ to account for both diverging and converging waves. As in Sect. 3.8, the following convention is adopted: Whenever there is a choice of sign, the upper one corresponds to diverging (or outgoing) waves, the lower to converging (or incoming) waves. The radius r_0 is normally selected to be the source radius or other distance where the waveform is known. Equation (4.280) is valid for $kr \gg 1$, where k is a relevant characteristic wave number. Now introduce the appropriate change of variables:

Spherical waves:

$$q = (r/r_0)p, \quad z = \pm r_0 \ln(r/r_0), \tag{4.281}$$

Cylindrical waves:

$$q = (r/r_0)^{1/2}p, \quad z = \pm 2(\sqrt{r} - \sqrt{r_0})\sqrt{r_0}, \tag{4.282}$$

in terms of which Eq. (4.280) becomes, with $dz/dr = \pm(r_0/r)^m$,

$$\frac{\partial q}{\partial z} = \frac{\beta q}{\rho_0 c_0^3} \frac{\partial q}{\partial \tau}. \tag{4.283}$$

Comparison with Eq. (4.28) reveals that the solutions of that equation (for plane waves) are also solutions of Eq. (4.283) (for spherical and cylindrical waves) if in the former we replace p by the scaled dependent variable q, and x by the stretched coordinate z, and interpret the retarded time according to its definition following Eq. (4.280). Note that the transformation yields $z > 0$ for both diverging ($r > r_0$) and converging ($r < r_0$) waves.

Consider, for example, use of the transformation in Eq. (4.281) to describe spherical waves in the preshock region. For a plane wave with source condition $p = f(t)$ at $x = 0$, the general solution is Eq. (4.29). The corresponding solution for a spherical wave with source condition $p = f(t)$ at $r = r_0$ is given by Eq. (4.29) with p replaced by q and x by z:

$$p(r, \tau) = \frac{r_0}{r} f\left(\tau \pm \frac{\beta p r}{\rho_0 c_0^3} \ln \frac{r}{r_0}\right). \tag{4.284}$$

The quasilinear solution for second-harmonic generation, corresponding to Eq. (4.54), is

$$p_2(r, \tau) = \pm \frac{\beta p_0^2 \omega r_0^2}{2\rho_0 c_0^3} \frac{\ln(r/r_0)}{r} \sin 2\omega\tau. \tag{4.285}$$

For the diverging wave, a maximum amplitude of $\beta p_0^2 \omega r_0 / 2e\rho_0 c_0^3$ is achieved at $r/r_0 = e$.

The shock formation distance is calculated in a similar way. If the source waveform is $f(t) = p_0 \sin \omega t$, the shock formation distance for the plane wave is given by Eq. (4.22), $\bar{x} = (\beta \varepsilon k)^{-1}$. The shock formation distance for a spherical wave is defined in terms of the stretched coordinate z by $\bar{z} = (\beta \varepsilon k)^{-1}$. Via the transformation in Eq. (4.281), the radial distance $r = \bar{r}$ where shock formation occurs is

$$\bar{r} = r_0 \exp(\pm 1/\beta \varepsilon k r_0). \tag{4.286}$$

For the diverging wave, the shock formation distance can be considerably farther than the corresponding distance for a plane wave [although note that for $\beta \varepsilon k r_0 \gg 1$ ($\bar{x} \ll r_0$), expansion of Eq. (4.286) yields $\bar{r} \simeq r_0 \pm \bar{x}$, and the plane-wave result is recovered]. Despite spreading losses, however, the result implies that shock formation is inevitable in the absence of dissipation. For the converging wave, it must be kept in mind that the model is valid only for $kr \gg 1$. In the same manner, the shock formation distance for a cylindrical wave that is sinusoidal at $r = r_0$ is found to be

$$\bar{r} = r_0 \left(1 \pm \frac{1}{2\beta \varepsilon k r_0} \right)^2 . \tag{4.287}$$

After shocks are formed, plane-wave solutions obtained on the basis of weak shock theory [e.g., Eqs. (4.172), (4.173), and (4.179)] may also be transformed for application to spherical and cylindrical waves via Eqs. (4.281) and (4.282). For a specific illustration of this technique, consider the N-wave solution given by Eq. (4.172). For the peak amplitude p_{sh}, we find

$$\text{Spherical N wave:} \quad p_{sh} = \frac{r_0}{r} \frac{p_0}{\sqrt{1 \pm b r_0 \ln(r/r_0)}} , \tag{4.288}$$

$$\text{Cylindrical N wave:} \quad p_{sh} = \sqrt{\frac{r_0}{r}} \frac{p_0}{\sqrt{1 \pm 2b(\sqrt{r} - \sqrt{r_0})\sqrt{r_0}}} . \tag{4.289}$$

For spherical spreading, an outgoing N wave decays at large distance as $r^{-1}(\ln r)^{-1/2}$, whereas the cylindrical N wave decays as $r^{-3/4}$ (Landau, 1945; Landau and Lifshiftz, 1987).

Similar transformations are available for more general types of spreading. For example, Eq. (3.60) describes the propagation of quasi-plane waves in a duct having a slowly varying cross-sectional area $A(x)$:

$$\frac{\partial p}{\partial x} + \frac{A'}{2A} p = \pm \frac{\beta p}{\rho_0 c_0^3} \frac{\partial p}{\partial \tau} , \tag{4.290}$$

where $A' = dA/dx$ and $\tau = t \mp (x - x_0)/c_0$. [Notice the similarity to Eq. (12.58) for nonlinear propagation along ray tubes in inhomogeneous media.] The appropriate transformation is

$$q = (A/A_0)^{1/2} p, \quad z = \pm \int_{x_0}^{x} (A_0/A)^{1/2} dx, \tag{4.291}$$

where $A_0 = A(x_0)$. Substitution into Eq. (4.290) again yields Eq. (4.283). Note that for $A \propto x^{2m}$, Eqs. (4.290) and (4.291) reduce to Eqs. (4.280)–(4.282) with r replaced by x.

We conclude with the example of propagation in an exponential horn. The horn cross-sectional area is given by

$$A = A_0 e^{2(x-x_0)/h}, \tag{4.292}$$

where h is the distance over which the diameter changes by the factor e. The stretched coordinate z in this case is, from Eq. (4.291),

$$z = \pm h(1 - e^{-(x-x_0)/h}), \tag{4.293}$$

and the corresponding shock formation distance, for $p = p_0 \sin \omega t$ at $x = x_0$, is

$$\bar{x} = x_0 + h \ln \left(\frac{\beta \varepsilon k h}{\beta \varepsilon k h \mp 1} \right). \tag{4.294}$$

Note that for propagation in the $+x$ direction (for which A increases with x), $\bar{x}$ increases toward infinity as the value of $\beta \varepsilon k h$ decreases toward unity, and therefore no shock formation is predicted for $\beta \varepsilon k h < 1$. The absence of shock formation is due to a flare rate so rapid (small h) that waveform steepening, which is proportional to the amplitude of the signal, cannot keep pace with the decrease in amplitude due to spreading. In contrast, note that Eq. (4.286) predicts that shock formation in a diverging spherical wave is inevitable (in the absence of dissipation). Although the shock-free limit for an exponential horn may violate the assumption of slowly varying $A(x)$, analogous phenomena are encountered in nonlinear ray theory (due mainly to variations in the ambient properties of the medium) and are referred to as waveform freezing (see Sects. 12.2.2 and 12.2.4).

When thermoviscous dissipation is included in model equations that account for spreading, no variable transformation is available that recovers the corresponding form of the equation for plane waves. For example, including thermoviscous dissipation in Eq. (4.280) produces the generalized Burgers equation given by Eq. (3.58). When Eqs. (4.281) and (4.282) are used to transform the generalized Burgers equation, the dissipation term acquires a coefficient that depends on distance [see Eq. (11.28)], and numerical techniques are normally required to obtain uniformly valid solutions.

4.6.2 Intensity and Absorption

Sections 4.3–4.5 have demonstrated that nonlinear distortion can markedly increase the absorption of sound by the fluid. One way to quantify the enhanced absorption is to define a finite-amplitude absorption coefficient α_f in terms of the intensity I of the wave [see, for example, Naugol'nykh (1958) and Rudnick (1958)]:

$$\alpha_f = -\frac{1}{2I}\frac{dI}{dx},$$ (4.295)

where plane waves have been assumed. For small signals, the intensity decays as $I = I_0 e^{-2\alpha x}$, where I_0 is the intensity at the source and α is the ordinary (amplitude) absorption coefficient; in this case, α_f reduces to α. To utilize Eq. (4.295) for finite-amplitude waves, we must first find a way to calculate their intensity.

The assumption of plane progressive waves is continued throughout the analysis that follows. The transformations discussed in Sect. 4.6.1 may be used to extend the results to other one-dimensional waves. For nonplanar waves, replace dI/dx in Eq. (4.295) with $\nabla \cdot \mathbf{I}$.

Intensity as defined here is the time average of the energy flux pu:

$$I = \frac{1}{t_{av}} \int_0^{t_{av}} pu \, dt,$$ (4.296)

where t_{av} is a suitable averaging time, for example, the period for a periodic wave or the duration for a transient. Since the wave motion is assumed to be progressive, particle velocity in the integrand may be replaced by $p/\rho_0 c_0$, and we obtain $I = p_{rms}^2/\rho_0 c_0$, where p_{rms} is the root-mean-square pressure. Finding the intensity thus reduces to finding p_{rms}^2. The calculation is made first for lossless waves and then extended to waves for which weak shock theory is valid.

For illustration, we again revisit the monofrequency source problem. Equation (4.46) gives the source signal, and Eqs. (4.175) the solution. Because of the symmetry of this particular wave, the averaging time need be only half a period, that is, $0 \le \omega\tau \le \pi$. Equation (4.296) thus becomes

$$I = \frac{2I_0}{\pi} \int_0^\pi P^2 d(\omega\tau).$$ (4.297)

First restrict attention to the shock-free region $\sigma \le 1$. Because for this case the relation between $\omega\tau$ and Φ is one-to-one [the limits $\omega\tau = (0, \pi)$ correspond to the limits $\Phi = (0, \pi)$], substitution from the second of Eqs. (4.175) into Eq. (4.297) yields

$$I = \frac{2I_0}{\pi} \int_0^\pi \sin^2\Phi(1 - \sigma\cos\Phi)\, d\Phi.$$ (4.298)

Integration produces $\pi/2$ for the first term in the integrand, zero for the second term. The result is

$$I = I_0 = \frac{p_0^2}{2\rho_0 c_0}.$$ (4.299)

The interpretation of this simple result is that no energy is lost during the transfer of energy from the fundamental to the higher-harmonic components (Rudenko and Soluyan, 1977), *as long as shocks have not yet formed*. Substitution of Eq. (4.299) in Eq. (4.295) of course confirms that $\alpha_f = 0$ in the shock-free region.

A frequency-domain calculation of the intensity is also of interest. If the Fubini solution, Eq. (4.49), is substituted in Eq. (4.297), and note is taken of the orthogonality of the series elements $\sin n\omega\tau$, one finds

$$I = I_0 \sum_{n=1}^{\infty} B_n^2(\sigma), \qquad (4.300)$$

where the harmonic amplitudes are given by $B_n = (2/n\sigma) J_n(n\sigma)$. The series sums to unity (Watson, 1944), and Eq. (4.299) is recovered, as indeed is required by Parseval's theorem. Although of theoretical interest, Eq. (4.300) is also of practical use as a test for computational models.

Next extend the analysis to the region where shocks are present, $\sigma > 1$ (Blackstock, 1990). As shown in Sect. 4.4.3, the range $\omega\tau = (0, \pi)$ now corresponds to $\Phi = (\Phi_{sh}, \pi)$.[8] Equation (4.298) thus becomes

$$I = \frac{2I_0}{\pi} \int_{\Phi_{sh}}^{\pi} \sin^2\Phi(1 - \sigma\cos\Phi)\,d\Phi, \qquad (4.301)$$

where Φ_{sh} is given by the second of Eqs. (4.177). The integral is easily evaluated:

$$I = \frac{I_0}{\pi}(\pi - \Phi_{sh} + \sin\Phi_{sh}\cos\Phi_{sh} + \tfrac{2}{3}\sigma\sin^3\Phi_{sh}). \qquad (4.302)$$

An alternative expression, found by using the relations $\Phi_{sh} = \sigma P_{sh}$ and $\sin\Phi_{sh} = P_{sh}$ [see Eqs. (4.177)], is

$$\frac{I}{I_0} = 1 - \frac{P_{sh}}{\pi}(\sigma - \cos\sigma P_{sh} - \tfrac{2}{3}\sigma P_{sh}^2). \qquad (4.303)$$

In the preshock region, where $P_{sh} = 0$, Eq. (4.303) reduces to $I = I_0$, in agreement with Eq. (4.299). After shocks form, the intensity decreases monotonically with distance. Two approaches may be used to find the intensity in the sawtooth region ($\sigma > 3$). First, the asymptotic expression $P_{sh} = \pi/(1 + \sigma)$ [Eq. (4.178)] may be substituted in Eq. (4.303). The cosine term becomes $\cos[\pi\sigma/(1 + \sigma)] = \cos[\pi - \pi/(1+\sigma)] = -\cos P_{sh} = -1 + \tfrac{1}{2}P_{sh}^2 \cdots$. Several terms on the right-hand side of Eq. (4.303) then cancel, and the leading remaining term is

---

[8] The integration range $\omega\tau = (0, \pi)$ may be broken into the segments $\omega\tau = (0, \omega\tau_{sh})$ and $\omega\tau = (\omega\tau_{sh}, \pi)$. Integration of P^2 over the first segment (i.e., the shock) yields nothing because in weak shock theory the shock rise time is zero ($\omega\tau_{sh} = 0+$). However, if the model were to include finite rise time, integration over the shock would yield a finite contribution.

$$\frac{I}{I_0} = \frac{2\pi^2}{3(1+\sigma)^2}, \qquad \sigma > 3. \tag{4.304}$$

The same result may be obtained by substituting the expression for the sawtooth waveform, Eq. (4.180), in Eq. (4.297) and integrating directly. Deep in the sawtooth region ($\sigma \gg 1$), Eq. (4.304) reduces to

$$I = \frac{\pi^2 \rho_0 c_0^3}{3\beta^2 k^2 x^2}, \qquad \sigma \gg 1. \tag{4.305}$$

Acoustical saturation is indicated, since I has become independent of the source intensity I_0.

Next, to find the absorption α_f, use Eq. (4.302) or (4.303) in Eq. (4.295). The calculation is tedious, but the final result is simple:

$$\alpha_f \bar{x} = \frac{\frac{2}{3} P_{sh}^3}{\pi - \sigma P_{sh} + P_{sh} \cos \sigma P_{sh} + \frac{2}{3}\sigma P_{sh}^3}. \tag{4.306}$$

A simpler-appearing alternative expression, found by recognizing that the denominator of Eq. (4.306) is $\pi I / I_0$ [see Eq. (4.302)], is

$$\alpha_f \bar{x} = \frac{2 P_{sh}^3 I_0}{3\pi I}. \tag{4.307}$$

Equations (4.303) and (4.306) are simple, exact algebraic relations, which, with the help of the first of Eqs. (4.177), are easily graphed. Curves of I/I_0 and $\alpha_f \bar{x}$ are shown on the top in Fig. 4.14. On the bottom is a plot of $\alpha_f x$; the same result was found by Dalecki et al. (1991), who made their calculation in the frequency domain (see also Fig. 15.3). The physical explanation is as follows: Since no dissipation takes place unless shocks are present, the region $\sigma < 1$ is one in which the intensity is constant and α_f is zero. The rapid growth of the shocks in the region $1 < \sigma < \pi/2$ (see Fig. 4.9) initiates the decay of the intensity and concurrent rise of the absorption. Beyond $\sigma = \pi/2$, the shocks lose amplitude. The intensity continues to drop, but at an ever-decreasing rate (the weaker the shock, the less the dissipation). After reaching a peak at $\sigma = 1.933$ ($\alpha_f \bar{x} = 0.2734$), the absorption begins to decline, in step with the decreasing rate of intensity decline.

For values of $\sigma \gg 1$, an asymptotic expression for the absorption is

$$\alpha_f = [\sigma/(1+\sigma)]x^{-1} \sim x^{-1}, \tag{4.308}$$

a result also found by Carstensen et al. (1982).

Finally, recall that weak shock theory is limited to shocks that are not too weak. In line with comments in Sect. 4.4.4, we note that the limit of validity of our calculation is reached when α_f becomes of order α.

Fig. 4.14 Intensity and
absorption of a plane wave
(Blackstock, 1990).

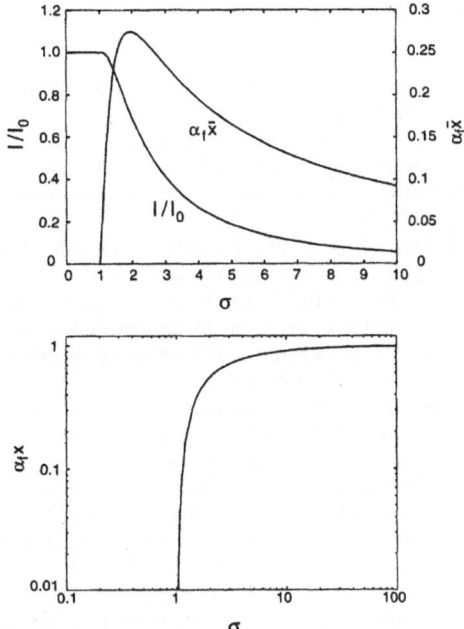

The absorption coefficient defined by Eq. (4.295) is of particular interest in biomedical applications of ultrasound, because the rate at which the temperature T increases as a result of dissipation of the acoustic energy in a plane wave is

$$\frac{dT}{dt} = -\frac{1}{\rho_0 c_m}\frac{dI}{dx}, \tag{4.309}$$

where c_m is the heat capacity of the medium (e.g., tissue) per unit mass. See Sect. 15.3.3. Elimination of dI/dx between Eqs. (4.295) and (4.309) yields

$$\frac{dT}{dt} = \frac{2\alpha_f I}{\rho_0 c_m}. \tag{4.310}$$

For the case of an initially sinusoidal wave, substitution of Eq. (4.307) into Eq. (4.310) gives

$$\frac{dT}{dt} = \frac{4 I_0 P_{sh}^3}{3\pi \rho_0 c_m \bar{x}}. \tag{4.311}$$

The rate at which the temperature increases is thus proportional to the cube of the shock amplitude. This result is similar to that given by Eq. (4.161).

References

Abramowitz, M., and Stegun, I. A., eds. (1970). *Handbook of Mathematical Functions* (Dover, New York).

Allen, C. H. (1950). Finite amplitude distortion in a spherically diverging sound wave in air. Ph.D. dissertation, Pennsylvania State University.

Bateman, H. (1915). Some recent researches on the motions of fluids. *Monthly Weather Review* **43**, 163–170.

Berktay, H. O. (1965). Possible exploitation of non-linear acoustics in underwater transmitting applications. *J. Sound Vib.* **2**, 435–461.

Bethe, H. A. (1942). The theory of shock waves for an arbitrary equation of state. Office of Scientific Research and Development, Report 545. NTIS Document No. PB 032189. Available from U.S. Department of Commerce, National Technical Information Service, Springfield, VA 22161 [(703) 487–4650].

Beyer, R. T. (1984). *Nonlinear Acoustics in Fluids* (Van Nostrand Reinhold, New York).

Blackstock, D. T. (1962). Propagation of plane sound waves of finite amplitude in nondissipative fluids. *J. Acoust. Soc. Am.* **34**, 9–30.

Blackstock, D. T. (1964a). On plane, spherical, and cylindrical sound waves of finite amplitude in lossless fluids. *J. Acoust. Soc. Am.* **36**, 217–219.

Blackstock, D. T. (1964b). Thermoviscous attenuation of plane, periodic, finite-amplitude sound waves. *J. Acoust. Soc. Am.* **36**, 534–542.

Blackstock, D. T. (1966a). Convergence of the Keck–Beyer perturbation solution for plane waves of finite amplitude in a viscous fluid. *J. Acoust. Soc. Am.* **39**, 411–413.

Blackstock, D. T. (1966b). Connection between the Fay and Fubini solutions for plane sound waves of finite amplitude. *J. Acoust. Soc. Am.* **39**, 1019–1026.

Blackstock, D. T. (1972). Nonlinear acoustics (theoretical). In *American Institute of Physics Handbook*, 3rd ed., D. E. Gray, ed. (McGraw-Hill, New York), Chap. 3n, pp. 3-183–3-205.

Blackstock, D. T. (1982). Spectral interactions in nonlinear acoustics—A review. In *Nonlinear Deformation Waves*, IUTAM Proceedings, U. Nigul and J. Engelbrecht, eds. (Springer, Berlin), pp. 301–315.

Blackstock, D. T. (1983). Propagation of a weak shock followed by a tail of arbitrary waveform. *Proceedings, 11th Int. Cong. Acoustics, Paris, France*, Vol. 1, pp. 305–308.

Blackstock, D. T. (1990). On the absorption of finite-amplitude sound. In *Frontiers of Nonlinear Acoustics: 12th ISNA*, M. F. Hamilton and D. T. Blackstock, eds. (Elsevier, London), pp. 119–124.

Borisov, A. A., Borisov, Al. A., Kutateladze, S. S., and Nakoryakov, V. E. (1983). Rarefaction shock wave near the critical liquid-vapour point. *J. Fluid Mech.* **126**, 59–73.

Burgers, J. M. (1940). Application of a model system to illustrate some points of the statistical theory of free turbulence. *Proc. Acad. Sci. Amsterdam* **43**, 2–12.

Burgers, J. M. (1948). A mathematical model illustrating the theory of turbulence. In *Advances in Applied Mechanics*, R. von Mises and T. von Kármán, eds. (Academic Press, New York), Vol. 1, pp. 171–199, in particular pp. 173, 181–184.

Carstensen, E. L., McKay, N. D., Dalecki, D., and Muir, T. G. (1982). Absorption of finite amplitude ultrasound in tissues. *Acustica* **51**, 116–123.

Cole, J. D. (1951). On a quasi-linear parabolic equation occurring in aerodynamics. *Q. Appl. Math.* **9**, 225–236.

Courant, R., and Friedrichs, K. O. (1948). *Supersonic Flow and Shock Waves* (Interscience, New York), pp. 116–172.

Cramer, M. S. (1989). Negative nonlinearity in selected fluorocarbons. *Phys. of Fluids A*, **1**, 1894–1897.

Dalecki, D., Carstensen, E. L., Parker, K. J., and Bacon, D. R. (1991). Absorption of finite amplitude focused ultrasound. *J. Acoust. Soc. Am.* **89**, 2435–2447.

Darvennes, C. M., Hamilton, M. F., Naze Tjøtta, J., and Tjøtta, S. (1991). Effects of absorption on the nonlinear interaction of sound beams. *J. Acoust. Soc. Am.* **89**, 1028–1036.

Earnshaw, S. (1860). On the mathematical theory of sound. *Trans. Roy. Soc.* (London) **150**, 133–148.

Fay, R. D. (1931). Plane sound waves of finite amplitude. *J. Acoust. Soc. Am.* **3**, 222–241.

Fenlon, F. H. (1972). An extension of the Bessel–Fubini series for a multiple-frequency cw acoustic source of finite amplitude. *J. Acoust. Soc. Am.* **51**, 284–289.

Fenlon, F. H. (1973). Derivation of the multiple frequency Bessel-Fubini series via Fourier analysis of the preshock time waveform. *J. Acoust. Soc. Am.* **53**, 1752–1754.

Fubini, E. (1935). Anomalie nella propagazione di ande acustiche de grande ampiezza. *Alta Frequenza* **4**, 530–581. English translation: Beyer (1984), 118–177.

Gol'dberg, Z. A. (1956). Second approximation acoustic equations and the propagation of plane waves of finite amplitude. *Sov. Phys. Acoust.* **2**, 346–350.

Gol'dberg, Z. A. (1957a). Certain second-order quantities in acoustics. *Sov. Phys. Acoust.* **3**, 157–162.

Gol'dberg, Z. A. (1957b). On the propagation of plane waves of finite amplitude. *Sov. Phys. Acoust.* **3**, 340–347.

Hayes, W. D. (1958). The basic theory of gasdynamic discontinuities. In *Fundamentals of Gas Dynamics (High Speed Aerodynamics and Jet Propulsion*, Vol. 3), Howard W. Emmons, ed. (Princeton University Press, Princeton, N.J.), pp. 416–481.

Hopf, E. (1950). The partial differential equation $u_t + uu_x = \mu u_{xx}$. *Comm. Pure and Appl. Math.* **3**, 201–230. (The equation appearing in that title, as printed by the journal, has a misprint. The correct form of the title is given here.)

Hugoniot, H. (1887, 1889). Mémoire sur la propagation du mouvement dans les corps et spécialement dans les gaz parfaits. *J. l'école polytech.* (Paris) **57**, 3–97, and *J. l'école polytech.* (Paris) **58**, 1–125. English translation of Sections 90–92, 110–112, 123–127 of 1889 article: Beyer (1984), 77–89.

Keck, W., and Beyer, R. T. (1960). Frequency spectrum of finite amplitude ultrasonic waves in liquids. *Phys. Fluids* **3**, 346–352.

Khokhlov, R. V., Naugol'nykh, K. A., and Soluyan, S. I. (1964). Waves of moderate amplitudes in absorbing media. *Acustica* **14**, 248–253.

Lagerstrom, P. A., Cole, J. D., and Trilling, L. (1949, 1950). *Problems in the Theory of Viscous Compressible Fluids* [Guggenheim Aeronautical Laboratory, California Institute of Technology (GALCIT)]. Prepared under the Office of Naval Research Contract N6-ONR-244—Task Order VIII. (Various U.S. university libraries have cataloged copies; it may also be available from the National Technological Information Service with reference to the report number ATI-55958.) The 1950 edition is identical to the first edition except for the addition of material appended at the end and was distributed by the Durand Reprinting Committee, California Institute of Technology.

Lambrakis, K. C., and Thompson, P. A. (1972). Existence of real fluids with a negative fundamental derivative Γ. *Phys. of Fluids* **15**, 933–935.

Landau, L. (1945). On shock waves at large distances from the place of their origin. *J. Phys. U.S.S.R.* **9**, 496–500.

Landau, L. D., and Lifshitz, E. M. (1987). *Fluid Mechanics*, 2nd ed. (Pergamon Press, New York).

Lighthill, M. J. (1950). The energy distribution behind decaying shocks.—I. Plane waves. *Phil. Mag.*, Ser. 7, **41**, 1101–1128.

Lighthill, M. J. (1956). Viscosity effects in sound waves of finite amplitude. In *Surveys in Mechanics*, G. K. Batchelor and R. M. Davies, eds. (Cambridge University Press, Cambridge, England), pp. 250–351.

Lindsay, R. B. (1974). *Physical Acoustics* (Hutchinson & Ross, Stroudsburg, Pa.).

Mendousse, J. S. (1953). Nonlinear dissipative distortion of progressive sound waves at moderate amplitudes. *J. Acoust. Soc. Am.* **25**, 51–54.

Moffett, M. B., Konrad, W. L., and Carlton, L. F. (1978). Experimental demonstration of the absorption of sound by sound in water. *J. Acoust. Soc. Am.* **63**, 1048–1051.

Morfey, C. L. (1993). Shock interactions as a source of backward-travelling acoustic waves. In *Advances in Nonlinear Acoustics: 13th ISNA*, H. Hobæk, ed. (World Scientific, London), pp. 167–172.

Morfey, C. L., and Sparrow, V. W. (1993). Plane compression front steepening in nonlinear media forms both a shock and a reflected wave. *J. Acoust. Soc. Am.* **93**, 3085–3088.

Naugol'nykh, K. A. (1958). On the absorption of sound waves of finite amplitude. *Sov. Phys. Acoust.* **4**, 115–124.

Naugol'nykh, K. A., Soluyan, S. I., and Khokhlov, R. V. (1963). Spherical waves of finite amplitude in a viscous thermally conducting medium. *Sov. Phys. Acoust.* **9**, 42–46.

Novikov, B. K., and Rudenko, O. V. (1976). Degenerate parametric amplification of sound. *Sov. Phys. Acoust.* **22**, 258–259.

Pierce, A. D. (1989). *Acoustics* (Acoustical Society of America, New York).

Poisson, S. D. (1808). Mémoire sur la théorie du son. *J. l'école polytech.* (Paris) **7**, 319–392. See Secs. 24–25, pp. 365–370. English translation of Sections 23–25: Beyer (1984), 23–28.

Rankine, W. J. M. (1870). On the thermodynamic theory of waves of finite longitudinal disturbance. *Phil. Trans. Roy. Soc.* **160**, 277–288, or Beyer (1984), 65–76.

Rayleigh, Lord (1910). Aerial plane waves of finite amplitude. *Proc. Roy. Soc.* **A 84**, 247–284, or *Scientific Papers of Lord Rayleigh* **V** (Dover, New York, 1964), or Lindsay (1974), 135–173.

Rogers, P. H. (1977). Weak-shock solution for underwater explosive shock waves. *J. Acoust. Soc. Am.* **62**, 1412–1419.

Rudenko, O. V., and Soluyan, S. I. (1977). *Theoretical Foundations of Nonlinear Acoustics* (Plenum, New York).

Rudnick, I. (1953). On the attenuation of a repeated sawtooth shock wave. *J. Acoust. Soc. Am.* **25**, 1012–1013.

Rudnick, I. (1958). On the attenuation of finite amplitude waves in a liquid. *J. Acoust. Soc. Am.* **30**, 564–567.

Schaffer, M. E. (1975). The suppression of sound with sound. Technical Report ARL-TR-75-64 (Applied Research Laboratories, The University of Texas at Austin, AD 778 868).

Shooter, J. A., Muir, T. G., and Blackstock, D. T. (1974). Acoustic saturation of spherical waves in water. *J. Acoust. Soc. Am.* **55**, 54–62.

Soluyan, S. I., and Khokhlov, R. V. (1961). Propagation of acoustic waves of finite amplitude in a dissipative medium. *Vestn. Mosk. Univ.* (Series III), *Fiz. Astron.* **3**, 52–61, or Beyer (1984), 193–206.

Stokes, G. G. (1848). On a difficulty in the theory of sound. *Phil. Mag.* (Series 3) **33**, 349–356, or Beyer (1984), 29–36.

Taylor, G. I. (1910). The conditions necessary for discontinuous motion in gases. *Proc. Roy. Soc.* **A 84**, 371–377.

Thompson, P. A. (1971). A fundamental derivative in gas dynamics. *Phys. of Fluids* **14**, 1843–1849.

Thompson, P. A., Carofano, G. C., and Kim, Y.-G. (1986). Shock waves and phase changes in a large-heat-capacity fluid emerging from a tube. *J. Fluid Mech.* **166**, 57–92.

Thompson, P. A., and Lambrakis, K. (1973). Negative shock waves. *J. Fluid Mech.* **60**, 187–208.

Thuras, A. L., Jenkins, R. T., and O'Neil, H. T. (1935). Extraneous frequencies generated in air carrying intense sound waves. *J. Acoust. Soc. Am.* **6**, 173–180.

Watson, G. N. (1944). *A Treatise on the Theory of Bessel Functions*, 2nd ed. (Cambridge University Press, Cambridge, England).

Webster, D. A., and Blackstock, D. T. (1977). Finite amplitude saturation of plane sound waves in air. *J. Acoust. Soc. Am.* **62**, 518–523.

Webster, D. A., and Blackstock, D. T. (1978). Asymptotic decay of periodic spherical waves in dissipative media. *J. Acoust. Soc. Am.* **64**, 533.

Whitham, G. D. (1952). The flow pattern of a supersonic projectile. *Commun. Pure Appl. Math.* **5**, 301–348.

Chapter 5
Dispersion

Mark F. Hamilton, Yurii A. Ilinskii, and Evgenia A. Zabolotskaya

Contents

5.1 Introduction

The dependence of phase speed on frequency is called dispersion. Dispersion in unbounded, homogeneous acoustical media is a weak effect that can often be neglected, in contrast to the strong dispersion of light waves in most optical media. For example, although dispersion accompanies the attenuation of sound due to viscosity and heat conduction [dispersion and attenuation are connected by the Kramers–Kronig relations; see, for example, Temkin (1990)], the dispersion is negligible and can normally be ignored. The propagation of finite amplitude sound in thermoviscous fluids is then described accurately by the Burgers equation. Dispersion due to molecular relaxation in air and seawater, however, alters wave profiles and postpones shock formation, but the effect is sufficiently weak that it can be taken into account by models similar to the Burgers equation.

    Section 5.2 is devoted to weak dispersion, which is analyzed with techniques similar to those described in Chap. 4 for finite amplitude sound in nondispersive media. Two examples of weak dispersion mechanisms are considered, relaxation in unbounded media and boundary layers in ducts. Section 5.3 focuses on two examples of acoustical media with strong dispersion, waveguides and bubbly

M. F. Hamilton · Yu. A. Ilinskii · E. A. Zabolotskaya
Department of Mechanical Engineering, The University of Texas at Austin, Austin, TX, USA

© The Author(s) 2024   149
M. F. Hamilton, D. T. Blackstock (eds.), *Nonlinear Acoustics*,
https://doi.org/10.1007/978-3-031-58963-8_5

liquids, the analyses of which require special methods. Propagation in horns, which is strongly dispersive at frequencies near cutoff and weakly dispersive at high frequencies, is not considered in this chapter. For more general discussion of nonlinear waves in dispersive media, the reader is referred to the textbook by Whitham (1974).

5.2   Weak Dispersion

When dispersion is sufficiently weak that the variation in phase speed as a function of frequency is a small percentage correction to a reference sound speed (say c_0), variations in the waveform, as a function of distance, are slow on the scale of a wavelength. It is then appropriate to consider the evolution of the waveform, due to both dispersion and nonlinearity, in a retarded time frame $\tau = t - x/c_0$. The same approach is followed when the combined effects of absorption and nonlinearity are taken into account by the Burgers equation [Eq. (3.54)]. We thus begin with the following generalized Burgers equation for plane waves (Blackstock, 1985):

$$\frac{\partial p}{\partial x} + L_\tau(p) = \frac{\beta p}{\rho_0 c_0^3} \frac{\partial p}{\partial \tau}, \tag{5.1}$$

where $p(x, \tau)$ is the sound pressure, ρ_0 is the ambient density, c_0 is the small-signal sound speed in some reference state, and β is the coefficient of nonlinearity [Eqs. (2.26)]. The linear, retarded time operator $L_\tau(p)$ describes the attenuation and dispersion properties of the medium. Analytic expressions for L_τ are known for several cases, for example, for attenuation and dispersion due to propagation in relaxing media or due to boundary-layer effects in ducts. These two examples are considered in Sects. 5.2.1 and 5.2.2, respectively. Geometrical effects due to spherical or cylindrical spreading, or due to diffraction in directional beams, can be taken into account by including the appropriate linear terms in Eq. (5.1) (Sects. 3.8 and 3.9; see also Cleveland et al., 1996). In the present section, attention is devoted only to plane-wave propagation.

It is often convenient to express the operator $L_\tau(p)$ in the frequency domain. First expand the pressure in the Fourier series

$$p(x, \tau) = \frac{1}{2} \sum_{n=1}^{\infty} p_n(x) e^{j\omega_n \tau} + \text{c.c.}, \tag{5.2}$$

where $\omega_n = n\omega_0$, ω_0 is a reference frequency, and c.c. designates complex conjugates of the preceding terms. Since L_τ is a retarded time operator, its effect in Eq. (5.2) is to introduce complex multiplicative constants in the summation. We label these constants $\tilde{\alpha}_n$ and define them through the following operation (Blackstock, 1985):

$$L_\tau(e^{j\omega_n\tau}) = \tilde{\alpha}_n e^{j\omega_n\tau}. \tag{5.3}$$

Substitution of Eq. (5.2) into Eq. (5.1) thus yields the following coupled equations for the complex spectral amplitudes p_n (see, for example, Korpel, 1980; also, Sect. 11.2.1):

$$\frac{dp_n}{dx} + \tilde{\alpha}_n p_n = \frac{j\omega_n\beta}{4\rho_0 c_0^3} \left(\sum_{m=1}^{n-1} p_m p_{n-m} + 2 \sum_{m=n+1}^{\infty} p_m p_{m-n}^* \right), \tag{5.4}$$

where p_n^* is the complex conjugate of p_n. The real part of the complex coefficient $\tilde{\alpha}_n$ is associated with attenuation, the imaginary part with variation in phase speed:

$$\tilde{\alpha}_n = \alpha_n + j\omega_n(c_n^{-1} - c_0^{-1}), \tag{5.5}$$

where α_n is the attenuation coefficient and c_n the phase speed at frequency ω_n. This interpretation follows from recognition that Eqs. (5.2) and (5.4) yield linear solutions of the form $\exp(-\tilde{\alpha}_n x + j\omega_n\tau) = \exp[-\alpha_n x + j\omega_n(t - x/c_n)]$.

The first summation in Eq. (5.4) accounts for sum-frequency generation, the second for difference-frequency generation. The coupled equations can be solved either analytically by successive approximations or perturbation methods (e.g., to investigate second-harmonic generation; see Sect. 10.2), or numerically by finite-difference methods (e.g., to study waveform distortion and shock formation; see Sect. 11.2.1). For numerical integration with a finite number of harmonics ($1 \le n \le N$), replace ∞ by N in Eqs. (5.2) and (5.4).

5.2.1 Relaxation

Relaxation refers to the finite time required for a medium to establish equilibrium in a new thermodynamic state produced by a change in one or more of the state variables (Landau and Lifshitz, 1987). Several internal processes may contribute to the time required to establish equilibrium—for example, chemical reaction, phase transition, and molecular vibration. In air it is mainly the vibration of oxygen and nitrogen molecules, and in seawater it is the dissociation of boric acid and magnesium sulfate molecules. Relaxation is accompanied by energy dissipation, and because the behavior of a given internal process depends on the relation of its *relaxation time* to the frequency of an acoustical disturbance, relaxation also introduces dispersion.

The operator for attenuation and dispersion in a thermoviscous fluid with multiple independent relaxation mechanisms is obtained by combining Eqs. (11-6.3b) and (11-6.5) of Pierce (1989):

$$L_\tau(p) = -\frac{\delta}{2c_0^3}\frac{\partial^2 p}{\partial \tau^2} - \sum_i \frac{c_i'}{c_0^2}\int_{-\infty}^{\tau}\frac{\partial^2 p}{\partial \tau'^2}e^{-(\tau-\tau')/t_i}\,d\tau', \qquad (5.6)$$

where δ is the sound diffusivity for the thermoviscous fluid [Eq. (3.42)], t_i is the relaxation time for the ith relaxation process ($1/2\pi t_i$ is the *relaxation frequency*), and c_i' is the corresponding net increase in phase speed as frequency varies from zero to infinity. Equation (5.6) is valid to lowest order in the diffusivity δ and the phase speed increments c_i'.

Interpretation of L_τ is best revealed by combining Eqs. (5.3) and (5.6) to obtain

$$\tilde{\alpha}_n = \frac{\delta\omega_n^2}{2c_0^3} + \frac{\omega_n^2}{c_0^2}\sum_i\frac{c_i't_i}{1+j\omega_n t_i}. \qquad (5.7)$$

Substitution of Eq. (5.7) into Eq. (5.5) yields, to first order in c_i',

$$\alpha_n = \frac{\delta\omega_n^2}{2c_0^3} + \frac{\omega_n}{c_0^2}\sum_i\frac{c_i'\omega_n t_i}{1+(\omega_n t_i)^2}, \qquad c_n = c_0 + \sum_i\frac{c_i'(\omega_n t_i)^2}{1+(\omega_n t_i)^2}. \qquad (5.8)$$

The phase speed c_n increases monotonically with frequency from the *equilibrium sound speed* c_0 at $\omega_n = 0$ to the *frozen sound speed* $c_\infty = c_0 + \sum c_i'$ at $\omega_n = \infty$. At acoustical frequencies that are low in comparison with each of the relaxation frequencies ($\omega_n^2 \ll 1/t_i^2$ for all i), the attenuation coefficient reduces to $\alpha_n = (\delta + 2c_0\sum c_i't_i)\omega_n^2/2c_0^3$, which exhibits the properties of thermoviscous attenuation with a diffusivity that is increased by $\delta' = 2c_0\sum c_i't_i$. At high frequencies, viscosity and heat conduction are the dominant dissipation mechanisms, and the attenuation coefficient reduces in the limit $\omega_n \to \infty$ to the classical thermoviscous attenuation coefficient $\alpha_n^{tv} = \delta\omega_n^2/2c_0^3$.

To investigate the combined effects of relaxation and nonlinearity, it is helpful to ignore thermoviscous dissipation and to consider only one relaxation process, $i = r$. Thus, let $\delta = 0$, $t_i = t_r$, and $c_i' = c' = c_\infty - c_0$. It is common practice to characterize the dispersion of a monorelaxing fluid by the parameter $m = (c_\infty^2 - c_0^2)/c_0^2$, which may be expressed as $m = 2c'/c_0$ to lowest order in c'. These modifications of Eqs. (5.7) and (5.8) yield

$$\tilde{\alpha}_n = \frac{m\omega_n^2 t_r}{2c_0(1+j\omega_n t_r)}, \qquad \alpha_n = \frac{m\omega_n^2 t_r}{2c_0(1+\omega_n^2 t_r^2)}, \qquad c_n = c_0 + \frac{mc_0\omega_n^2 t_r^2}{2(1+\omega_n^2 t_r^2)}. \qquad (5.9)$$

The frequency dependences of α_1 and c_1 are shown in Fig. 5.1.

Beginning with a more common form of the evolution equation for propagation in a monorelaxing fluid, rather than Eqs. (5.1) and (5.6), we shall derive a solution for stationary waves (Polyakova et al., 1962; see also Rudenko and Soluyan, 1977). The alternative form is obtained by substituting $\tilde{\alpha}_n$ from Eq. (5.9) into Eq. (5.4),

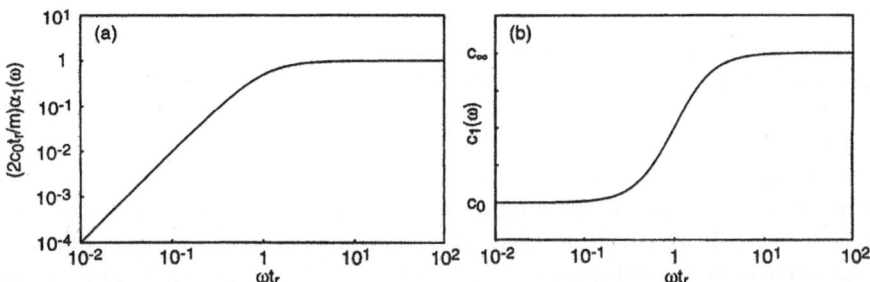

Fig. 5.1 Frequency dependence of the (a) attenuation coefficient and (b) phase speed for a monorelaxing fluid.

multiplying the resulting equation by $(1 + j\omega_n t_r)$, and transforming the result into the time domain:

$$\left(1 + t_r \frac{\partial}{\partial \tau}\right)\left(\frac{\partial p}{\partial x} - \frac{\beta p}{\rho_0 c_0^3}\frac{\partial p}{\partial \tau}\right) = \frac{m t_r}{2c_0}\frac{\partial^2 p}{\partial \tau^2}. \tag{5.10}$$

A stationary wave is one for which $\partial p / \partial x = 0$ in the retarded time frame τ, i.e., the waveform propagates without change in shape. With $\partial p / \partial x = 0$ and following one integration with respect to τ, Eq. (5.10) becomes

$$\left(p + \frac{m\rho_0 c_0^2}{2\beta}\right)\frac{dp}{d\tau} + \frac{p^2}{2t_r} = \text{const.} \tag{5.11}$$

Evaluation of the integration constant requires additional information about the waveform. We consider a pressure jump $2p_0$, from $-p_0$ ahead of the wave front to p_0 behind. If the fluid is quiet both far ahead of and far behind the wavefront, the boundary conditions become $p = -p_0$ and $dp/d\tau = 0$ at $\tau = -\infty$, and $p = p_0$ and $dp/d\tau = 0$ at $\tau = \infty$. These conditions combine to yield $p_0^2/2t_r$ for the integration constant. Applying these conditions to Eq. (5.11), separating variables, and integrating (with the remaining integration constant determined by requiring $p = 0$ at $\tau = 0$) yield

$$\tau = t_r \ln \frac{(1 + p/p_0)^{D-1}}{(1 - p/p_0)^{D+1}}, \tag{5.12}$$

where $D = m\rho_0 c_0^2/2\beta p_0$ measures the ratio of relaxation effects to nonlinear effects. In the limit of weak nonlinearity, Eq. (5.12) can be inverted to obtain an explicit dependence of pressure on time:

$$p = p_0 \tanh(\tau/2Dt_r), \quad D \gg 1. \tag{5.13}$$

The characteristic rise time of the pressure jump in this limit is given by $2Dt_r = m\rho_0c_0^2t_r/\beta p_0$, which is the same result that is obtained if the diffusivity δ in the corresponding stationary wave solution for thermoviscous fluids [Eq. (4.247)] is replaced by $\delta' = mc_0^2t_r$. In general, however, Eq. (5.12) cannot be inverted analytically.

Equation (5.12) describes a multivalued waveform for $D < 1$, and weak shock theory is needed to correct the solution (Sect. 4.4.1). The small-signal sound speed immediately behind and ahead of the shock, modeled as an instantaneous pressure jump, is the frozen sound speed $c_\infty = (1+m/2)c_0$. Thus, the shock must propagate at speed $dx_{sh}/dt = (1+m/2)c_0 + \beta(p_{sh} - p_0)/2\rho_0c_0$, where $x_{sh}(t)$ is the location of the shock, p_{sh} is the peak shock pressure, i.e., the value of p immediately behind the shock, and $-p_0$ is the value immediately ahead of the disturbance. For the waveform to be stationary in the retarded time frame $\tau = t - x/c_0$, we must also have $dx_{sh}/dt = c_0$. Equating the two expressions for the shock speed yields (Pierce, 1989)

$$p_{sh} = (1 - 2D)p_0, \quad 0 \le D \le 1. \tag{5.14}$$

To obtain the retarded arrival time of the shock, τ_{sh}, set $p = p_{sh}$ in Eq. (5.12):

$$\tau_{sh} = -t_r \ln[4D^{1+D}(1 - D)^{1-D}], \quad 0 \le D \le 1. \tag{5.15}$$

The wave profile is thus constructed for $0 \le D \le 1$ by setting $p = -p_0$ for $\tau \le \tau_{sh}$, and

$$\tau = \tau_{sh}, \qquad\qquad -p_0 \le p \le p_{sh}, \tag{5.16}$$

$$= t_r \ln \frac{(1 + p/p_0)^{D-1}}{(1 - p/p_0)^{D+1}}, \quad p_{sh} \le p \le p_0. \tag{5.17}$$

Calculated waveforms are shown in Fig. 5.2 for three values of D. For $D = 2$, there is no shock in the waveform, and the solution shown, based on Eq. (5.12), is similar

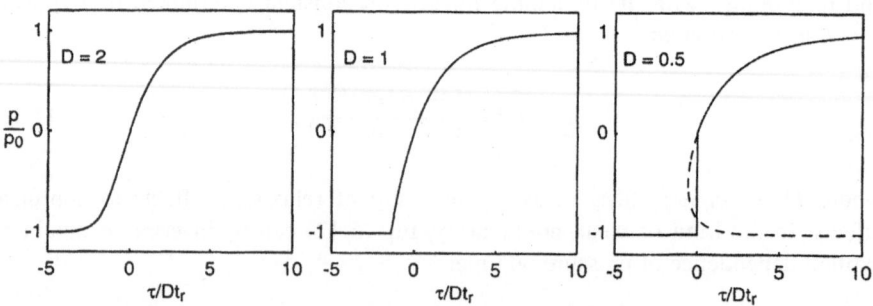

Fig. 5.2 Waveforms predicted for a stationary wave in a monorelaxing fluid.

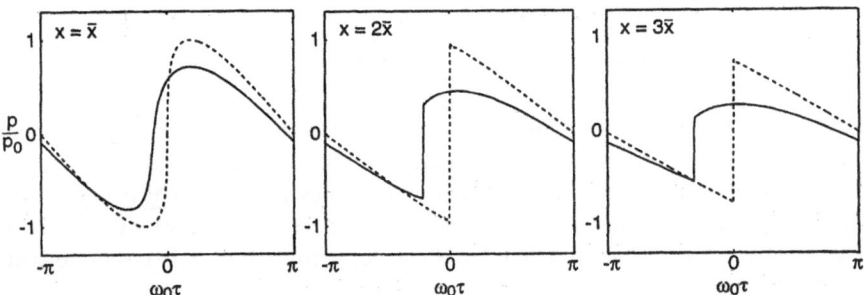

Fig. 5.3 Initially sinusoidal waveforms computed at increasing distances in a monorelaxing, thermoviscous fluid (solid curves), compared with waveforms in a purely thermoviscous fluid (no relaxation, dashed curves).

to the asymptotic result given by Eq. (5.13). For $D = 1$, the shock threshold, the solution is obtained from Eqs. (5.14)–(5.17), and the waveform has a discontinuity in slope at $\tau_{sh}/t_r = -\ln 4$, but not in amplitude ($p_{sh}/p_0 = -1$). A multivalued waveform is predicted by Eq. (5.12) for $D = 0.5$ (dashed curve); the corrected solution based on Eqs. (5.14)–(5.17) is given by the solid curve ($\tau_{sh} = p_{sh} = 0$).

The evolution of an initially sinusoidal waveform, $p = p_0 \sin \omega_0 t$ at $x = 0$, can be calculated numerically with Eq. (5.4) (see Sect. 11.2.1). Results are presented in Fig. 5.3 at various distances in a monorelaxing, thermoviscous fluid with $D = 0.5$, $\omega_0 t_r = 1$, and $\alpha_1^{tv}\bar{x} = 10^{-3}$, where $\bar{x} = \rho_0 c_0^3/\beta\omega_0 p_0$ is the plane-wave shock formation distance in an ideal fluid. Thermoviscous absorption is required for numerical stability when shock formation occurs. For comparison, results for $D = 0$ (no relaxation) are shown as dashed curves. The waveforms at $x = \bar{x}$ reveal that relaxation postpones shock formation, introduces asymmetry and attenuation, and causes the waveform to propagate faster than phase speed c_0. A shock exists in the waveform at $x = 2\bar{x}$, and the asymmetry is more pronounced. The rounded wave profiles immediately following the shocks at $x = 2\bar{x}$ and $x = 3\bar{x}$ are characteristic of the effect of relaxation.

Our analysis concludes with an asymptotic model for high-frequency ($\omega_0 t_i \gg 1$ for all i) propagation in a multirelaxing fluid, which generalizes the results obtained by Soluyan and Khokhlov (1962) for a monorelaxing fluid (see also Rudenko and Soluyan, 1977). To construct the evolution equation for this case, expand $\tilde{\alpha}_n$ in powers of the small quantities $(\omega_n t_i)^{-1}$ to obtain at leading order, with $\delta = 0$,

$$\tilde{\alpha}_n \simeq \frac{1}{c_0^2} \sum_i \left(\frac{1}{t_i} - j\omega_n\right) c_i', \quad \omega_0 t_i \gg 1. \tag{5.18}$$

Substitution into Eq. (5.4) and transformation back into the time domain yields

$$\frac{\partial p}{\partial x} + \frac{p}{c_0^2} \sum_i \frac{c_i'}{t_i} - \frac{1}{c_0^2} \frac{\partial p}{\partial \tau} \sum_i c_i' = \frac{\beta p}{\rho_0 c_0^3} \frac{\partial p}{\partial \tau}. \tag{5.19}$$

The third term on the left-hand side accounts for the increase in phase speed from c_0 to c_∞. This term can be eliminated by introducing the new retarded time $\tau' = t - x/c_\infty$, which to lowest order in c_i' becomes $\tau' = \tau + (\sum c_i')x/c_0^2$. Next, introducing the quantities $q = p \exp(x/l_r)$ and $z = l_r[1 - \exp(-x/l_r)]$, where $l_r = c_0^2(\sum c_i' t_i^{-1})^{-1}$ is a relaxation length scale, permits Eq. (5.19) to be rewritten as

$$\frac{\partial q}{\partial z} = \frac{\beta q}{\rho_0 c_0^3} \frac{\partial q}{\partial \tau'}. \tag{5.20}$$

Equation (5.20) is in the familiar form associated with the propagation of finite-amplitude sound in an ideal fluid [recall Eq. (5.1)], and it can be solved using the methods described in Chap. 4. For $x \ll l_r$, we have $q \simeq p$ and $z \simeq x$, in which case the effect of relaxation is negligible.

Note that $z \to l_r$ in the limit $x \to \infty$, and therefore waveform distortion is frozen at the stage corresponding to the distortion of sound at distance $x = l_r$ in an ideal fluid. The shock formation distance for a wave with amplitude p_0 and angular frequency ω_0 at the source is thus $x = \bar{x}_r$, where $z(x \equiv \bar{x}_r) = \bar{x}$:

$$\bar{x}_r = l_r \ln \left(\frac{1}{1 - R} \right). \tag{5.21}$$

Here $R = \bar{x}/l_r = (\rho_0 c_0/\beta p_0 \omega_0) \sum c_i' t_i^{-1}$ measures the ratio of relaxation effects to nonlinear effects, and for a monorelaxing fluid, with $l_r = 2 c_0 t_r/m$, we have $R = D/\omega_0 t_r$. Shock formation occurs only for $R < 1$, and for $R \ll 1$ Eq. (5.21) reduces to $\bar{x}_r = (1 + R/2)\bar{x}$.

5.2.2 Boundary Layers

A plane wave that propagates in a duct with rigid walls experiences dispersion and attenuation as a result of the thermoviscous boundary layer along the walls. Provided that the boundary layer is thin in comparison with the transverse dimension of the duct, the dispersion operator is (Blackstock, 1985)

$$L_\tau(p) = -\frac{\delta}{2 c_0^3} \frac{\partial^2 p}{\partial \tau^2} + b \sqrt{\frac{2}{\pi}} \int_0^\infty \frac{\partial p(x, \tau - \tau')}{\partial \tau} \frac{d\tau'}{\sqrt{\tau'}}, \tag{5.22}$$

where the first term accounts for free-stream thermoviscous losses, the boundary-layer parameter is $b = (C/4A)\sqrt{2\nu/c_0^2}[1 + (\gamma - 1)/\sqrt{Pr}]$, C is the perimeter of the duct and A its cross sectional area, ν is the kinematic viscosity, γ the ratio of specific heats, and Pr the Prandtl number. Combining Eqs. (5.3), (5.5), and (5.22) yields

$$\tilde{\alpha}_n = \frac{\delta\omega_n^2}{2c_0^3} + (1+j)b\sqrt{\omega_n}, \quad \alpha_n = \frac{\delta\omega_n^2}{2c_0^3} + b\sqrt{\omega_n}, \quad c_n = c_0 - \frac{bc_0^2}{\sqrt{\omega_n}}. \quad (5.23)$$

The restriction that the boundary layer be thin corresponds to the frequency restriction $\sqrt{\omega_n} \gg bc_0$, which was used to obtain the above relation for the phase speed. Note that in contrast to relaxation dispersion, tube wall dispersion produces phase speeds that are lower than the free-space phase speed c_0. For both tube wall and relaxation dispersion, however, the phase speed increases with frequency, and the classical thermoviscous attenuation coefficient $\alpha_n^{tv} = \delta\omega_n^2/2c_0^3$ is recovered at high frequencies.

Analytic solutions of Eq. (5.1) with L_τ given by Eq. (5.22), even without the free-stream losses, are not available. We therefore resort to numerical solutions of Eq. (5.4) (Sect. 11.2.1). The parameter $\tilde{\alpha}_n$ in Eqs. (5.23) was substituted into Eq. (5.4), with $(b\sqrt{\omega_0})\bar{x} = 0.5$ and $\alpha_1^{tv}\bar{x} = 10^{-3}$, to produce the waveforms in Fig. 5.4 (at the same distances as in Fig. 5.3) for the source condition $p = p_0 \sin\omega_0 t$ at $x = 0$. As with relaxation, thermoviscous absorption is required for numerical stability. Boundary-layer dispersion causes rounding of the positive portions and cusping of the negative portions of the waveforms (Coppens, 1971). The waveforms in Fig. 5.4 are typical of those reported in the literature for experiments performed with high-intensity sound in ducts (Webster and Blackstock, 1977).

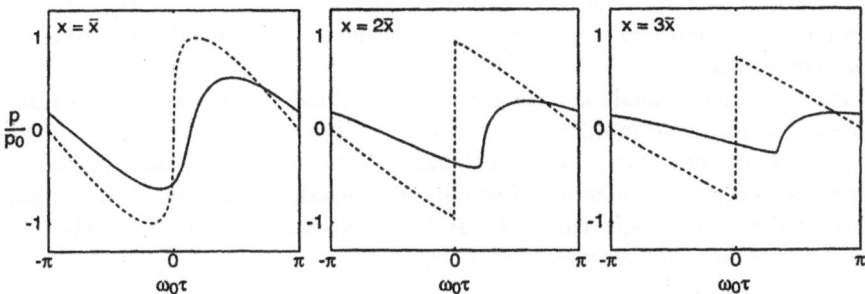

Fig. 5.4 Initially sinusoidal waveforms computed at increasing distances in a tube containing a thermoviscous fluid (solid curves), compared with waveforms in the absence of boundary-layer effects (dashed curves).

5.3   Strong Dispersion

When dispersion is strong in relation to nonlinearity, the large variations in phase speeds prevent efficient energy transfer among the interacting harmonic components. Shock formation is unlikely, and most of the energy tends to remain in the frequency components radiated by the source. Dispersion is often strong in waveguides and bubbly liquids, which are the subjects of Sects. 5.3.1 and 5.3.2, respectively.

5.3.1   Waveguides

Waveguides are formed in many ways. Perhaps the most common in acoustics are air-filled ducts with rigid walls, which are often used in laboratory experiments. Natural waveguides are formed by layered media consisting of fluids with different acoustical properties (e.g., shallow water channels bounded above by air and below by sediment). Waveguides are also formed by continuous variations in the properties of inhomogeneous media (e.g., underwater sound channels created by local sound speed minima).

We begin with the following second-order wave equation for a lossless, homogeneous fluid [Eq. (3.41) with $\delta = 0$]:

$$\nabla^2 p - \frac{1}{c_0^2} \frac{\partial^2 p}{\partial t^2} = -\frac{\beta}{\rho_0 c_0^4} \frac{\partial^2 p^2}{\partial t^2} - \left(\nabla^2 + \frac{1}{c_0^2} \frac{\partial^2}{\partial t^2} \right) \mathcal{L}, \qquad (5.24)$$

where $\mathcal{L} = \rho_0 u^2 / 2 - p^2 / 2\rho_0 c_0^2$ is the Lagrangian density, and u is the particle velocity. Equation (5.24) does not, by itself, account for dispersion. The dispersion relations are established by solving the eigenvalue problem for the normal modes, as discussed below.

Each waveguide mode can be decomposed into an angular spectrum of plane waves that propagate in different directions. Insight into the effect of waveguide dispersion on harmonic generation therefore follows from an understanding of harmonic generation due to the noncollinear interaction of just two of these plane waves (Naze Tjøtta and Tjøtta, 1987; Hamilton and Blackstock, 1988). Thus, consider the following primary wave field:

$$p_1 = p_a \sin(\omega_a t - \mathbf{k}_a \cdot \mathbf{r}) + p_b \sin(\omega_b t - \mathbf{k}_b \cdot \mathbf{r}), \qquad (5.25)$$

where $\mathbf{k}_a$ and $\mathbf{k}_b$ are wave vectors associated with plane waves at frequencies ω_a and ω_b, respectively, and $\mathbf{r} = (x, y, z)$. The angle θ between the directions of propagation is defined by $\cos\theta = (\mathbf{k}_a \cdot \mathbf{k}_b)/k_a k_b$, where $k_i = |\mathbf{k}_i| = \omega_i/c_0$ ($i = a, b$). Substitution of Eq. (5.25) into the right-hand side of Eq. (5.24), and making use of first-order relations to simplify second-order terms, yields the following inhomogeneous wave equation for the sum- or difference-frequency pressure $p_\pm$:

$$\nabla^2 p_\pm - \frac{1}{c_0^2}\frac{\partial^2 p_\pm}{\partial t^2} = \mp \beta_\pm(\theta)\frac{\omega_\pm^2 p_a p_b}{\rho_0 c_0^4}\cos[\omega_\pm t - (\mathbf{k}_a \pm \mathbf{k}_b)\cdot\mathbf{r}], \qquad (5.26)$$

where $\beta_\pm(\theta) = \beta - 2\sin^2(\theta/2) \pm 4(\omega_a\omega_b/\omega_\pm^2)\sin^4(\theta/2)$ is a modified coefficient of nonlinearity that depends on the interaction angle θ, $\beta = 1 + B/2A$, B/A is the ratio of coefficients in the equation of state (Chap. 2), and $\omega_\pm = \omega_a \pm \omega_b$ ($\omega_a > \omega_b$ is assumed). For $\omega_a \gg \omega_b$, the coefficient of nonlinearity reduces to $\beta_\pm(\theta) = \cos\theta + B/2A$, a form that lends itself to simple physical interpretation. The term $B/2A$ is a scalar effect due to nonlinearity in the pressure–density relation, whereas $\cos\theta$ is a vector effect that accounts for convection of one primary wave by the other.

The general behavior of the solution $p_\pm$ becomes evident following two observations. First, for $\theta \neq 0$, the particular solution of Eq. (5.26) (i.e., the *forced wave*) is proportional to the term on the right-hand side. The forced wave propagates with phase speed $\omega_\pm/|\mathbf{k}_a \pm \mathbf{k}_b|$ in the direction $\mathbf{k}_a \pm \mathbf{k}_b$, where $|\mathbf{k}_a \pm \mathbf{k}_b| = [k_\pm^2 \mp 4k_a k_b \sin^2(\theta/2)]^{1/2}$ and $k_\pm = k_a \pm k_b$. To the particular solution must be added a homogeneous solution (a *free wave*) to satisfy the source condition. For a planar source at which $p_\pm = 0$, the free wave propagates in the same direction as the forced wave, but with wave number $k_\pm$ and therefore phase speed $\omega_\pm/k_\pm$. When $\theta \neq 0$, the propagation speeds of the two waves differ, which causes the amplitude of $p_\pm$ to beat with spatial period $2\pi/||\mathbf{k}_a \pm \mathbf{k}_b| - k_\pm|$. The mechanical analogue of this phenomenon is the response of a mass-spring oscillator, which beats in time when driven at a frequency that does not coincide with its natural frequency. Increasing the interaction angle θ decreases the spatial periodicity, and the combination frequency generation becomes less efficient. For collinear interaction ($\theta = 0$ and thus $|\mathbf{k}_a \pm \mathbf{k}_b| = k_\pm$), there is synchronism between the forced and free waves, and the amplitude of the sum- and difference-frequency sound increases linearly with distance from the source. The beating of forced and free waves is a hallmark of dispersive nonlinear interactions in waveguides.

The second observation is that the amplitude of $p_\pm$ depends also on interaction angle in proportion to $\beta_\pm(\theta)$. However, the difference between $\beta_\pm(\theta)$ and β is significant only for large θ, in which case the interaction is very inefficient because of the phase mismatch discussed above. If we ignore the term containing $\mathcal{L}$ in Eq. (5.24), the corresponding form of Eq. (5.26) is identical except that $\beta_\pm(\theta)$ is replaced by β. Consequently, if the mode structure in a waveguide is such that significant harmonic interactions occur at small angles, the dominant nonlinear effects are characterized adequately by the Westervelt equation [Eq. (3.46) with $\delta = 0$]:

$$\nabla^2 p - \frac{1}{c_0^2}\frac{\partial^2 p}{\partial t^2} = -\frac{\beta}{\rho_0 c_0^4}\frac{\partial^2 p^2}{\partial t^2}. \qquad (5.27)$$

A sufficient condition for the validity of Eq. (5.27) is radiation from a source at frequencies well above cutoff for the modes excited by the source. Equation (5.27) is the starting point for the following analysis.

We begin the analysis of nonlinear interactions in waveguides by first reviewing briefly the linear solution of Eq. (5.27), i.e., with the right-hand side set to zero [see Morse and Ingard (1986) for a thorough discussion of linear waveguide theory]. The linear solution for a wave of frequency ω that propagates in a single mode may be written

$$p = \tfrac{1}{2} p_m \phi_m(y, z, \omega) e^{j(\omega t - k_m x)} + \text{c.c.}, \tag{5.28}$$

where p_m is the corresponding pressure amplitude, m is the mode number, and x is distance along the axis of the waveguide. The eigenfunctions $\phi_m(y, z, \omega)$ are solutions of

$$\left(\frac{\partial^2}{\partial y^2} + \frac{\partial^2}{\partial z^2} + \frac{\omega^2}{c_0^2} \right) \phi_m = k_m^2 \phi_m, \tag{5.29}$$

where the axial wave numbers k_m are eigenvalues determined by the boundary conditions along the walls of the waveguide. We assume that the walls are locally reacting, so that the boundary condition along the perimeter of the waveguide can be expressed in the form

$$-\frac{j \omega \rho_0 \phi_m}{\partial \phi_m / \partial n} = R_n(\omega) + j X_n(\omega), \tag{5.30}$$

where R_n is the real and X_n the imaginary part of the specific acoustic wall impedance, and n is the normal coordinate directed into the wall. For $R_n > 0$, the walls introduce losses, and the wave numbers k_m are complex. The phase speed $c_m(\omega)$ and attenuation coefficient $\alpha_m(\omega)$ are obtained from the real and imaginary parts of the axial wave number, respectively: $k_m = \omega/c_m - j\alpha_m$. For $R_n = 0$ the values of k_m^2 are real and decrease as m increases; wave propagation is lossless above the cutoff frequency defined by $k_m = 0$, and there is no wave propagation below cutoff. Eigenfunctions that satisfy Eqs. (5.29) and (5.30) are orthogonal:

$$\iint_S \phi_m(y, z, \omega) \phi_n(y, z, \omega) \, dy \, dz = \delta_{mn} N_m(\omega), \tag{5.31}$$

where N_m is a normalization factor and δ_{mn} is the Kronecker delta.

It is convenient to examine the nonlinearly generated harmonic components in terms of normal-mode solutions (Ostrovskii and Papilova, 1973; Zabolotskaya and Shvartsburg, 1988). We begin by considering a bifrequency primary wave field:

$$p_1 = \tfrac{1}{2} [p_a \phi_a(y, z, \omega_a) e^{j(\omega_a t - k_a x)} + p_b \phi_b(y, z, \omega_b) e^{j(\omega_b t - k_b x)}] + \text{c.c.}, \tag{5.32}$$

where p_a and p_b are real pressure amplitudes of the waves in modes a and b, respectively. Substitution of Eq. (5.32) into the right-hand side of Eq. (5.27) yields the following inhomogeneous wave equation for the sum- and difference-frequency pressures:

$$\left(\nabla^2 + \frac{\omega_\pm^2}{c_0^2}\right)p_\pm = \frac{\beta p_a p_b \omega_\pm^2}{2\rho_0 c_0^4}\phi_a(y, z, \omega_a)\phi_b^{(*)}(y, z, \omega_b)e^{j[\omega_\pm t - (k_a \pm k_b^{(*)})x]} + \text{c.c.},$$

(5.33)

where the notation $(*)$ indicates that the complex conjugate is to be taken only when performing calculations for the difference-frequency component. Making use of the orthogonality of the eigenfunctions, we may express the product $\phi_a \phi_b^{(*)}$ as an expansion of normal modes at the sum and difference frequencies:

$$\phi_a(y, z, \omega_a)\phi_b^{(*)}(y, z, \omega_b) = \sum_m a_m \phi_m(y, z, \omega_\pm),$$

(5.34)

where

$$a_m = \frac{1}{N_m(\omega_\pm)}\iint_S \phi_a(y, z, \omega_a)\phi_b^{(*)}(y, z, \omega_b)\phi_m(y, z, \omega_\pm)\,dydz.$$

(5.35)

Substitution of Eq. (5.34) into Eq. (5.33) yields for the particular solution

$$(p_\pm)_p = \frac{\beta p_a p_b \omega_\pm^2}{2\rho_0 c_0^4}\sum_m \frac{a_m \phi_m(y, z, \omega_\pm)}{k_m^2 - (k_a \pm k_b^{(*)})^2}e^{j[\omega_\pm t - (k_a \pm k_b^{(*)})x]} + \text{c.c.},$$

(5.36)

where k_m is evaluated at $\omega = \omega_\pm$. Equation (5.36) is the solution for the forced wave, the phase speed of which is $\omega_\pm/\text{Re}(k_a \pm k_b)$.

The complete solution is $p_\pm = (p_\pm)_p + (p_\pm)_h$, where $(p_\pm)_h$ is the homogeneous solution for the free wave that is needed to satisfy the source condition. Let the source be located at $x = 0$, and assume that no sum- or difference-frequency sound is radiated directly by the source. The related boundary condition is $(p_\pm)_h = -(p_\pm)_p$ at $x = 0$, and therefore $(p_\pm)_h$ equals the negative of Eq. (5.36), except with $k_a \pm k_b^{(*)}$ replaced by k_m in the exponential. The free waves thus propagate in different modes with different phase speeds $\omega_\pm/\text{Re}\,k_m$. The complete solution is

$$p_\pm = \frac{1}{2}\sum_m P_m(x)\phi_m(y, z, \omega_\pm)e^{j\omega_\pm t} + \text{c.c.},$$

(5.37)

where

$$P_m(x) = \frac{j\beta p_a p_b \omega_\pm^2 a_m}{2\rho_0 c_0^4(k_m + \Delta_m)}\frac{\sin(\Delta_m x)}{\Delta_m}e^{-j(k_m + \Delta_m)x},$$

(5.38)

and $\Delta_m = (k_a \pm k_b^{(*)} - k_m)/2$ is a complex parameter that vanishes in the absence of both attenuation and dispersion.

We note that the normal-mode solution for the second-order pressure $p_\pm$ is constructed with eigenfunctions that satisfy only a first-order relation for the boundary condition at the walls, Eq. (5.30). For the case of second-harmonic generation in a waveguide formed by two parallel planar surfaces, one rigid and the other pressure release, it has been shown that the second-order correction to the linear pressure-release boundary condition is negligible except for source frequencies very near cutoff (Hamilton and Zabolotskaya, 1991). Use of first-order boundary conditions is therefore consistent with use of Eq. (5.27), i.e., for source radiation at frequencies well above cutoff.

The beating phenomenon discussed above in relation to the noncollinear interaction of two plane waves is revealed clearly by the dependence of P_m on $\sin(\Delta_m x)$. Specifically, the distance between consecutive local minima along the axis of the waveguide is the dispersion length $\pi/|\mathrm{Re}\,\Delta_m|$. The extent to which efficient generation of sum- or difference-frequency sound is possible depends on whether $\mathrm{Re}\,\Delta_m \simeq 0$ is obtained for any given mode that is excited, i.e., for which free waves propagate at phase speeds close to that of the forced wave. Resonance thus occurs when the real part of the relation $k_m = k_a \pm k_b$ is satisfied, in which case there is no beating in that mode, and $|P_m|$ increases linearly with x when attenuation is negligible.

Consider, for example, sum- and difference-frequency generation in a rectangular duct with rigid walls. If the dimensions of the cross section are $a \times b$, the eigenfunctions may be written as $\phi_m = \cos(m_1\pi y/a)\cos(m_2\pi z/b)$, where m corresponds to the pair of mode indices (m_1, m_2), and the axial wave numbers are defined by $k_m^2 = (\omega/c_0)^2 - (m_1\pi/a)^2 - (m_2\pi/b)^2$. The simplest case that demonstrates the beating phenomenon is with $\phi_a = \cos(\pi y/a)$ and $\phi_b = 1$ for the primary waves (the latter is simply the plane-wave mode). The summation in Eq. (5.37) then reduces to the single term $(m_1, m_2) = (1, 0)$, and the sum- and difference-frequency sound is produced entirely in mode $\phi_{1,0} = \cos(\pi y/a)$, with $a_{1,0} = 1$ and $k_{1,0}^2 = (\omega_\pm/c_0)^2 - (\pi/a)^2$. Measurements obtained in an experiment corresponding to this case, with an air-filled duct, are shown in Fig. 5.5 (Hamilton and TenCate, 1987). The dashed curve is the prediction based on Eq. (5.38), and the solid curve is the result obtained when the contribution due to the Lagrangian density $\mathcal{L}$ is retained [recall the discussion preceding Eq. (5.27)]. Because of the simple structure of the primary wave field, a single interaction angle can be defined for this experiment: $\theta = \arcsin(\pi c_0/a\omega_a) = 58°$. Increasing ω_a decreases θ, and the dashed and solid curves in Fig. 5.5 come closer together.

Second-harmonic generation is a special case of the above analysis. To obtain the solution for the second-harmonic component at frequency $2\omega_a$, replace $\omega_\pm$ by $2\omega_a$, $p_a p_b$ by $p_a^2/2$, $\phi_a\phi_b^{(*)}$ by ϕ_a^2, $k_a \pm k_b^{(*)}$ by $2k_a$, and evaluate k_m at $\omega = 2\omega_a$ in all results for $p_\pm$. Equation (5.38) becomes

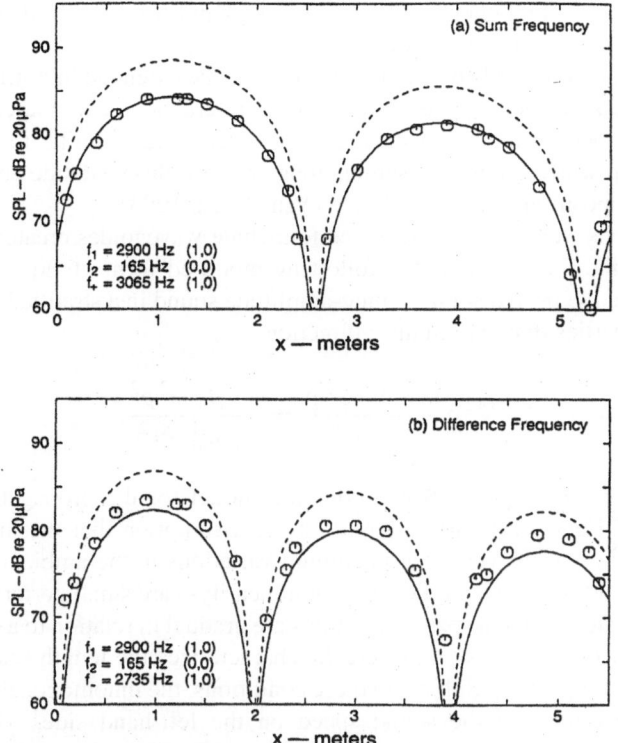

Fig. 5.5 Measurements of (**a**) sum- and (**b**) difference-frequency sound generated in an air-filled rectangular duct with dimensions $a = 7$ cm and $b = 3.8$ cm. The source levels were 120 dB (re 20 μPa) at $f_a = 2900$ Hz and 130 dB at $f_b = 165$ Hz ($f_1 \equiv f_a$ and $f_2 \equiv f_b$). The theoretical predictions (dashed and solid curves) are described in the text (Hamilton and TenCate, 1987).

$$P_m(x) = \frac{j\beta p_a^2 \omega_a^2 a_m}{\rho_0 c_0^4 (k_a + k_m/2)} \frac{\sin[(k_a - k_m/2)x]}{k_a - k_m/2} e^{-j(k_a + k_m/2)x}. \tag{5.39}$$

Examination of second-harmonic generation in a rectangular duct with rigid walls reveals that a primary wave in an arbitrary mode (m_1, m_2) interacts resonantly with the second-harmonic component that it generates in mode $(2m_1, 2m_2)$. More generally, the phase speed of the nth harmonic component (frequency $n\omega_a$) that is generated in mode (nm_1, nm_2) is the same for any n. Although nonresonant interactions also take place, entire families of harmonic components are thus in synchronism in rectangular ducts with rigid walls, and shock formation can occur (Hamilton and TenCate, 1988). However, this degeneracy is a special case. For example, replacing one of the walls with a pressure release surface destroys the synchronism (Hamilton and Zabolotskaya, 1991), although special choices of source frequencies and duct dimensions permit parametric amplification of selected spectral components (Ostrovskii and Papilova, 1973). In rigid circular ducts, resonance occurs only in the plane-wave mode.

The normal mode analysis described above can be extended in a straight-forward way to account for both higher-order spectral interactions and sources that excite arbitrary numbers of modes. A coupled first-order system similar to Eq. (5.4) can be derived, in which additional summations on the right-hand side account for the modal decomposition of the sound field (Van Doren, 1993).

The analysis can also be generalized to include waveguides created by inhomogeneous media. For example, the following modified form of Eq. (5.27) may be used to model the propagation of finite-amplitude sound in a stratified medium with ambient properties that vary in the z direction:

$$\nabla^2 p - \frac{1}{c^2(z)} \frac{\partial^2 p}{\partial t^2} = -\frac{\beta}{\rho_0 c_0^4} \frac{\partial^2 p^2}{\partial t^2}. \tag{5.40}$$

The linear form of Eq. (5.40) is a common model for the propagation of sound in the ocean. Equation (5.40) is based on the assumption that the inhomogeneity of the medium is weak (i.e., that maximum variations in the ambient properties—sound speed, density, and coefficient of nonlinearity—are small perturbations about a reference state), and that spatial variations are gradual in relation to a characteristic wavelength λ (i.e., $\lambda/L_z \ll 1$, where L_z characterizes the length scale associated with the inhomogeneity). Subject to these conditions, the inhomogeneity is manifest explicitly only through the sound speed on the left-hand side, which may be expressed as $c(z) = c_0 + c_1(z)$, with $|c_1|/c_0 \ll 1$. The values of ρ_0, c_0, and β are evaluated at a reference state in the fluid. For wave propagation in a direction perpendicular to the z axis, the analysis outlined by Eqs. (5.28) to (5.39) remains formally the same.

Finally, normal mode theory can be used to derive a nonlinear Schrödinger equation for the lossless ($R_n = 0$) propagation of a narrowband pulse (Zabolotskaya and Shvartsburg, 1988). The nonlinear Schrödinger equation is

$$j\frac{\partial A}{\partial x} + \frac{k_m''}{2} \frac{\partial^2 A}{\partial \tau_g^2} + \Lambda |A|^2 A = 0, \tag{5.41}$$

where $A(x, \tau_g)$ is a slowly varying amplitude modulation of a carrier wave that propagates in mode m at frequency ω, $\tau_g = t - x/c_g$ is a retarded time based on the group velocity $c_g = d\omega/dk_m$, and $k_m'' = d^2 k_m/d\omega^2$. The nonlinearity coefficient Λ depends on integrals over modes in which the carrier wave and its second harmonic propagate. For certain amplitude modulations, the effects of dispersion [taken into account by the second term in Eq. (5.41)] and nonlinearity are in balance, and the pulse envelope propagates without change in shape. Pulses of this type are referred to as envelope solitons, and they are well-known stationary wave solutions of Eq. (5.41) (Whitham, 1974; Ablowitz and Segur, 1981).

5.3.2 Bubbly Liquids

Common bubbly liquids of interest in acoustics include the near-surface layers in the ocean and flow in pipes. Of recent interest is sonoluminescence, the radiation of light from a collapsing bubble in a sound field. The presence of even small concentrations of bubbles in a liquid dramatically increases the compressibility and thus reduces the sound speed, and the resonant oscillations of the bubbles give rise to dispersion. In addition, nonlinearity due to the bubbles can exceed, by orders of magnitude, the nonlinearity due to the liquid alone. Analysis of finite-amplitude sound in bubbly liquids is quite complicated, and a variety of theoretical models have been proposed in this area. In this section we adopt the approach developed by Zabolotskaya and Soluyan (1973), which is based on the assumption that bubbles provide the dominant source of nonlinearity. The nonlinearity is taken into account by the Rayleigh–Plesset equation for a single bubble in an incompressible liquid.

We begin by deriving the linear wave equation for the propagation of sound through a bubbly liquid. Let the liquid contain a spatially uniform distribution of identical bubbles that are small compared with a wavelength. Define $\rho = \rho_0 + \rho'$ to be the density of the mixture, $\rho_l = \rho_{l0} + \rho'_l$ the density of the liquid, $\rho_g = \rho_{g0} + \rho'_g$ the density of the gas, and $V = V_0 + v$ the bubble volume, where the subscript 0 designates the equilibrium value. If N is the number of bubbles per unit volume in the equilibrium state, the corresponding volume fraction of liquid displaced by the bubbles is NV_0, and therefore

$$\rho_0 = NV_0\rho_{g0} + (1 - NV_0)\rho_{l0}. \tag{5.42}$$

Away from equilibrium, the density of the mixture is given by

$$\rho_0/\rho = NV + (1 - NV_0)\rho_{l0}/\rho_l. \tag{5.43}$$

When the volume fraction occupied by the gas is small ($NV \ll 1$), we have $\rho_0 \simeq \rho_{l0} \gg \rho_{g0}$, which permits the linear approximation of Eq. (5.43) to be written

$$\frac{\rho'}{\rho_0} = \frac{p}{\rho_0 c_0^2} - Nv, \tag{5.44}$$

where p is the sound pressure and c_0 is the small-signal sound speed in the liquid (i.e., $p = c_0^2 \rho'_l$ for $N = 0$). It is straightforward to combine Eq. (5.44) with the linearized continuity and momentum equations for a fluid with ambient density ρ_0 to obtain the inhomogeneous wave equation

$$\nabla^2 p - \frac{1}{c_0^2}\frac{\partial^2 p}{\partial t^2} = -\rho_0 N \frac{\partial^2 v}{\partial t^2}. \tag{5.45}$$

Solution of Eq. (5.45) requires an additional relation between the sound pressure and the bubble volume. Although Eq. (5.45) was derived by retaining only terms that are linear in p and v, it is hereafter assumed that the dominant contribution to finite-amplitude effects on the propagation of sound is due to nonlinearity in the relation $v = v(p)$ governing the bubble dynamics. To simplify the analysis, we assume that the bubbles are spherical, that their motions do not influence each other (i.e., that they are well separated), that they pulsate in their lowest (i.e., radially symmetric) mode, and that they are surrounded by an incompressible liquid (i.e., they do not radiate sound themselves). The pulsation of each bubble is then described by the Rayleigh-Plesset equation (Young, 1989):

$$R\ddot{R} + \tfrac{3}{2}\dot{R}^2 + 4v\frac{\dot{R}}{R} = \frac{P_g - P}{\rho_0},\tag{5.46}$$

where R is the bubble radius, the dots indicate differentiation with respect to time, v is the kinematic viscosity, P_g is the gas pressure in the bubble, and P is the total pressure in the mixture. Use of the adiabatic gas law $P_g/P_0 = (V_0/V)^\gamma$, where γ is the ratio of specific heats and P_0 is the ambient pressure in the mixture ($P = P_0 + p$), together with the relation $V = (4\pi/3)R^3$, permits the following expansion of Eq. (5.46) to be obtained to quadratic order in v (but to first order in loss terms):

$$\ddot{v} + \delta\omega_0\dot{v} + \omega_0^2 v + \eta p = av^2 + b(2v\ddot{v} + \dot{v}^2),\tag{5.47}$$

where $\delta = 4v/\omega_0 R_0^2$ is the viscous damping coefficient, $\omega_0^2 = 3\gamma P_0/\rho_0 R_0^2$, R_0 is the bubble radius at the equilibrium volume $V_0 = (4\pi/3)R_0^3$, $\eta = 4\pi R_0/\rho_0$, $a = (\gamma + 1)\omega_0^2/2V_0$, and $b = 1/6V_0$.

The first nonlinear term, av^2, is associated with the adiabatic gas law, and the second, $b(2v\ddot{v} + \dot{v}^2)$, is associated with the dynamic response of the bubble. In the absence of nonlinearity, Eq. (5.47) describes the forced response of a harmonic oscillator with undamped natural frequency ω_0. The quality factor for the bubble resonance is $Q = 1/\delta$. To include damping due to acoustic radiation by the bubble, compressibility of the liquid must be taken into account (Prosperetti, 1987), which introduces the additional loss factor $-R_0\dddot{v}/c_0$ on the left-hand side of Eq. (5.47) (Il'inskii and Zabolotskaya, 1992).

Equations (5.45) and (5.47) are solved simultaneously for the sound pressure in the mixture. The most straightforward way to accomplish this task is via the method of successive approximations. We illustrate this approach by considering the problem of second-harmonic generation (Zabolotskaya, 1976). Thus, let

$$p = \tfrac{1}{2}(p_1 e^{j\omega t} + p_2 e^{j2\omega t}) + \text{c.c.},\quad v = \tfrac{1}{2}(v_1 e^{j\omega t} + v_2 e^{j2\omega t}) + \text{c.c.},\tag{5.48}$$

where p_n and v_n are of order ε^n in terms of the acoustic Mach number. At order ε, substitution of Eqs. (5.48) into Eqs. (5.45) and (5.47) gives

$$\left(\nabla^2 + \frac{\omega^2}{c_0^2}\right)p_1 = \rho_0 N \omega^2 v_1, \tag{5.49}$$

$$(\omega_0^2 - \omega^2 + j\delta\omega_0\omega)v_1 = -\eta p_1. \tag{5.50}$$

Combining these two equations to eliminate v_1 yields

$$\left(\nabla^2 + \frac{\omega^2}{\tilde{c}_1^2}\right)p_1 = 0, \tag{5.51}$$

in which $\tilde{c}_1$ is defined by setting $n = 1$ in the following:

$$\frac{c_0^2}{\tilde{c}_n^2} = 1 + \frac{\mu C}{1 - n^2\omega^2/\omega_0^2 + j\delta n\omega/\omega_0}. \tag{5.52}$$

Here $\mu = N V_0$ is the volumetric void fraction in the equilibrium state, and $C = \rho_0 c_0^2/\gamma P_0$ is the ratio of the compressibility of the adiabatic gas $(1/\gamma P_0)$ to that of the liquid $(1/\rho_0 c_0^2)$. This ratio is $C = 1.54 \times 10^4$ for air bubbles in water at atmospheric pressure.

Noting that a plane wave solution of Eq. (5.51) is $\exp[-j(\omega/\tilde{c}_1)x]$, we can obtain the phase speed c_n and attenuation coefficient α_n from Eq. (5.52):

$$c_n(\omega) = 1/\operatorname{Re}\tilde{c}_n^{-1}, \quad \alpha_n(\omega) = -n\omega \operatorname{Im}\tilde{c}_n^{-1}. \tag{5.53}$$

In Fig. 5.6 are shown the (dimensionless) phase speed c_1 and attenuation coefficient α_1 as functions of frequency for $\mu C = 1$, both with ($\delta = 0.1$, solid curves) and without ($\delta = 0$, dashed curves) damping. Well below the bubble resonance frequency ($\omega^2 \ll \omega_0^2$), the phase speed reduces to $c_1 = c_0/\sqrt{1 + \mu C}$, which is independent of frequency. The low frequency phase speed corresponds to the sound speed in a liquid where the compressibility has been increased by the presence

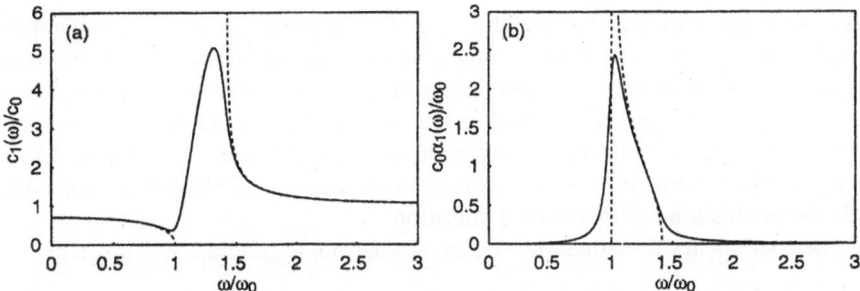

Fig. 5.6 (a) Phase speed and (b) attenuation coefficient for sound in a bubbly liquid, with ($\delta = 0.1$, solid curves) and without ($\delta = 0$, dashed curves) viscous damping.

of bubbles that pulsate in phase with the pressure. In the high frequency limit
($\omega \to \infty$), the response time of the bubbles is substantially longer than the period
of the acoustic oscillation. The bubble motion is effectively "frozen," and the phase
speed approaches the small-signal sound speed in the liquid alone ($c_1 \to c_0$).
The region of large variations in the phase speed, in the neighborhood of the
bubble resonance frequency, is characterized by high attenuation. In the absence
of viscous damping, the model predicts only evanescent waves (no propagation) in
the frequency range $1 < \omega/\omega_0 < \sqrt{1 + \mu C}$, with $\alpha_1 = 0$ outside this range.

For a distribution of equilibrium bubble radii, with $\mathcal{N}(R_0)\,dR_0$ bubbles per unit
volume having radii between R_0 and $R_0 + dR_0$, the void fraction μ is replaced by
$(4\pi/3)\mathcal{N}(R_0)R_0^3\,dR_0$, and integration yields in place of Eq. (5.52)

$$\frac{c_0^2}{\tilde{c}_n^2} = 1 + \frac{4\pi}{3}C \int_0^\infty \frac{\mathcal{N}(R_0)R_0^3\,dR_0}{1 - n^2\omega^2\kappa^2 R_0^2 + j\delta n\omega\kappa R_0}, \tag{5.54}$$

where $\kappa = (C/3)^{1/2}/c_0$. A distribution of bubble sizes broadens the resonances in
Fig. 5.6 (see Commander and Prosperetti, 1989).

We now develop the equations for second-harmonic generation. Substitution of
Eqs. (5.48) into Eqs. (5.45) and (5.47) yields at order ε^2

$$\left(\nabla^2 + \frac{4\omega^2}{c_0^2}\right)p_2 = 4\rho_0 N\omega^2 v_2, \tag{5.55}$$

$$(\omega_0^2 - 4\omega^2 + j2\delta\omega_0\omega)v_2 = -\eta p_2 + \tfrac{1}{2}(a - 3b\omega^2)v_1^2. \tag{5.56}$$

Equation (5.50) is used to express v_1^2 in terms of p_1^2, after which Eqs. (5.55) and
(5.56) can be combined into a single inhomogeneous wave equation for the second-
harmonic pressure:

$$\left(\nabla^2 + \frac{4\omega^2}{\tilde{c}_2^2}\right)p_2 = \beta_2(\omega)\frac{2\omega^2}{\rho_0 c_0^4}p_1^2, \tag{5.57}$$

where $\tilde{c}_2$ is given by Eq. (5.52) with $n = 2$. The reason for introducing the notation
$\beta_2(\omega)$ is to provide a basis for comparison with second-harmonic generation in a
liquid without bubbles. In the absence of bubbles, and when nonlinearities in the
equations of motion and state are taken into account, second-harmonic generation is
governed (in the quasilinear approximation) by Eq. (5.57) with $\beta_2 \equiv 1 + B/2A$ (and
$\tilde{c}_2 \equiv c_0$). Here, however, we have ignored nonlinearity in the liquid and considered
only the nonlinearity due to bubble pulsation.

The coefficient of nonlinearity for second-harmonic generation is given by

$$\beta_2(\omega) = \frac{\mu C^2(\gamma + 1 - \omega^2/\omega_0^2)}{2(1 - 4\omega^2/\omega_0^2 + j2\delta\omega_0\omega)(1 - \omega^2/\omega_0^2 + j\delta\omega_0\omega)^2}, \tag{5.58}$$

Fig. 5.7 Second-harmonic
coefficient of nonlinearity for
a mixture of air bubbles in
water, with ($\delta = 0.1$, solid
curve) and without ($\delta = 0$,
dashed curve) viscous
damping.

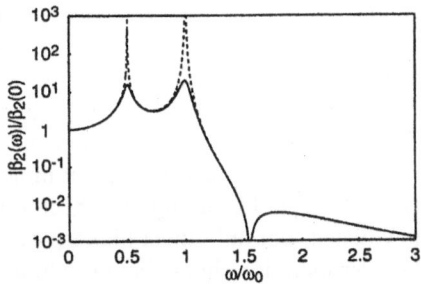

where ω is the frequency of the primary wave. The magnitude of Eq. (5.58) as a
function of frequency is shown in Fig. 5.7, for $\gamma = 1.4$, both with and without
viscous damping. There are two resonances and one antiresonance associated with
second-harmonic generation. At the higher resonance ($\omega/\omega_0 = 1$), the frequency
of the primary wave matches the bubble resonance frequency, and at the lower
resonance ($\omega/\omega_0 = 0.5$), the second-harmonic frequency coincides with the bubble
resonance. The antiresonance at $\omega/\omega_0 = \sqrt{\gamma + 1}$ occurs when the nonlinearity
in the Rayleigh-Plesset equation offsets the nonlinearity in the adiabatic gas law.
Note that the low-frequency limit, $\beta_2(0) = (\gamma + 1)\mu C^2/2$, can itself be very
large. For example, a void fraction of only 0.001% ($\mu = 10^{-5}$) yields $\beta_2(0) \sim$
2.8×10^3 for air bubbles in water, which is several orders of magnitude greater
than the coefficient of nonlinearity for the water alone (i.e., $\beta = 3.5$). However, for
$\sqrt{\gamma + 1} < \omega/\omega_0 < 3$ in this example, $|\beta_2|$ is comparable to β. For a distribution of
bubble sizes, the right-hand side of Eq. (5.58) becomes an integral with μ replaced
by $(4\pi/3)\mathcal{N}(R_0)R_0^3\,dR_0$, as in Eq. (5.54).

To complete the analysis of second-harmonic generation, boundary conditions
must be imposed on Eqs. (5.51) and (5.57). We assume plane wave propagation in
the $+x$ direction and require that $p_1 = p_0$ and $p_2 = 0$ at $x = 0$. For simplicity, we
also ignore viscous damping ($\delta = 0$) and assume that neither the primary nor the
secondary wave is evanescent. Equation (5.51) yields $p_1 = p_0 e^{-jk_1 x}$ ($k_n = n\omega/c_n$,
$n = 1, 2$), and substitution into Eq. (5.57) leads to the second-harmonic solution

$$p_2 = \frac{jp_0^2[(\gamma + 1)\omega_0^2/\omega^2 - 1]}{6\gamma P_0(1 - \omega^2/\omega_0^2)}\sin[(k_2 - 2k_1)x/2]e^{-j(k_2+2k_1)x/2}. \tag{5.59}$$

Although β_2 is proportional to the void fraction μ [recall Eq. (5.58)], the maximum
amplitude of p_2 is not. The reason is that the effect of nonlinearity is offset by
dispersion. To illustrate this point we consider the leading term in an expansion of
the wave number mismatch for small μ:

$$k_2 - 2k_1 = \frac{3\mu C\omega^3}{c_0\omega_0^2(1 - 4\omega^2/\omega_0^2)(1 - \omega^2/\omega_0^2)} + O(\mu^2). \tag{5.60}$$

At distances much shorter than the dispersion length $2\pi/|k_2 - 2k_1|$ and for μ sufficiently small, Eq. (5.59) yields $|p_2| = |\beta_2|p_0^2\omega x/2\rho_0 c_0^3$, which corresponds to the solution of Eq. (5.57) with $\tilde{c}_1 = \tilde{c}_2 = c_0$, i.e., in the absence of dispersion. Thus, whereas the initial rate of second-harmonic generation increases with μ, the dispersion length is reduced and the region of growth is terminated ever closer to the source. It must be kept in mind that Eq. (5.59) is valid only if the relation $|p_2| \ll |p_1|$ holds throughout the nonlinear interaction region. When this condition is violated, not only must higher-order spectral interactions be taken into account, but consideration should be given to including higher-order terms in Eqs. (5.45) and (5.47).

In addition to second-harmonic generation, a variety of phenomena with ana-logues in nonlinear optics have been investigated on the basis of Eqs. (5.45) and (5.47), such as parametric amplification, phase conjugation, stimulated Raman scattering, and active spectroscopy (Bunkin et al., 1986).

We conclude by deriving an evolution equation for low-frequency waves whose dominant frequency components satisfy the relation $\omega^2 \ll \omega_0^2$. Noting that the first nonlinear term in Eq. (5.47) dominates the second at low frequencies, we may write

$$v = -\frac{\eta}{\omega_0^2}p - \frac{\delta}{\omega_0}\dot{v} - \frac{1}{\omega_0^2}\ddot{v} + \frac{a}{\omega_0^2}v^2, \quad \omega^2 \ll \omega_0^2. \tag{5.61}$$

The second and third terms on the right-hand side of Eq. (5.61) are small in comparison with the first at low frequencies [e.g., see Eq. (5.50)], and therefore the approximate relation $v = -\eta p/\omega_0^2$ may be substituted into the second through fourth terms (which are all small relative to the first) to obtain

$$v = -\frac{\eta}{\omega_0^2}p + \frac{\delta\eta}{\omega_0^3}\dot{p} + \frac{\eta}{\omega_0^4}\ddot{p} + \frac{a\eta^2}{\omega_0^6}p^2. \tag{5.62}$$

Substitution of Eq. (5.62) into Eq. (5.45) gives

$$\nabla^2 p - \frac{1}{c_{00}^2}\frac{\partial^2 p}{\partial t^2} = -\frac{\mu\eta\rho_0}{\omega_0^4 V_0}\left(\frac{a\eta}{\omega_0^2}\frac{\partial^2 p^2}{\partial t^2} + \delta\omega_0\frac{\partial^3 p}{\partial t^3} + \frac{\partial^4 p}{\partial t^4}\right), \tag{5.63}$$

where $c_{00}^2 = c_0^2/(1 + \mu C)$ is recognized as the low-frequency limit of Eq. (5.52). Finally, for progressive plane waves we may introduce the slow scales (Sect. 3.7) $\tau = t - x/c_{00}$ and $x_1 = \varepsilon x$ in Eq. (5.63) to obtain, at leading order,

$$\frac{\partial p}{\partial x} = \frac{\beta_0 p}{\rho_0 c_{00}^3}\frac{\partial p}{\partial \tau} + a'\frac{\partial^3 p}{\partial \tau^3} + b'\frac{\partial^2 p}{\partial \tau^2}, \tag{5.64}$$

where $\beta_0 = (\gamma + 1)\mu C^2/2(1 + \mu C)^2$, $a' = \mu C^2 R_0^2/6c_{00}^3(1 + \mu C)^2$, and $b' = 4\nu a'/R_0^2$. The second term on the right accounts for dispersion, and the third

Fig. 5.8 B/A for a mixture
of air bubbles in water, taking
only bubble nonlinearity into
account (dashed curve), and
including liquid and gas
nonlinearities (solid curve).

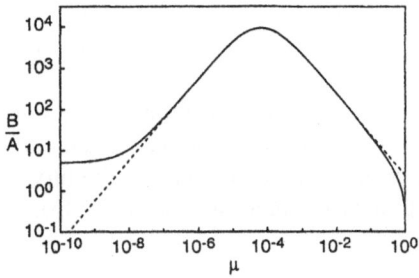

accounts for attenuation. Equation (5.64) is in the form of Eq. (5.1), and it is known
as the Korteweg–deVries–Burgers equation.

Solutions for steady shock waves have been derived from equations in the form of
Eq. (5.64), and the predictions have been verified experimentally in bubbly liquids
(Kuznetsov et al., 1978). Without attenuation ($b' = 0$), Eq. (5.64) reduces to the
KdV (Korteweg–deVries) equation, which admits solutions for solitons (Ablowitz
and Segur, 1981). Alternatively, in the absence of dispersion ($a' = 0$), Eq. (5.64)
reduces to the Burgers equation. See van Wijngaarden (1972) for further discussion.

Finally, via comparison of Eqs. (5.1) and (5.64), we identify β_0 as the low-
frequency coefficient of nonlinearity. As noted above, nonlinearity in our model
derives solely from the pressure–density relation for the mixture, Eq. (5.44), where
the nonlinearity enters through Eq. (5.47). We may therefore write $\beta_0 = B/2A$,
because for plane waves the "1" that appears in the expression $\beta = 1 + B/2A$ is
due to the nonlinearity in the continuity equation (Hamilton and Blackstock, 1988),
which we have ignored. We thus identify

$$\frac{B}{A} = \frac{(\gamma + 1)\mu C^2}{(1 + \mu C)^2} \tag{5.65}$$

as the effective nonlinearity parameter, which yields a maximum value of ($\gamma +$
1)$C/4$ at $\mu C = 1$, and $B/A \propto \mu$ for $\mu C \ll 1$, $B/A \propto \mu^{-1}$ for $\mu C \gg 1$. The
dashed curve in Fig. 5.8 is Eq. (5.65) evaluated for air bubbles in water.

Equation (5.65) does not yield the proper limiting values for $\mu = 0$ (pure liquid,
with $B/A = 5.0$ for water) or $\mu = 1$ (pure gas, with $B/A = 0.4$ for air) because
we have not taken the nonlinearity of the liquid alone into account, and the volume
fraction of gas was assumed to be small. These additional nonlinearities can be
included by using Eq. (2.24) to obtain

$$\frac{B}{A} = \frac{(\gamma + 1)\mu C^2 + (1 - \mu)[2 + (B/A)_l]}{[1 + \mu(C - 1)]^2} - 2, \tag{5.66}$$

where $(B/A)_l$ is the nonlinearity parameter for the liquid. Equation (5.66), which is
graphed as the solid curve in Fig. 5.8 (see also Everbach, 1989), reduces to $(B/A)_l$
for $\mu = 0$ and $(\gamma - 1)$ for $\mu = 1$ [Eq. (2.11)], as required. Comparison of the

two curves in Fig. 5.8 reveals that the approximate Eq. (5.65) is valid in the range $10^{-7} \lesssim \mu \lesssim 10^{-1}$. Equation (5.65) is recovered in general for $(B/A)_l = 0$, $C \gg 1$, and $\mu \ll 1$.

We conclude by noting that for $\mu \ll 1$, the analysis in the present section can be generalized to include all quadratic nonlinearities associated with propagation through the liquid phase of the mixture by adding the right-hand side of Eq. (5.24) to the right-hand side of Eq. (5.45).

References

Ablowitz, M. J., and Segur, H. (1981). *Solitons and the Inverse Scattering Transform* (Society for Industrial and Applied Mathematics, Philadelphia).

Blackstock, D. T. (1985). Generalized Burgers equation for plane waves. *J. Acoust. Soc. Am.* **77**, 2050–2053.

Bunkin, F. V., Kravtsov, Yu. A., and Lyakhov, G. A. (1986). Acoustic analogues of nonlinear-optics phenomena. *Sov. Phys. Usp.* **29**, 607–619.

Cleveland, R. O., Hamilton, M. F., and Blackstock, D. T. (1996). Time-domain modeling of finite-amplitude sound in relaxing fluids. *J. Acoust. Soc. Am.* **99**, 3312–3318.

Commander, K. W., and Prosperetti, A. (1989). Linear pressure waves in bubbly liquids: Comparison between theory and experiments. *J. Acoust. Soc. Am.* **85**, 732–746.

Coppens, A. B. (1971). Theoretical study of finite-amplitude traveling waves in rigid-walled ducts: Behavior for strengths precluding shock formation. *J. Acoust. Soc. Am.* **49**, 306–318.

Everbach, E. C. (1989). Tissue composition determination via measurement of the acoustic nonlinearity parameter. Ph.D. thesis, Yale University, Chap. 3.

Hamilton, M. F., and Blackstock, D. T. (1988). On the coefficient of nonlinearity β in nonlinear acoustics. *J. Acoust. Soc. Am.* **83**, 74–77.

Hamilton, M. F., and TenCate, J. A. (1987). Sum and difference frequency generation due to noncollinear wave interaction in a rectangular duct. *J. Acoust. Soc. Am.* **81**, 1703–1712.

Hamilton, M. F., and TenCate, J. A. (1988). Finite amplitude sound near cutoff in higher-order modes of a rectangular duct. *J. Acoust. Soc. Am.* **84**, 327–334.

Hamilton, M. F., and Zabolotskaya, E. A. (1991). Nonlinear propagation of sound in a liquid layer between a rigid and a free surface. *J. Acoust. Soc. Am.* **90**, 1048–1055.

Il'inskii, Yu. A., and Zabolotskaya, E. A. (1992). Cooperative radiation and scattering of acoustic waves by gas bubbles in liquids. *J. Acoust. Soc. Am.* **92**, 2837–2841.

Korpel, A. (1980). Frequency approach to nonlinear dispersive waves. *J. Acoust. Soc. Am.* **67**, 1954–1958.

Kuznetsov, V. V., Nakoryakov, V. E., Pokusaev, B. G., and Shreiber, I. R. (1978). Propagation of perturbations in a gas-liquid mixture *J. Fluid Mech.* **85**, 85–96.

Landau, L. D., and Lifshitz, E. M. (1987). *Fluid Mechanics*, 2nd ed. (Pergamon, New York).

Morse, P. M., and Ingard, K. U. (1986). *Theoretical Acoustics* (Princeton University Press, Princeton, N.J.), Chap. 9.

Naze Tjøtta, J., and Tjøtta, S. (1987). Interaction of sound waves. Part I: Basic equations and plane waves. *J. Acoust. Soc. Am.* **82**, 1425–1428.

Ostrovskii, L. A., and Papilova, I. A. (1973). Nonlinear mode interaction and parametric amplification in acoustic waveguides. *Sov. Phys. Acoust.* **19**, 45–50.

Pierce, A. D. (1989). *Acoustics: An Introduction to Its Physical Principles and Applications* (Acoustical Society of America, New York).

Polyakova, A. L., Soluyan, S. I., and Khokhlov, R. V. (1962). Propagation of finite disturbances in a relaxing medium. *Sov. Phys. Acoust.* **8**, 78–82.

Prosperetti, A. (1987). The equation of bubble dynamics in a compressible liquid. *Phys. Fluids* **30**, 3626–3628.

Rudenko, O. V., and Soluyan, S. I. (1977). *Theoretical Foundations of Nonlinear Acoustics* (Plenum, New York).

Soluyan, S. I., and Khokhlov, R. V. (1962). Finite amplitude acoustic waves in a relaxing medium. *Sov. Phys. Acoust.* **8**, 170–175.

Temkin, S. (1990). Attenuation and dispersion of sound in bubbly fluids via the Kramers–Kronig relations. *J. Fluid Mech.* **211**, 61–72.

Van Doren, T. W. (1993). Propagation of finite amplitude sound in multiple waveguide modes. Ph.D. dissertation, The University of Texas at Austin, Sec. 3–1.

van Wijngaarden, L. (1972). One-dimensional flow in liquids containing small gas bubbles. *Ann. Rev. Fluid Mech.* **4**, 369–396.

Webster, D. A., and Blackstock, D. T. (1977). Finite-amplitude saturation of plane sound waves in air. *J. Acoust. Soc. Am.* **62**, 518–523.

Whitham, G. B. (1974). *Linear and Nonlinear Waves* (Wiley, New York).

Young, F. R. (1989). *Cavitation* (McGraw-Hill, New York), Chap. 2.

Zabolotskaya, E. A. (1976). Acoustic second-harmonic generation in a liquid containing uniformly distributed air bubbles. *Sov. Phys. Acoust.* **21**, 569–571.

Zabolotskaya, E. A., and Shvartsburg, A. B. (1988). Propagation of finite amplitude waves in a waveguide. *Sov. Phys. Acoust.* **34**, 493–495.

Zabolotskaya, E. A., and Soluyan, S. I. (1973). Emission of harmonic and combination-frequency waves by air bubbles. *Sov. Phys. Acoust.* **18**, 396–398.

Chapter 6
Radiation Pressure and Acoustic Levitation

Taylor G. Wang and Chun P. Lee

Contents

6.1 Introduction

Acoustic radiation pressure was first studied by Lord Rayleigh (1902, 1905) as an acoustic counterpart of the pressure induced by an electromagnetic wave. Since then this subject has been studied by many researchers, with differing results (see a review by Beyer, 1978). The confusion arises mainly because the phenomenon is such a subtle nonlinear effect that sometimes the question must be *very carefully* posed in order to obtain a unique answer. In this chapter we shall present the subject starting from first principles (Lee and Wang, 1993). We shall use Eulerian coordinates, which will turn out to be much more straightforward than the more traditional use of Lagrangian coordinates.

When acoustic radiation pressure acts on the surface of a small object, it imposes a net force on the object called the acoustic radiation force. In practice, this force is very weak in comparison to gravity. However, with the accessibility of a microgravity environment provided by Space flights in recent years, intense interest has arisen in processing of materials without possible contamination by a crucible, i.e., "containerless processing." Acoustic radiation force is best suited

T. G. Wang · C. P. Lee
Center for Microgravity Research and Applications, Vanderbilt University, Nashville, TN, USA

© The Author(s) 2024   175
M. F. Hamilton, D. T. Blackstock (eds.), *Nonlinear Acoustics*,
https://doi.org/10.1007/978-3-031-58963-8_6

for this purpose. The containerless handling of materials using this force is called "acoustic levitation."

6.2 Radiation Pressure

6.2.1 General

6.2.1.1 Equation of State

We start by introducing a model for the adiabatic equation of state of the fluid:

$$P - P_0 = Q_0 \left[\left(\frac{\rho}{\rho_0} \right)^\gamma - 1 \right], \tag{6.1}$$

where P_0 is the ambient pressure, $P = P_0 + p$ is the total pressure, p is the acoustic pressure, ρ_0 is the ambient density, $\rho = \rho_0 + \rho'$, where ρ' is the excess density, γ is a constant, and $Q_0 = \rho_0 c_0^2 / \gamma$, in which c_0 is the small-signal speed of sound. For an ideal gas, γ is just the ratio of specific heats, in which case $c_0^2 = \gamma P_0 / \rho_0$, $Q_0 = P_0$, and Eq. (6.1) reduces to $P / P_0 = (\rho / \rho_0)^\gamma$. For liquids, γ is an empirical parameter, and Eq. (6.1) is known as the Tait equation [see Eq. (2.12)]. For water, $\gamma \approx 7$, and $Q_0 \approx 3000$ bars (Beyer, 1974).

6.2.1.2 Acoustic Radiation Stress Tensor

The equations of motion for an ideal fluid are the momentum equation

$$\rho \left(\frac{\partial u_i}{\partial t} + u_j \frac{\partial u_i}{\partial x_j} \right) = -\frac{\partial P}{\partial x_i}, \tag{6.2}$$

and the continuity equation

$$\frac{\partial \rho}{\partial t} + \frac{\partial (\rho u_j)}{\partial x_j} = 0, \tag{6.3}$$

where u_i is the ith component of the particle velocity field vector, and x_i is a component of the position vector. Equations (6.2) and (6.3) combine to give

$$\frac{\partial (\rho u_i)}{\partial t} + \frac{\partial (\rho u_i u_j)}{\partial x_j} = -\frac{\partial P}{\partial x_i}. \tag{6.4}$$

For acoustic oscillations, we can define the time average over one cycle, denoted by $\langle \ \rangle$. On averaging Eq. (6.4), the first term on the left side vanishes in steady state, such that

$$\frac{\partial S_{ij}}{\partial x_j} = 0, \tag{6.5}$$

where

$$S_{ij} = -\langle P \rangle \delta_{ij} - \langle \rho u_i u_j \rangle \tag{6.6}$$

is called the acoustic radiation stress tensor, which was first derived by Brillouin (1938, 1964), and δ_{ij} is the Kronecker delta. It does not make any difference in Eq. (6.5) if we replace P by $P - P_0$ in Eq. (6.6). Since u_i is first order, with $\rho = \rho_0 + \rho'$, Eq. (6.6) becomes, at second order,

$$S_{ij} = -\langle P - P_0 \rangle \delta_{ij} - \rho_0 \langle u_i u_j \rangle. \tag{6.7}$$

6.2.1.3   Mean Eulerian Excess Pressure

The quantity $\langle P - P_0 \rangle$ in Eq. (6.7) is a "mean excess pressure," and in general it is nonzero at finite amplitudes. It is Eulerian because it is evaluated at a fixed point in space, as opposed to the Lagrangian pressure. Since a sound field in an inviscid fluid is irrotational, we may write $\mathbf{u} = \nabla \phi$, where ϕ is the velocity potential. Equation (6.2) then becomes

$$\nabla \left[\frac{\partial \phi}{\partial t} + \tfrac{1}{2} |\nabla \phi|^2 \right] = -\frac{\nabla P}{\rho}. \tag{6.8}$$

If T is the temperature, and s and w are the entropy per unit mass and the enthalpy per unit mass of the fluid, respectively, then $dw = T\,ds + dP/\rho$. For an adiabatic process, $\nabla w = \nabla P/\rho$, and Eq. (6.8) becomes

$$w = -\frac{\partial \phi}{\partial t} - \tfrac{1}{2} |\nabla \phi|^2 + C' \tag{6.9}$$

after being integrated in space, where C' is constant in space but can depend on time. The pressure P can be expanded in a Taylor series in w as follows:

$$P = P_0 + \left(\frac{\partial P}{\partial w} \right)_{s,0} w + \frac{1}{2} \left(\frac{\partial^2 P}{\partial w^2} \right)_{s,0} w^2 + \cdots , \tag{6.10}$$

where the subscript $s, 0$ means "evaluated at constant entropy and at equilibrium." Since $(\partial w/\partial P)_s = 1/\rho$, we have $(\partial P/\partial w)_s = \rho$, and $(\partial^2 P/\partial w^2)_s = (\partial \rho/\partial w)_s = (\partial \rho/\partial P)_s (\partial P/\partial w)_s = \rho/c^2$, where the fundamental relation $(\partial P/\partial \rho)_s = c^2$ was used. We now let each of these quantities take its equilibrium value, and Eq. (6.10) becomes

$$P = P_0 + \rho_0 \left(-\frac{\partial \phi}{\partial t} - \tfrac{1}{2}|\nabla \phi|^2 + C' \right)$$

$$+ \frac{1}{2}\frac{\rho_0}{c_0^2} \left(-\frac{\partial \phi}{\partial t} - \tfrac{1}{2}|\nabla \phi|^2 + C' \right)^2 + \cdots. \qquad (6.11)$$

In linear acoustics, $C' = 0$ and Eq. (6.9) reduces to $w = p/\rho_0 = -\partial \phi/\partial t$. At second order, a finite value of C' is sometimes needed for the solution to satisfy a constraint. We therefore take C' to be a second-order quantity. By time-averaging Eq. (6.11), at second order we find

$$\langle P - P_0 \rangle = \frac{1}{2}\frac{\rho_0}{c_0^2} \left\langle \left(\frac{\partial \phi}{\partial t} \right)^2 \right\rangle - \tfrac{1}{2}\rho_0 \langle |\nabla \phi|^2 \rangle + C, \qquad (6.12)$$

where $C = \rho_0 \langle C' \rangle$ is a constant in both space and time. At second order, ϕ in the quadratic terms on the right-hand side can be replaced with linear relations. Substituting $\mathbf{u} = \nabla \phi$ and $\partial \phi/\partial t = -p/\rho_0$ into Eq. (6.12), we obtain for the mean Eulerian excess pressure

$$\langle P - P_0 \rangle = \langle P^E - P_0 \rangle = \frac{1}{2\rho_0 c_0^2} \langle p^2 \rangle - \tfrac{1}{2}\rho_0 \langle \mathbf{u} \cdot \mathbf{u} \rangle + C = \langle V \rangle - \langle K \rangle + C,$$
$$(6.13)$$

where $\langle V \rangle = \langle p^2 \rangle/2\rho_0 c_0^2$ and $\langle K \rangle = \rho_0 \langle \mathbf{u} \cdot \mathbf{u} \rangle/2$ are the time-averaged potential and kinetic energy densities, respectively, and P has been equated to P^E to emphasize that the pressure is Eulerian.

6.2.1.4 Mean Lagrangian Excess Pressure

In contrast to its Eulerian counterpart, the mean Lagrangian excess pressure is what a fluid particle experiences during acoustic vibration. The superscript L shall be used to denote Lagrangian. Let $\boldsymbol{\xi}$ be the displacement of a fluid particle from its equilibrium position $\mathbf{a}$. As we shall see, we require only a first-order relation between $\boldsymbol{\xi}$ and the fluid particle velocity, at which order there is no difference between the Lagrangian velocity and the Eulerian velocity. We therefore write $\mathbf{u} = \partial \boldsymbol{\xi}/\partial t$. An arbitrary Lagrangian quantity $q^L(\mathbf{a}, t)$ is related to the corresponding Eulerian quantity $q^E(\mathbf{x}, t)$, at second order, by $q^L(\mathbf{a}, t) = [q^E(\mathbf{x}, t)]_{\mathbf{x}=\mathbf{a}} + \boldsymbol{\xi} \cdot [\nabla q^E(\mathbf{x}, t)]_{\mathbf{x}=\mathbf{a}}$. We thus obtain $P^L = P^E + \boldsymbol{\xi} \cdot \nabla P^E$, or

$$\langle P^L - P_0 \rangle = \langle P^E - P_0 \rangle + \langle \boldsymbol{\xi} \cdot \nabla P^E \rangle. \qquad (6.14)$$

The last term on the right can be evaluated using first-order relations. The linearized form of Eq. (6.2) is $\rho_0 \partial \mathbf{u}/\partial t = -\nabla P$, or $\rho_0 \partial^2 \boldsymbol{\xi}/\partial t^2 = -\nabla P^E$. Integration by parts yields $\langle \boldsymbol{\xi} \cdot \nabla P^E \rangle = -\rho_0 \langle \boldsymbol{\xi} \cdot \partial^2 \boldsymbol{\xi}/\partial t^2 \rangle = \rho_0 \langle \partial \boldsymbol{\xi}/\partial t \cdot \partial \boldsymbol{\xi}/\partial t \rangle = \rho_0 \langle \mathbf{u} \cdot \mathbf{u} \rangle$, and thus $\langle \boldsymbol{\xi} \cdot \nabla P^E \rangle = 2\langle K \rangle$. Substituting both the last relation and Eq. (6.13) into Eq. (6.14), we obtain for the mean Lagrangian excess pressure

$$\langle P^L - P_0 \rangle = \langle V \rangle + \langle K \rangle + C = \langle E \rangle + C, \tag{6.15}$$

where $\langle E \rangle = \langle V \rangle + \langle K \rangle$ is the total mean energy density of the wave. Equations (6.15) and (6.13) are referred to as Langevin's first and second relations, respectively (Beissner, 1986).

To understand the significance of $\langle P^L - P_0 \rangle$, consider the x component of the stress defined in Eq. (6.7):

$$S_{xx} = -\langle P - P_0 \rangle - \rho_0 \langle u^2 \rangle, \tag{6.16}$$

where $\langle P - P_0 \rangle$ is the mean Eulerian excess pressure, and u is the acoustic particle velocity in the x direction. Using Eq. (6.13) and letting $\langle K_x \rangle = \rho_0 \langle u^2 \rangle/2$, $\langle K_y \rangle = \rho_0 \langle v^2 \rangle/2$, and $\langle K_z \rangle = \rho_0 \langle w^2 \rangle/2$ for the velocity components u, v, and w in the x, y, and z directions, respectively, we may rewrite Eq. (6.16) as

$$- S_{xx} = \langle V \rangle - \langle K_y \rangle - \langle K_z \rangle + \langle K_x \rangle + C. \tag{6.17}$$

If the motion is only in the x direction, then $\langle K_y \rangle = \langle K_z \rangle = 0$, and after letting $\langle K_x \rangle = \langle K \rangle$, comparison with Eq. (6.15) yields $-S_{xx} = \langle P^L - P_0 \rangle$ for Eq. (6.17). From the 1-D form of Eq. (6.5), $\partial S_{xx}/\partial x = 0$, it follows that $\partial \langle P^L - P_0 \rangle/\partial x = 0$, such that

$$\langle P^L - P_0 \rangle = \text{constant.} \tag{6.18}$$

Equation (6.18) is not generally true in a 2- or 3-D case. The 1-D case is interesting historically, since it has been studied most because of its simplicity. In practice, however, 1-D motion is a severe restriction, and the inadequacy of this approach has been pointed out by Beissner (1985, 1986). For example, oblique reflection of a plane wave, and even normal reflection of a diffracting sound beam, cannot be analyzed with 1-D equations. Therefore, the usefulness of Eq. (6.18) is somewhat limited in practice.

6.2.1.5   Rayleigh and Langevin Radiation Pressures

In electromagnetism, the radiation pressure is the pressure experienced by a material surface when it is illuminated by a light wave. An acoustic radiation pressure is therefore the mean excess pressure experienced by a material surface in a sound field. A mean pressure is defined throughout the fluid, but an acoustic radiation pressure is defined on a material surface only. The mean Eulerian and Lagrangian excess pressures are different expressions of the same physical entity. In contrast, the Rayleigh and Langevin radiation pressures are defined for different physical situations.

Beyer (1950, 1974, 1978) defined the Rayleigh radiation pressure as the difference between the mean pressure at a reflecting or absorbing wall in the presence of a sound wave and the pressure that would have existed if the fluid were at rest at the same mean density. He defined the Langevin radiation pressure as the difference between the mean pressure at a reflecting or absorbing wall and the pressure in the unperturbed fluid behind the wall, with the fluid being in contact with the two sides of the wall. The first definition identifies the Rayleigh radiation pressure as $\langle P^L - P_0 \rangle$ on a wall in a fluid with a fixed mean volume. The second definition identifies the Langevin radiation pressure as $\langle P^L - P_0 \rangle$ on a wall in a fluid that is in contact with the ambient unperturbed fluid. Gol'dberg (1971) defined the Rayleigh radiation pressure as the mean excess pressure on the wall which depends on the elasticity of the fluid and therefore on γ, and the Langevin radiation pressure as that which has no such dependence. For the first definition, there is no contact between the sound field and the unperturbed medium. For the second definition, contact is permitted such that C is set to zero to ensure "zero perturbation at infinity."

In most cases the definitions of Beyer and Gol'dberg are equivalent. However, the radiation pressure can depend on γ, whereas the mean volume of a confined fluid is not necessarily fixed (Sect. 6.2.2.2). Gol'dberg's definitions are therefore simpler. As we shall see, even if $C = 0$, the radiation pressure can depend implicitly on γ through its dependence on c_0, in which case Gol'dberg's definitions become somewhat vague (Sect. 6.2.2.3). Also, as long as there is a constraint, contact between the sound wave and the unperturbed medium does not always lead to the result expected by Gol'dberg, which is $C = 0$ (Lee and Wang, 1993). If the constraint is not related to conservation of mass or volume, then C need not be related to γ. Consequently, we modify Gol'dberg's definitions as follows: The radiation pressure is Langevin if it depends only on the sound waves, and it is Rayleigh if it depends on the sound waves plus a constraint. It takes a finite value of C to satisfy the constraint. Therefore equivalently, the radiation pressure is of Rayleigh type for $C \neq 0$ and of Langevin type for $C = 0$.

For the general 3-D case, we should consider the acoustic radiation stress tensor instead of the radiation pressure. If the surface is rigid, however, we can retain the notion of a radiation pressure. If the surface is absorbing, then in the region where the particle velocity is nearly normal to the surface, it is still convenient to speak of the radiation pressure.

6.2.2   Some Classical Problems

6.2.2.1   Plane Wave Incident on a Perfectly Reflecting Wall

This is the original problem posed by Rayleigh (1902, 1905). Consider a plane wave traveling in the x direction and being reflected at $x = 0$ by a rigid plane wall. The incident and reflected waves combine to produce a standing wave with acoustic pressure

$$p = A \cos kx \sin \omega t \tag{6.19}$$

and particle velocity

$$u = -\frac{A}{\rho_0 c_0} \sin kx \cos \omega t, \tag{6.20}$$

where A is the pressure amplitude, ω is the angular frequency, and $k = \omega/c_0$. Substituting Eqs. (6.19) and (6.20) into Eq. (6.13), we find $\langle V \rangle = A^2 \cos^2 kx/4\rho_0 c_0^2$ and $\langle K \rangle = A^2 \sin^2 kx/4\rho_0 c_0^2$, and thus

$$\langle P^E - P_0 \rangle = \frac{A^2}{4\rho_0 c_0^2} \cos 2kx + C = \langle E \rangle \cos 2kx + C = 2\langle E_i \rangle \cos 2kx + C, \tag{6.21}$$

where $\langle E \rangle = \langle V \rangle + \langle K \rangle = A^2/4\rho_0 c_0^2$ is the energy density of the standing wave, and $\langle E_i \rangle = \langle E \rangle/2$ is that of either the incident or the reflected wave. Conservation of mass requires that the mean excess density per wavelength be zero. The Eulerian excess density is given by

$$\langle \rho^E - \rho_0 \rangle = \left(\frac{\partial \rho}{\partial P}\right)_{s,0} \langle P^E - P_0 \rangle + \frac{1}{2}\left(\frac{\partial^2 \rho}{\partial P^2}\right)_{s,0} \langle (P^E - P_0)^2 \rangle + \cdots, \tag{6.22}$$

where $(\partial \rho/\partial P)_{s,0} = 1/c_0^2$ and $(\partial^2 \rho/\partial P^2)_{s,0} = -(\gamma - 1)/\rho_0 c_0^4$ follow from Eq. (6.1), $\langle P^E - P_0 \rangle$ is given by Eq. (6.21), and $\langle (P^E - P_0)^2 \rangle = \langle p^2 \rangle = (1/2)A^2 \cos^2 kx$ from Eq. (6.19). Integrating Eq. (6.22) in x over $2\pi/k$ and setting the left-hand side equal to zero, we have $C = (\gamma - 1)\langle E \rangle/2$. Equation (6.21) becomes

$$\langle P^E - P_0 \rangle = \left(\frac{\gamma - 1}{2} + \cos 2kx\right)\langle E \rangle = (\gamma - 1 + 2\cos 2kx)\langle E_i \rangle. \tag{6.23}$$

After substituting the value of C in Eq. (6.15) we find

$$\langle P^{\mathrm{L}} - P_0 \rangle = \left(\frac{\gamma + 1}{2} \right) \langle E \rangle = (\gamma + 1) \langle E_i \rangle. \tag{6.24}$$

Note that $\langle P^{\mathrm{E}} - P_0 \rangle$ in Eq. (6.23) depends on x, whereas $\langle P^{\mathrm{L}} - P_0 \rangle$ in Eq. (6.24) is constant. At $x = 0$, however, Eqs. (6.23) and (6.24) agree, as they should because the two representations must coincide at locations where there is no motion. The common value is the radiation pressure $\langle \Delta P \rangle$ experienced by the wall, and it agrees with Rayleigh's result (1905):

$$\langle \Delta P \rangle = \left(\frac{\gamma + 1}{2} \right) \langle E \rangle = (\gamma + 1) \langle E_i \rangle. \tag{6.25}$$

Because C is nonzero, Eq. (6.25) describes the Rayleigh radiation pressure. From Eq. (6.22), we also find that

$$\frac{\langle \rho^{\mathrm{E}} - \rho_0 \rangle}{\rho_0} = - \left(\frac{\gamma - 3}{8} \right) \left(\frac{A}{\rho_0 c_0^2} \right)^2 \cos 2kx, \tag{6.26}$$

which is sinusoidal in x and has a spatial average of zero. The Lagrangian counterpart is $\langle \rho^{\mathrm{L}} \rangle = \langle \rho^{\mathrm{E}} \rangle + \langle \xi \, \partial \rho^{\mathrm{E}} / \partial x \rangle$, where $u = \partial \xi / \partial t$ [see Eq. (6.14)]. On the right-hand side, $\langle \rho^{\mathrm{E}} \rangle$ is given by Eq. (6.26), $\xi = -(A/\rho_0 c_0 \omega) \sin kx \sin \omega t$ from Eq. (6.20), and $\partial \rho^{\mathrm{E}} / \partial x = c_0^{-2} \partial p / \partial x = -(kA/c_0^2) \sin kx \sin \omega t$ from Eq. (6.19), such that $\langle \xi \, \partial \rho^{\mathrm{E}} / \partial x \rangle = (A^2 / 2\rho_0 c_0^4) \sin^2 kx$. The result is

$$\frac{\langle \rho^{\mathrm{L}} - \rho_0 \rangle}{\rho_0} = \frac{1}{4} \left(\frac{A}{\rho_0 c_0^2} \right)^2 - \left(\frac{\gamma - 1}{8} \right) \left(\frac{A}{\rho_0 c_0^2} \right)^2 \cos 2kx. \tag{6.27}$$

It is noted that, unlike $\langle \rho^{\mathrm{E}} - \rho_0 \rangle / \rho_0$, $\langle \rho^{\mathrm{L}} - \rho_0 \rangle / \rho_0$ has a nonzero spatial average, equal to $(A/2\rho_0 c_0^2)^2$. The spatial average can be nonzero because the time average is taken by following an oscillating particle, and therefore the procedure is biased by the oscillation velocity, which controls the distribution of time spent by the particle along the path it travels during one period of oscillation. Again, Eqs. (6.26) and (6.27) agree at $x = 0$, where there is no particle motion.

6.2.2.2 Plane Wave Incident on a Perfectly Absorbing Wall

This problem is a controversial one. It differs from Rayleigh's original problem in that the wall is perfectly absorbing. The system consists of a closed tube with a transducer at one end that generates a plane wave traveling down the tube to the other end, which is an absorbing wall. Let the acoustic pressure and particle velocity be

$$p = A \sin(kx - \omega t) \tag{6.28}$$

and

$$u = \frac{A}{\rho_0 c_0} \sin(kx - \omega t). \tag{6.29}$$

Substituting Eqs. (6.28) and (6.29) into Eqs. (6.13) and (6.15), we find

$$\langle P^{\mathrm{E}} - P_0 \rangle = C \tag{6.30}$$

and

$$\langle P^{\mathrm{L}} - P_0 \rangle = \langle E_i \rangle + C, \tag{6.31}$$

since $\langle V \rangle = \langle K \rangle = A^2/4\rho_0 c_0^2 = \langle E_i \rangle/2$. A perfectly absorbing wall offers no reaction to the acoustic particle velocity on its surface. One can further consider the wall being pushed back a little by the radiation pressure and held at an equilibrium position by an opposing spring force. This is how Chu and Apfel (1984) defined a "perfectly absorbing" wall, in response to the comments of Nyborg and Rooney (1984) concerning an earlier paper by Chu and Apfel (1982). In Eq. (6.22), the right-hand side is evaluated using Eq. (6.30) for the first term and Eq. (6.28) for the second. On the left-hand side, conservation of mass is still applicable, but we must allow the fluid volume to change. Let the length of the tube be L in the absence of sound, and in the presence of sound, let the wall be displaced by ΔL. For a pure progressive wave there is no preferred spatial reference point, and the Eulerian density $\langle \rho^{\mathrm{E}} \rangle$ should be independent of x. Conservation of mass requires $\rho_0 L = \langle \rho^{\mathrm{E}} \rangle (L + \Delta L)$, or

$$\langle \rho^{\mathrm{E}} - \rho_0 \rangle = -\rho_0 \left(\frac{\Delta L/L}{1 + \Delta L/L} \right). \tag{6.32}$$

Substitution into Eq. (6.22) yields, following some rearrangement, an expression for C:

$$C = \left(\frac{\gamma - 1}{2} \right) \langle E_i \rangle - \rho_0 c_0^2 \left(\frac{\Delta L/L}{1 + \Delta L/L} \right). \tag{6.33}$$

Now let the absorbing wall be supported from behind by a force due to a spring with constant κ. Then $\Delta L/L$ is determined by the balance between $\langle P^{\mathrm{L}} - P_0 \rangle$ from Eq. (6.31) and the spring force:

$$\kappa \Delta L = \langle P^{\mathrm{L}} - P_0 \rangle S = [C + \langle E_i \rangle] S, \tag{6.34}$$

where S is the cross-sectional area of the tube. Combining Eqs. (6.33) and (6.34) to eliminate ΔL, keeping only terms up to second order, we find

$$C = \left[\frac{\gamma + 1}{2(1 + \kappa_0/\kappa)} - 1\right]\langle E_i\rangle, \tag{6.35}$$

where $\kappa_0 = \rho_0 c_0^2 S/L$ is the effective spring constant associated with adiabatic compression of the fluid in the tube. Substituting Eq. (6.35) into Eqs. (6.30) and (6.31), we have

$$\langle P^E - P_0\rangle = \left[\frac{\gamma + 1}{2(1 + \kappa_0/\kappa)} - 1\right]\langle E_i\rangle \tag{6.36}$$

and

$$\langle P^L - P_0\rangle = \left[\frac{\gamma + 1}{2(1 + \kappa_0/\kappa)}\right]\langle E_i\rangle. \tag{6.37}$$

Since $\langle P^L - P_0\rangle$ in Eq. (6.37) is independent of x and $C \neq 0$, it is the Rayleigh radiation pressure experienced by the absorbing surface:

$$\langle \Delta P\rangle = \left[\frac{\gamma + 1}{2(1 + \kappa_0/\kappa)}\right]\langle E_i\rangle. \tag{6.38}$$

Substituting Eq. (6.35) into Eq. (6.34) and using the definition of κ_0 yields

$$\frac{\Delta L}{L} = \tfrac{1}{2}(\gamma + 1)\left(\frac{\kappa_0/\kappa}{1 + \kappa_0/\kappa}\right)\frac{\langle E_i\rangle}{\rho_0 c_0^2}. \tag{6.39}$$

Equations (6.38) and (6.39) were obtained by Chu and Apfel (1984). Even if there is a mean density change in the fluid, the radiation pressure can still be of the Rayleigh type.

Two limiting cases of Eqs. (6.36) and (6.37) are now considered. For $\kappa_0/\kappa \ll 1$, which corresponds to an immovable absorbing wall, the equations become

$$\langle P^E - P_0\rangle = \tfrac{1}{2}(\gamma - 1)\langle E_i\rangle \tag{6.40}$$

and

$$\langle \Delta P\rangle = \langle P^L - P_0\rangle = \tfrac{1}{2}(\gamma + 1)\langle E_i\rangle \tag{6.41}$$

(Rooney and Nyborg, 1972). For $\kappa_0/\kappa = 1$, which defines a *truly* perfectly absorbing wall because even the spring constants on the two sides match, we have

$$\langle P^E - P_0\rangle = \tfrac{1}{4}(\gamma - 3)\langle E_i\rangle \tag{6.42}$$

and

$$\langle \Delta P \rangle = \langle P^{\text{L}} - P_0 \rangle = \tfrac{1}{4}(\gamma + 1)\langle E_i \rangle \tag{6.43}$$

(Westervelt, 1950). The differences between Eqs. (6.40) and (6.41) and Eqs. (6.42) and (6.43) demonstrate the sensitivity of radiation pressure to small variations in the boundary conditions. Both sets of results are thus correct if properly interpreted.

Let us consider a variation of the problem. The above results remain the same if the wave exists in a tube. If a hole is opened on the side wall, however, then $\langle P^{\text{E}} - P_0 \rangle = 0$ is obtained near the hole. From Eq. (6.30) we thus have $C = 0$, and it follows that for all x

$$\langle P^{\text{E}} - P_0 \rangle = 0, \tag{6.44}$$

and from Eq. (6.31)

$$\langle \Delta P \rangle = \langle P^{\text{L}} - P_0 \rangle = \langle E_i \rangle, \tag{6.45}$$

which is a Langevin radiation pressure. Instead of a hole, one can insert instead a freely moving side piston that allows the inner and outer pressures to equalize, and the same results are obtained (Beyer, 1950, 1974, 1978).

6.2.2.3   Sound Beam Incident on a Partially Reflecting Interface

Consider a flat interface at $z = 0$ between fluids 1 and 2, which occupy the regions $z < 0$ and $z > 0$, respectively, have corresponding densities ρ_1 and ρ_2, and sound speeds c_1 and c_2. Let a sound beam propagate through medium 1 toward the interface at normal incidence, and for simplicity let it be described by a single term in its Fourier–Bessel series representation:

$$p_i = A J_0(k_r r) \exp[i(K_1 z - \omega t)], \tag{6.46}$$

where $k_1 = \omega / c_1$, k_r is the reciprocal of the characteristic width of the beam, and $K_1 = (k_1^2 - k_r^2)^{1/2}$ is the effective wave number in the z direction, with $K_1 \gg k_r$. In general, the beam is partially reflected and partially transmitted. Let the reflected beam be described by

$$p_r = B J_0(k_r r) \exp[-i(K_1 z + \omega t)], \tag{6.47}$$

and the transmitted beam by

$$p_t = D J_0(k_r r) \exp[i(K_2 z - \omega t)], \tag{6.48}$$

where $k_2 = \omega/c_2$, and $K_2 = (k_2^2 - k_r^2)^{1/2}$ is the effective wave number in the z direction, with $K_2 \gg k_r$. The corresponding z components of the acoustic particle velocities are

$$w_i = \frac{K_1 A}{\omega \rho_1} J_0(k_r r) \exp[i(K_1 z - \omega t)], \tag{6.49}$$

$$w_r = -\frac{K_1 B}{\omega \rho_1} J_0(k_r r) \exp[-i(K_1 z + \omega t)], \tag{6.50}$$

and

$$w_t = \frac{K_2 D}{\omega \rho_2} J_0(k_r r) \exp[i(K_2 z - \omega t)], \tag{6.51}$$

and the r components are

$$u_i = \frac{k_r A}{i \omega \rho_1} J_0'(k_r r) \exp[i(K_1 z - \omega t)], \tag{6.52}$$

$$u_r = \frac{k_r B}{i \omega \rho_1} J_0'(k_r r) \exp[-i(K_1 z + \omega t)], \tag{6.53}$$

and

$$u_t = \frac{k_r D}{i \omega \rho_2} J_0'(k_r r) \exp[i(K_2 z - \omega t)]. \tag{6.54}$$

Imposing the boundary condition that the acoustic pressures on the two sides of the interface be continuous, i.e., $p_i + p_r = p_t$, we find that $A + B = D$. Imposing another boundary condition, that the z components of the particle velocities on the two sides be continuous, i.e., $w_i + w_r = w_t$, we find that $A - B = ZD$, where $Z = \rho_1 c_1 / \rho_2 c_2$ is the ratio of the impedances of media 1 and 2, with terms of order $(k_r/k_1)^2$ or $(k_r/k_2)^2$ neglected. Thus

$$B = \left(\frac{1 - Z}{1 + Z}\right) A \tag{6.55}$$

and

$$D = \left(\frac{2}{1 + Z}\right) A. \tag{6.56}$$

In medium 1, the total acoustic pressure is $p_1 = p_i + p_r$, and the total particle velocity is given by $w_1 = w_i + w_r$ and $u_1 = u_i + u_r$. Similarly, in medium 2, $p_2 = p_t$, $w_2 = w_t$ and $u_2 = u_t$. Substituting these expressions, using the real parts of Eqs. (6.46)–(6.54), into Eqs. (6.13) and (6.15), and setting $C = 0$ since this must be true for the unperturbed interface far from the beam, we obtain in a straightforward way for medium 1

$$
\langle P_1^E - P_0 \rangle = \frac{AB \cos 2K_1 z}{\rho_1 c_1^2} [J_0(k_r r)]^2
$$

$$
+ \frac{1}{4\rho_1 c_1^2} \left(\frac{k_r}{k_1}\right)^2 \left[(A^2 + B^2)\{[J_0(k_r r)]^2 - [J_0'(k_r r)]^2\} \right.
$$

$$
\left. - 2AB \cos 2K_1 z \{[J_0(k_r r)]^2 + [J_0'(k_r r)]^2\} \right] \tag{6.57}
$$

and

$$
\langle P_1^L - P_0 \rangle = \frac{A^2 + B^2}{2\rho_1 c_1^2} [J_0(k_r r)]^2
$$

$$
+ \frac{1}{4\rho_1 c_1^2} \left(\frac{k_r}{k_1}\right)^2 \left[-(A^2 + B^2)\{[J_0(k_r r)]^2 + [J_0'(k_r r)]^2\} \right.
$$

$$
\left. + 2AB \cos 2K_1 z \{[J_0(k_r r)]^2 - [J_0'(k_r r)]^2\} \right], \tag{6.58}
$$

and for medium 2

$$
\langle P_2^E - P_0 \rangle = \frac{D^2}{4\rho_2 c_2^2} \left(\frac{k_r}{k_2}\right)^2 \{[J_0(k_r r)]^2 - [J_0'(k_r r)]^2\} \tag{6.59}
$$

and

$$
\langle P_2^L - P_0 \rangle = \frac{D^2}{4\rho_2 c_2^2} \left([J_0(k_r r)]^2 - \frac{1}{2}\left(\frac{k_r}{k_2}\right)^2 \{[J_0(k_r r)]^2 + [J_0'(k_r r)]^2\} \right). \tag{6.60}
$$

Equations (6.57)–(6.60) vanish for large $k_r r$, as expected for a beam.

We now consider the limiting forms of Eqs. (6.57)–(6.60) for $(k_r/k_1)^2 \ll 1$, $(k_r/k_2)^2 \ll 1$. In the region $k_r r \ll 1$, the disturbance is a plane wave. In the region $k_r r \gg 1$, the sound field vanishes. The solutions therefore approximate those for the idealized case of a collimated plane-wave beam. Using Eqs. (6.55) and (6.56), we obtain the following simplified results: for medium 1,

$$
\langle P_1^E - P_0 \rangle = 2\langle E_i \rangle \left(\frac{1-Z}{1+Z}\right) \cos 2K_1 z \tag{6.61}
$$

and

$$\langle P_1^L - P_0 \rangle = 2\langle E_i \rangle \left(\frac{1 + Z^2}{(1 + Z)^2} \right), \tag{6.62}$$

and for medium 2,

$$\langle P_2^E - P_0 \rangle = 0 \tag{6.63}$$

and

$$\langle P_2^L - P_0 \rangle = \frac{4Zn}{(1 + Z)^2} \langle E_i \rangle, \tag{6.64}$$

where $\langle E_i \rangle = A^2/2\rho_1 c_1^2$ is the energy density of the incident wave, and $n = c_1/c_2$. From Eqs. (6.62) and (6.64), the difference in mean Lagrangian excess pressures at the interface between media 1 and 2, $\langle \Delta P^L \rangle = \langle P_1^L - P_2^L \rangle$, is

$$\langle \Delta P^L \rangle = \frac{2\langle E_i \rangle}{(1 + Z)^2}(1 + Z^2 - 2nZ). \tag{6.65}$$

Beyer (1974) and Landau and Lifshitz (1987; Sect. 66) obtained the same result. Equation (6.65) explains the "acoustic fountain," in which the radiation pressure of a sound beam propagating from one fluid into another displaces the interface, leading to the formation of a jet (Hertz and Mende, 1939; see also Beyer, 1974). The direction in which the interface is displaced, i.e., into medium 1 or medium 2, depends on the combination of Z and n.

6.3   Acoustic Levitation

6.3.1   Acoustic Radiation Force on a Small Sphere

In a sound field, the radiation stress acting on the surface of a sphere leads to an acoustic radiation force. Consider the long-wavelength limit $kR \ll 1$, where R is the radius of the sphere. In practice, we typically have $R \gg \delta = (\nu/\omega)^{1/2}$, where δ is the viscous boundary-layer thickness of the sound field, and ν is the kinematic viscosity of the host medium. We may thus neglect viscosity. To calculate the force, one usually specifies, for example, whether the sphere is rigid or compressible. We shall consider the following practical cases: (1) a solid or liquid sphere in a gaseous medium, (2) a liquid sphere in a liquid medium, and (3) a gas bubble in a liquid medium.

6.3.1.1 Rigid Sphere in a Gaseous Medium

Since liquid or solid is approximately 1000 times more dense than gas, we may consider the sphere as an immovable mass. Moreover, since the acoustic impedances of the sphere material and the gas are so mismatched, and acoustic resonance in the sphere is not possible in the limit $kR \ll 1$, we may consider the sphere to be rigid. King (1934) considered arbitrary densities for the sphere and the host medium. Since he did not also consider the compressibility of the sphere, his theory is valid only for a heavy rigid sphere in a gaseous medium.

Let the incident wave be described by $p_i = p_{i0} \exp(-i\omega t)$, where

$$p_{i0} = A \sin kz, \tag{6.66}$$

and therefore $z = 0$ is the location of a pressure node. Let the sphere be located on the z axis, at $z = Z$. Introduce a spherical coordinate system (r, θ) with its origin at the center of the sphere, and with the direction $\theta = 0$ aligned with the positive z axis. For a point (r, θ), we have $z = Z + r \cos \theta$. Substitution into Eq. (6.66), and using the identity

$$\exp(ikr \cos \theta) = \sum_{n=0}^{\infty} (2n + 1)i^n j_n(kr) P_n(\cos \theta) \tag{6.67}$$

(Abramowitz and Stegun, 1965), we have

$$p_{i0} = \sum_{n=0}^{\infty} (2n + 1)A_n j_n(kr) P_n(\cos \theta), \tag{6.68}$$

where

$$A_n = \frac{A}{2i} i^n [e^{ikZ} - (-1)^n e^{-ikZ}], \tag{6.69}$$

j_n is the spherical Bessel function, and P_n is the Legendre polynomial. Let the wave scattered from the sphere's surface S be described by $p_r = p_{r0} \exp(-i\omega t)$, where p_{r0} can be expanded as

$$p_{r0} = \sum_{n=0}^{\infty} B_n h_n^{(1)}(kr) P_n(\cos \theta), \tag{6.70}$$

in which $h_n^{(1)}$ is the spherical Hankel function of the first kind, which for large kr, together with the factor $\exp(-i\omega t)$, describes a spherical wave propagating outward from the center. The total sound pressure $p = p_0 \exp(-i\omega t)$, where $p_0 = p_{i0} + p_{r0}$, has a zero normal derivative on S, and thus the normal component of the particle velocity at the rigid surface is zero. We therefore have

$$
\left(\frac{\partial p_{i0}}{\partial r}\right)_{r=R} = -\left(\frac{\partial p_{r0}}{\partial r}\right)_{r=R}.
\tag{6.71}
$$

Substituting Eqs. (6.68) and (6.70) into Eq. (6.71) and equating terms with the same n, we find

$$
B_n = -\frac{(2n+1)j_n'(kR)}{h_n^{(1)'}(kR)} A_n.
\tag{6.72}
$$

From Eqs. (6.68)–(6.70) and (6.72), we thus obtain for the total sound pressure at $r = R$

$$
p_0 = \sum_{n=0}^{\infty}(2n+1)\left[j_n(kR) - \frac{j_n'(kR)h_n^{(1)}(kR)}{h_n^{(1)'}(kR)}\right] A_n P_n(\cos\theta).
\tag{6.73}
$$

In the limit of small kR, Eq. (6.73) reduces to

$$
p_0 = \left[1 - \frac{(kR)^2}{2}\right] A \sin kZ + \tfrac{3}{2}(kR)A\cos kZ\, P_1(\cos\theta)
$$
$$
- \tfrac{5}{9}(kR)^2 A \sin kZ\, P_2(\cos\theta)
\tag{6.74}
$$

at order $(kR)^2$. The tangential component of the particle velocity at $r = R$ is $u_\theta = u_{\theta0}\exp(-i\omega t)$, where $u_{\theta0} = (i\omega\rho_0 R)^{-1}(\partial p_0/\partial\theta)$. Equation (6.74) yields

$$
u_{\theta0} = \frac{3i}{2}\frac{A}{\rho_0 c_0}\cos kZ\sin\theta - \frac{5i}{3}\frac{(kR)A}{\rho_0 c_0}\sin kZ\cos\theta\sin\theta.
\tag{6.75}
$$

At the rigid surface, the normal component of the velocity vanishes. The sound field given by Eqs. (6.74) and (6.75) can be substituted into Eq. (6.13) to evaluate $\langle P^{\mathrm{E}} - P_0\rangle$ at $r = R$:

$$
\langle P^{\mathrm{E}} - P_0\rangle = \frac{A^2}{4\rho_0 c_0^2}[\sin^2 kZ + \tfrac{3}{2}(kR)\sin 2kZ\cos\theta
$$
$$
- \tfrac{9}{4}\cos^2 kZ\sin^2\theta + \tfrac{5}{2}(kR)\sin 2kZ\sin^2\theta\cos\theta].
\tag{6.76}
$$

In deriving Eq. (6.76) we have set $C = 0$ in Eq. (6.13), because the system is open, with a sphere scattering an infinite plane wave into an unbounded space. As far as the radiation pressure is concerned, there is no constraint on the system such as conservation of volume or mass, and consequently there is no need for a nonzero value of C.

The integral over S of the normal component of the stress in Eq. (6.7) leads to the acoustic radiation force. Since the normal component of the velocity is zero, the stress in Eq. (6.7) becomes $S_{ij}n_j = -\langle P^E - P_0 \rangle n_i$, where $\mathbf{n} = -\mathbf{e}_r$. By axisymmetry, the force acting on the sphere has only a z component. With $\mathbf{n} \cdot \mathbf{e}_z = -\cos\theta$, the force is given by

$$F_z = -\int_S \langle P^E - P_0 \rangle \cos\theta \, dS. \tag{6.77}$$

Substituting Eq. (6.76) and the relation $dS = 2\pi R^2 \sin\theta \, d\theta$, and integrating from $\theta = 0$ to $\theta = \pi$, we find

$$F_z = -\frac{5\pi}{6} \frac{A^2 k R^3}{\rho_0 c_0^2} \sin 2kZ \tag{6.78}$$

(King, 1934). The force thus pushes the sphere toward the pressure node at $Z = 0$. For small kZ, such that $\sin 2kZ \approx 2kZ$, the force acts like a spring that keeps the sphere at the pressure node. The potential well U defined by $F_z = -dU/dZ$ is

$$U = -\frac{5\pi}{12} \frac{A^2 R^3}{\rho_0 c_0^2} \cos 2kZ. \tag{6.79}$$

For a progressive wave $p_i = p_{i0} \exp(-i\omega t)$, where $p_{i0} = A \exp(ikz)$, the corresponding force is

$$F_z = \frac{11\pi}{18} \frac{A^2 k^4 R^6}{\rho_0 c_0^2} \tag{6.80}$$

(King, 1934), which is of order $(kR)^3$ times the force produced by the standing wave, given by Eq. (6.78). For a progressive plane wave, the force is due to the deflection of incident-wave momentum. Measurements of the acoustic radiation force on a sphere, reported by Rudnick (1977) and by Leung et al. (1981), are consistent with King's theory for a standing wave [Eq. (6.78); see Fig. 6.1].

For the purpose of levitation, two standing waves are sometimes needed. If the frequencies are different, the total force is the vector sum of the forces due to the two waves acting independently, because time averaging eliminates any coupling between them. If the frequencies are the same, however, then the coupling is finite, and the force is affected. Consider two pressure waves $p_{ix} = p_{ix0} \exp(-i\omega t)$ and $p_{iy} = p_{iy0} \exp(-i\omega t)$ that produce corresponding vibrations in the x and y directions, where $p_{ix0} = A_x \sin kx$, $p_{iy0} = A_y \exp(i\psi) \sin ky$, A_x and A_y are pressure amplitudes, and ψ is the phase difference. The origin is located at a pressure node. Let $\alpha = A_y/A_x$ ($0 < \alpha \leq 1$), and let the sphere be centered at (X, Y). With the sphere expected to remain near $X = Y = 0$, the potential well

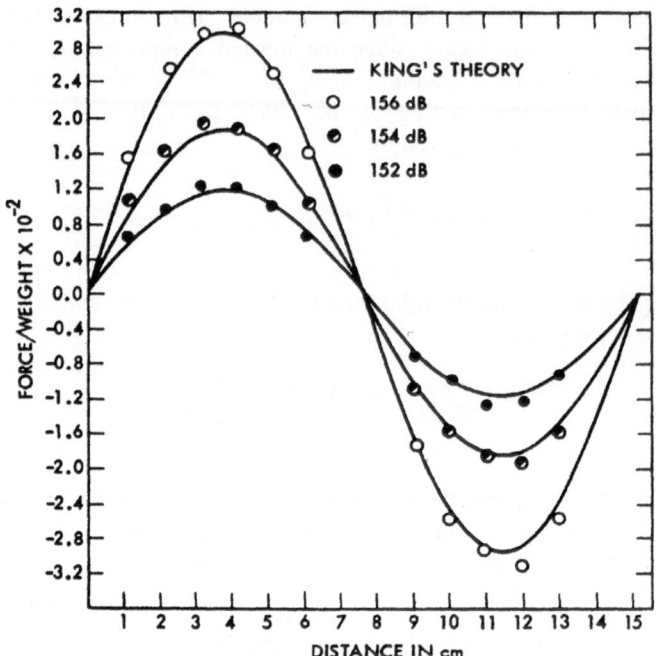

Fig. 6.1 Force/weight of a sphere vs. sphere position for a plane standing wave at three sound pressure levels, compared with King's theory. The sound pressure levels were measured using a microphone attached on an end wall, where the pressure antinode was located. Second and higher harmonics of the fundamental mode were generated by nonlinearity and observed on a spectrum analyzer, giving the microphone output signal a sawtooth appearance. In an attempt to restore the wave to a sinusoidal form, a second harmonic was introduced by the speaker into the chamber to cancel the second harmonic generated in the fluid (from Leung et al., 1981, Fig. 4, p. 1764).

U, related to the force $\mathbf{F}$ by $\mathbf{F} = -\nabla U$, is given approximately by Lee and Wang (1988a):

$$U = \frac{5\pi}{6} \frac{A_x^2 k^2 R^3}{\rho_0 c_0^2} (X^2 + \alpha^2 Y^2 + \tfrac{4}{5}\alpha XY \cos\psi). \tag{6.81}$$

If $\psi = \pi/2$, the potential well is just the sum of the contributions from the two waves. Otherwise, the two waves are coupled. The contours of the well are in general elliptical, oriented at an angle depending on α and ψ. If the sphere is instead a liquid drop, it will be flattened along its major axis. Only for the special case $\psi = \pi/2$ and $\alpha = 1$ are the contours circular.

6.3.1.2   Liquid Sphere in an Immiscible Liquid Medium

Yosioka and Kawasima (1955) considered the general case of a compressible fluid sphere in a compressible fluid medium. For a drop in a liquid, the compressibility and density of the drop are comparable with those of the medium, so that the vibration of the drop must be taken into account. Therefore, the normal velocity on the surface S is not required to be zero. The force is obtained by integrating the normal component of the radiation stress tensor in Eq. (6.7) over S. Alternatively, we shall derive the results using the integral momentum equation (Westervelt, 1957). By integrating Eq. (6.5) over the space between S and a much larger spherical surface S_0 concentric with the sphere, and using Gauss's theorem, we obtain a surface integral of $S_{ij}n_j$ over $S' = S + S_0$, equated to zero, where n_j is the normal unit vector on the surface S' pointing away from the enclosed space. Since the integral of $-S_{ij}n_j$ over S is the force F_i acting on the sphere, we have, using Eq. (6.7),

$$F_i = \int_S (\langle P^E - P_0 \rangle \delta_{ij} + \rho_0 \langle u_i u_j \rangle) n_j \, dS$$

$$= -\int_{S_0} (\langle P^E - P_0 \rangle \delta_{ij} + \rho_0 \langle u_i u_j \rangle) n_j \, dS, \qquad (6.82)$$

where $\mathbf{n} = -\mathbf{e}_r$ on S and $\mathbf{n} = \mathbf{e}_r$ on S_0. We now calculate the force in terms of far-field relations. Let $\phi_i = \phi_{i0} \exp(-i\omega t)$ be the velocity potential of the incident wave, and $\phi = \phi_0 \exp(-i\omega t)$ be that of the total wave. At large kr, the velocity potential approaches the form

$$\phi_0 = \phi_{i0} + \frac{f(\theta)}{r} \exp(ikr), \qquad (6.83)$$

where $f(\theta)$ is determined by the boundary conditions on S. Using Eq. (6.83), and setting $C = 0$ in Eq. (6.12) [for the same reason given just after Eq. (6.76)], we can evaluate $\langle P^E - P_0 \rangle$. Substituting $\langle P^E - P_0 \rangle$ and $\mathbf{u} = \nabla \phi$ into Eq. (6.82), we obtain

$$F_z = -\pi \rho_0 k^2 \int_{-1}^{1} |f(\theta)|^2 \cos\theta \, d(\cos\theta)$$

$$- \pi \rho_0 r^2 \int_{-1}^{1} \mathrm{Re}\left(\frac{\partial \phi_{i0}}{\partial r} \frac{\partial \phi_{i0}^*}{\partial z}\right) d(\cos\theta)$$

$$- \frac{\pi \rho_0 r^2}{2} \int_{-1}^{1} (k^2 |\phi_{i0}|^2 - |\nabla \phi_{i0}|^2) \cos\theta \, d(\cos\theta)$$

$$- \pi \rho_0 r^2 k^2 \int_{-1}^{1} \mathrm{Re}\left(\phi_{i0}^* \frac{f(\theta)}{r} \exp(ikr)\right) \cos\theta \, d(\cos\theta)$$

$$+ \pi \rho_0 r^2 k \int_{-1}^{1} \mathrm{Im}\left(\frac{\partial \phi_{i0}^*}{\partial z} \frac{f(\theta)}{r} \exp(ikr)\right) d(\cos\theta), \qquad (6.84)$$

where Re and Im denote real and imaginary parts. The incident wave is given by Eq. (6.66) and expanded as in Eqs. (6.68) and (6.69). The scattered wave is given by Eq. (6.70). The sound wave inside the sphere is given by $p_t = p_{t0} \exp(-i\omega t)$, where

$$p_{t0} = \sum_{n=0}^{\infty} D_n j_n(kr) P_n(\cos\theta). \qquad (6.85)$$

The boundary condition that the acoustic pressures be continuous at $r = R$ gives

$$A_n(2n+1) j_n(kR) + B_n h_n^{(1)}(kR) = D_n j_n(k_1 R), \qquad (6.86)$$

where $k_1 = \omega/c_1$, in which c_1 is the speed of sound in the fluid sphere. The condition that the velocities be continuous at $r = R$ leads to

$$\lambda\sigma[A_n(2n+1) j_n'(kR) + B_n h_n^{(1)'}(kR)] = D_n j_n'(k_1 R), \qquad (6.87)$$

where $\lambda = \rho_1/\rho_0$, $\sigma = c_1/c_0$, and ρ_1 is the density of the sphere. When both the sphere and the medium are liquids, we expect that $\lambda \sim 1$ and $\sigma \sim 1$. We thus have $kR \sim k_1 R \ll 1$. Using Eqs. (6.86) and (6.87) to solve for B_n leads to

$$B_n = -\frac{A_n(2n+1)[j_n(kR) j_n'(k_1 R) - \lambda\sigma j_n'(kR) j_n(k_1 R)]}{h_n^{(1)}(kR) j_n'(k_1 R) - \lambda\sigma h_n^{(1)'}(kR) j_n(k_1 R)}. \qquad (6.88)$$

Substituting Eq. (6.88) into Eq. (6.86) to solve for D_n gives

$$D_n = -\frac{i\lambda\sigma A_n(2n+1)}{(kR)^2[h_n^{(1)}(kR) j_n'(k_1 R) - \lambda\sigma h_n^{(1)'}(kR) j_n(k_1 R)]}, \qquad (6.89)$$

where we have made use of the Wronskian relation $(j_n h_n' - j_n' h_n)(x) = i/x^2$. The force can be evaluated from the radiation stress tensor on S, making use of the expression for the field on the inside surface with D_n given by Eq. (6.89).

In our case, the far field is given by the incident wave represented by Eq. (6.68), and the asymptotic form of the scattered wave represented by Eq. (6.70). Using the scattered potential $\phi_{r0} = p_{r0}/i\omega\rho_0$ for the second term on the right-hand side of Eq. (6.83), with the scattered acoustic pressure p_{r0} given by Eq. (6.70), we obtain for large kr

$$f(\theta) = \frac{1}{i\omega\rho_0 k} \sum_{n=0}^{\infty} (-i)^{n+1} B_n P_n(\cos\theta). \qquad (6.90)$$

Expanding B_n in Eq. (6.88) to lowest order in kR, using $\lambda \sim 1$, $\sigma \sim 1$, and $kR \sim k_1 R \ll 1$, we find that B_0 and B_1 are of order $(kR)^3 A$, and for $n > 1$, $B_n \sim$

$(kR)^{2n+1}A$. We then have

$$
f(\theta) = \frac{AkR^3}{\rho_0 c_0} \left[\left(\frac{1-\lambda}{1+2\lambda} \right) \cos kZ \cos \theta + \frac{i}{3} \left(1 - \frac{1}{\lambda \sigma^2} \right) \sin kZ \right]. \tag{6.91}
$$

Substituting Eq. (6.91), and Eq. (6.66) with $z = Z + r \cos \theta$, into Eq. (6.84), using $\phi_{i0} = p_{i0}/i\omega\rho_0$, and the radiation condition to eliminate the dependence on r, we find

$$
F_z = -\frac{\pi A^2 k R^3}{\rho_0 c_0^2} \left\{ \left[\frac{\lambda + \frac{2}{3}(\lambda - 1)}{1 + 2\lambda} \right] - \frac{1}{3\lambda \sigma^2} \right\} \sin 2kZ, \tag{6.92}
$$

where the first term inside the curly braces comes from the interaction of the incident wave with the dipole scattered field, and the second term from the interaction with the monopole scattered field. The latter effect depends on the compressibility of the sphere. In the limit $\lambda \to \infty$, Eq. (6.92) reduces to King's formula, Eq. (6.78). In the $\sigma\lambda$ plane, the curve $\sigma(\lambda)$ decreases monotonically from infinity at $\lambda = 0$ to zero at $\lambda = \infty$, passing through the point $(1,1)$ in between. Unlike the case with King's formula, the force predicted by Eq. (6.92) can point toward either a pressure node or an antinode, depending on whether the point (λ, σ) lies above or below this curve, respectively (Yosioka and Kawasima, 1955). Crum (1971) measured the acoustic pressure required to hold different small droplets in an immiscible liquid medium at different positions, and found good agreement with Eq. (6.92) (Fig. 6.2). For a progressive plane wave $p_i = p_{i0} \exp(-i\omega t)$, where $p_{i0} = A \exp(ikz)$, the force is evaluated in a similar fashion after Eq. (6.84) is simplified by conservation of energy (Lee and Wang, 1984):

$$
F_z = \frac{2\pi A^2 k^4 R^6}{\rho_0 c_0^2 (1 + 2\lambda)^2} \left\{ \left(\lambda - \frac{1+2\lambda}{3\lambda\sigma^2} \right)^2 + \frac{2}{9}(1 - \lambda)^2 \right\}, \tag{6.93}
$$

which agrees with the result obtained by Yosioka and Kawasima, and which reduces to King's result, Eq. (6.80), in the limit $\lambda \to \infty$. Equation (6.93) is of order $(kR)^3$ times its counterpart for the standing wave, Eq. (6.92). The force always points in the direction of the wave propagation, as expected because it is due to the deflection of wave momentum.

Gor'kov (1962) calculated the force in terms of the local incident wave. His formula is valid for any wave except a progressive plane wave. His separate result for the latter agrees with Eq. (6.93). For any other wave, let

$$
f_1 = 1 - \frac{1}{\lambda\sigma^2} \tag{6.94}
$$

and

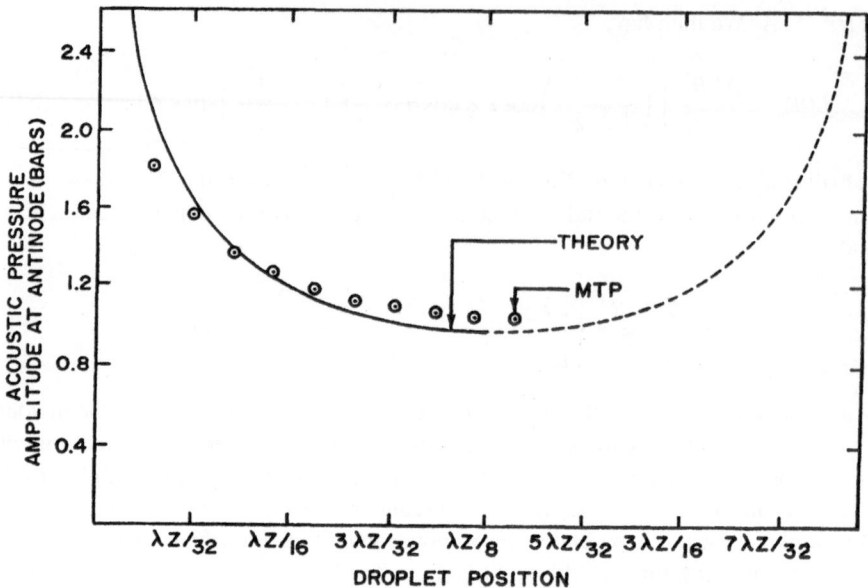

Fig. 6.2 Acoustic pressure amplitude vs. droplet position for paraldehyde droplets in water. The solid line is the theoretical prediction [Crum (1971) derived Eq. (6.92) differently] for $\lambda = \rho_1/\rho_0 = 0.994$, $\sigma = c_1/c_0 = 0.776$, and $k = 0.81$ cm^{-1}. The droplet position is the displacement from the pressure antinode at the left. MTP stands for minimum trapping pressure (from Crum, 1971, Fig. 3, p. 161).

$$f_2 = \frac{2(\lambda - 1)}{2\lambda + 1}. \tag{6.95}$$

The force is given by $\mathbf{F} = -\nabla U$, where

$$U = 2\pi R^3 \left(\frac{f_1}{3\rho_0 c_0^2} \langle p_i^2 \rangle - \frac{f_2 \rho_0}{2} \langle \mathbf{u}_i \cdot \mathbf{u}_i \rangle \right), \tag{6.96}$$

and where p_i and $\mathbf{u}_i$ are, respectively, the incident acoustic pressure and particle velocity at the position of the center of the sphere. For a plane standing wave, Eq. (6.96) agrees with Eq. (6.92). Marston (1980) showed that a drop can be levitated, deformed, and excited into oscillations in an immiscible liquid host using an amplitude-modulated wave.

6.3.1.3 Bubble in a Liquid Medium

The possible resonance between the wave and the monopole oscillation of the bubble is even more important than the bubble compressibility (Yosioka and

Kawasima, 1955; Eller, 1968). The density of gas is much less than that of liquid, but the speed of sound in gas is only a few times less, and thus $\lambda \ll 1$ but $\sigma \sim 1$. The incident wave is given by Eq. (6.66). The analysis begins with Eqs. (6.82)–(6.90). Let us expand B_n in Eq. (6.88) to lowest order in kR. The condition $(kR)^2 = 3\lambda\sigma^2$ in the denominator $\{-kR + i[1 - 3\lambda\sigma^2/(kR)^2]\}/3\sigma$ for B_0 in Eq. (6.88) is associated with the monopole oscillation resonance. Thus $B_0 = -[\sigma(kR/\sigma)^3 A \sin kZ]/\{\sigma(kR/\sigma)^3 + i[3\lambda - (kR/\sigma)^2]\} \sim O(A)$, but for $n > 1$ we have $B_n \sim (kR)^{2n+1} A$, and B_n can be ignored. We then have

$$
f(\theta) = \frac{kR^3 A}{\rho_0 c_0} \left\{ \cos kZ \cos\theta + \frac{\sin kZ}{\sigma^2 \left\{ \sigma \left(\frac{kR}{\sigma}\right)^3 + i \left[3\lambda - \left(\frac{kR}{\sigma}\right)^2 \right] \right\}} \right\}. \tag{6.97}
$$

Equations (6.97) and (6.66) are substituted into Eq. (6.84) to obtain the force:

$$
F_z = \frac{\pi A^2 kR^3}{\rho_0 c_0^2} \left\{ \frac{\frac{1}{\sigma^2}\left[3\lambda - \left(\frac{kR}{\sigma}\right)^2\right]}{\sigma^2 \left(\frac{kR}{\sigma}\right)^6 + \left[3\lambda - \left(\frac{kR}{\sigma}\right)^2\right]^2} \right\} \sin 2kZ \tag{6.98}
$$

(Wu and Du, 1990; Löfstedt and Putterman, 1991). The corresponding result of Yosioka and Kawasima (1955) differs by a factor of $\sigma^2/(kR)^2$ because of an algebraic error. Equation (6.98) agrees with the more simple theory of Eller (1968), except that the latter is singular at the monopole resonance. Using $\rho_1 c_1^2 = 3\alpha P_0$, where here $\alpha = \gamma$ or $\alpha = 1$ for an adiabatic or isothermal bubble, respectively, the condition $(kR)^2 = 3\lambda\sigma^2$ yields the monopole frequency $\omega_0 = (3\alpha P_0/\rho_0 R^2)^{1/2}$ [see Eq. (5.47)]. For a given bubble, the force points toward the pressure antinode for $\omega < \omega_0$, and toward the node for $\omega > \omega_0$. Alternatively, for a given frequency ω, the bubble travels toward the pressure antinode for $R < R_0$, and toward the node for $R > R_0$, where $R_0 = (3a P_0/\rho_0 \omega^2)^{1/2}$. The bubble oscillates in phase with the sound if excited below resonance (Prosperetti, 1986). When the pressure rises, the bubble is pushed toward the pressure node while it is compressed, and vice versa when the pressure falls. The second half-cycle is more important because for the same force per unit volume, the bubble has a larger volume. On average, the bubble is thus pushed toward the pressure antinode. The bubble oscillates 180° out of phase with the sound if excited above resonance, and the opposite happens. For a progressive plane wave $p_i = p_{i0} \exp(-i\omega t)$, where $p_{i0} = A \exp(ikz)$, the force is

$$
F_z = \frac{2\pi A^2 k^4 R^6}{\rho_0 c_0^2 \sigma^4 \left\{ \sigma^2 \left(\frac{kR}{\sigma}\right)^6 + \left[3\lambda - \left(\frac{kR}{\sigma}\right)^2 \right]^2 \right\}}, \tag{6.99}
$$

which is always positive and does not vanish at the monopole resonance.

6.3.1.4   Sphere in a Gaseous Medium with a Temperature Profile

Investigations involving nonuniform temperature are motivated by the need to heat and cool samples in containerless processing. For this purpose, a chamber with a nonuniform temperature can be used (Trinh et al., 1986; Robey et al., 1987; Collas and Barmatz, 1987). An intense focused light beam can also be used to heat a sample in a chamber at room temperature. For the latter scheme, the force is perturbed by the thermal profile around the sphere (Lee and Wang, 1984, 1988b). Without the profile, the sphere prevents the air in its volume from oscillating, and at the same time the sphere experiences a force. The obstruction can be considered correlated to the force. If the air around the sphere is heated, the lighter hot air vibrates more than the ambient air, reducing the obstruction and therefore the force. Heating can even reverse the force, sending the sphere to the wall. The heated air near the poles ($\theta = 0, \pi$) is at the stagnation points of the vibration and cannot move. Only the air around the equator ($\theta = \pi/2$) can affect the force. However, acoustic streaming (Nyborg, 1965; Lee and Wang, 1990; see also Chap. 7) tends to drive the heated air from the equator toward the poles (Oran et al., 1979; Leung and Wang, 1985). If the streaming is strong enough to clear the heated air from the equator, the thermal effect on the force is negligible (Lee and Wang, 1988b). Another effect of strong acoustic streaming is the enhancement of heat transfer from the sample (Lee and Wang, 1988b; Gopinath and Mills, 1993).

It was observed by Leung and Wang (1985) that for such a sphere, the percentage change in the force is negligible at high sound pressure levels. From the observed profile in Fig. 6.3, most of the heated air was pushed from the equator toward the poles by the streaming. If hot air suddenly appears around the sphere—for example, due to release of latent heat when the sample is cooled to its freezing point—then streaming cannot immediately clear the heated air from the equator, and the force can be weakened or even reversed in direction.

6.3.2   Acoustic Viscous Torque

The Rayleigh torque is the moment of the radiation stress on an asymmetric object (Rayleigh, 1882). Two sound waves of the same frequency, vibrating in perpendicular directions, can produce a different kind of torque on an axisymmetric object. A fluid particle moves in a circular orbit in the compound wave field. A simple model is based on the assumption that if the fluid particle moves in the counterclockwise direction, the object will be induced to rotate in the clockwise direction, like a cogwheel (Wang et al., 1977; Wang and Kanber, 1978). Consider two waves of equal pressure amplitudes A, causing a circular but irrotational particle velocity of amplitude $u_a \sim A/\rho_0 c_0$. Inside the viscous boundary layer of thickness $\delta = (\nu/\omega)^{1/2}$ lining the sphere's surface, the velocity field is subjected to the no-slip boundary condition and becomes rotational. The gyration of a fluid particle inside the viscous boundary layer, around an approximately closed loop, produces

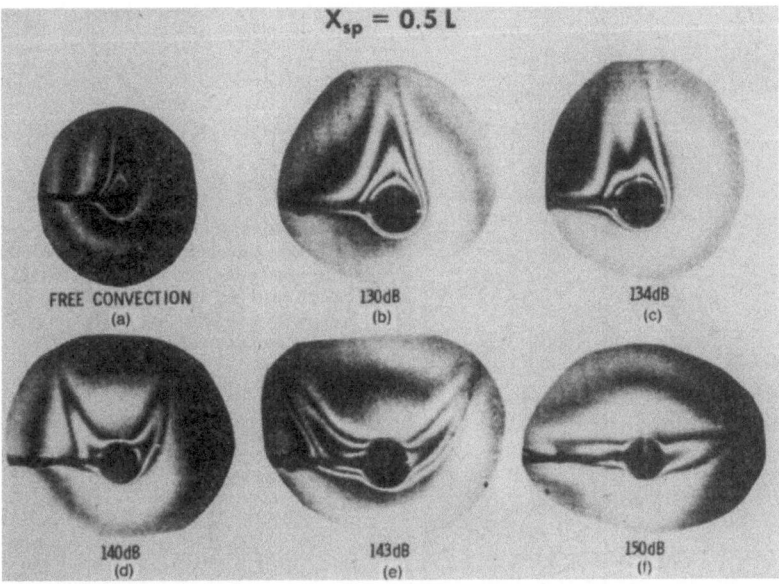

Fig. 6.3 Holograms of a heated sphere at equilibrium at the pressure node in a horizontal acoustic vibration at various sound pressure levels (from Leung and Wang, 1985, Fig. 9, p. 1689).

a tangential stress of order $\rho v u_a/\delta$ on the side of the loop near the surface. It also produces an opposite stress times a factor of order $R/(R + a)$, where $a \sim u_a/\omega$ is the radius of the gyration, on the side of the loop far from the surface. The two effects add up to produce a net stress of order $\mu u_a^2/\delta \omega R$. There is thus a torque of order $\pi R^2 \delta A^2/\rho_0 c_0^2$ on the object, which agrees with the prediction of a more thorough analysis by Busse and Wang (1981):

$$T = 3\sqrt{2}\pi R^2 \delta \frac{A_x A_y}{\rho_0 c_0^2} \sin \psi. \tag{6.100}$$

The torque vanishes when $\psi = 0$, and it is maximized when $\psi = \pi/2$. This simplified model thus provides a shortcut to the answer. However, the sphere actually rotates in the same direction as a fluid particle, according to the more rigorous theory and also experimental observation, and in contradiction to the cogwheel model, because the acoustic streaming inside the boundary layer is opposite to that outside (Lee and Wang, 1989). The more general theory was tested for the case of a circular disc, and it agrees quite well with experiment (Figs. 6.4 and 6.5).

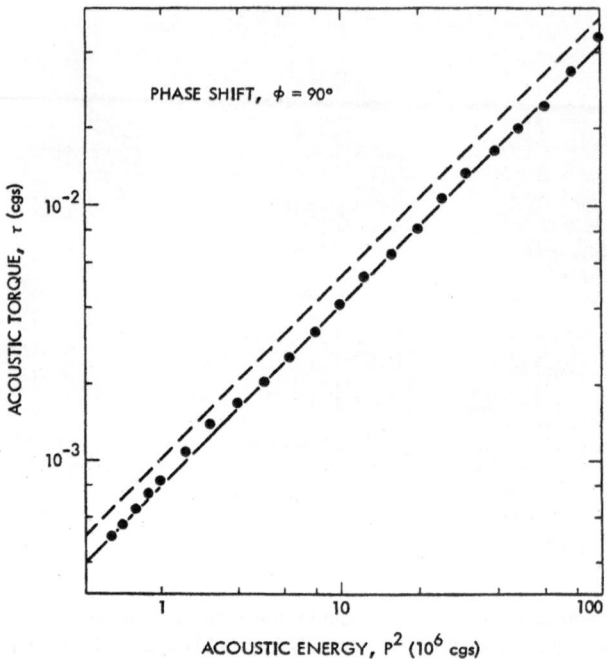

Fig. 6.4 Acoustic torque on a circular disc vs. acoustic energy. The points are the experimental data, the solid line represents the model $T = 2^{1/2}\delta S(A_x A_y/2\rho_0 c_0^2)\sin\omega t$ after normalization, with S the surface area of the disc, and the dashed line the rigorous theory (from Busse and Wang, 1981, Fig. 1, p. 1634).

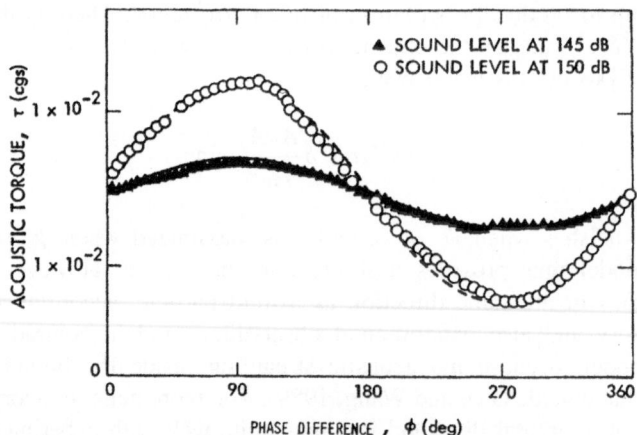

Fig. 6.5 Acoustic torque on a circular disc vs. phase difference between the x and y components of the sound field. The points are experimental data. The curves are normalized values from the model $T = 2^{1/2}\delta S(A_x A_y/2\rho_0 c_0^2)\sin\omega t$ (from Busse and Wang, 1981, Fig. 2, p. 1635).

6.3.3   Resonance Frequency Shift

The introduction of even a small sample into a resonant chamber causes the resonance frequency to shift (Leung et al., 1982; Barmatz et al., 1983). A decrease in the sound power and a corresponding decrease in the force, when coupled with the finite response time of the wave field, can lead to positional instabilities and consequently loss of the sample from the potential well (Rudnick and Barmatz, 1990). The shift is a linear effect. The sample reduces the volume of the chamber, which corresponds to an increase in the resonance frequency. It also forces the sound to travel a greater distance, corresponding to a reduction in the resonance frequency. For a chamber of volume V and length L in the z direction, let the incident sound wave be $p_i = A \sin(n\pi z/L) \exp(-i\omega t)$. With a sphere at $z = Z$, the resonance frequency shift is (Leung et al., 1982)

$$\frac{\delta\omega}{\omega_n} = \frac{4\pi R^3/3}{V}\{[\tfrac{5}{4} - \tfrac{229}{360}(k_n R)^2]\cos 2k_n Z - [\tfrac{1}{4} + \tfrac{67}{360}(k_n R)^2]\}, \qquad (6.101)$$

where $k_n = n\pi/L$, and $\omega_n = k_n c_0$ is the resonance frequency for the empty chamber. Measurements of the resonance-frequency shift are compared with Eq. (6.101) in Fig. 6.6.

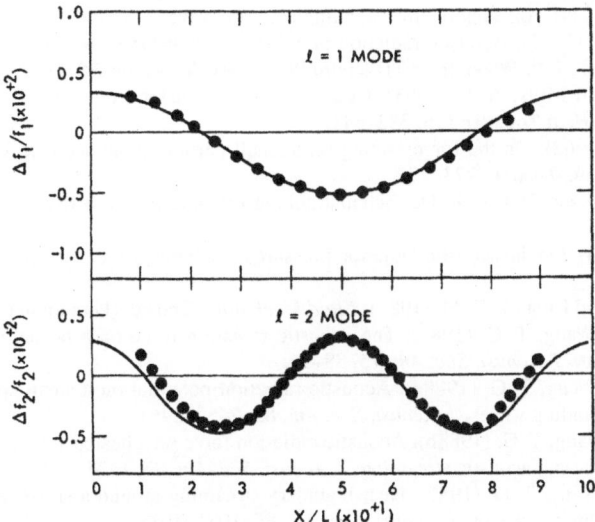

Fig. 6.6 Resonance-frequency shift vs. position for a sphere of radius 0.5 in. Circles represent measurement, and curves represent theory (from Leung et al., 1982, Fig. 3, p. 618).

References

Abramowitz, M., and Stegun, I. A., eds. (1965). *Handbook of Mathematical Functions* (Dover, New York).

Barmatz, M., Allen, J. L., and Gasper, M. (1983). Experimental investigation of the scattering effects of a sphere in a cylindrical resonant chamber. *J. Acoust. Soc. Am.* **73**, 725–732.

Beissner, K. (1985). On the time-average acoustic pressure. *Acustica* **51**, 1–4.

Beissner, K. (1986). Two concepts of acoustic radiation pressure. *J. Acoust. Soc. Am.* **79**, 1610–1612.

Beyer, R. T. (1950). Radiation pressure in a sound wave. *Am. J. Phys.* **18**, 25–29.

Beyer, R. T. (1974). *Nonlinear Acoustics.* Naval Ship Systems Command, Department of the Navy.

Beyer, R. T. (1978). Radiation pressure—the history of a mislabeled tensor. *J. Acoust. Soc. Am.* **63**, 1025–1030.

Brillouin, L. (1938). *Les Tenseurs en Mécanique et en Elasticité* (Masson et Cie, Paris). English translation from French by Brennan, R. O. (1964): *Tensors in Mechanics and Elasticity* (Academic Press, New York).

Busse, F. H., and Wang, T. G. (1981). Torque generated by orthogonal acoustic waves—theory. *J. Acoust. Soc. Am.* **69**, 1634–1638.

Chu, B. T., and Apfel, R. E. (1982). Acoustic radiation pressure produced by a beam of sound. *J. Acoust. Soc. Am.* **72**, 1673–1687.

Chu, B. T., and Apfel, R. E. (1984). Response to the comments of Nyborg and Rooney [*J. Acoust. Soc. Am.* **75**, 263–264 (1984)]. *J. Acoust. Soc. Am.* **75**, 1003–1004.

Collas, P., and Barmatz, M. (1987). Acoustic radiation force on a particle in a temperature gradient. *J. Acoust. Soc. Am.* **81**, 1327–1330.

Crum, L. A. (1971). Acoustic force on a liquid droplet in an acoustic stationary wave. *J. Acoust. Soc. Am.* **50**, 157–163.

Eller, A. (1968). Force on a bubble in a standing acoustic wave. *J. Acoust. Soc. Am.* **43**, 170–171.

Gol'dberg, Z. A. (1971). Acoustic radiation pressure. In *High-Intensity Ultrasonic Fields* (L. D. Rozenberg, ed., J. S. Wood, trans.) (Plenum Press, New York), pp. 75–133.

Gopinath, A., and Mills, A. F. (1993). Convective heat transfer from a sphere due to acoustic streaming. *J. Heat Transfer* **115**, 332–341.

Gor'kov, L. P. (1962). On the forces acting on a small particle in an acoustical field in an ideal fluid. *Sov. Phys.–Dokl.* **6**, 773–775.

Hertz, G., and Mende, H. (1939). Der Schallstrahlungsdruck in Flüssigkeiten. *Z. Physik* **114**, 354–367.

King, L. V. (1934). On the acoustic radiation pressure on spheres. *Proc. Roy. Soc. (London)* **A147**, 212–240.

Landau, L. D., and Lifshitz, E. M. (1987). *Fluid Mechanics*, 2nd ed. (Pergamon Press, New York).

Lee, C. P., and Wang, T. G. (1984). The acoustic radiation force on a heated (or cooled) rigid sphere—theory. *J. Acoust. Soc. Am.* **75**, 88–96.

Lee, C. P., and Wang, T. G. (1988a). Acoustic radiation potential on a small sphere due to two orthogonal standing waves. *J. Acoust. Soc. Am.* **83**, 2459–2461.

Lee, C. P., and Wang, T. G. (1988b). Acoustic radiation force on a heated sphere including effects of heat transfer and acoustic streaming. *J. Acoust. Soc. Am.* **83**, 1324–1331.

Lee, C. P., and Wang, T. G. (1989). Near-boundary streaming around a small sphere due to two orthogonal standing waves. *J. Acoust. Soc. Am.* **85**, 1081–1088.

Lee, C. P., and Wang, T. G. (1990). Outer acoustic streaming. *J. Acoust. Soc. Am.* **88**, 2367–2375.

Lee, C. P., and Wang, T. G. (1993). Acoustic radiation pressure. *J. Acoust. Soc. Am.* **94**, 1099–1109.

Leung, E., Jacobi, N., and Wang, T. (1981). Acoustic radiation force on a rigid sphere in a resonance chamber. *J. Acoust. Soc. Am.* **70**, 1762–1767.

Leung, E., Lee, C. P., Jacobi, N., and Wang, T. G. (1982). Resonance frequency shift of an acoustic chamber containing a rigid sphere. *J. Acoust. Soc. Am.* **72**, 615–620.

Leung, E., and Wang, T. (1985). Force on a heated sphere in a horizontal plane acoustic standing wave field. *J. Acoust. Soc. Am.* **77**, 1686–1691.

Löfstedt, R., and Putterman, S. (1991). Theory of long wavelength acoustic radiation pressure. *J. Acoust. Soc. Am.* **90**, 2027–2033.

Marston, P. L. (1980). Shape oscillation and static deformation of drops and bubbles driven by modulated radiation stresses—theory. *J. Acoust. Soc. Am.* **67**, 15–26.

Nyborg, W. L. (1965). Acoustic streaming. In *Physical Acoustics*, Vol. IIB, W. P. Mason, ed. (Academic Press, New York), pp. 266–331.

Nyborg, W. L., and Rooney, J. A. (1984). Comments on "Acoustic radiation pressure produced by a beam of sound" [*J. Acoust. Soc. Am.* **72**, 1673–1687 (1982)], *J. Acoust. Soc. Am.* **75**, 263–264.

Oran, W. A., Witherow, W. K., Ross, B. B., and Rush, J. E. (1979). Some limitations on processing materials in acoustic levitation devices. In *Ultrasonics Symposium Proceedings* (IEEE, New York), pp. 482–486.

Prosperetti, A. (1986). Physics of acoustic cavitation. In *Frontiers in Physical Acoustics*, D. Sette, ed. (North-Holland, New York), pp. 145–188.

Rayleigh, Lord (1882). On an instrument capable of measuring the intensity of aerial vibrations. *Phil. Mag.* **14**, 186.

Rayleigh, Lord (1902). On the pressure of vibrations. *Phil. Mag.* **3**, 338–346.

Rayleigh, Lord (1905). On the momentum and pressure of gaseous vibrations, and on the connexion with the virial theorem. *Phil. Mag.* **10**, 364–374.

Robey, J. L., Trinh, E. H., and Wang, T. G. (1987). Acoustic force measurement in a dual-temperature resonant chamber. *J. Acoust. Soc. Am.* **82**, 1039–1044.

Rooney, J. A., and Nyborg, W. L. (1972). Acoustic radiation pressure in a traveling plane wave. *Am. J. Phys.* **40**, 1825–1830.

Rudnick, I. (1977). Measurements of the acoustic radiation pressure on a sphere in a standing wave field. *J. Acoust. Soc. Am.* **62**, 20–22.

Rudnick, J., and Barmatz, M. (1990). Oscillational instabilities in single-mode acoustic levitators. *J. Acoust. Soc. Am.* **87**, 81–92.

Trinh, E., Robey, J., Jacobi, N., and Wang, T. G. (1986). Dual-temperature acoustic levitation and sample transport apparatus. *J. Acoust. Soc. Am.* **79**, 604–612.

Wang, T. G., and Kanber, H. (1978). Nonlinear acoustic torque in intense sound field. *J. Acoust. Soc. Am.* **64**, Suppl. 1, S14–S15.

Wang, T. G., Kanber, H., and Rudnick, I. (1977). First order torques and solid body spinning velocities in intense sound fields. *Phys. Rev. Lett.* **38**, 128–130.

Westervelt, P. J. (1950). The mean pressure and velocity in a plane acoustic wave in a gas. *J. Acoust. Soc. Am.* **22**, 319–327.

Westervelt, P. J. (1957). Acoustic radiation pressure. *J. Acoust. Soc. Am.* **29**, 26–29.

Wu, J., and Du, G. (1990). Acoustic radiation force on a small compressible sphere in a focused beam. *J. Acoust. Soc. Am.* **87**, 997–1003.

Yosioka, K., and Kawasima, Y. (1955). Acoustic radiation pressure on a compressible sphere. *Acustica* **5**, 167–173.

Chapter 7
Acoustic Streaming

Wesley L. Nyborg

Contents

7.1   Introduction

In a region of fluid where a sound field exists, the pressure and velocity vary with time, but, in general, the temporal averages of these quantities are not zero. The average over time of the velocity is called *acoustic streaming*, and is the subject of this chapter. It was probably in studies of plates vibrating in air that the first observations of this acoustical phenomenon were made. Faraday (1831) found currents of air to rise at points of maximum vibration amplitude on such plates, and to descend at displacement nodes. Since then, the circulatory motions characteristic of acoustic streaming have been observed and investigated in many situations. Some of these involve vibrating objects or the interaction of sound fields with boundaries, while others involve beams of sound (especially ultrasound) propagating in a large space, or in other situations where boundaries are less important.

W. L. Nyborg
Department of Physics, University of Vermont, Burlington, VT, USA

© The Author(s) 2024
M. F. Hamilton, D. T. Blackstock (eds.), *Nonlinear Acoustics*,
https://doi.org/10.1007/978-3-031-58963-8_7

The subject was reviewed in detail by the author (Nyborg, 1965) in an earlier publication. In the present chapter, the intent is to discuss basic aspects of the theory, and to recognize developments during the past few decades. It has not been possible to be all-inclusive. For a more comprehensive treatment of the older literature, the interested reader is referred to the 1965 review cited above. For discussions of other or more recent findings, the reviews by Riley (1965), by Rudenko and Soluyan (1977), and by Lighthill (1978a, 1978b) are highly recommended.

In Sect. 7.2, the general theory is taken up, including the significance of effective force density and Reynolds stresses in driving the streaming. In Sect. 7.3, solutions of the streaming equations are discussed, both for streaming arising from ultrasound beams propagating in homogeneous media under free-field conditions and for streaming arising from boundary layers set up at the surfaces of walls or other solid objects. Applications of acoustic streaming are described briefly in Sect. 7.4.

7.2   General Considerations

7.2.1   Basic Equations

We assume a homogeneous isotropic fluid in which the pressure, density, and velocity at any point are given instantaneously by $P(x, y, z, t)$, $\rho(x, y, z, t)$, and $\mathbf{u}(x, y, z, t)$, respectively. Considering a small element dv of fluid, we suppose that the only forces acting on dv are surface stresses due to the elasticity and viscosity of the fluid. The dynamic equation for the fluid element is then

$$\mathbf{f} = \rho \left[\frac{\partial \mathbf{u}}{\partial t} + (\mathbf{u} \cdot \nabla)\mathbf{u} \right], \tag{7.1}$$

where $\mathbf{f}$ is the net force per unit volume due to stresses. For a Newtonian viscous fluid, $\mathbf{f}$ is given by

$$\mathbf{f} = -\nabla P + [\mu_B + (4/3)\mu]\nabla\nabla \cdot \mathbf{u} - \mu\nabla \times \nabla \times \mathbf{u}, \tag{7.2}$$

where μ and μ_B are, respectively, the coefficients of shear and bulk viscosity for the fluid. Both μ and μ_B are assumed here to be independent of space and time, although it is known that μ_B, especially, can be very much a function of frequency. By using the continuity equation

$$\frac{\partial \rho}{\partial t} + \nabla \cdot \rho\mathbf{u} = 0, \tag{7.3}$$

together with Eqs. (7.1) and (7.2) one obtains

$$\frac{\partial(\rho\mathbf{u})}{\partial t} - \mathbf{F}' = -\nabla P + [\mu_B + (4/3)\mu]\nabla\nabla\cdot\mathbf{u} - \mu\nabla\times\nabla\times\mathbf{u}, \tag{7.4a}$$

where

$$-\mathbf{F}' = \rho(\mathbf{u}\cdot\nabla)\mathbf{u} + \mathbf{u}\nabla\cdot\rho\mathbf{u}. \tag{7.4b}$$

It has been noted by Lighthill (1978a, 1978b) that $\mathbf{F}'$ is equal to the negative divergence of a momentum flux tensor, i.e., the ith component may be written, in index notation, as $F_i' = -\partial(\rho u_i u_j)/\partial x_j$.

7.2.2   Flow Produced by an External Force Field

An equation analogous to Eq. (7.4a) applies to the flow produced in a viscous fluid by an external force field, in the absence of sound; for this application, it is necessary only to replace $\mathbf{F}'$ in Eq. (7.4a) by $\mathbf{F}_e$, the latter representing the external force per unit volume. If the terms in Eq. (7.4a) involving $\partial/\partial t$ and $\nabla\cdot\mathbf{u}$ are set equal to zero, the resulting equation is

$$\mathbf{F}_e = \nabla P + \mu\nabla\times\nabla\times\mathbf{u} = \nabla P - \mu\nabla^2\mathbf{u}, \tag{7.5}$$

and becomes a governing equation for steady flow of an incompressible viscous fluid produced by an external force field. Solutions of Eq. (7.5) can be readily applied to textbook examples (Nyborg, 1975) of laminar flow in tubes and channels.

A general solution of Eq. (7.5) gives the steady velocity $\mathbf{u}$ at any observation point (x, y, z) in an unbounded fluid in terms of a volume integral:

$$\mathbf{u} = \frac{1}{8\pi\mu}\int\left\{\frac{\mathbf{F}_e}{r} + \frac{(\mathbf{F}_e\cdot\mathbf{r})\mathbf{r}}{r^3}\right\}dv. \tag{7.6}$$

Here $\mathbf{F}_e$ is the external force per unit volume at the location (x', y', z') of any volume element dv, and $\mathbf{r}$ is the vector position of the observation point relative to the volume element (Lighthill, 1978a, 1978b). Equation (7.6) is a generalization of an expression given by Lamb (1945) for the velocity resulting from a "point-force" (see also Happel and Brenner, 1991); if the only component of the external force is F_{ex}, in the x direction, and if it exists only in a single volume element centered at the origin, the velocity $\mathbf{u}$ at any position $\mathbf{r}$ is given by

$$\mathbf{u} = \frac{1}{8\pi\mu}\left\{\frac{\mathbf{i}F_{ex}}{r} + \frac{(\mathbf{F}_e\cdot\mathbf{r})\mathbf{r}}{r^3}\right\}dv, \tag{7.7}$$

where $\mathbf{i}$ is the unit vector along x. Along the x axis, the corresponding component u is given by

$$u = F_{ex}dv/4\pi\mu|x|, \tag{7.8}$$

where $|x|$ denotes the (unsigned) magnitude of x. Equation (7.8) shows that on the axis, u is always positive, is infinite at the point where the force is applied, and decreases monotonically with distance from that point.

The singularity (at $x = 0$) in Eq. (7.8) can be avoided by considering the force to be distributed over a finite volume. For example, suppose $\mathbf{F}_e$ is uniform and equal to $\mathbf{i}F_{ex}$ everywhere in a thin disc of radius a and thickness ε centered at the origin, its axis being along x. Then the velocity component u on the axis is found to be

$$u = \frac{\varepsilon F_{ex}a^2}{4\mu r'}, \quad r' = (a^2 + x^2)^{1/2}. \tag{7.9}$$

7.2.3 Interpretation of Terms in Eqs. (7.4)

We return now to the general equations of motion, Eqs. (7.4). From Eq. (7.2), the right-hand side of Eq. (7.4a) is equal to the net force applied to the element, per unit volume. This force is entirely a result of stresses at the surface of the element applied by the surrounding fluid; the term $-\nabla P$ gives the surface force exerted by elastic stresses, and the other terms give the corresponding force from viscous stresses.

7.2.4 Time-Averaging

To obtain $\langle \mathbf{u} \rangle$, the steady part of the flow, terms in Eqs. (7.4) are averaged with respect to time. Since the average over time of $\partial(\rho\mathbf{u})/\partial t$ must be zero, the temporal average of the left-hand side of Eq. (7.4a) becomes $\langle -\mathbf{F}' \rangle$. For convenience, $\mathbf{F}$ is defined as $\langle \mathbf{F}' \rangle$ and we have from Eq. (7.4b) that

$$\mathbf{F} = -\langle \rho(\mathbf{u} \cdot \nabla)\mathbf{u} + \mathbf{u}\nabla \cdot \rho\mathbf{u} \rangle. \tag{7.10}$$

Thus $\mathbf{F}$ is the negative of the time-averaged rate of increase of momentum per unit volume in an element of fluid. It also follows from an earlier comment that the component form of Eq. (7.10) is $F_i = -\partial\langle\rho u_i u_j\rangle/\partial x_j$, where $\langle\rho u_i u_j\rangle$ is referred to as the Reynolds stress (Lighthill, 1978a, 1978b).

It is often appropriate to consider the time-averaged velocity $\langle \mathbf{u} \rangle$ as flow of an incompressible fluid, so that, approximately,

$$\nabla \cdot \langle \mathbf{u} \rangle = 0. \tag{7.11}$$

When this condition is assumed, the equation that results from time-averaging Eq. (7.4a) can be written in a form analogous to Eq. (7.5), with $\langle P \rangle$ and $\langle \mathbf{u} \rangle$ replacing P and $\mathbf{u}$:

$$\mathbf{F} = \nabla \langle P \rangle + \mu \nabla \times \nabla \times \langle \mathbf{u} \rangle = \nabla \langle P \rangle - \mu \nabla^2 \langle \mathbf{u} \rangle. \tag{7.12}$$

The vector field $\mathbf{F}$ (a time-averaged quantity) corresponds to $\mathbf{F}_e$ in Eq. (7.5) and is therefore analogous to an external field of steady force in its relation to the fields of time-averaged pressure and velocity.

In the above discussion, the operator $\langle \cdot \rangle$ represents an average with respect to time at a fixed point in space, i.e., an *Eulerian average*. Defined operationally, the symbol $\langle \mathbf{u} \rangle$ denotes the average at a point Q of velocity measurements $\mathbf{u}_i$ on a series of infinitesimal "particles" (volume elements of fluid) passing through Q. Later, reference will be made to the *Lagrangian average* $\mathbf{U}$ of the velocity; this quantity is defined at Q as the time-averaged velocity of a particle during a short time interval immediately after passing through Q. An approximate expression for transforming $\langle \mathbf{u} \rangle$ to $\mathbf{U}$ is discussed in Sect. 7.3.1. Rudenko and Soluyan (1977) define the force quantity $\rho \mathbf{F}$, with units of force per unit mass, and derive an equation analogous to Eq. (7.12), but in which the fluid velocity on the right-hand side is $\mathbf{U}$ rather than $\langle \mathbf{u} \rangle$.

7.2.5 Other Forms of the General Equations

Equation (7.12), with $\mathbf{F}$ given by Eq. (7.10), is a governing equation for determining the quantities $\langle P \rangle$ and $\langle \mathbf{u} \rangle$. An alternative form of the equation is obtained by operating on all terms in Eq. (7.12) with the curl operator ($\nabla \times$); the term involving $\nabla \langle P \rangle$ then vanishes. Setting $\mathbf{R}$ equal to the vorticity $\nabla \times \mathbf{u}$, one obtains

$$\mu \nabla^2 \langle \mathbf{R} \rangle = -\nabla \times \mathbf{F}. \tag{7.13}$$

Equation (7.13) has the form of the well-known equation of Poisson, and the quantity $\nabla \times \mathbf{F}$ has been called the "vorticity source strength" (Medwin and Rudnick, 1953). It has been shown (Nyborg, 1965) that the time-averaged torque about the center of a spherical fluid element is $-(I/2\rho)\nabla \times \mathbf{F}$, where I is the moment of inertia of the element about its center. Since the flow $\langle \mathbf{u} \rangle$ is essentially that of an incompressible fluid, for which $\nabla \cdot \langle \mathbf{u} \rangle$ is zero, it is possible in some situations to express $\langle \mathbf{u} \rangle$ in terms of the Stokes stream function. Streamlines of the flow $\langle \mathbf{u} \rangle$ then correspond to paths along which the stream function is constant.

7.3   Solutions

7.3.1   Method of Successive Approximations

We now make approximations to the pressure, density, and velocity appropriate for a steady-state "sound field" on which is superposed a steady flow, the latter being absent when there is no sound. For this purpose, we represent the excess pressure $P - P_0$, excess density $\rho - \rho_0$ and fluid velocity $\mathbf{u}$ at any point in series form as

$$P - P_0 = p_1 + p_2 + \cdots,$$
$$\rho - \rho_0 = \rho_1 + \rho_2 + \cdots, \tag{7.14}$$
$$\mathbf{u} = \mathbf{u}_1 + \mathbf{u}_2 + \cdots.$$

Here p_1, ρ_1, and $\mathbf{u}_1$ are first-order approximations to the steady-state values of pressure, density, and velocity; they vary sinusoidally in time with angular frequency ω, and thus represent the "sound field." Second-order approximations to the solutions yield correction terms to be added to p_1, ρ_1, and $\mathbf{u}_1$. These corrections include time-independent quantities as well as terms with frequency 2ω; of these, it is only the former, the time-independent terms, that are treated here. In the remainder of this chapter, the symbols p_2, ρ_2, and $\mathbf{u}_2$ will be used to represent *only the time-independent parts* of the corresponding second-order acoustic quantities, i.e., the second-order approximations to $\langle P \rangle$, $\langle \rho \rangle$ and $\langle \mathbf{u} \rangle$, respectively. In particular, $\mathbf{u}_2$ is the second-order approximation to the velocity of acoustic streaming, the subject of this chapter. (Although p_2, ρ_2, and $\mathbf{u}_2$ are time-averaged quantities, the enclosing bracket, $\langle \cdot \rangle$, is omitted for convenience.)

In the method of successive approximations under discussion, one substitutes the series expressions for pressure, density, and velocity from Eqs. (7.14) into Eq. (7.12) [with $\mathbf{F}$ given by Eq. (7.10)], then forms equations by selecting terms of like order (see also Sect. 10.2). Terms of the type ∇p_n are of order n, while terms of the type $\rho_l(\mathbf{u}_m \cdot \nabla)\mathbf{u}_n$ are of order $(l + m + n)$. Beginning with terms of zero order, we find that there is only one in Eq. (7.12), namely, ∇P_0. The equation formed by selecting all zero-order terms is simply a statement that $\nabla P_0 = 0$, i.e., that P_0 is constant; this is a reasonable result, since external forces, such as gravity, have been assumed absent.

One obtains no equation from first-order terms in Eq. (7.12), since the time-average (over many cycles) of a quantity varying sinusoidally with time is zero. It is from second-order terms in Eq. (7.12) that we obtain an equation that is of special relevance to the topic of this chapter. This equation [which involves the desired approximation ($\mathbf{u}_2$) to the streaming velocity $\langle \mathbf{u} \rangle$] can be written as

$$\mathbf{F}_2 = \nabla p_2 - \mu \nabla^2 \mathbf{u}_2, \tag{7.15a}$$

where

$$\mathbf{F}_2 = -\rho_0 \langle (\mathbf{u}_1 \cdot \nabla)\mathbf{u}_1 + \mathbf{u}_1(\nabla \cdot \mathbf{u}_1) \rangle. \tag{7.15b}$$

Equation (7.15a) is of exactly the same form as Eq. (7.12), but here $\langle \mathbf{F} \rangle$, $\langle P \rangle$ and $\langle \mathbf{u} \rangle$ are replaced by second-order approximations to them, $\mathbf{F}_2$, p_2 and $\mathbf{u}_2$. It is seen in Eq. (7.15b) that the vector $\mathbf{F}_2$, the "effective force density," can be calculated when $\mathbf{u}_1$ is known.

As explained in Sect. 7.2.4, the notation $\langle \mathbf{q} \rangle$ represents the *Eulerian average* of any field quantity $\mathbf{q}$. That is, $\langle \mathbf{q} \rangle$ is the average over a suitable time interval of $\mathbf{q}(Q, t)$, where the latter is the instantaneous value of $\mathbf{q}$ at a fixed point Q. When movement occurs, as in a sound field, a continuous series of particles (i.e., volume elements of the fluid) passes through Q, $\mathbf{q}(Q, t)$ is the value of $\mathbf{q}$ for that particle that happens to be at Q at time t, and $\langle \mathbf{q} \rangle$ is the average of $\mathbf{q}$ for this series of particles.

For a selected particle whose location $Q'(t)$ varies with time but is in the vicinity of Q, the temporal average should be of $\mathbf{q}(Q', t)$, where the latter gives $\mathbf{q}$ at time t for the selected particle *wherever it may be*. From the Taylor series, we may write, approximately,

$$\mathbf{q}[Q'(t)] = \mathbf{q}(Q, t) + [(\xi \cdot \nabla)\mathbf{q}]_{Q,t}, \tag{7.16}$$

where ξ is the instantaneous vector distance from Q to Q'. We now let $\mathbf{q}$ be the velocity $\mathbf{u}$, average each term in the equation with respect to time, and evaluate the terms to second order, obtaining

$$\mathbf{U} = \mathbf{u}_2 + \left\langle \left(\int \mathbf{u}_1 \, dt \right) \cdot \nabla \mathbf{u}_1 \right\rangle; \tag{7.17}$$

here ξ_1 has been equated to $\int \mathbf{u}_1 \, dt$. The second term on the right-hand side of Eq. (7.17) is referred to as the *velocity transform*, or the *Stokes drift*, and has been discussed extensively by Westervelt (1953a, 1953b). In experimental studies of acoustic streaming, $\mathbf{u}_2$ would seem to be relevant to measurements with hot-wire anemometry (Starritt et al., 1989) and $\mathbf{U}$ to observed motions of small indicator particles that follow the fluid motion. The quantity $\mathbf{U}$ is also approximately equal to the *mass transport velocity*, a quantity whose integral $\int \mathbf{U} \cdot d\mathbf{S}$ over a surface is equal to M/ρ_0, where M is the time-averaged rate at which mass flows across the surface (Nyborg, 1965). Finally, we note that p_2 is the Eulerian radiation pressure, labeled P^E in Chap. 6. To calculate the Lagrangian radiation pressure, labeled P^L in Chap. 6, replace $\mathbf{U}$ by P^L, $\mathbf{u}_2$ by p_2, and $\nabla \mathbf{u}_1$ by ∇p_1 in Eq. (7.17).

Much of the older published theoretical work on acoustic streaming is based on the second-order approximation described in this section. In this approximation, the flow pattern is independent of the first-order velocity amplitude A, and the streaming speed is proportional to A^2. It was recognized by early investigators that these features apply only when the amplitude and the associated streaming speeds are relatively low. At higher values of these quantities, the flow pattern sometimes changes considerably (Ingard and Labate, 1950; Andres and Ingard, 1953; Raney

et al., 1954; Skavlem and Tjøtta, 1955). In Sect. 7.3.2.2, experiments are described
for which theory capable of dealing with higher streaming speeds is required.

7.3.2 Forces and Flows Produced by a Beam of Ultrasound

7.3.2.1 Linear Propagation in a Large Space

Experimental results have been reported by Starritt et al. (1989, 1991) for acoustic
streaming arising from a focused beam of ultrasound propagating through a tank of
water. In such a situation, the dimensions of the tank are typically large compared
to those of the region where the intensity is highest, that is, the focal zone. Hence
the flow in that region is probably not very dependent on the boundaries; it is thus
instructive to model the situation as one involving flow in an unbounded medium.

For simplicity, the focal region is considered to be cylindrical (Fig. 7.1), and
the sound field in that region is initially assumed to be an attenuated plane wave
traveling in the positive x direction. The first-order velocity can be written as (the
real part of) the vector $\mathbf{i}u_1$, where u_1 is given by

$$u_1 = u_a \exp(j\omega t), \quad u_a = u_{a0} \exp[-(\alpha + jk)x]. \tag{7.18}$$

Here u_{a0} is the velocity amplitude at $x = 0$, α is the attenuation coefficient, and k
is the (real) propagation constant, equal to ω/c_0, where c_0 is the small-signal speed
of sound. Replacing $\mathbf{u}_1$ by $\mathbf{i}u_1$ in Eq. (7.15b), one finds that $\mathbf{F}_2$ can be expressed
as $\mathbf{i}F_{2x}$, where F_{2x} is equal to $-2\rho_0\langle u_1 \partial u_1/\partial x \rangle$. Evaluating the expression,[1] one

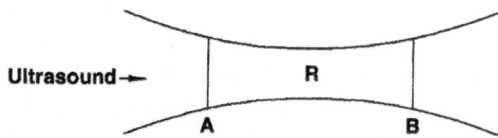

Fig. 7.1 Focal zone (region R) of ultrasound beam, propagating in water away from boundaries.
Under conditions of nonlinear propagation, the wave distortion and energy dissipation are assumed
to occur primarily in this zone.

---

[1] In evaluating the time average, it is necessary, in principle, to use the real parts of u_1 and $\partial u_1/\partial x$.
However, the step of extracting real parts can be avoided by using a property of complex quantities:
Consider two functions given by the real parts of

$$G_1 = g_1 \exp(j\omega t), \quad G_2 = g_2 \exp(j\omega t),$$

where g_1 and g_2 are functions of space and may be complex; the time-averaged product of the
functions is given by

$$\langle (\text{Re } G_1)(\text{Re } G_2) \rangle = (1/2)\text{Re}[g_1 g_2^*],$$

where g_2^* is the complex conjugate of g_2.

obtains

$$F_{2x} = \alpha \rho_0 u_{a0}^2 e^{-2\alpha x}. \tag{7.19}$$

For a beam with circular cross section of radius a, the time-averaged transmitted acoustic power W is the product of $\pi a^2 \rho_0 c_0 / 2$ and $u_{a0}^2 e^{-2\alpha x}$; hence, for such a beam, an alternative for Eq. (7.19) is

$$F_{2x}(x) = \frac{2\alpha W(x)}{\pi a^2 c_0}. \tag{7.20}$$

In seeking an expression for $\mathbf{u}_2$ we recall the analogy between Eqs. (7.5) and (7.15a), and use the result in Eqs. (7.9). Replacing F_{ex} by F_{2x} from Eq. (7.20), we can obtain an expression for the axial velocity at a given distance from an elemental disc in the beam, resulting from the "effective force" produced by ultrasound in the disc. For this purpose, the position of a given elemental "force disc" is designated as x_1, its thickness as dx_1 (replacing ε), and the distance from the disc as $|x - x_1|$. The axial streaming velocity at x is then

$$u_2(x, x_1) = GW(x_1)\frac{dx_1}{r_1}, \tag{7.21a}$$

where G and r_1 are given by

$$G = \frac{\alpha}{2\pi \mu c_0}, \quad r_1 = [a^2 + (x - x_1)^2]^{1/2}. \tag{7.21b}$$

In Eq. (7.21a), the product $GW dx_1$ can be written as $-\Delta W / 4\pi \mu c_0$, where the quantity $-\Delta W$, equal to $2\alpha W(x_1)\, dx_1$, is the loss in acoustic power that results from absorption in the disc. Hence Eq. (7.21a) can be written as

$$u_2(x) = -\frac{\Delta W}{4\pi \mu c_0 r_1}. \tag{7.22}$$

At points (on the axis) within the disc, where $x_1 = x$ and $r_1 = a$, the velocity reduces to u_d, given by

$$u_d = GW(x)\frac{dx_1}{a} = \frac{|-\Delta W|}{4\pi \mu c_0 a}. \tag{7.23}$$

It is worth noting that the expression on the right of Eq. (7.23) resembles that for the limiting speed of an object moving through a viscous fluid in response to a steady force. Thus $\Delta W / c_0$ is the acoustic radiation force on a nonreflecting object that absorbs acoustic power in the amount ΔW, while $4\pi \mu a$ is comparable to the drag coefficient for a solid or fluid sphere of radius a.

For applications to medical diagnostic ultrasound, we calculate G and u_d for water and a frequency of 3.5 MHz. From Herzfeld and Litovitz (1959), the absorption coefficient α for water at 20 °C is 0.31 Np/m at 3.5 MHz. Choosing μ as 10^{-3} Pa·s and c_0 as 1500 m/s leads to 0.033 N^{-1} for G. If a is 1 mm, dx' is 0.1 mm, and W is 0.1 W, the velocity u_d is 0.33 mm/s.

To estimate the streaming velocity on the axis of an ultrasound beam, one can suppose the beam to be divided into a large number of thin discs and superpose their contributions at points of interest. At the center of a focal region, we assume further, for simplicity, that the contributions come primarily from a cylindrical region of radius a and length l. The axial velocity at the center of such a "force cylinder" can be calculated by carrying out an integration, with the right-hand side of Eq. (7.21a) as the integrand. Choosing $x = 0$, one obtains

$$u_{2x}(0) = \int_{-l/2}^{l/2} GW(x_1)(a^2 + x_1^2)^{-1/2} dx_1. \tag{7.24}$$

If the attenuation in the region is small and independent of x, the product GW can be assumed constant for purposes of the integration, and the total power loss ΔW_c equated to $2\alpha l W$; one then obtains

$$u_{2x}(0) = u_{2c}\Phi, \quad u_{2c} = \frac{\Delta W}{4\pi\mu ac_0},$$

$$\Phi = \frac{a}{l}\ln\left(\frac{\chi + 1}{\chi - 1}\right), \quad \chi = \sqrt{1 + (2a/l)^2}. \tag{7.25}$$

When l is small ($l \ll a$), the nondimensional factor Φ in Eqs. (7.25) reduces to unity and the velocity $u_{2x}(0)$ to u_{2c}, in agreement with Eq. (7.23). As l increases, Φ decreases, though rather slowly. Specifically, when a, W, and other parameters have the values assumed above and l/a is equated to 2, 4, 6, 8, and 10, one obtains for Φ the values 0.82, 0.72, 0.61, 0.52, and 0.46, respectively. Hence, for a force cylinder of length up to ten times the diameter, the streaming velocity at its center can be estimated as u_{2c}, within a factor of about 2. When l is large ($l \gg a$), Φ approaches $(2a/l)\ln(l/a)$.

As explained earlier (Sects. 7.2.4 and 7.3.1), the quantity u_{2x} is an Eulerian average of the velocity. To obtain the corresponding Lagrangian average U_x, one utilizes Eq. (7.17); under the conditions considered above, the velocity transform is 0.014 mm/s, and hence U_x is essentially equal to u_{2x}.

If the streaming speed near the source transducer (or near any other solid boundary) were to be considered, it would be necessary to take into account the (usually nonslip) boundary condition there. Blake (1971) and Lighthill (1978a, 1978b) have shown how this can be done with image techniques.

Examining the expression for G in Eq. (7.21b), one finds that it is proportional to the ratio α/μ. From classical theory of linear acoustics (Rayleigh, 1945), if the attenuation α_c is entirely a result of absorption arising from shear viscosity of the fluid, the ratio is independent of μ and is given by

$$\frac{\alpha_c}{\mu} = \frac{8\pi^2 f^2}{3\rho_0 c_0^3} = \frac{2k^2}{3\rho_0 c_0}. \tag{7.26}$$

When attenuation arises from other causes, such as volume relaxation or scattering, the ratio α/μ is increased over the value given in Eq. (7.26). For example, the ratio α/μ for water is more than three times that expected from Eq. (7.26) (Hall, 1948), and that for biological suspensions can be much greater (Carstensen and Schwan, 1959; Dunn et al., 1969; Duck, 1990; NCRP, 1992).

7.3.2.2 Nonlinear Propagation in a Large Space

It was shown by Starritt et al. (1989, 1991) that when conditions are such that a propagating wave undergoes nonlinear distortion, the speed of acoustic streaming can be greatly increased. Figure 7.2 shows streaming velocity along the axis of a focused ultrasound beam, as measured with a hot-wire anemometer in water. A 3.5-MHz transducer, with 19 mm diameter and 9.5 cm focal length, was used to generate the beam in a $50 \times 20 \times 15$-cm water bath containing distilled water. The two upper curves are for pulsed modes, and the lower one is for a continuous mode of operation. For all of these, the total acoustic power was 100 mW, but the (spatial and temporal) peak positive pressure (p^+) varied from one mode of operation to another. For curve c, p^+ was 0.2 MPa, for curve b, it was 1.4 MPa, and for curve a, it was 3.4 MPa.

It is seen in Fig. 7.2 that the streaming velocity is small near the transducer and rises to a rather broad maximum a little beyond the focal distance. According to Eqs. (7.21) to (7.25), the streaming velocity should depend only on the time-averaged power and hence should be the same for all three curves. Instead, there is a considerable increase in streaming speed when the pulsed mode is used, and the peak positive pressure is increased. From curves b and c, an increase of p^+ by a factor of 7 (1.4 MPa/0.2 MPa) causes, approximately, a doubling of velocity; from curves a and b, an increase of p^+ by only a factor of 2.4 (3.4 MPa/1.4 MPa) causes nearly a tripling of the velocity.

Fig. 7.2 Variation of measured streaming velocity with distance from transducer. Frequency, 3.5 MHz; total acoustic power, 100 mW. Curve a: pulsed, $p^+ = 3.4$ MPa. Curve b: pulsed, $p^+ = 1.4$ MPa. Curve c: continuous, $p^+ = 0.2$ MPa. (Adapted from Starritt et al., 1989).

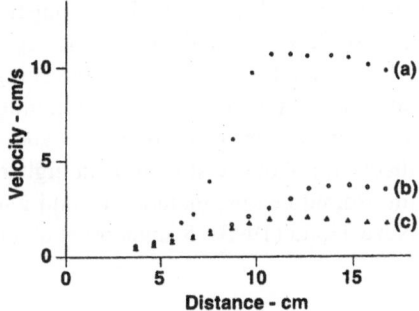

As shown by the investigators, the observed enhancement of streaming can be understood by considering the increased absorption associated with wave distortion that occurs at the higher values of p^+. In water and similar media, a distorted wave, whose spectrum contains not only a fundamental frequency but also a series of harmonics, is absorbed more strongly than a monofrequency wave at the fundamental frequency, since the absorption coefficient increases with frequency. The effect is qualitatively as if the attenuation coefficient α were increased without any change in the shear viscosity coefficient μ; hence, according to Eqs. (7.21)–(7.25), there should be an increase in the acoustic streaming speed u_2.

As a method for taking nonlinearity into account, Starritt et al. (1991) calculated the equivalent of F_{2x} in Eq. (7.20) (expressed by them as a pressure gradient) by summing energy losses in the components of a waveform that had acquired a sawtooth spectral distribution. [Their procedure was equivalent to replacing $\alpha W(x)$ in Eq. (7.20) by a sum of terms of the type $\alpha_n W_n$, one for each of the harmonics.] Among their results was an estimate, for a device of relatively high acoustic output used in medical diagnostic ultrasound, that in water, F_{2x} is 175 times greater than it would have been in the absence of harmonic generation. This enhanced value of F_{2x} occurs, especially, in and near the focal region.

If one knew the components of $\mathbf{F}_2$ everywhere, one might calculate the streaming velocity by evaluating an integral analogous to that in Eq. (7.6). The simplified treatment in Sect. 7.3.2.1, in which Eq. (7.24) is used, enables one to make estimates under conditions where the acoustic amplitudes are relatively low. At higher amplitudes, use of Eq. (7.24) is made difficult by the need to know G and W as functions of x under nonlinear conditions. A special limiting situation applies if the region of energy dissipation and of F_{2x} is very localized, as in the region R of Fig. 7.1. The streaming velocity in that region will then be given approximately by the simple expression in Eqs. (7.25), where Φ is of the order of unity and ΔW_c is the total loss of acoustic power from the beam during transit through the lossy region.

Rudenko and Soluyan (1977) and Lighthill (1978a, 1978b) have pointed out the inadequacy of solutions based on the second-order approximation in dealing with flows where the Reynolds number is high and $\mathbf{u}_2$ is not small compared to $\mathbf{u}_1$. For example, in such flows, the term $(\mathbf{u}_2 \cdot \nabla)\mathbf{u}_2$, or $(\mathbf{U} \cdot \nabla)\mathbf{U}$ should be included, as well as $(\mathbf{u}_1 \cdot \nabla)\mathbf{u}_1$. When this is done, a solution such as that for the axial flow from a "force disc" [Eq. (7.22)] is no longer symmetric about the origin; the velocity for positive x is larger than that for negative x at the same distance from the origin. Also, when the Reynolds number is high, the flow may become turbulent. Parabolic approximations of nonlinear streaming equations for flows produced by directional sound beams are given by Gusev and Rudenko (1979). Others who have dealt with theory for acoustic streaming at higher speeds, especially in applications to focused ultrasound beams, include Wu and Du (1993), Mitome (1993a, 1993b), Tjøtta and Naze Tjøtta (1993), Kamakura et al. (1995), and Hamilton et al. (1995).

7.3.2.3   Propagation in a Bounded Space

When an ultrasound beam travels in a tube (or in any bounded space), the presence of the boundaries may significantly affect the acoustic streaming velocity, even at points near the axis. This time-independent flow must be circulatory when it occurs in a bounded region containing a fixed mass of fluid. When the dimensions of the tube (or, more generally, those of the bounded region) are large compared to the effective dimensions of the beam, the return flow is distributed over a volume that is large compared to that occupied by the beam and has little effect on flow near the axis. If, however, the ratio of tube dimensions to beam dimensions is not large, the axial streaming velocity is dependent on all of these dimensions.

A number of special cases involving beam propagation in a channel have been treated (neglecting effects of nonlinear propagation), and are discussed in the earlier review by Nyborg (1965). These include the acoustic streaming produced by an attenuated beam traveling between parallel walls, in a circular tube (Eckart, 1948), and in a conical tube (Johnsen and Tjötta, 1957). Solutions are given there for a beam traveling in a tube with open ends (in which case the region is not simply connected) and with closed ends; it is commonly assumed that the beam does not extend laterally to the wall(s) of the tube or channel in which it travels and, hence, that the beam does not interact tangentially with boundaries.

Eckart's theory for a beam traveling in a fluid inside a tube with closed ends ("quartz wind") generated considerable interest, especially because of its finding that measurements of acoustic streaming can be used to determine the bulk viscosity μ_B of the fluid. This is suggested by the expressions for u_2 in Eqs. (7.21) (although the latter are for a beam traveling in unbounded space) when it is noted that the constant G is proportional to α/μ and that theory for absorption in a viscous fluid (Herzfeld and Litovitz, 1959) relates α to the linear combination $(\mu_B + 4\mu/3)$. By measurements of streaming, determinations of α, and hence of μ_B, were made for a series of liquids by Liebermann (1949) and for a series of gases by Medwin (1954). Piercy and Lamb (1954) made further measurements on liquids using a modified arrangement in which flow was produced and measured in a capillary "side-arm" attached to a larger vessel containing a fluid of interest through which the ultrasound was caused to propagate.

7.3.3   Forces and Flows Caused by Boundary Layers

7.3.3.1   General Considerations

In the applications described in Sect. 7.3.2, the effective force $\mathbf{F}_2$ was produced primarily in the interior of a homogeneous fluid, well away from boundaries. In an equally important set of applications, the distributions of $\mathbf{F}_2$ (or of $\nabla \times \mathbf{F}_2$) produced near boundaries are significant and, in fact, may control the streaming. This is true of streaming produced by sound waves that graze solid surfaces, as well as those produced near vibrating objects.[2]

7.3.3.2 Force Density for Plane Traveling Wave that Grazes Wall

We consider an attenuated plane wave as in Eq. (7.18), except that a rigid boundary exists in the plane $z = 0$, which presents a nonslip condition so that both components of the velocity are zero there. To derive an expression for the first-order velocity that matches this condition, we return to Eqs. (7.3) and (7.4) and approximate them to first order. The resulting equations are then combined with a first-order approximation to the equation of state, $p_1 = c_0^2 \rho_1$, to obtain an equation in which $\mathbf{u}_1$ is the only dependent variable. When (as is commonly true) α is much less than k, this equation can be written as

$$\frac{2\nabla\nabla\cdot\mathbf{u}_1}{(k-j\alpha)^2} + 2\mathbf{u}_1 = \frac{j\nabla\times\nabla\times\mathbf{u}_1}{\kappa^2}, \tag{7.27}$$

where the constants k, κ, α, and b are defined by

$$k = \frac{\omega}{c_0}, \quad \kappa^2 = \frac{\omega\rho_0}{2\mu}, \quad \alpha = \frac{bk^3}{4\kappa^2}, \quad \mu b = \mu_B + \tfrac{4}{3}\mu. \tag{7.28}$$

It can be seen that Eq. (7.27) is satisfied by either an irrotational vector function $\mathbf{u}_a$ that is a solution of

$$\nabla\nabla\cdot\mathbf{u}_a = \nabla^2\mathbf{u}_2 = -(k-j\alpha)^2\mathbf{u}_a \tag{7.29}$$

or a divergence-free vector function $\mathbf{u}_b$ that is a solution of

$$-\nabla\times\nabla\times\mathbf{u}_b = 2j\kappa^2\mathbf{u}_b = m^2\mathbf{u}_b, \tag{7.30}$$

where $m = (1+j)\kappa$. For $\kappa \gg k \gg \alpha$, an approximate solution $\mathbf{u}_s$ for the traveling-wave situation, formed from the sum of $\mathbf{u}_a$ and $\mathbf{u}_b$, has x and z components given (omitting the time factor), respectively, by

$$u_s = u_{a0}e^{-(\alpha+jk)x}(1 - e^{-mz}), \quad w_s = -\left(\frac{\alpha+jk}{m}\right)u_s. \tag{7.31}$$

---

[2] In Sect. 7.3.3 the analysis is simplified in that effects of thermal diffusivity and of nonlinear propagation are ignored. Qi (1993) and Qi et al. (1995) present a much more detailed analysis of the acoustic boundary layer, and show corrections to be made for thermal diffusivity when these are significant.

Examining Eqs. (7.31), we see that, as required, u_s and w_s are both zero at $z = 0$. Also, at distances from the wall equal to a few multiples of κ^{-1}, u_s approaches the value given as u_1 in Eqs. (7.18). The distance κ^{-1} has been called the *ac boundary-layer thickness*, and a sheetlike region of this thickness just inside the wall has been called the *ac boundary layer*.

Substituting $\mathbf{u}_s$ for $\mathbf{u}_1$ in Eq. (7.15b), and making much use of indicated approximations, one finds that the x component F_{2x} of $\mathbf{F}_2$ is much larger than the z component F_{2z} and that

$$F_{2x} = F_{2xa} + F_{2xb}, \tag{7.32a}$$

where

$$F_{2xa} = \alpha \rho_0 u_{a0}^2 e^{-2\alpha x}, \quad F_{2xb} = \tfrac{1}{2} F_{2xa}\left[\frac{k}{\alpha} f_1(n) + f_2(n)\right],$$

$$f_1(n) = C + S - e^{-2n}, \quad f_2(n) = e^{-2n} - 3C + S, \tag{7.32b}$$

$$C = e^{-n}\cos n, \quad S = e^{-n}\sin n, \quad n = \kappa z.$$

The force density field F_{2xa} is the same as for an unbounded plane traveling wave, Eq. (7.19). It is independent of z and, if $2\alpha x$ is small, not very dependent on x.

In contrast, F_{2xb} is very z-dependent, as is evident from the plots in Fig. 7.3. It is concentrated in a region near the boundary at $z = 0$, being negligible for n greater than, say, 5. When k is much larger than α, as is commonly true, the term $f_2(n)$ can be neglected in the expression for F_{2xb}. Then, since the maximum value of $f_1(n)$ is 0.46, the ratio of the maximum of F_{2xb} to F_{2xa} is $0.23k/\alpha$. In water at 20°C for a

Fig. 7.3 Plots of $f_1(n)$, $f_2(n)$, and $f_3(n)$ vs. the nondimensional quantity $n = \kappa z$, based on Eqs. (7.32) and (7.34); these apply when a plane traveling wave grazes a boundary. The force density component F_{2xb} is proportional to a linear combination of $f_1(n)$ and $f_2(n)$, while the streaming velocity component u_{2b} is proportional to $f_3(n)$.

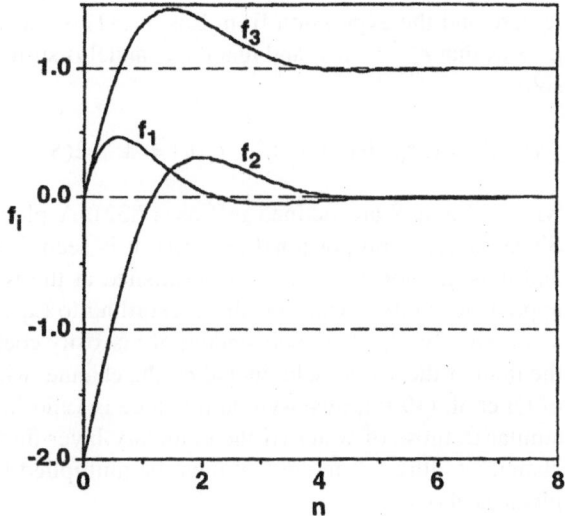

frequency of 1 MHz, the ratio k/α is 1.66×10^5; it is evident that in the boundary layer, the force density F_{2x} comes almost entirely from F_{2xb}.

7.3.3.3 Flow from Plane Traveling Wave in Channel with Open Ends

We now use the results obtained for F_{2x} in Eqs. (7.32) to treat the situation where a plane wave travels along the x direction in a channel between two parallel walls, one at $z = 0$ and the other at $z = h$. Near the wall at $z = 0$, the force distribution is given by Eqs. (7.32); near the wall at $z = h$, from symmetry, a similar boundary layer exists. On the basis of Eqs. (7.15), the vector $\mathbf{F}_2$ is considered to have only an x component, given by F_{2x}. If the channel has open ends, we can take ∇p_2 to be zero; then $\nabla^2 \mathbf{u}_2$ has only one nonzero component, $\nabla^2 u_2$. Also, if αx is small in a region of interest, the factor $e^{-2\alpha x}$ differs little from unity there, and F_{2x}, being nearly independent of x, is essentially a function of z alone. With the flow considered to be along x, and having speed depending only on z, $\nabla^2 u_2$ becomes $\partial^2 u_2/\partial z^2$, and Eq. (7.15a) reduces to

$$-\mu \frac{\partial^2 u_2}{\partial z^2} = F_{2x} = F_{2xa} + F_{2xb}, \qquad (7.33)$$

where F_{2xa} and F_{2xb} are given in Eqs. (7.32) for the region $0 < z < h/2$, and are given by similar expressions [in which $(h - z)$ replaces z] for the region $h/2 < z < h$. It is instructive to determine the flow resulting from F_{2xa} and F_{2xb} separately. Because of symmetry, it is sufficient to find results for the flow in the region $0 < z < h/2$.

The streaming velocity u_{2b} associated with the force density generated in the boundary layer is obtained from a solution of Eq. (7.33) in which F_{2xa} is set equal to zero and the expression from Eqs. (7.32) is substituted for F_{2xb}. Assuming that $k \gg \alpha$, that $e^{-2\alpha x} \simeq 1$, and that the channel is sufficiently wide, one obtains for the solution

$$u_{2b}(z) = (u_{a0}^2/4c_0)f_3(n), \quad f_3(n) = 1 + 2(S - C) + e^{-2n}, \quad n = \kappa z, \qquad (7.34)$$

where C and S are defined in Eqs. (7.32). A plot of $f_3(n)$ is shown in Fig. 7.3. Since $u_{2b}(z)$ is proportional to $f_3(n)$, it is seen from this plot that $u_{2b}(0)$ is zero, and thus the nonslip condition is satisfied at the walls. At large values of κz, f_3 approaches unity asymptotically. According to Eqs. (7.34), the asymptotic value of u_{2b} is simply $u_{a0}^2/4c_0$, independent of viscosity coefficients and other properties of the fluid or the sound field, including the channel width h. According to the analysis of Qi et al. (1995), this asymptotic value is valid in a fluid with thermal properties similar to those of water (if the boundary-layer thickness is small compared to the channel width), but in general must be multiplied by a factor; for air, the factor is given as about 1.7.

Similarly, to obtain the streaming velocity u_{2a} associated with the force density generated in a free-traveling wave, the expression for F_{2xa} from Eqs. (7.32) is substituted into Eq. (7.33). As before, the attenuation is considered small in the region of interest, so that $e^{-2\alpha x} \simeq 1$ and F_{xa} is essentially constant. If it is assumed that a nonslip condition exists on the streaming velocity at $z = 0$ and at $z = h$, the solution of Eq. (7.33) is

$$u_{2a}(z) = C_1 n_1 (1 - n_1), \quad n_1 = \frac{z}{h}, \quad C_1 = \frac{\alpha \rho_0 u_{a0}^2 h^2}{2\mu}. \tag{7.35}$$

It is seen that the flow pattern $u_{2a}(z)$ is parabolic, with maximum speed $C/4$ occurring at $z = h/2$. The total streaming velocity u_2 in the half-channel ($0 < z < h/2$) is a superposition of (1) the velocity $u_{2b}(z)$ given by Eqs. (7.34), associated with the boundary layer at $z = 0$, and (2) a parabolic velocity distribution $u_{2a}(z)$ given by Eqs. (7.35), associated with the free-traveling wave. We have, then, for the total streaming velocity in this half-channel,

$$u_2 = (u_{a0}^2/4c_0) f_3(n) + C_1 n_1 (1 - n_1), \tag{7.36}$$

where C_1 is given in Eqs. (7.35). Because of symmetry, Eq. (7.36) applies with z measured from either wall.

As before, u_2 is the Eulerian average of the velocity; it gives the average at a point as it would be measured with a small hot-wire anemometer, for example. To proceed from u_2 to the Lagrangian average U, the velocity components u_s and w_s from Eq. (7.31) are used for $\mathbf{u}_1$ in Eq. (7.17), with the result

$$U = \frac{3u_{a0}^2}{4c_0} \left[f_4(n) + \left(\frac{4\phi k^2 h^2}{9} \right) n_1 (1 - n_1) \right], \quad f_4(n) = 1 - 2C + e^{-2n}, \tag{7.37}$$

in which $k \gg \alpha$ is assumed. In Eqs. (7.37), the absorption coefficient α, which appears in the expression for C_1 in Eqs. (7.35), is taken to be $\phi \alpha_c$, where α_c [given by α in Eqs. (7.28) when $\mu_B = 0$, $b = 4/3$] is the classical result that applies if the absorption is from shear viscosity alone. (The factor ϕ is 3 or more for water, is much larger for some liquids and suspensions, and is somewhat greater than unity for air in the megahertz frequency range.)

7.3.3.4 Flow from Plane Traveling Wave in Channel with Closed Ends

In the preceding discussion, the time-averaged pressure gradient $\partial p_2/\partial x$ is considered to be zero. This corresponds to an assumption that the ends of the channel are open; no restraints to the flow exist there, the time-averaged pressure being the same at the two ends. In contrast, if the ends are closed[3] a pressure gradient is set up. Under the assumptions made here, the traveling wave extends laterally throughout the channel (unlike the situations discussed in Sect. 7.3.2.3). The pressure gradient has only an x component $\partial p_2/\partial x$, and Eq. (7.15a) can now be written in the form

$$K - F_{xb} = \mu\frac{\partial^2 u_{2x}}{\partial z^2};\tag{7.38}$$

here the force density F_{xb} is given in Eqs. (7.32b), and K, equal to the difference $(\partial p_2/\partial x - F_{xa})$, is essentially constant in a region of interest (if $e^{-2\alpha x} \simeq 1$ in that region). A solution of Eq. (7.38) is obtained and the constant K is determined by imposing the condition that the net flow across any cross section is zero. This condition is

$$\int_0^{h/2} U_x(z)\,dz = 0.\tag{7.39}$$

7.3.3.5 Other Examples of Flow from Boundary Layers

Boundary-layer acoustic streaming occurs in situations where there is relative oscillatory motion between a fluid and a boundary, with a component tangential to the boundary. Theory for the regularly spaced vortices produced by a standing wave in a channel between parallel walls was derived by Rayleigh (1883, 1945) in what appears to have been the first successful analysis of acoustic streaming. Further development of the theory, including a correction, has been described by Raney et al. (1954).

Considerable attention has been given to the streaming produced near a solid cylinder in oscillation relative to a surrounding fluid (Ingard and Labate, 1950; Andres and Ingard, 1953; Skavlem and Tjøtta, 1955; Holtsmark et al., 1954; Raney et al., 1954; Westervelt, 1955). In its simplest form, which applies at low amplitudes, the flow pattern is independent of amplitude and consists of four symmetric circulations. More generally, the pattern depends on the amplitude of oscillation, as well as on the cylinder radius and the boundary-layer thickness κ^{-1}. At the higher amplitudes, a jetlike fluid motion occurs, outward from the cylinder along the axis of oscillation (Riley, 1965; Stuart, 1966; Davidson and Riley, 1972; Bertelsen et al., 1973).

Streaming near a solid sphere in the field of a point source of sound was treated theoretically by Wang (1982), who showed that the pattern changes markedly with distance between sphere and source; it reduces to that near a sphere in a uniform

---

[3] Here it is supposed that the ends of the channel are closed with acoustical absorbing materials or other terminations that are barriers to the acoustic streaming but do not significantly perturb the first-order field.

field when the distance is large. The streaming near a sphere at higher Reynolds numbers was taken up by Amin and Riley (1990). An explanation was offered by Nyborg (1994) of differences that had appeared in the literature on theory for the streaming near a sphere.

Other aspects of the streaming near a solid sphere in oscillating fields have been treated by Lee and Wang (1989, 1990; see also Sect. 6.3.1.4). For example, they developed theory for the streaming around a small solid sphere in the field of two orthogonal standing waves, including an expression for the steady torque exerted on the sphere. The streaming near an oscillating elliptic cylinder was treated by Davidson and Riley (1972).

As shown by an example in Sect. 7.3.3.3, the velocity parallel to a boundary often increases quickly from its boundary value to a "limiting" value, then rises much more slowly. Simplified methods can sometimes be used to calculate the limiting velocity. It was shown by Longuet-Higgins (1953) that for two-dimensional fields satisfying required conditions, the limiting velocity U_L is given by

$$U_L = -\frac{3}{8\omega}\frac{\partial u_{a0}^2}{\partial x}, \tag{7.40}$$

where x is a coordinate (which may be curvilinear) that is tangent to the boundary in the region of interest, and u_{a0} is the amplitude of the first-order velocity component along x, as it would be in the absence of viscosity. For example, on an oscillating cylinder of radius a, when x measures arc length along the periphery (in a plane perpendicular to the axis), u_{a0} can be written as $A\sin(x/a)$ and U_L is $(3A^2/2\omega a)\sin(2x/a)$, as given by Schlichting (1932, 1955).

The Longuet-Higgins result in Eq. (7.40) was generalized by Nyborg (1958, 1965) and further generalized by Lee and Wang (1989). One application is to the streaming produced by a small hemispherical sound source on a solid plane boundary (which can represent a vibrating gas bubble resting on a boundary); for this, an estimate of the limiting velocity was found to be

$$U_L = \frac{Q^2}{4\pi\omega r^5}, \tag{7.41}$$

where Q is the maximum during a cycle of volume flow through the hemispherical surface of the source and r is distance from the source.

7.3.3.6 Acoustic Streaming in Special Situations

Special features of acoustic streaming theory were explored by Powell (1982) as they occur in a medium characterized by nonlinear viscoelasticity, and by Whitworth (1990) as they occur in a rotating fluid.

7.4 Applications

Numerous investigations have shown that applications of sound, including ultra-sound, can bring about changes in structures or in processes if acoustic streaming is brought to bear on them. In this brief account, the emphasis is on applications that have been reported since an earlier review (Nyborg, 1965).

Applications of ultrasound in liquids often depend on phenomena occurring at gas-liquid interfaces, and on the activity of vibrating and/or cavitating gas bubbles. Since Elder's basic investigation of the microstreaming produced by a vibrating bubble resting on a solid boundary (Elder, 1959), studies have been made on the consequences of such motion. Gould (1966) demonstrated enhancement of heat transfer into liquid from a solid surface by the presence of a vibrating bubble. In later work, he showed (Gould, 1974) that the rate of gas transport between a vibrating bubble and the surrounding liquid was altered considerably when surface waves were present on the bubble, evidently because of the associated acoustic streaming.

The acoustic streaming near a gas bubble in a sound field has been studied theoretically by Statnikov (1968) and by Davidson and Riley (1971). Theories for its influence on mass transport have been advanced by Kapustina and Statnikov (1968), Davidson (1971), and Church (1988).

Using a frequency of about 20 kHz, Rooney (1970, 1972) demonstrated that the viscous shear associated with microstreaming near a single resonant bubble caused hemolysis of red cells in saline suspension. Williams et al. (1970) showed that hemolysis was produced in a similar way by microstreaming generated by a vibrating wire whose tip was of the same curvature as the bubble surface. Further studies on cellular effects produced by microstreaming, especially in the lower ultrasonic frequency range, have been reviewed by Williams (1983).

Miller (1987) has reviewed findings from related studies at higher frequencies, extending into the 1–10 MHz frequency range. Included are numerous investigations of plant leaves or other plant tissues in which gas-filled channels exist intracellularly. In other studies, hydrophobic Nuclepore® membranes are used; when these membranes are immersed in an aqueous liquid, air is trapped in the pores, providing approximately cylindrical gas bodies of fairly uniform dimensions, a few microns in diameter and depth. When a strip of this membrane is immersed in a suspension of particles, such as biological cells, and exposed to ultrasound while viewed through a microscope, microstreaming is observed near each gas-filled pore. Biological effects of the motion have included cell destruction, aggregation of platelets, and release of ATP from red cells.

Shiran et al. (1990) cite studies that show the influence of bubble-associated microstreaming on reactions that occur at a solid surface. They also give evidence that the pattern that appears in the Sarvazyan method of mapping ultrasound fields results from enhancement of the reaction rate by microstreaming from small bubbles.

Dunn (1985) observed inactivation of Chinese hamster V79 cells irradiated with a beam of 3-MHz ultrasound at elevated temperature, and attributed the biological change to viscous shear associated with acoustic streaming. Pohl et al. (1995) found increased agglutination of human red cells subjected to an ultrasound beam of 2-MHz or 10-MHz frequency, and proposed acoustic streaming as a cause. While the above experiments by Dunn and by Pohl et al. were done in vitro with cell suspensions in laboratory containers, Frizzell et al. (1986) made observations on blood vessels in tissue of living mice during exposure to 1-MHz ultrasound. In these in vivo exposures, vigorous small-scale streaming was observed in a vessel near an "obstacle" deliberately formed by indenting the vessel wall with a small rod applied externally. Under some conditions, continuation of the acoustic exposure led to thrombus formation near the obstacle.

Experimental and theoretical studies of heat transport from a cylinder or sphere serving as heat source, and of the increased heat transport resulting from acoustic streaming generated by vibrating the object, have been made by Davidson (1973) and by Lee and Wang (1990). Boraker et al. (1992) obtained significant improvements in the sensitivity and reaction time of the well known ELISA techniques used in immunology by applying low-frequency microstreaming to the surfaces at which the biochemical reactions occur.

Applications to the transporting of materials in small-scale devices led Moroney et al. (1990) to study acoustic streaming produced in fluids by traveling flexural waves in thin membranes. In a summary of research activities in Japan, Mitome (1993a) describes applications of acoustic streaming to ultrasonic motors and to ultrasonic cleaning. Qi and Brereton (1995) discuss mechanisms for removal of small impurities from a surface by using focused ultrasound to produce streaming at the surface.

Lee and Wang (1989) have used orthogonal standing waves to exert torque on a small sphere via the acoustic streaming generated at its surface. An application of this technique is to the manipulation of objects in gravity-free environments (Sect. 6.3).

In studies on hearing, Békésy (1960) observed that eddies are produced in the cochlea when sound is being received. Lighthill (1992) has pointed out that boundary layers are established at the surface of the basilar membrane. As a result, acoustic streaming is expected in the cochlea, resembling in some ways the channel flow discussed in Sect. 7.3, but with novel features because of the nature of basilar membrane motion.

Several recent publications describe roles for acoustic streaming in medical applications. Wu et al. (1994) showed that when ultrasound passes through a liquid immediately before impinging on bone tissue, the heating produced in the bone is reduced by acoustic streaming in the liquid. Stavros and Dennis (1993) and Nightingale et al. (1995) have investigated a technique for distinguishing cysts from solid masses during use of ultrasound breast imaging to evaluate lesions; the technique depends on findings that streaming motions produced by the ultrasound can be observed during ultrasonic examination of liquid-filled cysts.

References

Amin, N., and Riley, N. (1990). Streaming from a sphere due to a pulsating force. *J. Fluid Mech.* **210**, 459–473.

Andres, J. M., and Ingard, U. (1953). Acoustic streaming at low Reynolds numbers. *J. Acoust. Soc. Am.* **25**, 932–938.

Békésy, G. von (1960). *Experiments in Hearing* (McGraw-Hill New York).

Bertelsen, A., Svardal, A., and Tjøtta, S. (1973). Nonlinear streaming effects associated with oscillating cylinders. *J. Fluid Mech.* **59**, 493–511.

Blake, J. R. (1971). A note on the image system for a stokeslet in a no-slip boundary. *Proc. Cambridge Phil. Soc.* **70**, 303–310.

Boraker, D. K., Bugbee, S. J., and Reed, B. A. (1992). Acoustic probe-based ELISA. *J. Immunological Methods* **155**, 91–94.

Carstensen, E. L., and Schwan, H. P. (1959). Acoustic properties of hemoglobin solutions. *J. Acoust. Soc. Am.* **31**, 305–311.

Church, C. C. (1988). A method to account for acoustic micro-streaming when predicting bubble growth rates produced by rectified diffusion. *J. Acoust. Soc. Am.* **84**, 1758–1764.

Davidson, B. J. (1971). Mass transfer due to cavitation microstreaming. *J. Sound Vib.* **17**, 261–270.

Davidson, B. J. (1973). Heat transfer from a vibrating cylinder. *Int. J. Heat and Mass Transfer* **16**, 1703–1727.

Davidson, B. J., and Riley, N. (1971). Cavitation microstreaming. *J. Sound Vib.* **15**, 217–233.

Davidson, B. J., and Riley, N. (1972). Jets induced by oscillatory motion. *J. Fluid Mech.* **53**, 287–303.

Duck, F. A. (1990). *Physical Properties of Tissues* (Academic Press, Ltd., London).

Dunn, F. (1985). Cellular inactivation by heat and shear. *Radiation and Environmental Biophysics* **24**, 131–139.

Dunn, F., Edmonds, P. D., and Fry, W. J. (1969). Absorption and dispersion of ultrasound in biological media. In *Biological Engineering*, H. P. Schwann, ed. (McGraw-Hill, New York), pp. 205–332.

Eckart, C. (1948). Vortices and streams caused by sound waves. *Phys. Rev.* **73**, 68–76.

Elder, S. A. (1959). Cavitation microstreaming. *J. Acoust. Soc. Am.* **31**, 54–64.

Faraday, M. (1831). On a peculiar class of acoustical figures. *Phil. Trans. R. Soc. London* **121**, 299–327.

Frizzell, L. A., Miller, D. L., and Nyborg, W. L. (1986). Ultrasonically induced intravascular streaming and thrombus formation adjacent to a pipette. *Ultrasound Med. Biol.* **12**, 217–221.

Gould, R. K. (1966). Heat transfer across a solid-liquid interface in the presence of acoustic streaming. *J. Acoust. Soc. Am.* **40**, 219–225.

Gould, R. K. (1974). Rectified diffusion in the presence of, and absence of, acoustic streaming. *J. Acoust. Soc. Am.* **54**, 1740–1746.

Gusev, V. E., and Rudenko, O. V. (1979). Nonsteady quasi-one-dimensional acoustic streaming in unbounded volumes with hydrodynamic nonlinearity. *Sov. Phys. Acoust.* **25**, 493–497.

Hall, L. (1948). The origin of ultrasonic absorption in water. *Phys. Rev.* **73**, 775–781.

Hamilton, M. F., Il'inskii, Yu. A., and Zabolotskaya, E. A. (1995). Acoustic streaming at high Reynolds numbers in focused sound beams. *J. Acoust. Soc. Am.* **97**, 3376(A).

Happel, J., and Brenner, H. (1991). *Low Reynolds Number Hydrodynamics* (Kluwer Academic Publishers, Dordrecht), Section 3.4.

Herzfeld, K. F., and Litovitz, T. A. (1959). *Absorption and Dispersion of Ultrasonic Waves* (Academic Press, New York).

Holtsmark, J., Johnsen, I., Sikkeland, T., and Skavlem, S. (1954). Boundary layer flow near a cylindrical obstacle in an oscillating, incompressible fluid. *J. Acoust. Soc. Am.* **26**, 26–39.

Ingard, U., and Labate, S. (1950). Acoustic circulation effects and the nonlinear impedance of orifices. *J. Acoust. Soc. Am.* **22**, 211–218.

Johnsen, I., and Tjötta, S. (1957). Eine theoretische und experimentelle untersuchung über den quarzwind. *Acustica* **7**, 7–16.

Kamakura, T., Kazuhisa, M., Kumamoto, Y., and Breazeale, M. A. (1995). Acoustic streaming induced in focused Gaussian beams. *J. Acoust. Soc. Am.* **97**, 2740–2746.

Kapustina, O. A., and Statnikov, Y. G. (1968). Influence of acoustic microstreaming on the mass transfer in a gas bubble-liquid system. *Sov. Phys. Acoust.* **13**, 327–329.

Lamb, H. (1945). *Hydrodynamics* (Dover Publications, New York), Section 340.

Lee, C. P., and Wang, T. G. (1989). Near-boundary streaming around a small sphere due to two orthogonal standing waves. *J. Acoust. Soc. Am.* **85**, 1081–1088.

Lee, C. P., and Wang, T. G. (1990). Outer acoustic streaming. *J. Acoust. Soc. Am.* **88**, 2367–2375.

Liebermann, L. N. (1949). The second viscosity of liquids. *Phys. Rev.* **75**, 1415–1422.

Lighthill, J. (1978a). Acoustic streaming. *J. Sound Vib.* **61**, 391–418.

Lighthill, J. (1978b). *Waves in Fluids* (Cambridge University Press, Cambridge, U.K.).

Lighthill, J. (1992). Acoustic streaming in the ear itself. *J. Fluid Mech.* **239**, 551–606.

Longuet-Higgins, M. S. (1953). Mass transport in water waves. *Phil. Trans. R. Soc.* **A245**, 535–581.

Medwin, H. (1954). An acoustic streaming experiment in gases. *J. Acoust. Soc. Am.* **26**, 332–341.

Medwin, H., and Rudnick, I. (1953). Surface and volume sources of vorticity in acoustic fields. *J. Acoust. Soc. Am.* **25**, 538–540.

Miller, D. L. (1987). A review of the ultrasonic bioeffects of microsonation, gas-body activation, and related cavitation-like phenomena. *Ultrasound Med. Biol.* **13**, 443–470.

Mitome, H. (1993a). Acoustic streaming and nonlinear acoustics research activities in Japan. In *Advances in Nonlinear Acoustics*, 13th International Symposium on Nonlinear Acoustics, H. Hobæk, ed. (World Scientific, Singapore), pp. 43–54.

Mitome, H. (1993b). Enhancement of driving force of acoustic streaming due to nonlinear attenuation of ultrasound. In *Ultrasonics International 93 Conference Proceedings* (Butterworth-Heinemann Ltd., Linacre House, Jordan Hill, Oxford OX2 8DP), pp. 439–442.

Moroney, R. M., White, R. M., and Howe, R. T. (1990). Fluid motion produced by ultrasonic Lamb waves. *Proceedings of IEEE 1990 Ultrasonics Symposium*, Honolulu, Hawaii, December 4–7, pp. 355–358.

NCRP (1992). Exposure Criteria for Medical Diagnostic Ultrasound. I. Criteria Based on Thermal Mechanisms. National Council of Radiation Protection and Measurements Report No. 113 (NCRP Publications, Bethesda, MD).

Nightingale, K. R., Kornguth, P. J., Walker, W. F., McDermott, B. A., and Trahey, G. E. (1995). A novel ultrasonic technique for differentiating cysts from solid lesions: Preliminary results in the breast. *Ultrasound Med. Biol.* **21**, 745–751.

Nyborg, W. L. (1958). Acoustic streaming near a boundary. *J. Acoust. Soc. Am.* **30**, 329–339.

Nyborg, W. L. (1965). Acoustic streaming. In *Physical Acoustics*, Vol. 2B, W. P. Mason, ed. (Academic Press, New York).

Nyborg, W. L. (1975). *Intermediate Biophysical Mechanics* (Cummings Publishing Co., Menlo Park, CA).

Nyborg, W. L. (1994). Acoustic streaming near a rigid sphere. *J. Acoust. Soc. Am.* **95**, 556.

Piercy, J. E., and Lamb, J. (1954). Acoustic streaming in liquids. *Proc. R. Soc.* **A226**, 43–50.

Pohl, E. E., Rosenfeld, E. H., Pohl, P., and Millner, R. (1995). Effects of ultrasound on agglutination and aggregation of human erythrocytes *in vitro*. *Ultrasound Med. and Biol.* **21**, 711–719.

Powell, R. L. (1982). Acoustic streaming reversal by a finite amplitude sound wave in a non-Newtonian fluid. *J. Acoust. Soc. Am.* **71**, 1118–1123.

Qi, Q. (1993). Effect of compressibility on acoustic streaming near a rigid boundary for a plane traveling wave. *J. Acoust. Soc. Am.* **94**, 1090–1098.

Qi, Q., and Brereton, G. J. (1995). Mechanisms of removal of micron-sized particles by high-frequency ultrasonic waves. *IEEE Trans. Ultrason. Ferroelectr. Freq. Control* **42**, 619–629.

Qi, Q., Johnson, R. E., and Harris, J. G. (1995). Boundary layer attenuation and acoustic streaming accompanying plane-wave propagation in a tube. *J. Acoust. Soc. Am.* **97**, 1499–1509.

Raney, W. P., Corelli, J. C., and Westervelt, P. J. (1954). Acoustical streaming in the vicinity of a cylinder. *J. Acoust. Soc. Am.* **26**, 1006–1014.

Rayleigh, Lord (1883). On the circulation of air observed in Kundt's tubes, and on some allied acoustical problems. *Phil. Trans. R. Soc.* **A175**, 1–21.

Rayleigh, Lord (1945). *Theory of Sound* (Dover Publications, New York).

Riley, N. (1965). Oscillating viscous flows. Mathematika **12**, 161–175.

Rooney, J. A. (1970). Hemolysis near an ultrasonically pulsating bubble. *Science* **169**, 869–871.

Rooney, J. A. (1972). Shear as a mechanism for sonically induced biological effects. *J. Acoust. Soc. Am.* **52**, 1718–1724.

Rudenko, O. V., and Soluyan, S. I. (1977). *Theoretical Foundations of Nonlinear Acoustics* (translation from the Russian by R. T. Beyer) (Consultants Bureau, New York).

Schlichting, H. (1932). Berechnung ebener periodischer Grenzschichtstromungen. *Phys. unserer Z.* **33**, 327–335.

Schlichting, H. (1955). *Boundary Layer Theory* (McGraw-Hill, New York).

Shiran, M. B., Quan, K. M., Watmough, D. J., Abdellatif, K., Mallard, J. R., Marshall, D., and Gregory, D. W. (1990). Some of the factors involved in the Sarvazyan method for recording ultrasound field distributions with special reference to the application of ultrasound in physiotherapy. *Ultrasonics* **28**, 411–414.

Skavlem, S., and Tjøtta, S. (1955). Steady rotational flow of an incompressible viscous fluid enclosed between two coaxial cylinders. *J. Acoust. Soc. Am.* **27**, 26–33.

Starritt, H. C., Duck, F. A., and Humphrey, V. F. (1989). An experimental investigation of streaming in pulsed diagnostic ultrasound beams. *Ultrasound Med. Biol.* **15**, 363–373.

Starritt, H. C., Duck, F. A., and Humphrey, V. F. (1991). Forces acting in the direction of propagation in pulsed ultrasound fields. *Phys. Med. Biol.* **36**, 1465–1474.

Statnikov, Y. G. (1968). Microstreaming about a gas bubble in a liquid. *Sov. Phys. Acoust.* **13**, 398–399.

Stavros, A. T., and Dennis, M. A. (1993). The ultrasound of breast pathology. In *Percutaneous Breast Biopsy*, S. H. Parker and W. E. Jobe, eds. (Raven Press, New York), Chapter 11, pp. 111–115.

Stuart, J. T. (1966). Double boundary layers in oscillatory viscous flow. *J. Fluid Mech.* **24**, 673–687.

Tjøtta, S., and Naze Tjøtta, J. (1993). Acoustic streaming in ultrasound beams. In *Advances in Nonlinear Acoustics*, 13th International Symposium on Nonlinear Acoustics, H. Hobæk, ed. (World Scientific, Singapore), pp. 601–606.

Wang, C.-Y. (1982). Acoustic streaming of a sphere near an unsteady source. *J. Acoust. Soc. Am.* **71**, 580–584.

Westervelt, P. J. (1953a). The theory of steady rotational flow generated by a sound field. *J. Acoust. Soc. Am.* **25**, 60–67.

Westervelt, P. J. (1953b). Hydrodynamic flow and Oseen's approximation. *J. Acoust. Soc. Am.* **25**, 951–953.

Westervelt, P. J. (1955). Acoustic streaming near a small obstacle. *J. Acoust. Soc. Am.* **27**, 379.

Whitworth, G. (1990). Acoustic streaming in a rotating fluid. *J. Acoust. Soc. Am.* **88**, 1960–1963.

Williams, A. R. (1983). *Ultrasound: Biological Effects and Potential Hazards* (Academic Press, New York).

Williams, A. R., Hughes, D. E., and Nyborg, W. L. (1970). Hemolysis near a transversely oscillating wire. *Science* **169**, 871–873.

Wu, J., and Du, G. (1993). Acoustic streaming generated by a focused Gaussian beam and finite amplitude tonebursts. *Ultrasound Med. Biol.* **19**, 167–176.

Wu, J., Winkler, A. J., and O'Neill, T. P. (1994). Effect of acoustic streaming on ultrasonic heating. *Ultrasound Med. and Biol.* **20**, 195–201.

Chapter 8
Sound Beams

Mark F. Hamilton

Contents

8.1   Introduction

Medical ultrasound, sonar, acoustic microscopy, and nondestructive testing are some of the areas in which intense directional sound beams are encountered. At high acoustic intensities, and for propagation in real fluids, a proper model of directional acoustic radiation must take into account the combined effects of diffraction, absorption, and nonlinearity. The Khokhlov–Zabolotskaya–Kuznetsov (KZK) equation (Zabolotskaya and Khokhlov, 1969; Kuznetsov, 1971), presented in Sect. 8.2, accounts consistently for all three effects, and it provides the theoretical foundation for the present chapter. In Sect. 8.3, general integral expressions are derived for second-harmonic, sum-frequency, and difference-frequency generation in weakly nonlinear, axisymmetric sound beams. Solutions of these integrals are obtained for radiation from sources with Gaussian amplitude shading, both unfocused and focused. Asymptotic expressions are developed for the parametric array and for self-demodulation produced by radiation from a circular piston. In Sect. 8.4, numerical results are presented for strongly nonlinear piston radiation,

M. F. Hamilton
Department of Mechanical Engineering, The University of Texas at Austin, Austin, TX, USA

© The Author(s) 2024   231
M. F. Hamilton, D. T. Blackstock (eds.), *Nonlinear Acoustics*,
https://doi.org/10.1007/978-3-031-58963-8_8

both unfocused and focused, illustrating shock formation and other effects not taken into account by the quasilinear analysis in Sect. 8.3.

8.2 Parabolic Wave Equation

For consistency, all analyses in the present chapter are based on the KZK parabolic wave equation (Bakhvalov et al., 1987), Eq. (3.65), which accounts for nonlinearity, absorption due to viscosity and heat conduction, and diffraction in directional sound beams:

$$\frac{\partial^2 p}{\partial z \partial \tau} = \frac{c_0}{2} \nabla_\perp^2 p + \frac{\delta}{2c_0^3} \frac{\partial^3 p}{\partial \tau^3} + \frac{\beta}{2\rho_0 c_0^3} \frac{\partial^2 p^2}{\partial \tau^2}. \tag{8.1}$$

Here p is the sound pressure, z is the coordinate along the axis of the beam (which is assumed to propagate in the $+z$ direction), $\tau = t - z/c_0$ is retarded time, c_0 is the small-signal sound speed, δ is the diffusivity of sound corresponding to thermoviscous absorption, β is the coefficient of nonlinearity, and ρ_0 is the ambient density. The operator $\nabla_\perp^2$ is the Laplacian in the (x, y) plane, perpendicular to the axis of the beam. We shall consider only axisymmetric sound beams, for which $\nabla_\perp^2 = \partial^2/\partial r^2 + r^{-1}(\partial/\partial r)$, where $r^2 = x^2 + y^2$. For plane waves $\nabla_\perp^2 p = 0$, and Eq. (8.1) reduces to the Burgers equation, Eq. (3.54). The effect of the parabolic approximation on the accuracy with which Eq. (8.1) models diffraction is discussed in Sect. 8.3.2 in connection with radiation from a circular piston. Although Eq. (8.1) is written in terms of sound pressure, it is consistent to use the linear plane-wave impedance relation $p = \rho_0 c_0 u_z$ to express the KZK equation and its solutions in terms of u_z, the z component of the particle velocity vector. See Sect. 3.9 for further discussion of Eq. (8.1).

8.3 Quasilinear Theory

General integral expressions are derived below for second-harmonic, sum-frequency and difference-frequency generation due to weak finite-amplitude radiation from axisymmetric sources. The expressions are obtained using the method of successive approximations. We begin by restricting our attention to second-harmonic genera-tion by a sound beam radiated at a single angular frequency ω. A quasilinear solution of the form

$$p = p_1 + p_2 \tag{8.2}$$

is assumed, where p_1 is the linear solution (first approximation) of Eq. (8.1) for the pressure at frequency ω, and p_2 is a small correction to p_1 (the second approximation, for $|p_2| \ll |p_1|$ throughout the nonlinear interaction region) at the second-harmonic frequency 2ω. Now write

$$p_n(r, z, \tau) = \frac{1}{2j} q_n(r, z) e^{jn\omega\tau} + \text{c.c.}, \quad n = 1, 2, \tag{8.3}$$

where q_1 and q_2 are complex pressure amplitudes, and c.c. denotes the complex conjugate of preceding terms. Equation (8.3) reduces to $p_n = q_n \sin n\omega\tau$ when q_n is real. Substitution of Eqs. (8.2) and (8.3) into Eq. (8.1) yields the following quasilinear system of equations for q_1 and q_2 (see, e.g., Sect. 10.2):

$$\frac{\partial q_1}{\partial z} + \frac{j}{2k} \nabla_\perp^2 q_1 + \alpha_1 q_1 = 0, \quad \frac{\partial q_2}{\partial z} + \frac{j}{4k} \nabla_\perp^2 q_2 + \alpha_2 q_2 = \left(\frac{\beta k}{2\rho_0 c_0^2}\right) q_1^2, \tag{8.4}$$

where $\alpha_n = \delta n^2 \omega^2 / 2c_0^3$ is the thermoviscous attenuation coefficient at frequency $n\omega$, and $k = \omega/c_0$. Arbitrary absorption laws may be used hereafter to define α_1 and α_2 in Eqs. (8.4) (see Sect. 5.2, where dispersion is also taken into account). For the source condition, we take

$$p(r, 0, t) = \frac{1}{2j} q_1(r, 0) e^{j\omega t} + \text{c.c.}, \tag{8.5}$$

where $q_1(r, 0)$ is an arbitrary distribution that takes into account both amplitude and phase shading. It is thus assumed that the source does not radiate at the second-harmonic frequency, i.e., $q_2(r, 0) = 0$.

Both of Eqs. (8.4) are linear, the second because q_1^2 is determined independently by the first. Integral forms of these equations that are more convenient for analysis can therefore be constructed with Green's functions (see, e.g., Morse and Ingard, 1968, pp. 319–322, for a discussion of Green's functions of the Helmholtz equation). We define the Green's function $G_n(r, z|r', z')$ here, at frequency $n\omega$, to be the solution of the inhomogeneous equation

$$\frac{\partial G_n}{\partial z} + \frac{j}{2nk} \nabla_\perp^2 G_n + \alpha_n G_n = \frac{1}{2\pi r} \delta(r - r') \delta(z - z'), \tag{8.6}$$

where the right-hand side is the three-dimensional Dirac delta function in cylindrical coordinates, for axial symmetry. Primed coordinates correspond to locations of source points. To solve Eq. (8.6), we employ the Hankel transform pair

$$\tilde{f}(\kappa) = \int_0^\infty f(r) J_0(\kappa r) r \, dr, \quad f(r) = \int_0^\infty \tilde{f}(\kappa) J_0(\kappa r) \kappa \, d\kappa, \tag{8.7}$$

where J_0 is the zeroth-order Bessel function. Transformation of Eq. (8.6) yields

$$\frac{\partial \tilde{G}_n}{\partial z} - \left(\frac{j\kappa^2}{2nk} - \alpha_n\right)\tilde{G}_n = \frac{J_0(\kappa r')}{2\pi}\delta(z - z'), \qquad (8.8)$$

the solution of which is

$$\tilde{G}_n(\kappa, z|r', z') = \exp\left[\left(\frac{j\kappa^2}{2nk} - \alpha_n\right)(z - z')\right]\frac{J_0(\kappa r')}{2\pi}H(z - z'), \qquad (8.9)$$

where $H(x)$ is the Heaviside unit step function ($H = 0$ for $x < 0$ and $H = 1$ for $x > 0$). The physical significance of the step function is that only nonlinear interactions between the source plane and a given field point (i.e., for $z' < z$ and therefore $H = 1$) contribute significantly to the field at that point. Mathematically, this is a consequence of the parabolic approximation, which leads to Eq. (8.1) as a first-order differential equation in z that describes only forward propagating waves. Inversion of Eq. (8.9) using the second of Eqs. (8.7) gives

$$G_n(r, z|r', z') = \frac{jnk}{2\pi(z - z')}J_0\left(\frac{nkrr'}{z - z'}\right)$$

$$\times \exp\left[-\alpha_n(z - z') - \frac{jnk(r^2 + r'^2)}{2(z - z')}\right], \qquad (8.10)$$

where the dependence on $H(z - z')$ is now suppressed.

Solutions of Eqs. (8.4) are obtained by integrating over the product of the Green's function and the appropriate source function to sum up the contributions from all source points. For q_1, the source function is $q_1(r', 0)$, and the integration is performed over the surface elements $dS' = 2\pi r' dr'$ in the plane $z' = 0$. The source function for q_2 is the right-hand side of the second of Eqs. (8.4), which is proportional to the volume distribution $q_1^2(r', z')$, and the integration is performed over the volume elements $dV' = 2\pi r' dr' dz'$. We thus obtain

$$q_1(r, z) = 2\pi \int_0^\infty q_1(r', 0)G_1(r, z|r', 0)r' dr', \qquad (8.11)$$

$$q_2(r, z) = \frac{\pi\beta k}{\rho_0 c_0^2}\int_0^z\int_0^\infty q_1^2(r', z')G_2(r, z|r', z')r' dr'dz'. \qquad (8.12)$$

The source condition $q_2(r, 0) = 0$ establishes the lower integration limit $z' = 0$ in Eq. (8.12), and the (suppressed) step function $H(z - z')$ establishes the upper limit $z' = z$. On the basis of Eq. (8.12), second-harmonic generation in the quasilinear approximation is interpreted as a sound field radiated by a volume distribution of *virtual sources* whose strengths are proportional to $q_1^2(r, z)$.

The same procedure may be followed to obtain integral expressions for sum- and difference-frequency generation. Begin by replacing Eqs. (8.3) with

$$p_1(r, z, \tau) = \frac{1}{2j}[q_{1a}(r, z)e^{j\omega_a\tau} + q_{1b}(r, z)e^{j\omega_b\tau}] + \text{c.c.}, \tag{8.13}$$

$$p_2(r, z, \tau) = \frac{1}{2j}[q_{2a}(r, z)e^{j2\omega_a\tau} + q_{2b}(r, z)e^{j2\omega_b\tau}$$

$$+ q_+(r, z)e^{j\omega_+\tau} + q_-(r, z)e^{j\omega_-\tau}] + \text{c.c.}, \tag{8.14}$$

which describe a sound beam radiated at the two primary frequencies ω_a and ω_b (with $q_{2a} = q_{2b} = q_+ = q_- = 0$ at $z = 0$), leading to the nonlinear generation of sum and difference frequencies $\omega_\pm = \omega_a \pm \omega_b$ (with $\omega_a > \omega_b$ assumed) in addition to the second harmonics $2\omega_a$ and $2\omega_b$. The solutions q_{1a} and q_{1b} for the primary beams are given by Eq. (8.11), q_{2a} and q_{2b} by Eq. (8.12), and

$$q_\pm(r, z) = \pm\frac{\pi\beta k_\pm}{\rho_0 c_0^2}\int_0^z\int_0^\infty q_{1a}(r', z')q_{1b}^{(*)}(r', z')G_\pm(r, z|r', z')r'\, dr'dz', \tag{8.15}$$

where $G_\pm$ is given by Eq. (8.10) with nk replaced by $k_\pm = \omega_\pm/c_0$ and α_n replaced by $\alpha_\pm$. The superscript $(*)$ signifies that the complex conjugate of q_{1b} is taken only for calculation of the difference-frequency component.

At this point, one may substitute Eq. (8.11) into Eqs. (8.12) and (8.15) to express q_2 and $q_\pm$ as quadruple integrals that depend directly on the source functions. This is the approach taken by Berntsen et al. (1984), who use these forms of the solutions for numerical calculations and to obtain general asymptotic expressions for the far field. The asymptotic form of q_2 is presented below only for specific source functions, Gaussian (Sect. 8.3.1) and uniform piston (Sect. 8.3.2). Here, we conclude by deriving the general asymptotic form assumed by q_1 in the far field.

Let the source have nominal radius a, in which case $q_1(r', 0) \simeq 0$ for $r' > a$. Also let the location of the field point satisfy $z \gg z_0$, where the quantity $z_0 = \frac{1}{2}ka^2$ is referred to as the *Rayleigh distance*, which marks the transition between the near- and far-field regions of the primary beam. Following the substitution of Eq. (8.10) into Eq. (8.11), the resulting phase term $-jkr'^2/2z$ is at most $z_0/z \ll 1$ in magnitude at source points $r' \sim a$, and it can thus be ignored. Comparison of the remaining integral with the first of Eqs. (8.7) yields

$$q_1(\theta, z) \simeq jkz^{-1}e^{-\alpha_1 z}\tilde{q}_1(k\tan\theta, 0)\exp(-\tfrac{1}{2}jkz\tan^2\theta), \tag{8.16}$$

where θ is the angle with respect to the axis of the beam, defined by $\tan\theta = r/z$. The value of q_1 in the far field is therefore determined by the Hankel transform of the source distribution [with $\kappa = k\tan\theta$ in Eqs. (8.7)]. A similar result is obtained from the solution of the Helmholtz equation (Morse and Ingard, 1968, p. 377). Equation (8.16) reveals that the amplitude of the primary beam decays as $z^{-1}e^{-\alpha_1 z}$ in the far field because of spherical spreading and absorption. The phase term $-\frac{1}{2}jkz\tan^2\theta$ accounts for the spherical curvature of the wavefronts, within the framework of the parabolic approximation (see footnote in Sect. 8.3.3). We define a corresponding far-field directivity function by

$$D_1(\theta) = \frac{\tilde{q}_1(k\tan\theta, 0)}{\tilde{q}_1(0, 0)}, \tag{8.17}$$

which is normalized such that $D_1(0) = 1$.

8.3.1 Gaussian Sources

Sources with Gaussian amplitude shading are frequently used in models of finite-amplitude sound beams because solutions in closed form can often be obtained in the quasilinear approximation. Owing to their simplicity, the solutions provide insight into the main features of the combined effects of diffraction, absorption, and nonlinearity.

We begin by considering second-harmonic generation and let the source function be defined by

$$q_1(r, 0) = p_0 \exp[-(r/a)^2], \tag{8.18}$$

where p_0 is the peak source pressure, and a is the effective source radius. Substitution of Eq. (8.18) into Eq. (8.11) yields for the linear solution

$$q_1(r, z) = \frac{p_0 e^{-\alpha_1 z}}{1 - jz/z_0} \exp\left[-\frac{(r/a)^2}{1 - jz/z_0}\right], \tag{8.19}$$

where $z_0 = \frac{1}{2}ka^2$. For $z \ll z_0$, Eq. (8.19) reduces to $q_1 \simeq p_0 \exp[-\alpha_1 z - (r/a)^2]$, which describes a collimated beam of planar wavefronts having the same transverse amplitude distribution as at the source. The far-field solution is found from Eq. (8.16), with the directivity function given by Eq. (8.17):

$$D_1(\theta) = \exp[-\tfrac{1}{4}(ka)^2 \tan^2\theta]. \tag{8.20}$$

Larger values of ka—i.e., larger ratios of source dimension to radiated wavelength—produce narrower beams. The axial amplitude predicted by Eq. (8.19) with $\alpha_1 = 0$ is shown in Fig. 8.1, which displays a monotonically decreasing function of distance. The axial phase of q_1 approaches 90° for $z/z_0 \gg 1$.

The quasilinear solution for the second-harmonic pressure is obtained by substituting Eq. (8.19) into Eq. (8.12):

$$q_2(r, z) = \frac{j P_2 e^{-\alpha_2 z + j(2\alpha_1 - \alpha_2)z_0}}{1 - jz/z_0} \exp\left[-\frac{2(r/a)^2}{1 - jz/z_0}\right]$$
$$\times \{E_1[j(2\alpha_1 - \alpha_2)z_0] - E_1[j(2\alpha_1 - \alpha_2)(z_0 - jz)]\}, \tag{8.21}$$

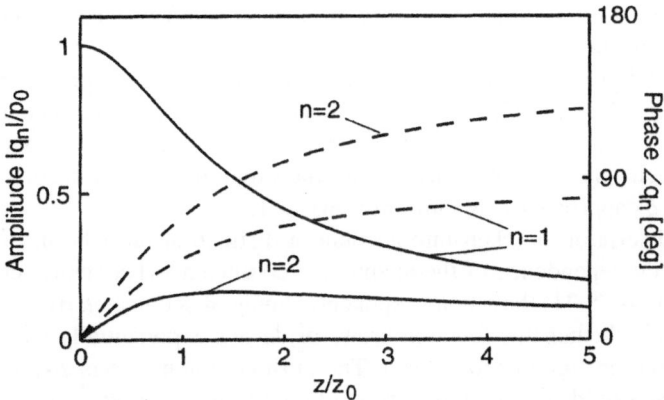

Fig. 8.1 Predicted propagation curves for the axial amplitudes (solid curves) and phases (dashed curves) for the fundamental ($n = 1$) and second-harmonic ($n = 2$) components in a Gaussian beam, with no absorption.

where $P_2 = \beta p_0^2 k^2 a^2 / 4\rho_0 c_0^2$. The function $E_1(\xi) = \int_\xi^\infty u^{-1} e^{-u}\, du$ is the exponential integral, which arises frequently in quasilinear solutions for diffracting sound beams. Two useful asymptotic properties of the exponential integral are $E_1(\xi) \simeq \xi^{-1} e^{-\xi}$ for $|\xi| \gg 1$, and $E_1(\xi) \simeq -(C + \ln \xi)$ for $|\xi| \ll 1$, where $C = 0.577$ is Euler's constant. Comparison of Eqs. (8.19) and (8.21) reveals $q_2(r) \propto q_1^2(r)$, which indicates that the second-harmonic beamwidth is narrower at all ranges z than the width of the primary beam by a factor of $1/\sqrt{2}$. Harmonic generation is most efficient along the z axis, where the primary beam is most intense.

Consider now the case for which absorption is negligible and set $\alpha_1 = \alpha_2 = 0$, whereby Eq. (8.21) reduces to

$$q_2(r, z) = \frac{j P_2 \ln(1 - jz/z_0)}{1 - jz/z_0} \exp\left[-\frac{2(r/a)^2}{1 - jz/z_0}\right]. \tag{8.22}$$

In the collimated near field of the primary beam ($z \ll z_0$), expansion of Eq. (8.22) yields $q_2 \simeq (p_0^2 \beta k z / 2\rho_0 c_0^2) \exp[-2(r/a)^2]$, which describes the same linear growth rate as for second-harmonic generation in plane wave fields close to the source [see Eq. (4.54)]. For $z \gg z_0$, Eq. (8.22) becomes

$$q_2(\theta, z) \simeq -P_2 \frac{\ln(z/z_0) - j\pi/2}{z/z_0} D_1^2(\theta) \exp(-jkz \tan^2 \theta). \tag{8.23}$$

The directivity of the second-harmonic component is equal to the square of the directivity of the primary beam. The eventual decay rate associated with the second-harmonic component is $(z_0/z) \ln(z/z_0)$, which is the dependence for second-harmonic generation by a spherical primary wave [see Eq. (4.285)], and which is not attained until distances for which $\ln(z/z_0) \gg 1$. Whereas the far field in linear acoustics is at distances $z/z_0 \gg 1$, the far field of the second-harmonic component, in the absence of absorption, is determined by $\ln(z/z_0) \gg 1$ (Garrett et al., 1984; Berntsen et al., 1984). The axial amplitude and phase predicted by Eq. (8.22) are shown in Fig. 8.1. In the absence of absorption, the second-harmonic component is generated mainly within the near field of the primary beam ($z < z_0$), beyond which spherical spreading reduces the strength of the interaction. The phase advance of q_2 approaches $180°$ for $\ln(z/z_0) \gg 1$.

When absorption is taken into account, a distinction must be made regarding the frequency dependence of the attenuation coefficients (Darvennes et al., 1991). Note that in Eq. (8.21) the second exponential integral becomes $E_1[(2\alpha_1 - \alpha_2)z]$ for $z \gg z_0$, and whether this function grows or decays exponentially with increasing z depends on the sign of $(2\alpha_1 - \alpha_2)$. The most common case is $2\alpha_1 < \alpha_2$, which applies to any fluid for which α is proportional to ω^m with $m > 1$ (e.g., classical thermoviscous fluids, for which $m = 2$). The relation $2\alpha_1 > \alpha_2$ is obtained for $m < 1$.

First let $2\alpha_1 < \alpha_2$, in which case the second exponential integral in Eq. (8.21) increases exponentially for large z and dominates the first, to give

$$q_2(\theta, z) \simeq -\frac{P_2 z_0}{(\alpha_2 - 2\alpha_1)} \frac{e^{-2\alpha_1 z}}{z^2} D_1^2(\theta) \exp(-jkz \tan^2 \theta). \qquad (8.24)$$

We thus obtain $q_2(\theta, z) \propto q_1^2(\theta, z)$ in the far field [recall Eqs. (8.16) and (8.17)]; i.e., the second-harmonic component depends only on the local properties of the primary beam. The reason for this dependence is that the amplitude of the virtual source distribution for second-harmonic generation is asymptotically proportional to $z^{-2} e^{-2\alpha_1 z}$ (because it is proportional to q_1^2). In contrast, the second-harmonic sound that is radiated by these virtual sources decays asymptotically as $z^{-1} e^{-\alpha_2 z}$, i.e., as a freely propagating spherical wave. For $2\alpha_1 < \alpha_2$, the virtual sources are thus attenuated less rapidly than the sound they radiate, and therefore the behavior of the second-harmonic field is determined in the far field by the behavior of neighboring virtual sources—i.e., by q_1^2.

Now take $2\alpha_1 > \alpha_2$, in which case for z sufficiently large the second exponential integral in Eq. (8.21) can be ignored in comparison with the first. It is seen by inspection that the amplitude of the second-harmonic component is asymptotically proportional to $z^{-1} e^{-\alpha_2 z} D_1^2(\theta)$. The z dependence of the second-harmonic component is the same as that of a freely propagating spherical wave at frequency 2ω, because in this case the virtual source distribution is attenuated more rapidly than the sound it radiates.

To conclude, we consider sum- and difference-frequency generation due to radiation from a bifrequency Gaussian source defined by

$$q_{1a}(r, 0) = p_{0a} \exp[-(r/a)^2], \qquad q_{1b}(r, 0) = p_{0b} \exp[-(r/b)^2], \qquad (8.25)$$

where the pressure amplitudes p_{0a} and p_{0b}, and the corresponding source radii a and b, may differ. For brevity here we ignore absorption, in which case Eq. (8.15) yields

$$q_\pm(r, z) = (jP_\pm/f_\pm) \exp(-k_\pm^2 r^2/2f_\pm)$$

$$\times \left[E_1\left(\frac{k_a k_b (z_{0a} \mp z_{0b})^2 k_\pm^2 r^2}{2f_\pm(g_\pm \mp jf_\pm k_\pm z)} \right) - E_1\left(\frac{k_a k_b (z_{0a} \mp z_{0b})^2 k_\pm^2 r^2}{2f_\pm g_\pm} \right) \right],$$

$$(8.26)$$

where $P_\pm = \beta k_\pm^2 z_{0a} z_{0b} p_{0a} p_{0b}/2\rho_0 c_0^2$, $f_\pm(z) = k_a z_{0a} + k_b z_{0b} - jk_\pm z$, $g_\pm(z) = k_\pm^2 z_{0a} z_{0b} - j(k_b z_{0a} + k_a z_{0b})k_\pm z$, $z_{0a} = \frac{1}{2}k_a a^2$, $z_{0b} = \frac{1}{2}k_b b^2$, $k_a = \omega_a/c_0$, and $k_b = \omega_b/c_0$. Equation (8.26) reduces on axis to

$$q_\pm(0, z) = (jP_\pm/f_\pm) \ln(1 \mp jk_\pm z f_\pm/g_\pm), \tag{8.27}$$

and the far-field directivity is given by the product of the primary beam directivities, $D_{1a}(\theta)D_{1b}(\theta)$. Further discussion of far-field properties, in relation to difference-frequency radiation from a parametric array, may be found in Sect. 8.3.4. A generalization of Eq. (8.26) is available for sources that are also displaced and steered (Darvennes and Hamilton, 1990), which permits the analysis of nonlinear effects produced by sound beams whose axes intersect at nonzero angles. Radiation of sum- and difference-frequency sound from the nonlinear interaction region formed by the intersection of two noncollinear primary beams is referred to as scattering of sound by sound. See Berntsen et al. (1989) for a general analysis of this problem, and for a review of earlier related theories and experiments.

8.3.2 Piston Sources

Here the amplitude distribution for the source is taken to be

$$q_1(r, 0) = p_0 H(a - r). \tag{8.28}$$

Equation (8.28) is used to model the vibration of a circular piston having velocity amplitude $u_0 = p_0/\rho_0 c_0$. Radiation from pistons, linear as well as nonlinear, is more difficult to investigate analytically than radiation from Gaussian sources. Specifically, closed-form solutions of Eq. (8.11) are available only on axis and in the far field, and only asymptotic approximations of Eqs. (8.12) and (8.15) are available in closed form.

Following substitution of Eq. (8.28) into Eq. (8.11), the latter may be rewritten in the following form (Naze Tjøtta and Tjøtta, 1980):

$$q_1(r, z) = p_0 e^{-\alpha_1 z} \left[1 - e^{-jz_0/z} + 2e^{-jz_0/z} \int_0^{rz_0/az} e^{-jz\xi^2/z_0} J_1(2\xi)\, d\xi \right],$$

$$(8.29)$$

where $z_0 = \frac{1}{2}ka^2$, and J_1 is the first-order Bessel function. Equation (8.29) is particularly convenient for numerical calculations because r appears only in the upper integration limit. Note that a singularity exists at $z = 0$, in the neighborhood of which the solution is invalid, and the source condition (8.28) is not recovered. [In contrast, the Gaussian source condition (8.18) is recovered from Eq. (8.19).] The effect of the singularity is more easily understood from analysis of the axial solution, for which the integral in Eq. (8.29) vanishes and the remaining terms may be combined as

$$q_1(0, z) = j2p_0 \sin(z_0/2z) \exp(-\alpha_1 z - jz_0/2z). \tag{8.30}$$

It can be shown by expanding the corresponding solution of the Helmholtz equation for radiation from a baffled circular piston that Eq. (8.30) is in agreement for $z/a \gtrsim (ka)^{1/3}$ (Kunitsyn and Rudenko, 1978), i.e., at distances typically beyond just a few source radii. As z increases, the magnitude of Eq. (8.30) oscillates between 0 and $2p_0 e^{-\alpha_1 z}$, passing through zeros at $z = z_0/2m\pi$ ($m = 1, 2, \ldots$). Near-field oscillations are associated with rapid phase variations, which create poor conditions for efficient harmonic generation prior to the last axial null ($m = 1$). The last maximum is at $z \simeq z_0/\pi$, and for $z \gg z_0$ the amplitude decays as $z^{-1}e^{-\alpha_1 z}$. Propagation in the far field is described by Eq. (8.16), with the directivity function given by

$$D_1(\theta) = \frac{2J_1(ka \tan \theta)}{ka \tan \theta}. \tag{8.31}$$

The exact, more familiar directivity function for small-signal radiation from a circular piston in a rigid baffle is obtained by replacing $\tan \theta$ with $\sin \theta$ in Eq. (8.31). The extent to which $\tan \theta \simeq \sin \theta$—e.g., for $\theta \lesssim 20°$—is indicative of the maximum angle at which the parabolic approximation is valid. See Naze Tjøtta and Tjøtta (1980) for more detailed comparison of the lossless, linear solution of Eq. (8.1) with the corresponding solution of the Helmholtz equation.

The quasilinear solution for the second-harmonic component involves, in general, a quadruple integral that must be evaluated numerically. Therefore, direct numerical integration of the second of Eqs. (8.4), or even Eq. (8.1), is normally the method used to model second-harmonic generation in the field of a piston. This is the procedure followed in Sect. 8.4, where axial propagation curves and beam patterns are presented for both unfocused and focused circular pistons. Here we restrict our attention to far-field asymptotic solutions for the second-harmonic component (Garrett et al., 1984; Berntsen et al., 1984), which shall assist interpretation of the numerical and experimental results in Sect. 8.4.

For $\alpha_2 > 2\alpha_1$ one obtains

$$q_2(\theta, z) \simeq -\frac{\beta p_0^2 k^3 a^4}{8\rho_0 c_0^2 (\alpha_2 - 2\alpha_1)} \frac{e^{-2\alpha_1 z}}{z^2} D_1^2(\theta) \exp(-jkz \tan^2 \theta), \qquad (8.32)$$

which is observed to be identical in form to Eq. (8.24), and the discussion of the latter in Sect. 8.3.1 applies directly to the present case as well. For $z_0 \ll z \ll L_a$, where $L_a = \min(\frac{1}{2}\alpha_1^{-1}, \alpha_2^{-1})$ is an absorption length, the field point is beyond the near field of the primary beam but prior to where absorption becomes important, and the following far-field result for $\alpha_1 = \alpha_2 = 0$ is applicable:

$$q_2(\theta, z) \simeq -\frac{\beta p_0^2 k^3 a^4}{8\rho_0 c_0^2} \exp(-jkz \tan^2 \theta)$$
$$\times \frac{1}{z}\left\{\left[\ln\left(\frac{2z}{\nu z_0}\right) - j\frac{\pi}{2}\right] D_1^2(\theta) + D_{2f}(\theta)\right\}. \qquad (8.33)$$

Here $\nu = e^C = 1.78$, and $D_{2f}(\theta)$ is an integral that must be evaluated numerically, and that is independent of z and is of the same order as $D_1^2(\theta)$ in magnitude. Therefore, the product directivity $D_1^2(\theta)$ does not accurately represent the beam pattern until distances for which $\ln(z/z_0) \gg 1$, where the contribution due to $D_{2f}(\theta)$ may be ignored. The most striking contribution due to $D_{2f}(\theta)$ is the appearance of side lobes where nulls are predicted by $D_1^2(\theta)$, the effect of which is that twice as many side lobes appear in the second-harmonic field as in the primary beam. The additional side lobes associated with $D_{2f}(\theta)$ are sometimes referred to as *fingers*; they become less significant as z increases, but this relative decay rate is logarithmic and therefore extremely slow. The earliest clear observations of fingers were reported by Lockwood et al. (1973) and by Moffett (1979), prior to when the theoretical explanation of this phenomenon became available (Berntsen et al., 1984). The influence of $D_{2f}(\theta)$ is discussed further in Sect. 8.4 in relation to Figs. 8.5b and 8.6.

8.3.3   Focused Sources

Focused sound beams can be produced in several ways—for example, by using a curved source, by placing a lens in front of a planar source, or by using a phased array. In each case, the intent is to generate wavefronts that are approximately spherical. The radius of curvature d of these wavefronts is called the focal length of the source.

We consider first the phase distribution that is required in the source plane $z = 0$ to produce a spherical wave that, in the limit of geometrical acoustics (very high frequencies), converges at the axial point $z = d$. Begin by writing the small-signal pressure of the converging spherical wave in the form $p = (d/R)p_0 f[t + (R - d)/c_0]$, where $f(t)$ is the time dependence of the signal at the source, $R = \sqrt{r^2 + (d - z)^2}$ is distance from the focal point, and we have $p = p_0 f(t)$ at $(r, z) = (0, 0)$. The appropriate source distribution in the plane $z = 0$ is thus obtained with $R = \sqrt{r^2 + d^2}$. When the source radius a is small in comparison with the focal length ($a \ll d$), and the source is therefore excited only at points $r \ll d$, we may retain only the leading terms in the binomial expansion of the expression for R to obtain $R \simeq d + r^2/2d$. The effect of the term $r^2/2d$ is greater on the phase than on the amplitude of p, and it is consistent within the parabolic approximation to replace the amplitude coefficient d/R by unity, and the phase term $(R - d)/c_0$ by $r^2/2c_0 d$ (Lucas and Muir, 1982). These approximations account to dominant order for the spherical curvature of wavefronts in the source plane $z = 0$.

To convert an unfocused source condition to one with focal length d, thus replace t according to the transformation[1]

$$t \rightarrow t + r^2/2c_0 d. \tag{8.34}$$

A simple transformation rule can be developed for the special case of Gaussian beams, as follows (Novikov et al., 1987). Noting that the unfocused Gaussian source condition is obtained by combining Eq. (8.18) with the time factor $e^{j\omega t}$ in Eq. (8.5), introduce the phase term $j\omega r^2/2c_0 d$ to obtain

$$q_1(r, 0) = p_0 \exp(-r^2/a^2 + jkr^2/2d). \tag{8.35}$$

Now rewrite Eq. (8.35) as

$$q_1(r, 0) = p_0 \exp(-r^2/\tilde{a}^2), \quad \tilde{a}^2 = \frac{a^2}{1 - jG}, \tag{8.36}$$

where $G = z_0/d = ka^2/2d$. Comparison with Eq. (8.18) reveals that the Gaussian beam solutions presented in Sect. 8.3.1 for radiation from monofrequency sources can be modified for application to focused Gaussian beams merely by replacing a everywhere by $\tilde{a}$ (including the factors of a in the definitions of z_0 and P_2). To account for focusing in Gaussian beam solutions for sum- and difference-frequency generation [for example, Eq. (8.26)], define the two focusing gains $G_a = k_a a^2/2d_a$ and $G_b = k_b b^2/2d_b$ to form expressions similar to the second of Eqs. (8.36), and replace a by $\tilde{a}$, but replace b by $\tilde{b}$ for the sum frequency and b by $\tilde{b}^*$ for the difference frequency because of the complex conjugate that appears in Eq. (8.15).

For example, consider transformation of the lossless solutions associated with second-harmonic generation in a Gaussian beam. Equation (8.19) with $\alpha_1 = 0$ and

---

[1] Defocused sources are modeled by using instead the transformation $t \rightarrow t - r^2/2c_0 d$, i.e., by replacing d with $-d$. Note that in Eq. (8.16) we have $-\frac{1}{2}jkz\tan^2\theta = -j\omega r^2/2c_0 z$, and the phase term thus describes a diverging spherical wavefront with radius of curvature z, centered at the source.

Eq. (8.22) thus become

$$q_1(r, z) = \frac{p_0}{1 - (1 + jG^{-1})z/d} \exp\left[-\frac{(1 - jG)(r/a)^2}{1 - (1 + jG^{-1})z/d}\right], \tag{8.37}$$

$$q_2(r, z) = \frac{jP_2}{(1 - jG)} \frac{\ln[1 - (1 + jG^{-1})z/d]}{1 - (1 + jG^{-1})z/d} \exp\left[-\frac{2(1 - jG)(r/a)^2}{1 - (1 + jG^{-1})z/d}\right]. \tag{8.38}$$

In the focal plane ($z = d$), the solutions reduce to

$$q_1(r, d) = jGp_0 \exp[-(Gr/a)^2 - jG(r/a)^2], \tag{8.39}$$

$$q_2(r, d) = j\left(\frac{\ln G + j\pi/2}{1 + jG^{-1}}\right)P_2 \exp[-2(Gr/a)^2 - j2G(r/a)^2]. \tag{8.40}$$

The relation $|q_1(0, d)| = Gp_0$ is obtained for the primary beam at the geometric focus, and therefore G is referred to as the focusing gain. The transverse distribution of q_1 in the focal plane is characterized by a radius a/G, in comparison with the radius a at the source. The width of the primary beam in the focal plane is thus reduced by a factor of G. As with unfocused Gaussian sources, the second-harmonic beam is everywhere narrower than the primary beam by a factor of $\sqrt{2}$.

Axial amplitudes and phases described by Eqs. (8.37) and (8.38) for $G = 10$ are displayed in Fig. 8.2. The primary beam experiences a 180° phase shift during passage through the focus, and the second-harmonic component experiences more nearly a 270° phase shift. As G increases and the limit of geometrical acoustics is approached, the phase shift in the primary beam tends toward a 180° step function

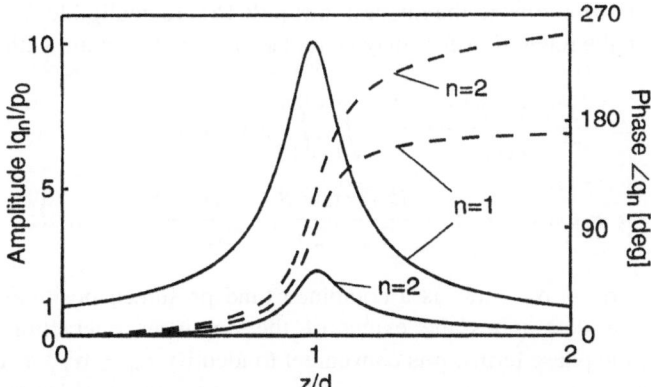

Fig. 8.2 Predicted propagation curves for the axial amplitudes (solid curves) and phases (dashed curves) for the fundamental ($n = 1$) and second-harmonic ($n = 2$) components in a focused Gaussian beam, with no absorption.

at the focus, whereas the phase advance of the second-harmonic component varies more slowly.

The corresponding quasilinear theory for focused pistons, with comparison to experiment, is provided by Lucas and Muir (1983) for second-harmonic generation and by Lucas et al. (1983) for difference-frequency generation.

8.3.4 Parametric Array

A common model of difference-frequency generation referred to as the parametric array is analyzed in the present section. The primary and secondary waves are described in general by Eqs. (8.13) and (8.14). We consider neighboring primary frequencies $\omega_a \simeq \omega_b$ and sufficiently strong absorption that the nonlinear interaction is limited to the near field of the bifrequency source (Westervelt, 1963). To ensure the latter condition we require $\alpha_0 z_0 \gtrsim 1$, where α_0 and z_0 are the attenuation coefficient and Rayleigh distance, respectively, evaluated at the mean primary frequency $\omega_0 = \frac{1}{2}(\omega_a + \omega_b)$. Only far-field properties of the difference-frequency field shall be considered here, so for simplicity the primary beam is approximated by collimated plane waves radiated by a circular piston of radius a:

$$q_{1a}(r, z) = p_{0a} H(a - r) e^{-\alpha_a z}, \quad q_{1b}(r, z) = p_{0b} H(a - r) e^{-\alpha_b z}. \tag{8.41}$$

Substitute Eqs. (8.41) into Eq. (8.15) and make the following approximations. Since the nonlinear interaction is confined by absorption to the near field of the primary beam (i.e., the main contribution to the integral is in the region $z' < z_0$) and we are interested only in the far field of the difference-frequency radiation, replace the term $(z - z')^{-1} J_0[k_- r r'/(z - z')]$ in the Green's function by $z^{-1} J_0(k_- r r'/z)$. Next, since contributions to the integral in Eq. (8.15) are negligible for large z', the upper limit on the integral over z' may be replaced by ∞ to obtain, with $r = z \tan \theta$,

$$q_-(\theta, z) \simeq -\frac{j p_{0a} p_{0b} \beta k_-^2}{2 \rho_0 c_0^2} \frac{e^{-\alpha_- z}}{z} \int_0^\infty \int_0^a J_0(k_- r' \tan \theta)$$

$$\times \exp\left(-\alpha_T z' - \frac{j k_- z^2 \tan^2 \theta}{2(z - z')} - \frac{j k_- r'^2}{2(z - z')}\right) r' \, dr' dz', \tag{8.42}$$

where $\alpha_T = \alpha_a + \alpha_b - \alpha_-$ is a combined (and presumed positive) attenuation coefficient that determines the extent of the nonlinear interaction region. To approximate the phase terms, it is convenient to identify $L_a = \alpha_T^{-1}$ as the effective length (often referred to as the *absorption length*) of the parametric array. Since the integration across the beam is restricted to the domain $r' < a$, the term containing r'^2 may be ignored for $z - L_a \gg \frac{1}{2} k_- a^2$. Binomial expansion is used to approximate the term containing $\tan^2 \theta$, for small z'/z, by $-\frac{1}{2} j k_- (z + z') \tan^2 \theta$. The resulting integrals over z' and r' are now independent and yield

$$q_-(\theta, z) \simeq -\frac{j p_{0a} p_{0b} \beta k_-^2 a^2}{4 \rho_0 c_0^2 \alpha_T} \frac{e^{-\alpha_- z}}{z} D_W(\theta) D_A(\theta) \exp(-\tfrac{1}{2} j k_- z \tan^2 \theta), \quad (8.43)$$

where the Westervelt directivity D_W (Westervelt, 1963), which results from the integral over z', and the aperture factor D_A (Naze and Tjøtta, 1965), which results from the integral over r', are given by[2]

$$D_W(\theta) = \frac{1}{1 + j(k_-/2\alpha_T)\tan^2\theta}, \quad D_A(\theta) = \frac{2 J_1(k_- a \tan\theta)}{k_- a \tan\theta}. \quad (8.44)$$

In terms of L_a, we restate the main restrictions on Eq. (8.43) as follows: strong absorption (nominally $L_a < z_0$) and field points far from the nonlinear interaction region ($z \gg L_a$). Note that the axial amplitude of the difference-frequency pressure is proportional to the array length L_a and varies with distance as $z^{-1} e^{-\alpha_- z}$.

We now consider the angular dependence. For small values of $k_- a$, as is frequently the case for $\omega_a \simeq \omega_b$, D_A is a relatively weak function of θ compared with D_W, and the directivity of the difference-frequency field is determined mainly by the latter. Observe that $|D_W|$ is a monotonically decreasing function of θ, with a maximum at $\theta = 0$ and no side lobes. This radiation pattern is created by the end-fire character of the array. As they propagate along the z axis at speed c_0, the primary waves constitute a phased line array of virtual sources that pump energy resonantly and therefore most efficiently into difference-frequency sound that propagates in the same direction. The absence of side lobes results from the continuous, exponential amplitude taper due to the factor $e^{-\alpha_T z'}$ in Eq. (8.42).

The half-power angle related to D_W is defined by $(k_-/2\alpha_T)\tan^2\theta_{HP} = 1$. To within the accuracy of the parabolic approximation (i.e., for narrow beams), we may let $\tan\theta_{HP} \simeq \theta_{HP}$ to obtain

$$\theta_{HP} = \sqrt{2\alpha_T/k_-} = \sqrt{2/k_- L_a}. \quad (8.45)$$

Whereas the beamwidths of the primary waves are determined by $k_a a$ and $k_b a$ via Eq. (8.31), the beamwidth of the difference-frequency sound depends mainly on $k_- L_a$. The difference-frequency beamwidth is essentially independent of the source radius a provided that absorption restricts the nonlinear interaction to the near field of the primary beam, and the aperture factor can be ignored. Parametric radiation thus provides a means for producing a more directional sound beam, at a lower frequency and with much lower side lobes, than is possible by direct, small-signal radiation from the same source of radius a (the latter situation would produce a beam pattern given by D_A). In addition, for $\omega_a \simeq \omega_b$, a small percentage

---

[2] Westervelt (1963) actually obtained D_W with $(2k_-/\alpha_T)\sin^2(\theta/2)$ in place of $(k_-/2\alpha_T)\tan^2\theta$, while Naze and Tjøtta (1965) obtained D_A with $k_- a \sin\theta$ in place of $k_- a \tan\theta$. These differences from Eqs. (8.44), because of the parabolic approximation associated with Eq. (8.1), are negligible for small θ.

change in either primary frequency corresponds to a large percentage change in the difference frequency, which makes wideband radiation (at the difference frequency) possible using a narrowband transducer (at the primary frequencies). The trade-off in comparison with the use of conventional acoustic sources is the lower efficiency associated with parametric generation of sound. Bellin and Beyer (1962) reported the first experiment on the parametric array, which verified the main features of Westervelt's theory.

Measurements obtained by Muir and Willette (1972) are presented in Fig. 8.3. A circular piston of radius $a = 3.8$ cm radiated a bifrequency primary beam in water at $\omega_a/2\pi = 482$ kHz and $\omega_b/2\pi = 418$ kHz. For these parameters, the Rayleigh distance is $z_0 = 1.4$ m and the absorption length is $L_a \sim 100$ m. We thus have $L_a \gg z_0$, and therefore difference-frequency generation is dominated by nonlinear interaction in the far field of the primary beam, which is not taken into account by the Westervelt model. Nevertheless, even though $\omega_a/\omega_- = 7.5$, the beamwidths at ω_a (Fig. 8.3a) and at ω_- (Fig. 8.3b) are similar. Note also the absence of side lobes at the difference frequency. The beam pattern at ω_+ (Fig. 8.3c) is narrower than at the primary frequencies and is given approximately by the product $D_{1a}(\theta)D_{1b}(\theta)$. The corresponding axial propagation curves are shown in the right half of Fig. 8.3. Absorption ultimately takes a greater toll on the primary, second-harmonic, and sum-frequency waves, leaving the difference-frequency component as the lone survivor far from the source. Muir and Willette's numerical results, shown with the measurements as solid curves, account for nonlinear interaction only in the far field, and correspond to using Eq. (8.16) in place of Eqs. (8.41)

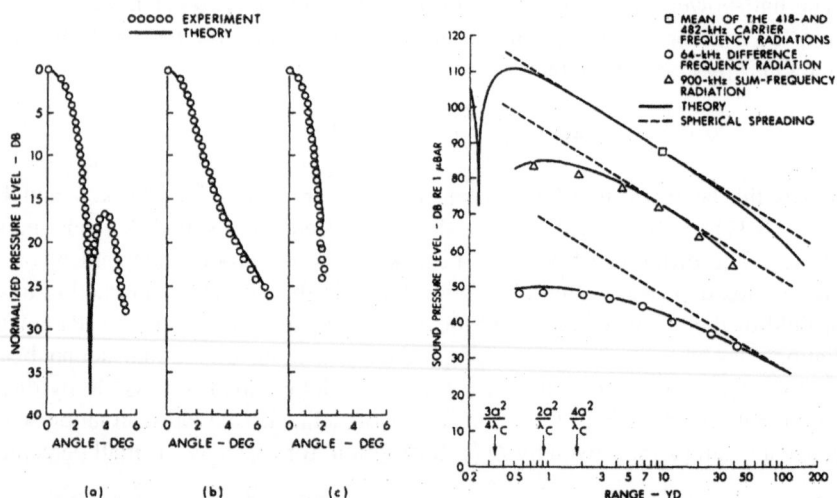

Fig. 8.3 Measurements of beam patterns produced by a parametric array in water at the (**a**) upper primary, (**b**) difference, and (**c**) sum frequencies, with corresponding propagation curves on the right (Muir and Willette, 1972; the solid curves are from their theoretical model).

for the primary beam. See also the article by Garrett et al. (1983) for comparison of experiment with theory based on direct numerical evaluation of Eq. (8.15).

We conclude by describing briefly the parametric receiving array, which employs an intense, collimated "pump" beam of frequency ω_a to determine the direction of propagation of a second signal with much lower frequency $\omega_b \ll \omega_a$. A receiver placed at distance L away from the source of the pump beam, along its axis, is used to detect the sum or difference frequency $\omega_\pm$ ($\simeq \omega_a$) produced by the noncollinear interaction of the two primary waves. It can be shown that the amplitude of $q_\pm$ at the receiver is approximately proportional to $D_b(\theta) = (\sin\Theta)/\Theta$, where $\Theta = k_b L \sin^2(\theta/2)$, and θ is the angle between the axis of the pump beam and the direction in which the low-frequency signal propagates (Berktay and Al-Temini, 1969; Zverev and Kalachev, 1970). Note that $D_b(\theta)$ is the directivity function at frequency ω_b for a continuous end-fire array of length L. The nonlinear interaction along the axis of the pump beam thus synthesizes the differential elements of the end-fire array. The main advantage is that the directivity is determined by the separation distance L between the source of the pump beam and the receiver, and therefore large effective apertures can in principle be constructed with only two relatively small transducers. An extensive theoretical and experimental investigation of parametric receiving arrays was performed by Truchard (1975a, 1975b).

The reader is referred also to a book by Novikov et al. (1987) that is devoted almost entirely to parametric arrays.

8.3.5 Self-demodulation

Here we consider the following transient source condition for a piston:

$$p(r, 0, t) = p_0 f(t) H(a - r), \quad f(t) = E(t) \sin[\omega_0 t + \phi(t)], \tag{8.46}$$

where the amplitude modulation $E(t)$ and phase modulation $\phi(t)$ are slowly varying functions of time (in comparison with $\sin\omega_0 t$). The instantaneous angular frequency of the carrier wave is $\Omega(t) = \omega_0 + d\phi/dt$. As in Sect. 8.3.4, we take the attenuation coefficient α_0 at frequency ω_0 to be sufficiently large ($\alpha_0 z_0 \gtrsim 1$, where $z_0 = \frac{1}{2}k_0 a^2$, $k_0 = \omega_0/c_0$) that the nonlinear interaction is confined to the near field of the beam. Under these conditions, an asymptotic result can be obtained for the axial waveform as follows (Averkiou et al., 1993).

Following the model of the parametric array, we approximate the primary beam as a collimated plane wave, and further assume that the exponential attenuation acts locally according to the instantaneous angular frequency $\Omega(\tau)$:

$$p_1(r, z, \tau) \simeq p_0 e^{-\alpha(\tau)z} E(\tau) \sin[\omega_0 \tau + \phi(\tau)] H(a - r), \tag{8.47}$$

where, for a thermoviscous fluid, we have $\alpha(\tau) = [\Omega(\tau)/\omega_0]^2\alpha_0$ for the time dependent attenuation coefficient. The frequency content of the secondary pressure p_2 is determined by p_1^2, which contains high frequencies associated with $E^2(\tau)\cos[2\omega_0\tau + 2\phi(\tau)]$ and low frequencies associated with $E^2(\tau)$. The high-frequency spectrum is absorbed more rapidly than the low-frequency spectrum, leaving only the contribution due to the latter in the far field. Concentrating on the latter, we write

$$p_1^2(r, z, \tau) \simeq \tfrac{1}{2}p_0^2 e^{-2\alpha(\tau)z} E^2(\tau) H(a - r) \tag{8.48}$$

and ignore the generation of higher frequencies. We now let $L_a = (2\alpha_0)^{-1}$ characterize the length of the nonlinear interaction region, assume $L_a < z_0$, and seek a solution for $z \gg L_a$.

The next assumption is that absorption of the nonlinearly generated low-frequency components is a relatively weak effect, which is justified if $E(t)$ and $\phi(t)$ are sufficiently slowly varying functions of time, corresponding to very low frequencies. The inhomogeneous wave equation for p_2 thus becomes, following integration of Eq. (8.1) with $\delta = 0$,

$$\frac{\partial p_2}{\partial z} - \frac{c_0}{2} \int_{-\infty}^{\tau} (\nabla_\perp^2 p_2)\, d\tau = \frac{\beta}{2\rho_0 c_0^3} \frac{\partial p_1^2}{\partial \tau}. \tag{8.49}$$

An axial Green's function solution for p_2 (in the time domain) can be constructed from Eq. (8.10) (which is expressed in the frequency domain) as follows: On axis ($r = 0$), for no absorption ($\alpha_n = 0$), and with harmonic frequency dependence nk replaced by ω/c_0, Eq. (8.10) becomes

$$G_\omega(0, z | r', z') = \frac{j\omega}{2\pi c_0(z - z')} \exp\left[-\frac{j\omega r'^2}{2c_0(z - z')} \right]. \tag{8.50}$$

Recalling the time convention $e^{j\omega\tau}$, from Fourier transform theory we associate the first factor of $j\omega$ in Eq. (8.50) with the derivative $\partial/\partial\tau$, and the argument of the exponential with the time delay $-r'^2/2c_0(z - z')$. Using Fourier transforms and the definition of the Green's function, one finds that the axial solution of Eq. (8.49) for arbitrary $p_1(r, z, \tau)$ is

$$p_2(0, z, \tau) = \frac{\beta}{2\rho_0 c_0^4} \frac{\partial^2}{\partial\tau^2} \int_0^z \int_0^\infty p_1^2\left[r', z', \tau - \frac{r'^2}{2c_0(z - z')} \right] \frac{r'\, dr'dz'}{z - z'}. \tag{8.51}$$

Now introduce the same approximations employed in Sect. 8.3.4. Take z to be sufficiently large that the phase term $r'^2/2c_0(z - z')$ can be ignored, replace $dz'/(z - z')$ by dz'/z, and extend the upper integration limit on z' to ∞. Substitution of Eq. (8.48) into Eq. (8.51) then results in elementary integrals over both r' and z', which yield

$$p_2(0, z, \tau) \simeq \frac{\beta p_0^2 a^2}{16 \rho_0 c_0^4 z} \frac{d^2}{d\tau^2} \frac{E^2(\tau)}{\alpha(\tau)}. \tag{8.52}$$

One of the derivatives comes from the right-hand side of Eq. (8.49), and the second is associated with propagation from the near field to the far field [the latter corresponds to the leading factor of jk in Eq. (8.16)]. For $\phi = \text{const}$ and therefore $\alpha(\tau) = \alpha_0$, Eq. (8.52) reduces to Berktay's (1965) solution. Moffett et al. (1970) reported the first measurements of self-demodulation, and demonstrated agreement (Moffett et al., 1971) with the waveform predicted by Berktay.

Finally, we may construct a complete solution $p = p_1 + p_2$ for the axial waveform, taking thermoviscous absorption into account, as follows. Begin with the linear axial solution of Eq. (8.1) corresponding to the source condition in the first of Eqs. (8.46), for arbitrary $f(t)$ (Frøysa et al., 1993):

$$p_1(0, z, \tau) = p_0[f(\tau) - f(\tau - a^2/2c_0 z)] * D(z, \tau), \tag{8.53}$$

where $D(z, \tau) = (c_0^3/2\pi \delta z)^{1/2} \exp(-c_0^3 \tau^2/2\delta z)$ is a thermoviscous dissipation function [see Eqs. (4.230)], and the asterisk indicates convolution with respect to τ. For $f(t) = \sin \omega t$, Eq. (8.53) reduces to Eq. (8.3) with q_1 given by Eq. (8.30). In what follows, $f(t)$ is given by the second of Eqs. (8.46). For $z > L_a$, propagation of the field p_2 may be described by linear theory. Assuming that p_2 is generated within a relatively compact volume directly in front of the source, we may combine Eqs. (8.52) and (8.53) to obtain the following solution for the axial waveform (Averkiou et al., 1993):

$$p(0, z, \tau) \simeq p_0 \left[f(\tau) - f(\tau - a^2/2c_0 z) + \frac{\beta p_0 a^2}{16 \rho_0 c_0^4 z} \frac{d^2}{d\tau^2} \frac{E^2(\tau)}{\alpha(\tau)} \right] * D(z, \tau). \tag{8.54}$$

Near the source ($z \sim L_a$), prior to where the asymptotic result for p_2 is valid, the dominant contribution to p is the primary wave p_1. Comparisons with numerical solutions of Eq. (8.1) show Eq. (8.54) to be accurate for $\alpha_0 z_0 > 1$.

Theory and experiment for the self-demodulation of a tone burst with center frequency $\omega_0/2\pi = 3.5$ MHz, radiated in glycerin from a piston of radius $a = 6.4$ mm, are compared in Fig. 8.4 (after Averkiou et al., 1993). Here $z_0/L_a \sim 30$, and the nonlinear interaction region is terminated well within the near field of the primary beam. The theory is given by Eq. (8.54) with $E(t) = \exp[-(\omega_0 t/25\pi)^{10}]$, $\phi(t) = 0$, and $z_0/\bar{z} = 1.6$, where $\bar{z} = \rho_0 c_0^3/\beta p_0 \omega_0$ is the lossless plane-wave shock formation distance at frequency ω_0. Observe how rapidly the tone burst is attenuated in comparison with the low-frequency, "self-demodulated" waveform. At $z/z_0 = 1.15$, virtually all that remains is the waveform predicted by Eq. (8.52). The corresponding frequency spectra in the right column reveal no significant second-harmonic generation.

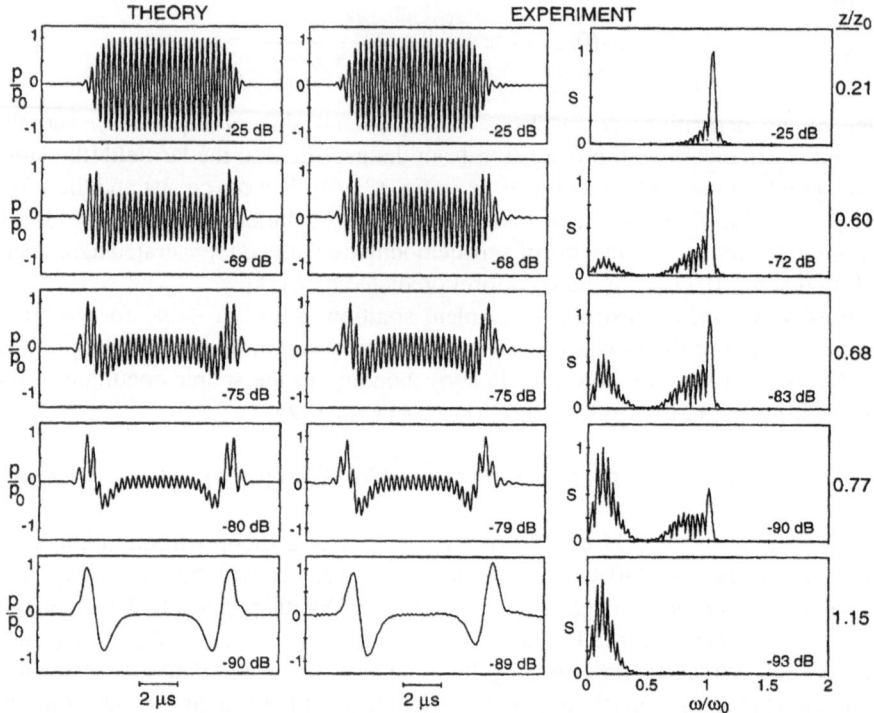

Fig. 8.4 Axial waveforms (theory and experiment, first two columns) and corresponding frequency spectra (experiment, third column) illustrating self-demodulation of a pulsed sound beam in glycerin (after Averkiou et al., 1993). Inset decibel values indicate levels relative to source.

8.4 Strong Nonlinearity

By strong nonlinearity we mean effects that cannot be modeled with quasilinear theory, for example, waveform distortion and shock formation. For these cases one usually resorts to numerical solutions of Eq. (8.1) (see Sect. 11.3.1). Before proceding to a discussion of the corresponding numerical results, we call attention to two analytically based techniques for solving the fully nonlinear form of the KZ equation, i.e., the lossless form of Eq. (8.1). Although Gaussian amplitude shading at the source is assumed in both approaches, each method may be extended to include other source distributions.

One method incorporates analytical techniques used in nonlinear geometrical acoustics (Hamilton et al., 1997). Coupled equations are obtained from an expansion of the sound pressure about the beam axis, and an approximate axial solution is derived for the preshock region of a Gaussian beam radiated by a monofrequency source. The solution is derived in the time domain and has a simple implicit analytic form. In the high-frequency, geometrical acoustics limit where effects of diffraction disappear, the solution reduces to the plane-wave solution of the KZ equation, i.e., the solution of the lossless Burgers equation. The solution is also expressed in the frequency domain as an explicit Fourier series, which in the limit of very high frequencies reduces to the Fubini solution. Good agreement is shown with numerical solutions of the KZ equation. The lossless axial forms of the quasilinear solutions (8.19) and (8.22) are recovered by straightforward expansion in powers of the acoustic Mach number. One consequence of the analysis is the following transcendental equation that approximates the shock formation distance $\bar{z}_G$ along the axis of a Gaussian beam in a lossless fluid, subject to the source condition associated with Eq. (8.18):

$$\tfrac{1}{4} \ln^2[1 + (\bar{z}_G/z_0)^2] + \arctan^2(\bar{z}_G/z_0) = (\bar{z}/z_0)^2, \tag{8.55}$$

where $\bar{z} = \rho_0 c_0^3/\beta p_0 \omega_0$ is the corresponding plane-wave shock formation distance. It can be seen that the limits $\bar{z}_G \to \bar{z}$ as $z_0 \to \infty$ ($a \to \infty$) and $\bar{z}_G \to z_0 e^{\bar{z}/z_0}$ as $z_0 \to 0$ ($a \to 0$) are obtained, consistent with the theory for plane and spherical waves, respectively. The effect of focusing was also included in the results described above.

The second method is based on combining the perturbation technique referred to as renormalization (Sect. 10.4) with what amounts to an application of weak shock theory (Frøysa and Coulouvrat, 1996). Although also restricted to Gaussian beams, this method is more general than the solution described in the previous paragraph in that it applies to pulses, both on and off axis, and shock formation is taken into account. Comparisons with numerical solutions reveal the technique to be accurate over a wide range of parameters, even well beyond the shock formation distance. The trade-off is an increase in complexity. Mainly, multiple numerical integrations in the corresponding quasilinear solution of the problem are required to perform the renormalization. An intricate procedure must also be followed to determine where to place the shocks, but similar difficulties would be faced in the first method if shocks were included.

We turn now to direct numerical solutions of Eq. (8.1). To facilitate computations, the following transformation is frequently employed (Hamilton et al., 1985):

$$P = \left(1 + \frac{z}{z_0}\right)\frac{p}{p_0}, \quad R = \frac{r/a}{1 + z/z_0}, \quad Z = \frac{z}{z_0}, \quad T = \omega_0 \tau - \frac{r^2/a^2}{1 + z/z_0}, \tag{8.56}$$

where a is a characteristic source radius, ω_0 is a reference frequency, $z_0 = \tfrac{1}{2}k_0 a^2$ ($k_0 = \omega_0/c_0$), and p_0 is a characteristic source amplitude. Substitution into Eq. (8.1) and integration with respect to T yields

$$\frac{\partial P}{\partial Z} = \frac{1}{4(1 + Z)^2} \int_{-\infty}^{T} (\nabla_R^2 P) \, dT + A\frac{\partial^2 P}{\partial T^2} + \frac{NP}{1 + Z}\frac{\partial P}{\partial T}, \tag{8.57}$$

where $\nabla_R^2 = \partial^2/\partial R^2 + R^{-1}(\partial/\partial R)$, $A = \alpha_0 z_0$ is a dimensionless attenuation coefficient (α_0 is the attenuation coefficient at frequency ω_0), and $N = z_0/\bar{z}$ is a dimensionless nonlinearity coefficient ($\bar{z}$ is again the lossless plane-wave shock formation distance). The transformation preserves the parabolic form of Eq. (8.1), but with range dependent coefficients. For $Z \gg 1$ we have $R \simeq \frac{1}{2}k_0 a \tan\theta$ and $T \simeq \omega_0\tau - k_0 r^2/2z$, and therefore constant values of R and T coincide with, respectively, the radial divergence and spherical wavefront curvature in the far field of a sound beam [recall Eq. (8.16)]. Equations (8.56) and (8.57) can be modified to accommodate the convergence and subsequent divergence of focused sound beams (Hart and Hamilton, 1988).

Equation (8.57) may be integrated directly in the time domain (Lee and Hamilton, 1995; see also Sect. 11.3.1.3), which is convenient for investigating pulsed sound beams. Effects of relaxation have also been included in time-domain numerical solutions of Eq. (8.57) (Cleveland et al., 1996). For sound beams radiated by monofrequency or bifrequency sources, frequency-domain algorithms are often more efficient for numerical computations. For these latter cases, solutions are sought in the form of a Fourier series, for example,

$$P(R, Z, T) = \frac{1}{2}\sum_{n=1}^{M} P_n(R, Z)e^{jnT} + \text{c.c.} \tag{8.58}$$

The series is truncated, for numerical reasons, at the Mth harmonic. Substitution into Eq. (8.57) yields the following coupled equations that can be solved numerically for the spectral amplitudes P_n (Naze Tjøtta et al., 1990; see also Sect. 11.3.1.2):

$$\frac{\partial P_n}{\partial Z} = \frac{\nabla_R^2 P_n}{j4n(1+Z)^2} - n^2 A P_n$$

$$+ \frac{jnN}{4(1+Z)}\left(\sum_{m=1}^{n-1} P_m P_{n-m} + 2\sum_{m=n+1}^{M} P_m P_{m-n}^*\right). \tag{8.59}$$

For fluids with attenuation coefficients that do not depend quadratically on frequency, $n^2 A$ can be replaced in Eqs. (8.59) by arbitrary coefficients $A_n + jD_n$, which account also for dispersion through D_n (see Sect. 5.2). The first comprehensive series of experiments performed for comparison with numerical solutions of coupled spectral equations based on the KZK equation is reported in a sequence of articles by Baker (1992), Baker and Humphrey (1992), and Baker et al. (1988).

Measurements of harmonic generation in sound beams radiated in water by unfocused circular pistons are compared in Fig. 8.5 with calculations based on Eqs. (8.59). The measurements in Fig. 8.5a (Averkiou and Hamilton, 1997) were obtained with a source of radius $a = 9.4$ mm that radiated at $\omega_0/2\pi = 2.25$ MHz ($A = 0.053$, $N = 0.97$), and the measured beam patterns at $z/z_0 = 2.9$ in Fig. 8.5b (TenCate, 1993) were obtained with a source of radius $a = 12.1$ mm

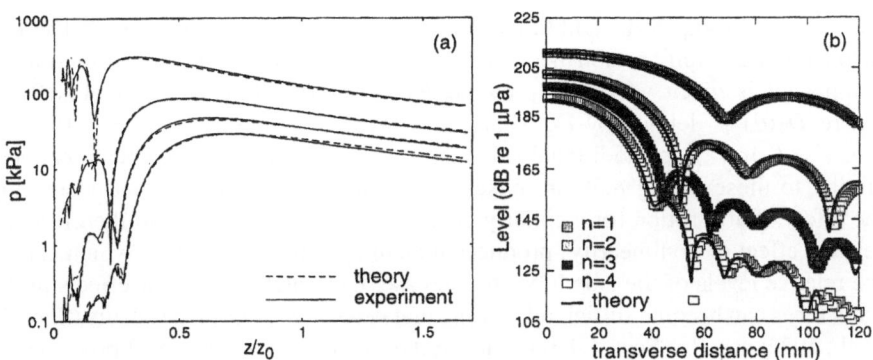

Fig. 8.5 (a) Axial propagation curves (Averkiou and Hamilton, 1997) and (b) beam patterns (TenCate, 1993) for harmonic generation produced by radiation from a circular piston in water.

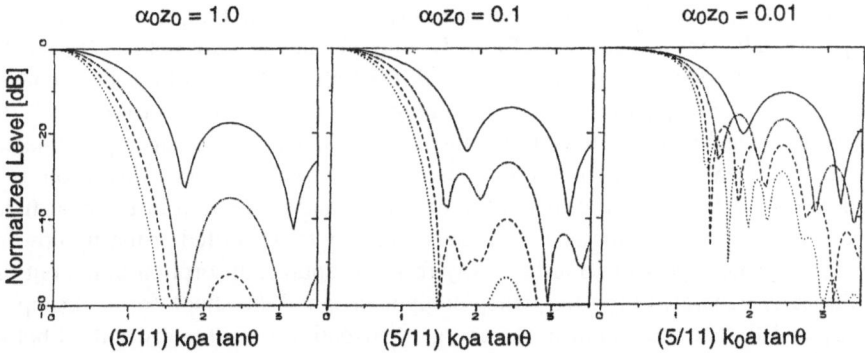

Fig. 8.6 Calculations showing the effect of absorption on harmonic beam patterns at $z = 10z_0$ in the field of a circular piston (Hamilton et al., 1985) for $n = 1$ (solid line), $n = 2$ (dash-dot-dashed line), $n = 3$ (dashed line), and $n = 4$ (dotted line).

radiating at 1 MHz ($A = 0.008$, $N = 0.42$). As noted in Sect. 8.3.2, efficient harmonic generation occurs mainly beyond the last axial minimum in the primary beam (Fig. 8.5a). In Fig. 8.5b, it can be seen that the nth-harmonic beam pattern has n times as many side lobes as the beam pattern at the source frequency (Berntsen et al., 1984; Hamilton et al., 1985). With increasing n, the harmonic components become more directional and exhibit greater side-lobe suppression. The additional side lobes are near-field effects that can be significant out to distances on the order of a hundred Rayleigh distances. Measurements made by TenCate (1993) reveal that the amplitudes of the additional side lobes in the second-harmonic beam pattern [corresponding to the function $D_{2f}(\theta)$ in Eq. (8.33)] decay as $z^{-1}e^{-\alpha_2 z}$.

Figure 8.6 shows numerical results (Hamilton et al., 1985) that demonstrate the effect of absorption on the relative levels of the additional side lobes in the field of a circular piston ($z/z_0 = 10$, $N = 1.5$). For strong absorption ($A = \alpha_0 z_0 = 1.0$, and thus $z = 10\alpha_0^{-1}$), the additional side lobes are completely attenuated. The far-field structure is fully established, and the directivity function for the nth-harmonic component is given very nearly by $D_n(\theta) = D_1^n(\theta)$ (Lockwood et al., 1973), where $D_1(\theta)$ is defined by Eq. (8.31). For weak absorption ($\alpha_0 z_0 = 0.01$, and thus $z = 0.1\alpha_0^{-1}$), the additional side lobes are very prominent, and beam patterns similar to those in Fig. 8.5b are observed. Finally, it can be seen that decreasing the effect of absorption (or, similarly, increasing the source level and therefore the relative effect of nonlinearity) produces flattening of the main lobes, which causes the relative levels of the side lobes to increase. Ultimately, nonlinear effects in the main lobes can become sufficiently strong that *acoustical saturation* occurs (Shooter et al., 1974; see also Sect. 4.4.3.4), and further increase in source level produces no increase in the axial pressure at a given distance.

The propagation of a high-intensity tone burst is demonstrated with numerical results in Fig. 8.7, for which the source condition at $z = 0$ is given by Eq. (8.46) with a Gaussian envelope defined by $E(t) = \exp[-(\omega_0 t/3\pi)^2]$, and with $\phi(t) = 0$, $A = 0.1$, and $N = 2.0$ (Lee and Hamilton, 1995). Frequency spectra are displayed adjacent to each waveform. By the end of the near field ($z/z_0 = 1$), the combined effects of nonlinearity and diffraction on the waveform have caused sharpening of the positive portions, rounding of the negative portions, and the formation of shock fronts. In addition, the peak positive pressures are approximately twice the peak negative pressures. This asymmetric waveform distortion is a hallmark of the combined effects of nonlinearity and diffraction in the near field. Waveform distortion in the near field causes energy to be shifted primarily upward in the frequency spectrum. Farther away from the source, absorption filters out the nonlinearly generated high-frequency components, and the shock fronts disappear ($z/z_0 = 10$). The maximum in the energy distribution is eventually shifted below the primary frequency as self-demodulation occurs ($z/z_0 = 30$–50). The waveform ultimately resembles $d^2 E^2/d\tau^2$ (but with some asymmetry) at $z/z_0 = 100$, where only frequencies associated with the pulse envelope characterize the energy spectrum. Parametric arrays with shocks exhibit comparable waveform distortion and spectral evolution (Naze Tjøtta et al., 1990).

Features similar to those observed in Figs. 8.5, 8.6 and 8.7 also appear in focused sound beams. Shown in Fig. 8.8a and b are comparisons of measurements (solid curves) with theory (dashed curves) for harmonic generation in the field of a focused piston with radius $a = 1.9$ cm and focal length $d = 16$ cm, radiating at 2.25 MHz in water with focusing gain $G = 10.6$, $\alpha_0 d = 0.025$, and $d/\bar{z} = 0.21$ (Averkiou and Hamilton, 1995). Also shown are measurements of the waveform (Fig. 8.8c) and corresponding frequency spectrum (Fig. 8.8d) for shock formation in a pulse at the geometric focus ($z = d$), with $d/\bar{z} = 0.34$ but otherwise under the same conditions (after Averkiou and Hamilton, 1997, with theory removed for clarity).

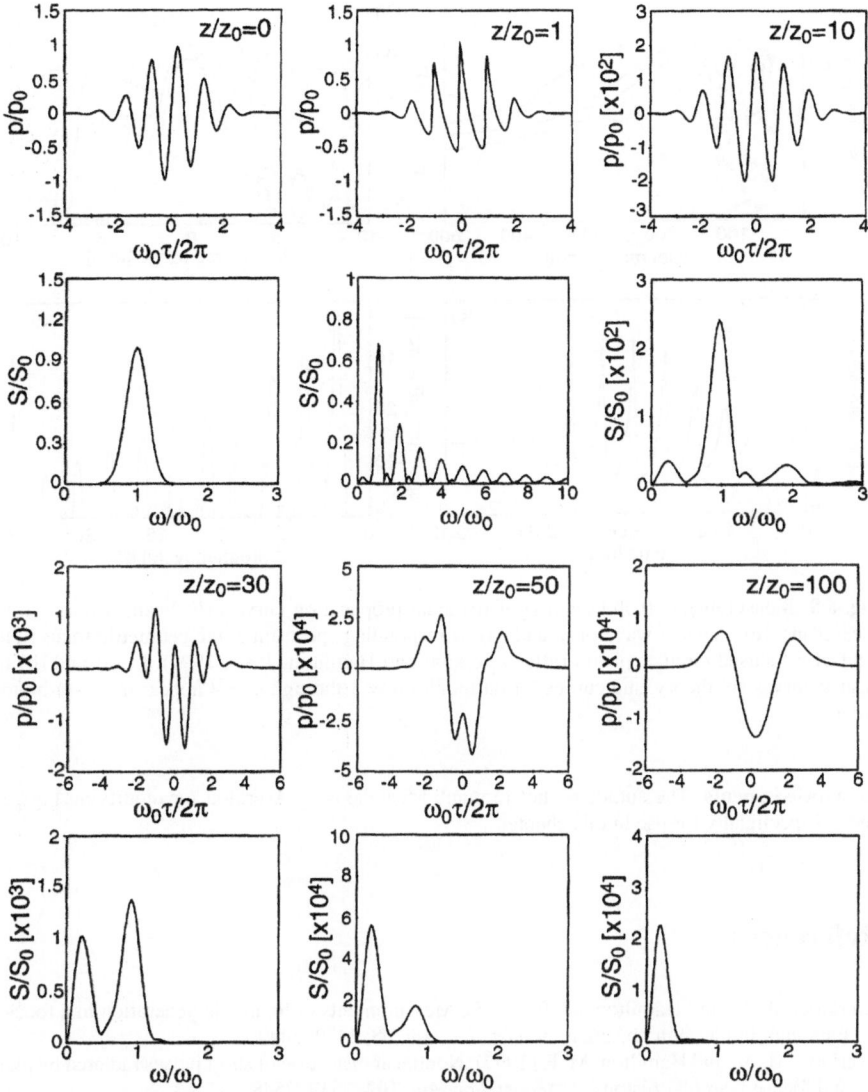

Fig. 8.7 Calculations showing waveforms and frequency spectra for a pulse along the axis of a sound beam radiated by a circular piston (Lee and Hamilton, 1995).

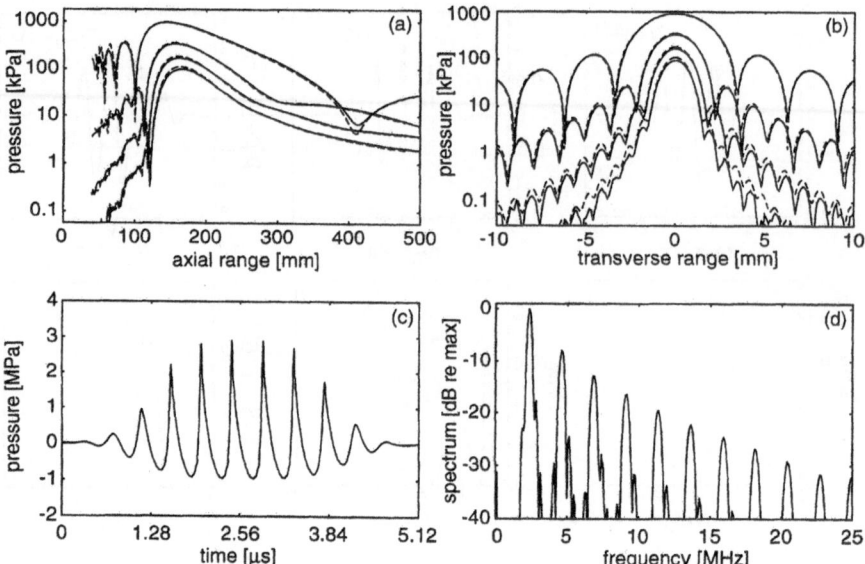

Fig. 8.8 Measurements (solid curves) of (**a**) axial propagation curves, (**b**) beam patterns in the focal plane, (**c**) transient waveform, and (**d**) corresponding spectrum at the geometric focus in the field of a focused circular piston (after Averkiou and Hamilton, 1995, 1997). In (a) and (b) the dashed curves are theory, and curves for harmonics $n = 1$ through $n = 4$ appear in top-to-bottom order.

Acknowledgments The author wishes to thank Michalakis A. Averkiou for modifying Figs. 8.4 and 8.8 specifically for use in this chapter.

References

Averkiou, M. A., and Hamilton, M. F. (1995). Measurements of harmonic generation in a focused finite-amplitude sound beam. *J. Acoust. Soc. Am.* **98**, 3439–3442.

Averkiou, M. A., and Hamilton, M. F. (1997). Nonlinear distortion of short pulses radiated by plane and focused circular pistons. *J. Acoust. Soc. Am.* **102**, 2539–2548.

Averkiou, M. A., Lee, Y.-S., and Hamilton, M. F. (1993). Self-demodulation of amplitude and frequency modulated pulses in a thermoviscous fluid. *J. Acoust. Soc. Am.* **94**, 2876–2883.

Baker, A. C. (1992). Nonlinear pressure fields due to focused circular apertures. *J. Acoust. Soc. Am.* **91**, 713–717.

Baker, A. C., Anastasiadis, K., and Humphrey, V. F. (1988). The nonlinear pressure field of a plane circular piston: Theory and experiment. *J. Acoust. Soc. Am.* **84**, 1483–1487.

Baker, A. C., and Humphrey, V. F. (1992). Distortion and high-frequency generation due to nonlinear propagation of short ultrasonic pulses from a plane circular piston. *J. Acoust. Soc. Am.* **92**, 1699–1705.

Bakhvalov, N. S., Zhileikin, Ya. M., and Zabolotskaya, E. A. (1987). *Nonlinear Theory of Sound Beams* (American Institute of Physics, New York).

Bellin, J. L. S., and Beyer, R. T. (1962). Experimental investigation of an end-fire array. *J. Acoust. Soc. Am.* **34**, 1051–1054.

Berktay, H. O. (1965). Possible exploitation of non-linear acoustics in underwater transmitting applications. *J. Sound Vib.* **2**, 435–461.

Berktay, H. O., and Al-Temini, C. A. (1969). Virtual arrays for underwater applications. *J. Sound Vib.* **9**, 295–307.

Berntsen, J., Naze Tjøtta, J., and Tjøtta, S. (1984). Nearfield of a large acoustic transducer. Part IV: Second harmonic and sum frequency radiation. *J. Acoust. Soc. Am.* **75**, 1383–1391.

Berntsen, J., Naze Tjøtta, J., and Tjøtta, S. (1989). Interaction of sound waves. Part IV: Scattering of sound by sound. *J. Acoust. Soc. Am.* **86**, 1968–1983.

Cleveland, R. O., Hamilton, M. F., and Blackstock, D. T. (1996). Time-domain modeling of finite-amplitude sound in relaxing fluids. *J. Acoust. Soc. Am.* **99**, 3312–3318.

Darvennes, C. M., and Hamilton, M. F. (1990). Scattering of sound by sound from two Gaussian beams. *J. Acoust. Soc. Am.* **87**, 1955–1964.

Darvennes, C. M., Hamilton, M. F., Naze Tjøtta, J., and Tjøtta, S. (1991). Effects of absorption on the nonlinear interaction of sound beams. *J. Acoust. Soc. Am.* **89**, 1028–1036.

Frøysa, K.-E., and Coulouvrat, F. (1996). A renormalization method for nonlinear pulsed sound beams. *J. Acoust. Soc. Am.* **99**, 3319–3328.

Frøysa, K.-E., Naze Tjøtta, J., and Tjøtta, S. (1993). Linear propagation of a pulsed sound beam from a plane or focusing source. *J. Acoust. Soc. Am.* **93**, 80–92.

Garrett, G. S., Naze Tjøtta, J., and Tjøtta, S. (1983). Nearfield of a large acoustic transducer. Part II: Parametric radiation. *J. Acoust. Soc. Am.* **74**, 1013–1020.

Garrett, G. S., Naze Tjøtta, J., and Tjøtta, S. (1984). Nearfield of a large acoustic transducer. Part III: General results. *J. Acoust. Soc. Am.* **75**, 769–779.

Hamilton, M. F., Khokhlova, V. A., and Rudenko, O. V. (1997). Analytical method for describing the paraxial region of finite amplitude sound beams. *J. Acoust. Soc. Am.* **101**, 1298–1308.

Hamilton, M. F., Naze Tjøtta, J., and Tjøtta, S. (1985). Nonlinear effects in the farfield of a directive sound source. *J. Acoust. Soc. Am.* **89**, 202–216.

Hart, T. S., and Hamilton, M. F. (1988). Nonlinear effects in focused sound beams. *J. Acoust. Soc. Am.* **84**, 1488–1496.

Kunitsyn, V. E., and Rudenko, O. V. (1978). Second-harmonic generation in the field of a piston radiator. *Sov. Phys. Acoust.* **24**, 310–313.

Kuznetsov, V. P. (1971). Equations of nonlinear acoustics. *Sov. Phys. Acoust.* **16**, 467–470.

Lee, Y.-S., and Hamilton, M. F. (1995). Time-domain modeling of pulsed finite-amplitude sound beams. *J. Acoust. Soc. Am.* **97**, 906–917.

Lockwood, J. C., Muir, T. G., and Blackstock, D. T. (1973). Directive harmonic generation in the radiation field of a circular piston. *J. Acoust. Soc. Am.* **53**, 1148–1153.

Lucas, B. G., and Muir, T. G. (1982). The field of a focusing source. *J. Acoust. Soc. Am.* **72**, 1289–1296.

Lucas, B. G., and Muir, T. G. (1983). Field of a finite-amplitude focusing source. *J. Acoust. Soc. Am.* **74**, 1522–1528.

Lucas, B. G., Naze Tjøtta, J., and Muir, T. G. (1983). Field of a parametric focusing source. *J. Acoust. Soc. Am.* **73**, 1966–1971.

Moffett, M. B. (1979). Measurement of fundamental and second harmonic pressures in the field of a circular piston source. *J. Acoust. Soc. Am.* **65**, 318–323.

Moffett, M. B., Westervelt, P. J., and Beyer, R. T. (1970). Large-amplitude pulse propagation—A transient effect. *J. Acoust. Soc. Am.* **47**, 1473–1474.

Moffett, M. B., Westervelt, P. J., and Beyer, R. T. (1971). Large-amplitude pulse propagation—A transient effect. II. *J. Acoust. Soc. Am.* **49**, 339–343.

Morse, P. M., and Ingard, K. U. (1968). *Theoretical Acoustics* (McGraw-Hill, New York).

Muir, T. G., and Willette, J. G. (1972). Parametric acoustic transmitting arrays. *J. Acoust. Soc. Am.* **52**, 1481–1486.

Naze, J., and Tjøtta, S. (1965). Nonlinear interaction of two sound beams. *J. Acoust. Soc. Am.* **37**, 174–175.

Naze Tjøtta, J., and Tjøtta, S. (1980). An analytical model for the nearfield of a baffled piston transducer. *J. Acoust. Soc. Am.* **68**, 334–339.

Naze Tjøtta, J., Tjøtta, S., and Vefring, E. (1990). Propagation and interaction of two collinear finite amplitude sound beams. *J. Acoust. Soc. Am.* **88**, 2859–2870.

Novikov, B. K., Rudenko, O. V., and Timoshenko, V. I. (1987). *Nonlinear Underwater Acoustics* (Acoustical Society of America, New York).

Shooter, J. A., Muir, T. G., and Blackstock, D. T. (1974). Acoustic saturation of spherical waves in water. *J. Acoust. Soc. Am.* **55**, 54–62.

TenCate, J. A. (1993). An experimental investigation of the nonlinear pressure field produced by a plane circular piston. *J. Acoust. Soc. Am.* **94**, 1084–1089.

Truchard, J. J. (1975a). Parametric acoustic receiving array. I. Theory. *J. Acoust. Soc. Am.* **58**, 1141–1145.

Truchard, J. J. (1975b). Parametric acoustic receiving array. II. Experiment. *J. Acoust. Soc. Am.* **58**, 1146–1150.

Westervelt, P. J. (1963). Parametric acoustic array. *J. Acoust. Soc. Am.* **35**, 535–537.

Zabolotskaya, E. A., and Khokhlov, R. V. (1969). Quasi-plane waves in the nonlinear acoustics of confined beams. *Sov. Phys. Acoust.* **15**, 35–40.

Zverev, V. A., and Kalachev, A. I. (1970). Modulation of sound by sound in the intersection of sound waves. *Sov. Phys. Acoust.* **16**, 204–208.

Chapter 9
Finite-Amplitude Waves in Solids

Andrew N. Norris

Contents

9.1 Introduction

The study of nonlinear acoustics of solids requires the introduction of new variables not used for fluids. This stems from the fact that a solid in motion is generally in a nonhydrostatic state of stress and cannot be characterized by pressure alone. The fundamental parameter is the local state of strain, and a careful definition of it is required for a discussion of nonlinear dynamics. Two general sources of nonlinearity can be recognized: the kinematic, or convective, nonlinearity that is independent of the material properties, and the inherent physical nonlinearity of the solid, as characterized by its constitutive behavior. As with fluids, both effects must be taken into account, although for highly nonlinear materials, such as rocks, the geometric nonlinearity can be insignificant.

The focus here will be on weakly nonlinear wave motion, for which the appropriate small parameter is the ratio of dynamic displacement to wavelength. To leading order, elastic waves propagate isentropically, although thermal losses through heat flux can be the dominant source of attenuation, at least in metals. Internal friction, akin to viscosity in fluids, is the primary energy loss mechanism for nonmetallic materials, and appropriate viscoelastic models are available (Kolsky, 1963). The leading-order adiabatic approximation is satisfactory for linear sinusoidal compressional waves of frequencies below about 1 GHz (Bland, 1969),

A. N. Norris
Department of Mechanical and Aerospace Engineering, Rutgers University, Piscataway, NJ, USA

M. F. Hamilton, D. T. Blackstock (eds.), *Nonlinear Acoustics*,
https://doi.org/10.1007/978-3-031-58963-8_9

and it is generally adequate except over regions of rapid change, as occurs at shock fronts, for example. Analysis of shocks and solutions of the fully nonlinear equations of motion, including effects of finite thermal conductivity, can be found in, for instance, Bland (1969). A major strand of research concerns the propagation of singular surfaces and acceleration waves. These are formally exact solutions for wavefronts of vanishing thickness, and their analysis stems from research by Hadamard (1903), and advanced by T. Y. Thomas, C. Truesdell, and others in the 1950s and 1960s (McCarthy, 1975). Useful reviews of waves in solids with an emphasis on nonlinearity are given by Zarembo and Krasil'nikov (1971) and by Thurston (1984).

9.2   Equations of Nonlinear Elastodynamics

Wave motion in solids is governed by the following momentum balance equation, which replaces Eq. (3.2):

$$\rho \frac{D\mathbf{u}}{Dt} = \nabla \cdot \boldsymbol{\sigma}. \tag{9.1}$$

Here, ρ is mass density, $\mathbf{u}$ is particle velocity, $\boldsymbol{\sigma} = \boldsymbol{\sigma}^{\mathrm{T}}$ is the stress tensor, also known as the Cauchy stress, and $(\nabla \cdot \boldsymbol{\sigma})_i = \partial \sigma_{ij}/\partial x_j$.[1] Nonlinear elasticity in solids is usually formulated in terms of a Lagrangian (or material) description, in contrast to that in fluids, which is normally considered in Eulerian (or spatial) coordinates. The distinction is that the current coordinate of a particle, which is $\mathbf{x}$, is displaced from its original or *natural* position $\mathbf{a}$ by the displacement $\mathbf{U} = \mathbf{x} - \mathbf{a}$. A laboratory sample is conveniently described in the unstressed, equilibrium state, which corresponds to the material coordinates $\mathbf{a}$. Thus, a typical nonlinear wave experiment in a solid that measures the transit time of a wave across a sample of a given *material* length provides data on the wave speed in material coordinates.

The connection between the current and material descriptions is through the deformation gradient tensor, defined as

$$\mathbf{F} = \frac{\partial \mathbf{x}}{\partial \mathbf{a}} = \mathbf{I} + \frac{\partial \mathbf{U}}{\partial \mathbf{a}}, \tag{9.2}$$

where $\mathbf{I}$ is the second-rank identity tensor, or $F_{ij} = \partial x_i/\partial a_j = \delta_{ij} + \partial U_i/\partial a_j$, where δ_{ij} is the Kronecker delta. The lengths of $d\mathbf{x}$ and $d\mathbf{a}$, corresponding to the same infinitesimal line element, are related by $dx^2 - da^2 = d\mathbf{x} \cdot d\mathbf{x} - d\mathbf{a} \cdot d\mathbf{a} = 2d\mathbf{a} \cdot (\mathbf{E} \cdot d\mathbf{a})$, where $\mathbf{E} = \mathbf{E}^{\mathrm{T}}$ is the Lagrangian, or Green, strain tensor, given by

---

[1] The summation convention is employed, which implies summation over repeated indices, e.g., $\partial \sigma_{ij}/\partial x_j = \partial \sigma_{i1}/\partial x_1 + \partial \sigma_{i2}/\partial x_2 + \partial \sigma_{i3}/\partial x_3$.

$$\mathbf{E} = \tfrac{1}{2}(\mathbf{F}^{\mathrm{T}} \cdot \mathbf{F} - \mathbf{I}), \quad \text{or} \quad E_{ij} = \frac{1}{2}\left(\frac{\partial U_i}{\partial a_j} + \frac{\partial U_j}{\partial a_i} + \frac{\partial U_k}{\partial a_i}\frac{\partial U_k}{\partial a_j}\right). \tag{9.3}$$

Mass conservation is taken into account by the relation $\rho \, dV = \rho_0 \, dV_0$ [rather than Eq. (3.1) for fluids], where dV and dV_0 are the volumes of infinitesimal boxes in the current and material coordinates, each containing the same particles, and ρ_0 is the constant density in the reference (unstressed) configuration. By definition, $dV = (\det \mathbf{F}) \, dV_0$, and hence

$$\frac{\rho}{\rho_0} = \frac{1}{\det \mathbf{F}}. \tag{9.4}$$

Equation (9.4) may be expressed explicitly in terms of the strain by writing $\det \mathbf{F} = (1 + 2I_E + 4II_E + 8III_E)^{1/2}$, where I_E, II_E, and III_E are the principal invariants of the strain tensor (Eringen and Suhubi, 1974):

$$\frac{\rho}{\rho_0} = [1 + 2\,\mathrm{tr}\,\mathbf{E} + 2(\mathrm{tr}\,\mathbf{E})^2 - 2\,\mathrm{tr}\,\mathbf{E}^2 + \tfrac{4}{3}(\mathrm{tr}\,\mathbf{E})^3 - 4(\mathrm{tr}\,\mathbf{E})\mathrm{tr}\,\mathbf{E}^2 + \tfrac{8}{3}\mathrm{tr}\,\mathbf{E}^3]^{-1/2}. \tag{9.5}$$

Either Eq. (9.4) or (9.5) eliminates density from consideration as a variable. It remains to express the stress in terms of the strain.

Equation (9.1) is cast in Eulerian coordinates, with the velocity $\mathbf{u} = \mathbf{u}(\mathbf{x}, t)$ implicitly a function of $\mathbf{x}$ and t. In the Lagrangian description, we consider the displacement as a function of $\mathbf{a}$ and t, i.e., $\mathbf{U} = \mathbf{U}(\mathbf{a}, t)$, and thus $\mathbf{u} = \partial \mathbf{U}/\partial t$. In order to transform Eq. (9.1) to Lagrangian coordinates, we introduce the non-symmetric tensor $\mathbf{P} = (\rho_0/\rho)\,\boldsymbol{\sigma} \cdot (\mathbf{F}^{-1})^{\mathrm{T}}$, known as the first Piola–Kirchhoff stress tensor, or sometimes the Lagrangian stress tensor. Substituting $\boldsymbol{\sigma} = (\rho/\rho_0)\,\mathbf{P} \cdot \mathbf{F}^{\mathrm{T}}$ and $D\mathbf{u}/Dt = \partial^2\mathbf{U}/\partial t^2$ into Eq. (9.1) and using the Euler–Piola–Jacobi identity $\nabla \cdot [(\rho/\rho_0)\mathbf{F}^{\mathrm{T}}] = \nabla \cdot |\mathbf{F}^{\mathrm{T}}/\det \mathbf{F}) = 0$ (Truesdell and Toupin, 1960, p. 246), we obtain $\rho_0 \partial^2 U_i/\partial t^2 = F_{jk}\partial P_{ik}/\partial x_j = \partial P_{ij}/\partial a_j$, or

$$\rho_0 \frac{\partial^2 \mathbf{U}}{\partial t^2} = \nabla_a \cdot \mathbf{P}, \tag{9.6}$$

where ∇_a denotes the gradient with respect to the material coordinates $\mathbf{a}$.

The equations of motion (9.6) can also be obtained from the Lagrangian density $\mathcal{L} = \tfrac{1}{2}\rho_0 u^2 - \rho_0 W$, where W is the specific strain energy of the elastic body per unit mass. The Euler–Lagrange equations for $\mathcal{L}$ yield Eq. (9.6) with $\mathbf{P} = \rho_0 \partial W/\partial \mathbf{F}$ [see, e.g., Eqs. (26.2) and (26.3) of Landau and Lifshitz, 1986]. For most materials, it is reasonable to assume that the strain energy depends upon the local stretching and volume change, which in turn are completely determined by the Green strain tensor $\mathbf{E}$. Hence $W = W(\mathbf{E})$, and it follows from the previous formula for $\mathbf{P}$ that

$$\mathbf{P} = \rho_0 \mathbf{F} \cdot \frac{\partial W}{\partial \mathbf{E}} \quad \Leftrightarrow \quad \boldsymbol{\sigma} = \rho \mathbf{F} \cdot \frac{\partial W}{\partial \mathbf{E}} \cdot \mathbf{F}^{\mathrm{T}}. \tag{9.7}$$

The strain energy is assumed to have the following expansion for small strains:

$$\rho_0 W = \frac{1}{2!} C_{ijkl} E_{ij} E_{kl} + \frac{1}{3!} C_{ijklmn} E_{ij} E_{kl} E_{mn} + \cdots, \tag{9.8}$$

and the symmetry of $\mathbf{E}$ implies that the second- and third-order moduli can be expressed using Voigt's notation: $C_{ijkl} = c_{IJ}$, $C_{ijklmn} = c_{IJK}$, where $I, J, K \in \{1, 2, 3, 4, 5, 6\}$ with the relationships $ij = 11, 22, 33, 23, 31, 12 \leftrightarrow I = 1, 2, 3, 4, 5, 6$. Equations (9.7) and (9.8) together imply that

$$P_{ij} = C_{ijkl} \frac{\partial U_k}{\partial a_l} + \frac{1}{2} M_{ijklmn} \frac{\partial U_k}{\partial a_l} \frac{\partial U_m}{\partial a_n} + \frac{1}{3} M_{ijklmnpq} \frac{\partial U_k}{\partial a_l} \frac{\partial U_m}{\partial a_n} \frac{\partial U_p}{\partial a_q} + \cdots, \tag{9.9}$$

where

$$M_{ijklmn} = C_{ijklmn} + C_{ijln}\delta_{km} + C_{jnkl}\delta_{im} + C_{jlmn}\delta_{ik}, \tag{9.10}$$

and Thurston (1984) gives an expression for the higher-order coefficients $M_{ijklmnpq}$. Note that $M_{ijklmn} \neq M_{jiklmn}$, which implies that the nonsymmetry of $\mathbf{P}$ is a second-order effect.

The number of second- and third-order moduli, at most 21 and 56, respectively, is much lower in the presence of *material symmetry*. Pure crystals display a symmetry associated with molecular arrangement, whereas man-made materials display *textured symmetry*; see Cowin and Mehrabadi (1995) for a complete account of elastic symmetries. We shall concentrate on isotropic solids, for which the strain energy has the expansion

$$\rho_0 W = \frac{\lambda}{2}(\mathrm{tr}\,\mathbf{E})^2 + \mu \,\mathrm{tr}\,\mathbf{E}^2 + \frac{\mathcal{C}}{3}(\mathrm{tr}\,\mathbf{E})^3 + \mathcal{B}(\mathrm{tr}\,\mathbf{E})\mathrm{tr}\,\mathbf{E}^2 + \frac{\mathcal{A}}{3}\mathrm{tr}\,\mathbf{E}^3 + \cdots, \tag{9.11}$$

where λ and μ are the Lamé moduli. The third-order moduli $\mathcal{A}$, $\mathcal{B}$, and $\mathcal{C}$ are those used by Landau and Lifshitz (1986), but there are many other notations, some of which are shown in Table 9.1. The isotropic moduli are

$$C_{ijkl} = \lambda \delta_{ij}\delta_{kl} + 2\mu I_{ijkl}, \tag{9.12}$$

where $I_{ijkl} = (\delta_{ik}\delta_{jl} + \delta_{il}\delta_{jk})/2$, and

$$C_{ijklmn} = 2\mathcal{C}\delta_{ij}\delta_{kl}\delta_{mn} + 2\mathcal{B}(\delta_{ij}I_{klmn} + \delta_{kl}I_{mnij} + \delta_{mn}I_{ijkl})$$

$$+ \frac{\mathcal{A}}{2}(\delta_{ik}I_{jlmn} + \delta_{il}I_{jkmn} + \delta_{jk}I_{ilmn} + \delta_{jl}I_{ikmn}). \tag{9.13}$$

Table 9.1 Relations between third-order elastic constants for isotropic solids.

Toupin and Bernstein (1961)	Murnaghan (1951)	Bland (1969)	Eringen and Suhubi (1974)		Standard, c_{IJK}	
$v_1 = 2C$	$l = B + C$	$\alpha = \frac{1}{3}C$	$l_E = \frac{1}{3}A + B + \frac{1}{3}C$	$c_{123} = 2C$	$c_{111} = 2A + 6B + 2C$	
$v_2 = B$	$m = \frac{1}{2}A + B$	$\beta = B$	$m_E = -A - 2B$	$c_{144} = B$	$c_{112} = 2B + 2C$	
$v_3 = \frac{1}{4}A$	$n = A$	$\gamma = \frac{1}{3}A$	$n_E = A$	$c_{456} = \frac{1}{4}A$	$c_{166} = \frac{1}{2}A + B$	

A solid is characterized by a positive shear modulus μ and a positive bulk modulus $K = \lambda + \frac{2}{3}\mu$, but the signs of the third-order moduli are not definite. An inviscid fluid is formally obtained by taking $\mu = 0$, $\lambda = A = \rho_0 c_0^2$, where c_0 is the small-signal sound speed in the fluid, and $A = 0$, $B = -A$, and $C = (A - B)/2$ (Kostek et al., 1993), where $B = \rho_0^2(\partial^2 P/\partial \rho^2)_0$, and P is pressure [see Eqs. (2.2)–(2.4) and Sect. 2.5].

The theory as presented ignores internal attenuation. The simultaneous effects of thermal and viscoelastic dissipation can be included by replacing $\mathbf{P}$ in Eq. (9.6) with $\mathbf{P} + \mathbf{D}$, where $\mathbf{D}$ is a viscous-like stress tensor defined by (see, e.g., Landau and Lifshitz, 1986)

$$D_{ij} = 2\eta(\dot{E}_{ij} - \tfrac{1}{3}\delta_{ij}\dot{E}_{kk}) + (\zeta + \chi)\delta_{ij}\dot{E}_{kk}, \tag{9.14}$$

where η is the shear viscosity coefficient, ζ is the bulk viscosity coefficient, and $\chi = \kappa T_0(\alpha_T K/c_l C_p)^2$, in which κ is thermal conductivity ($\eta, \zeta, \kappa > 0$), T_0 is ambient temperature, α_T is the thermal expansion coefficient, and C_p is the specific heat per unit volume at constant pressure. The dots in Eq. (9.14) indicate time derivatives.

9.3 Longitudinal and Transverse Plane Waves

A single equation of motion, in the absence of viscosity, is obtained by combining Eqs. (9.6) and (9.9):

$$\rho_0 \frac{\partial^2 U_i}{\partial t^2} = \frac{\partial^2 U_k}{\partial a_j \partial a_l}\left(C_{ijkl} + M_{ijklmn}\frac{\partial U_m}{\partial a_n} + M_{ijklmnpq}\frac{\partial U_m}{\partial a_n}\frac{\partial U_p}{\partial a_q} + \cdots\right). \tag{9.15}$$

Let $\mathbf{U} = (U, V, W)$ be a function of $a = a_1$ and t; then Eq. (9.15) yields[2] (Goldberg, 1961)

---

[2] For clarity in Eqs. (9.16)–(9.18), the subscripts on the displacement components represent partial differentiation with respect to the indicated quantities; for example, $U_{tt} = \partial^2 U/\partial t^2$.

$$U_{tt} - c_l^2 U_{aa} = \left(3c_l^2 + \frac{c_{111}}{\rho_0}\right) U_a U_{aa}$$

$$+ \left(c_t^2 + \frac{c_{166}}{\rho_0}\right)(V_a V_{aa} + W_a W_{aa}) + \cdots, \quad (9.16)$$

$$V_{tt} - c_t^2 V_{aa} = \left(c_t^2 + \frac{c_{166}}{\rho_0}\right)(U_a V_{aa} + V_a U_{aa}) + \cdots, \quad (9.17)$$

$$W_{tt} - c_t^2 W_{aa} = \left(c_t^2 + \frac{c_{166}}{\rho_0}\right)(U_a W_{aa} + W_a U_{aa}) + \cdots, \quad (9.18)$$

where c_l and c_t are the propagation speeds of linearized (small-signal) compressional and transverse elastic waves, respectively:

$$c_l = \sqrt{\frac{\lambda + 2\mu}{\rho_0}}, \quad c_t = \sqrt{\frac{\mu}{\rho_0}}. \quad (9.19)$$

The three equations of motion (9.16)–(9.18) reduce to one for purely longitudinal motion ($V = W = 0$), which can be written as (Thurston, 1984)

$$\frac{\partial^2 U}{\partial t^2} = c_l^2 \frac{\partial^2 U}{\partial a^2} g\left(\frac{\partial U}{\partial a}\right), \quad (9.20)$$

where

$$g(\xi) = 1 + \left(3 + \frac{c_{111}}{\rho_0 c_l^2}\right)\xi + \left(3 + \frac{3c_{111} + c_{1111}}{\rho_0 c_l^2}\right)\frac{\xi^2}{2!} + \cdots. \quad (9.21)$$

The one-dimensional equation (9.20) may be solved along characteristics as in Sect. 3.3.1. A wave traveling in the $+x$ direction has Riemann invariant $\frac{1}{2}(\lambda - u) = 0$ (λ here is not the Lamé constant!), where now

$$\lambda = -c_l \int_0^{\partial U/\partial a} g^{\frac{1}{2}}(\xi)\, d\xi. \quad (9.22)$$

The wave speed relative to the reference configuration is $c_{\text{ref}} = c_l g^{\frac{1}{2}}(\partial U/\partial a)$. The speed in actual space is $c + u$, where $c = c_{\text{ref}}(1 + \partial U/\partial a)$. By eliminating $\partial U/\partial a$ using Eq. (9.22) and $\lambda = u$, we find that $c + u = c_l + \beta u + \cdots$, with coefficient of nonlinearity

$$\beta = -\left(\frac{3}{2} + \frac{c_{111}}{2\rho_0 c_l^2}\right). \quad (9.23)$$

This agrees with Eq. (2.25) when the identity $c_{111} = 2\mathcal{A} + 6\mathcal{B} + 2\mathcal{C}$ (see Table 9.1) is used.

The nonlinear distortion of a wave is accompanied by the generation of harmonics of all orders for single-frequency input. A simple perturbation analysis of Eq. (9.20) shows that the source excitation $U(0, t) = U_0 \sin \omega t$, $V = W = 0$, produces a propagating second harmonic according to (Zarembo and Krasil'nikov, 1971)

$$U(a, \tau) = U_0 \sin \omega \tau + \frac{\beta}{4} \left(\frac{\omega U_0}{c_l} \right)^2 a \cos 2\omega \tau + \cdots, \tag{9.24}$$

where $\tau = t - a/c_l$. Note that the amplitude of the second harmonic is of order aU_0^2/λ_0^2, where λ_0 is the fundamental wavelength, and it grows in direct proportion to propagation distance. Equation (9.24) provides a practical means to determine β, and hence the TOE (third-order elasticity) constant c_{111}. Theoretical and experimental applications of harmonic generation techniques for measuring TOE constants of cubic crystals are reviewed by Breazeale and Philip (1984). One advantage of this approach is that it permits measurements to be made as a continuous function of temperature. The technique depends upon absolute amplitude measurements, which can be exacerbated by intrinsic attenuation in the sample. This is not a serious problem in crystals at the frequency ranges of practical interest, but it becomes a severe limitation for highly attenuative materials such as rock. With thermoviscous attenuation included [see Eq. (9.46)], the solution becomes (Zarembo and Krasil'nikov, 1971)

$$U(a, \tau) = U_0 e^{-\alpha a} \sin \omega \tau + \frac{\beta}{8\alpha} \left(\frac{\omega U_0}{c_l} \right)^2 (e^{-2\alpha a} - e^{-4\alpha a}) \cos 2\omega \tau + \cdots, \tag{9.25}$$

where $\alpha = \omega^2 (\frac{4}{3}\eta + \zeta + \chi)/2\rho_0 c_l^3$. The amplitude of the second harmonic is bounded in this case, having a maximum value of $\beta U_0^2 \omega^2/32\alpha c_l^2$ at distance $(\ln 2)/2\alpha$ from the source.

The analysis leading to Eq. (9.24) was based upon a truncation of the exact constitutive equation, and ignored the fact that coefficients of fourth order and higher could also generate second harmonics. A more precise examination of the problem by Thurston and Shapiro (1967) considered all higher orders, with expansions in powers of the acoustic Mach number at the source, $\varepsilon = \omega U_0/c_l$. They obtained definitive results relating experimental data to third-order coefficients, independent of fourth-order coefficients.

In addition to longitudinal waves, an isotropic elastic solid supports transverse, or shear, wave motion, for which the displacement is polarized perpendicular to the propagation direction. Small-signal transverse waves travel at speed c_t, which is always less than c_l. In fact, the inequality $c_t < c_l/\sqrt{2}$ holds for solids with positive Poisson's ratio (solids with negative Poisson's ratio are theoretically possible—e.g., origami—but are of little interest as far as wave motion is concerned). When the source displacement is not purely longitudinal—for example, $U(0, t) = U_0 \sin \omega t$ and $V(0, t) = V_0 \sin \omega t$—an additional contribution to the longitudinal second-harmonic component results from the term $V_a V_{aa}$ in Eq. (9.16). Generation of a longitudinal second-harmonic component with propagation speed c_l by a shear wave with propagation speed c_t is an asynchronous interaction that causes the amplitude of the former to beat with spatial period $l = \pi/(k_t - k_l)$, where $k_t = \omega/c_t$ and $k_l = \omega/c_l$ (see Fig. 5.5 for measurements of an analogous process, asynchronous harmonic generation of sound in a waveguide). Since l is of order one wavelength because of the disparity in propagation speeds, this interaction is very inefficient, and the resonant second-harmonic generation taken into account by Eqs. (9.24) and (9.25) dominates at distances $a \gg l$. Generalizations of Eqs. (9.24) and (9.25) that account for transverse source displacement are provided by Polyakova (1964).

Equations (9.16)–(9.18) do not support uncoupled transverse waves of the form $U = W = 0$. As just described, a first-order component V generates a second-order component U, and wave mixing occurs. Moreover, solution of Eq. (9.17) by perturbation methods shows that a first-order component V generates no second-order contribution to V, and consequently the quadratic nonlinearity for transverse waves is zero. The absence of transverse harmonics can be simply understood in terms of the isotropic material symmetry, which forbids quadratic terms for shear deformation (Norris, 1991).

The wave mixing process can, however, produce interesting three-wave interactions, which provides another means to measure combinations of TOE constants. Jones and Kobett (1963) calculated resonance conditions based on wave vector matching for oblique elastic-wave interactions (see also Landau and Lifshitz, 1986, Sect. 26), and experiments subsequently measured the amplitude of the wave generated from the interaction of two primary waves (Rollins et al., 1964). For example, the resonance condition for which two transverse plane waves having the same frequency generate a longitudinal second-harmonic wave is $|\mathbf{k}_t^a + \mathbf{k}_t^b| = 2k_l$, where the transverse-wave vectors indicate respective directions of propagation, and $|\mathbf{k}_t^a| = |\mathbf{k}_t^b| = k_t$. The resonance condition is satisfied when the angle formed by $\mathbf{k}_t^a$ and $\mathbf{k}_t^b$ is $\theta = 2\cos^{-1}(c_t/c_l)$. Zarembo and Krasil'nikov (1971) discuss three-wave interaction processes and provide a review of experimental results on the generation of transverse-wave harmonics in crystals.

9.4 Acoustoelasticity: Stress Dependence of the Wave Speeds

The most commonly used and the most precise method for determining TOE constants is based on the acoustoelastic effect, in which a static state of stress and strain applied to an elastic body changes the speeds of small-signal waves. The effect is relatively small (e.g., $\sim 10^{-5}$/MPa for aluminum), and it requires high-precision measurements, such as the "sing-around" technique in which the resonance frequency of a slab is measured. It also introduces the possibility of birefringence, whereby the degeneracy of the transverse waves is broken, and transverse waves polarized along two principal directions have slightly different speeds. Acoustoelasticity can also be used to measure an existing state of stress, such as residual stress (Pao et al., 1984). A summary of the experimental data on TOE constants of crystals can be found in the monograph by Thurston (1984).

The theory underlying acoustoelasticity is well developed, starting with the fundamental paper by Toupin and Bernstein (1961). Three states and their coordinates need to be distinguished: the natural, $\mathbf{a}$; the initial, $\mathbf{X}$; and the current, $\mathbf{x}$. For simplicity, suppose that the initial stress and strain are uniform and defined by the static displacements $\mathbf{U}^s$: $\mathbf{X} = \mathbf{a} + \mathbf{U}^s$. The acoustoelastic response is measured by the further "small on large" dynamic displacement $\mathbf{U}^d$: $\mathbf{x} = \mathbf{X} + \mathbf{U}^d$. The equation of motion for $\mathbf{U}^d$ follows from Eqs. (9.6) and (9.9) by linearization about the initial state. Also, it is useful to express the acoustoelastic equations in the initial coordinates, which, because of the infinitesimal nature of $\mathbf{U}^d$, coincide with the laboratory coordinates. The change of variable $\mathbf{a} \to \mathbf{X}$ is achieved by use of the chain rule, with the result

$$\rho_0 \frac{\partial^2 U_i^d}{\partial t^2} = B_{ijkl} \frac{\partial^2 U_k^d}{\partial X_j \partial X_l}, \tag{9.26}$$

where the effective elastic stiffnesses are

$$B_{ijkl} = C_{ijkl} + \delta_{ik} C_{jlqr}(\partial U_q^s/\partial X_r) + C_{rjkl}(\partial U_i^s/\partial X_r) + C_{irkl}(\partial U_j^s/\partial X_r)$$
$$+ C_{ijrl}(\partial U_k^s/\partial X_r) + C_{ijkr}(\partial U_l^s/\partial X_r) + C_{ijklmn}(\partial U_m^s/\partial X_n). \tag{9.27}$$

Equation (9.26) is a second-order (quadratic) approximation of Eqs. (9.6) and (9.9). The assumption of uniform initial stress and strain implies that the coefficients B_{ijkl} are constants.

Consider a plane wave propagating in the direction of the unit vector $\mathbf{n}$, $\mathbf{U}^d = \mathbf{U}^{d0} \sin \omega(t - \mathbf{n} \cdot \mathbf{X}/v)$, where the polarization $\mathbf{U}^{d0}$ (constant) satisfies, from Eq. (9.27),

$$\rho_0 v^2 U_i^{d0} = B_{ijkl} n_j n_l U_k^{d0}. \tag{9.28}$$

If the solid is assumed to be isotropic in its undeformed state, the eigenvalue equation (9.28) predicts one quasi-longitudinal wave in the direction of $\mathbf{n}$, and two quasi-transverse waves. With no loss in generality, let the coordinate axes coincide with the principal axes of static strain, $e_{ij}^s = \frac{1}{2}(\partial U_i^s/\partial X_j + \partial U_j^s/\partial X_i)$, and static stress, $\sigma_{ij}^s = C_{ijkl} e_{kl}^s$. If the propagation direction is aligned with one axis, say $\mathbf{n} = \mathbf{e}_1$, then the longitudinal and transverse modes are pure, with polarizations in the coordinate directions and corresponding propagation speeds given by

$$\rho_0 v_l^2 = \rho_0 c_l^2 + \sigma_{11}^s + (4\rho_0 c_l^2 + c_{111})e_{11}^s + c_{112}(e_{22}^s + e_{33}^s), \tag{9.29}$$

Table 9.2 Stress derivatives of longitudinal and transverse wave speeds in an isotropic solid [$K = \lambda + \frac{2}{3}\mu$, $E = 2\mu(1 + \nu)$, $\nu = \lambda/2(\lambda + \mu)$].

Stress	Mode	Propagation $\mathbf{n}$	Polarization	$\rho_0(dv^2/dp)_0$
Hydrostatic	Longitudinal	Arbitrary	$\parallel \mathbf{n}$	$-\frac{5-3\nu}{1+\nu} - \frac{1}{3K}(c_{111} + 2c_{112})$
Hydrostatic	Transverse	Arbitrary	$\perp \mathbf{n}$	$-\frac{3(1-\nu)}{1+\nu} - \frac{1}{3K}(c_{144} + 2c_{166})$
Uniaxial	Longitudinal	$\parallel$ stress	$\parallel \mathbf{n}$	$-1 - \frac{4(1-\nu)}{(1+\nu)(1-2\nu)} -$ $\frac{1}{E}(c_{111} - 2\nu c_{112})$
Uniaxial	Longitudinal	$\perp$ stress	$\parallel \mathbf{n}$	$\frac{4\nu(1-\nu)}{(1+\nu)(1-2\nu)} +$ $\frac{1}{E}[\nu c_{111} - (1 - \nu)c_{112}]$
Uniaxial	Transverse	$\parallel$ stress	$\perp \mathbf{n}$	$-\frac{2}{1+\nu} + \frac{1}{E}[\nu c_{144} - (1 - \nu)c_{166}]$
Uniaxial	Transverse	$\perp$ stress	$\parallel$ stress	$-\frac{1-\nu}{1+\nu} + \frac{1}{E}[\nu c_{144} - (1 - \nu)c_{166}]$
Uniaxial	Transverse	$\perp$ stress	$\perp$ stress	$\frac{2\nu}{1+\nu} - \frac{1}{E}(c_{144} - 2\nu c_{166})$

$$\rho_0 v_{t2}^2 = \rho_0 c_t^2 + \sigma_{11}^s + (2\rho_0 c_t^2 + c_{166})(e_{11}^s + e_{22}^s) + c_{144}e_{33}^s, \tag{9.30}$$

$$\rho_0 v_{t3}^2 = \rho_0 c_t^2 + \sigma_{11}^s + (2\rho_0 c_t^2 + c_{166})(e_{11}^s + e_{33}^s) + c_{144}e_{22}^s. \tag{9.31}$$

Acoustoelastic measurements are normally performed by varying the applied static stress according to a single parameter p, such as a hydrostatic pressurization, $\boldsymbol{\sigma}^s = -p\mathbf{I}$, or a uniaxial compression in the direction $\mathbf{m}$, $\boldsymbol{\sigma}^s = -p\mathbf{m} \times \mathbf{m}$. Some specific applications of the above formulae are listed in Table 9.2, which gives the dimensionless derivative $\rho_0(dv^2/dp)_0$ for states of hydrostatic pressurization and uniaxial compression (the subscript 0 on the derivative indicates evaluation at $p = 0$). In a typical experiment, the resonance frequency f is measured for a slab of thickness L_0 in the undeformed state, with the waves propagating in the direction normal to the faces. The wave speed is determined from the equation $v = 2Lf$, where L is the deformed length. In practice, it is simpler to measure the "natural" wave speed (Thurston and Brugger, 1964) $w = 2L_0 f$. The two speeds are related by $v/w = L/L_0$, or

$$\rho_0 \left(\frac{dv^2}{dp}\right)_0 = \rho_0 \left(\frac{dw^2}{dp}\right)_0 + 2\rho_0 v^2 n_i n_j \left(\frac{de_{ij}^s}{dp}\right)_0. \tag{9.32}$$

Values of the pressure derivatives of velocity (the first two in Table 9.2) are listed in Table 9.3 for a variety of materials. The values for rock (Berea sandstone) are noticeably larger than the rest, indicating a high degree of nonlinearity, although the measured values for rock display large spreads (Winkler and Liu, 1996).

The propagation speed of a transverse wave polarized in the 2 direction and traveling in the 1 direction is $v_{t2} \equiv v_{12}$. A pure transverse mode polarized in the 1 direction can also propagate in the 2 direction with speed v_{21}, where, according to

Table 9.3 Dimensionless dependence of sound speed on pressure for some common materials.

Material	$\rho_0(dv_l^2/dp)_0$	$\rho_0(dv_t^2/dp)_0$
Pyrex	−8.6	−2.84
Fused silica	−4.32	−1.42
Nickel-steel	2.84	1.55
Molybdenum	3.48	1.05
Alumina	4.46	1.12
Tungsten	4.58	0.70
Water	5.0	0
Niobium	6.18	0.29
Gold	6.4	0.90
Magnesium	6.89	1.47
Steel (Hecla)	7.45	1.46
Benzene	9.0	0
Armco-iron	9.3	5.7
Lucite*	10.3	1.6
Polystyrene	11.6	1.57
Aluminum	12.4	2.92
PMMA	15.0	3.0
Cemented glass beads*	288	84
Berea sandstone (A)*	1628	956

Data compiled from the literature by Johnson et al. (1994), except for those with asterisks, which are calculated from moduli reported by Winkler and Liu (1996)

Eq. (9.30),

$$\rho_0 v_{21}^2 - \rho_0 v_{12}^2 = \sigma_{22}^s - \sigma_{11}^s. \tag{9.33}$$

In a similar situation for an unstressed but anisotropic solid, the wave speeds satisfy $\rho_0 v_{12}^2 = C_{2121}$, $\rho_0 v_{21}^2 = C_{1212}$, which are identical because of the symmetry of the elastic moduli, $C_{2121} = C_{1212}$. The difference, Eq. (9.33), depends upon the principal stresses, and offers a means to distinguish stress-induced effects from those caused by intrinsic material anisotropy. For instance, a uniaxially stressed isotropic material does not act like a solid with transverse isotropy.

9.5 Sound Beams in Solids

The evolution of an initial disturbance with a well-defined direction of propagation can be described by a nonlinear parabolic equation similar to the Kuznetsov–Zabolotskaya–Khokhlov equation for sound beams of finite amplitude in thermo-viscous fluids (Sect. 3.9). KZK-type equations have been derived for longitudinal waves in isotropic solids (Zabolotskaya, 1986b) and for waves in anisotropic solids

(Zabolotskaya, 1986a; Norris and Kostek, 1993). However, beams of purely transverse motion in isotropic solids are undistorted in the second-order approximation. Zabolotskaya (1986b) included fourth-order elastic coefficients and found that the cubic nonlinearity generates a third harmonic that mixes with the fundamental to produce a transverse second-harmonic wave. Calculations for a linearly polarized Gaussian beam showed that the second harmonic is polarized in the direction perpendicular to the fundamental.

We illustrate the procedure for deriving a KZK-type equation for the simplest case, longitudinal wave motion in an isotropic solid. The following extension of the ordering procedure used in Sect. 3.9 for sound beams in fluids is employed. Introduce the retarded time $\tau = t - a_1/c_l$ and the scaled variables $\bar{a}_1 = \varepsilon a_1$, $(\bar{a}_2, \bar{a}_3) = \varepsilon^{1/2}(a_2, a_3)$, where $\varepsilon \ll 1$ is a characteristic acoustic Mach number. Internal energy loss is included, although it is assumed that the damping is small and scales as $(\eta, \zeta, \chi) = \varepsilon(\bar{\eta}, \bar{\zeta}, \bar{\chi})$. We assume the following forms for the displacements, velocities, and stresses: $(U_1, u_1, P_{11}, P_{22}, P_{33}) = \varepsilon(\bar{U}_1, \bar{u}_1, \bar{P}_{11}, \bar{P}_{22}, \bar{P}_{33})$, $(U_i, u_i, P_{1i}, P_{i1}) = \varepsilon^{3/2}(\bar{U}_i, \bar{u}_i, \bar{P}_{1i}, \bar{P}_{i1})$ for $i = 2, 3$, and $(P_{23}, P_{32}) = \varepsilon^2(\bar{P}_{23}, \bar{P}_{32})$.

The equations of motion (9.6) can be cast as the following quasi–first-order system:

$$\frac{\partial U_i}{\partial t} = u_i, \tag{9.34}$$

$$\rho_0 \frac{\partial u_i}{\partial t} = \frac{\partial P_{ij}}{\partial a_j} + \frac{\partial D_{ij}}{\partial a_j}, \tag{9.35}$$

$$\frac{\partial P_{ij}}{\partial t} = \frac{\partial P_{ij}}{\partial F_{kl}} \frac{\partial u_k}{\partial a_l}. \tag{9.36}$$

Substitution of the scaled variables into Eq. (9.34) gives simply $\partial \bar{U}_i/\partial \tau = \bar{u}_i$. The force balances in the three directions become, from Eq. (9.35) with D_{ij} given by Eq. (9.14),

$$\rho_0 \frac{\partial \bar{u}_1}{\partial \tau} + \frac{1}{c_l} \frac{\partial \bar{P}_{11}}{\partial \tau} = \varepsilon \left(\frac{\partial \bar{P}_{11}}{\partial \bar{a}_1} + \frac{\partial \bar{P}_{12}}{\partial \bar{a}_2} + \frac{\partial \bar{P}_{13}}{\partial \bar{a}_3} + \frac{(\frac{4}{3}\bar{\eta} + \bar{\zeta} + \bar{\chi})}{c_l^2} \frac{\partial^2 \bar{u}_1}{\partial \tau^2} \right)$$

$$+ O(\varepsilon^2), \tag{9.37}$$

$$\rho_0 \frac{\partial \bar{u}_2}{\partial \tau} + \frac{1}{c_l} \frac{\partial \bar{P}_{21}}{\partial \tau} - \frac{\partial \bar{P}_{22}}{\partial \bar{a}_2} = \varepsilon \left(\frac{\partial \bar{P}_{21}}{\partial \bar{a}_1} + \frac{\partial \bar{P}_{23}}{\partial \bar{a}_3} \right) + O(\varepsilon^2), \tag{9.38}$$

$$\rho_0 \frac{\partial \bar{u}_3}{\partial \tau} + \frac{1}{c_l} \frac{\partial \bar{P}_{31}}{\partial \tau} - \frac{\partial \bar{P}_{33}}{\partial \bar{a}_3} = \varepsilon \left(\frac{\partial \bar{P}_{31}}{\partial \bar{a}_1} + \frac{\partial \bar{P}_{32}}{\partial \bar{a}_2} \right) + O(\varepsilon^2). \tag{9.39}$$

An expression for the time rate of change of the stress is obtained from Eq. (9.9):

$$\frac{\partial P_{ij}}{\partial t} = C_{ijkl}\frac{\partial u_k}{\partial a_l} + M_{ijklmn}\frac{\partial U_m}{\partial a_n}\frac{\partial u_k}{\partial a_l} + \cdots,\tag{9.40}$$

which in terms of the scaled variables becomes

$$\frac{\partial \bar{P}_{11}}{\partial \tau} + \rho_0 c_l\frac{\partial \bar{u}_1}{\partial \tau} = \varepsilon\left[\rho_0 c_l^2\frac{\partial \bar{u}_1}{\partial a_1} + \lambda\left(\frac{\partial \bar{u}_2}{\partial a_2} + \frac{\partial \bar{u}_3}{\partial a_3}\right) - 2\beta\rho_0\bar{u}_1\frac{\partial \bar{u}_1}{\partial \tau}\right] + O(\varepsilon^2),\tag{9.41}$$

where β is the coefficient of nonlinearity for longitudinal waves, defined in Eq. (9.23). Subtraction of $1/c_l$ times Eq. (9.41) from the longitudinal force balance equation, Eq. (9.37), eliminates all $O(\varepsilon^0)$ terms. Next, differentiation of the result with respect to τ, and use of the relation[3] $\bar{P}_{11,1\tau} = -\rho_0 c_l\bar{u}_{1,1\tau} + O(\varepsilon)$ obtained from Eq. (9.37), yields the $O(\varepsilon)$ relation

$$2\frac{\partial^2\bar{u}_1}{\partial \bar{a}_1\partial \tau} + \frac{\lambda}{\lambda+2\mu}\Lambda_1 + \Lambda_2 - \frac{\beta}{c_l^2}\frac{\partial^2\bar{u}_1^2}{\partial \tau^2} - \frac{(\frac{4}{3}\bar{\eta}+\bar{\zeta}+\bar{\chi})}{\rho_0 c_l^3}\frac{\partial^3\bar{u}_1}{\partial \tau^3} = 0,\tag{9.42}$$

where $\Lambda_1 = \bar{u}_{2,2\tau} + \bar{u}_{3,3\tau}$ and $\Lambda_2 = -(\rho_0 c_l)^{-1}(\bar{P}_{12,2\tau} + \bar{P}_{13,3\tau})$. The 12 and 21 stress rates are symmetric to leading order and are given by $\bar{P}_{12,\tau} = \bar{P}_{21,\tau} = \mu(\bar{u}_{1,2} - c_l^{-1}\bar{u}_{2,\tau})$. Using similar relations for the 13 and 31 stresses, combined with the leading-order terms from the transverse force balances, Eqs. (9.38) and (9.39), we find that

$$\frac{c_t^2}{c_l^2}\Lambda_1 - \Lambda_2 = \frac{\mu}{\rho_0 c_l}\bar{\nabla}_\perp^2\bar{u}_1,\tag{9.43}$$

$$\Lambda_1 - \Lambda_2 = \frac{1}{\rho_0}(\bar{P}_{22,22} + \bar{P}_{33,33}),\tag{9.44}$$

where $\bar{\nabla}_\perp^2 = \partial^2/\partial\bar{a}_2^2 + \partial^2/\partial\bar{a}_3^2$. The leading-order terms in Eqs. (9.38) and (9.39) allow us to determine that $\bar{P}_{22,22}+\bar{P}_{33,33} = -(\lambda/c_l)\bar{\nabla}_\perp^2\bar{u}_1$, from which the relations $\Lambda_1 = -c_l\bar{\nabla}_\perp^2\bar{u}_1$ and $\Lambda_2 = -2(c_t^2/c_l)\bar{\nabla}_\perp^2\bar{u}_1$ follow.

Finally, we revert to the physical variables of interest by removing the dependence upon ε. Following elimination of Λ_1 and Λ_2, Eq. (9.42) thus reduces to

$$\frac{\partial^2 u}{\partial a\partial \tau} - \frac{c_l}{2}\nabla_\perp^2 u - \frac{\beta}{2c_l^2}\frac{\partial^2 u^2}{\partial \tau^2} - \frac{\delta}{2c_l^3}\frac{\partial^3 u}{\partial \tau^3} = 0,\tag{9.45}$$

where $u = u_1$ is the longitudinal particle velocity, $a = a_1$ is the coordinate along the nominal axis of the beam, $\nabla_\perp^2 = \partial^2/\partial a_2^2 + \partial^2/\partial a_3^2$ signifies the transverse

---

[3] Subscripts preceded by a comma represent partial differentiation with respect to the indicated scaled quantities; for example, $\bar{P}_{11,1\tau} = \partial^2\bar{P}_{11}/\partial\bar{a}_1\partial\tau$.

Laplacian, and $\delta = \rho_0^{-1}(\frac{4}{3}\eta + \zeta + \chi)$ is an acoustic diffusivity. Equation (9.45) is the longitudinal wave counterpart of the KZK equation in nonlinear acoustics, Eq. (3.65). For plane waves, set $\nabla_\perp^2 u = 0$ and integrate Eq. (9.45) with respect to τ to obtain the Burgers equation,

$$\frac{\partial u}{\partial a} - \frac{\beta}{c_l^2} u \frac{\partial u}{\partial \tau} - \frac{\delta}{2c_l^3} \frac{\partial^2 u}{\partial \tau^2} = 0, \qquad (9.46)$$

the acoustic counterpart of which is Eq. (3.54). Although Eqs. (9.45) and (9.46) are expressed in Lagrangian coordinates, whereas the corresponding KZK and Burgers equations for sound waves in fluids are expressed in Eulerian coordinates, this distinction is of higher order than the approximations leading to these model equations, and it may therefore be ignored. Equations (9.24) and (9.25) are simple perturbation solutions of Eq. (9.46).

References

Bland, D. R. (1969). *Nonlinear Dynamic Elasticity* (Blaisdell, Waltham, Mass.).

Breazeale, M. A., and Philip, J. (1984). Determination of third-order elastic constants from ultrasonic harmonic generation measurements. In *Physical Acoustics*, Vol. 17, W. P. Mason and R. N. Thurston, eds. (Academic Press, New York), pp. 1–60.

Cowin, S. C., and Mehrabadi, M. M. (1995). Anisotropic symmetries of linear elasticity. *Appl. Mech. Rev.* **48**, 247–285.

Eringen, A. C., and Suhubi, E. S. (1974). *Elastodynamics*, Vol. 1 (Academic Press, New York).

Gol'dberg, Z. A. (1961). Interaction of plane longitudinal and transverse elastic waves. *Sov. Phys. Acoust.* **6**, 306–310.

Hadamard, J. (1903). *Leçons sur la Propagation des Ondes et les Equations de l'Hydrodynamique* (Hermann, Paris).

Johnson, D. L., Kostek, S., and Norris, A. N. (1994). Nonlinear tube waves. *J. Acoust. Soc. Am.* **96**, 1829–1843.

Jones, G. L., and Kobett, D. (1963). Interaction of elastic waves in an isotropic solid. *J. Acoust. Soc. Am.* **35**, 5–10.

Kolsky, H. (1963). *Stress Waves in Solids* (Dover, New York).

Kostek, S., Sinha, B. K., and Norris, A. N. (1993). Third-order elastic constants for an inviscid fluid. *J. Acoust. Soc. Am.* **94**, 3014–3017.

Landau, L. D., and Lifshitz, E. M. (1986). *Theory of Elasticity*, 3rd ed. (Pergamon Press, Oxford).

McCarthy, M. F. (1975). Singular surfaces and waves. In *Continuum Physics*, Vol. 2, A. C. Eringen, ed. (Academic Press, New York), pp. 449–521.

Murnaghan, F. D. (1951). *Finite Deformations of an Elastic Solid* (Chapman and Hall, New York).

Norris, A. N. (1991). Symmetry conditions for third order elastic moduli and implications in nonlinear wave theory. *J. Elasticity* **25**, 247–257.

Norris, A. N., and Kostek, S. (1993). Nonlinear parabolic wave equations for solids. *Advances in Nonlinear Acoustics: 13th ISNA*, H. Hobæk, ed. (World Scientific, Singapore), pp. 463–471.

Pao, Y.-H., Sachse, W., and Fukuoka, H. (1984). Acoustoelasticity and ultrasonic measurements of residual stresses. In *Physical Acoustics*, Vol. 17, W. P. Mason and R. N. Thurston, eds. (Academic Press, New York), pp. 61–143.

Polyakova, A. L. (1964). Nonlinear effects in a solid. *Sov. Phys. Solid State* **6**, 50–54.

Rollins, F., Taylor, L. H., and Todd, P. H. (1964). Ultrasonic study of three-phonon interactions. II. Experimental results. *Phys. Rev.* **136A**, 597–601.

Thurston, R. N. (1984). Waves in solids. In *Mechanics of Solids*, Vol. 4, C. Truesdell, ed. (Springer–Verlag, Berlin), pp. 109–308.

Thurston, R. N., and Brugger, K. (1964). Third-order elastic constants and the velocity of small amplitude elastic waves in homogeneous stressed media. *Phys. Rev.* **133**, A1604–A1610.

Thurston, R. N., and Shapiro, M. J. (1967). Interpretation of ultrasonic experiments on finite-amplitude waves. *J. Acoust. Soc. Am.* **41**, 1112–1125.

Toupin, R. A., and Bernstein, B. (1961). Sound waves in deformed perfectly elastic materials. Acoustoelastic effect. *J. Acoust. Soc. Am.* **33**, 216–225.

Truesdell, C., and Toupin, R. A. (1960). The classical field theories. In *Handbuch der Physik*, Vol. III/1, S. Flügge, ed. (Springer, Berlin), pp. 226–793.

Winkler, K. W., and Liu, X. (1996). Measurements of third-order elastic constants in rocks. *J. Acoust. Soc. Am.* **100**, 1392–1398.

Zabolotskaya, E. A. (1986a). Nonlinear propagation of a sound beam in a crystal. *Sov. Phys. Acoust.* **32**, 36–37.

Zabolotskaya, E. A. (1986b). Sound beams in a nonlinear isotropic solid. *Sov. Phys. Acoust.* **32**, 296–299.

Zarembo, L. K., and Krasil'nikov, V. A. (1971). Nonlinear phenomena in the propagation of elastic waves. *Sov. Phys. Usp.* **13**, 778–797.

Chapter 10
Perturbation Methods

Jerry H. Ginsberg

Contents

10.1 Background

Perturbation techniques have proved to be an effective method for deriving analytical descriptions of finite-amplitude waves in fluids. Solutions obtained from perturbation methods have provided physical insights into underlying physical phenomena. They also provide benchmarks against which solutions obtained by computational methods may be measured. Several procedures are available; each involves a trade-off between ease of implementation and degree of generality. A comprehensive treatment may be found in the text by Nayfeh (1973).

The acoustic Mach number ε, defined as the ratio of the peak particle velocity at a representative location to the small-signal speed of sound, is the small parameter for the perturbation expansion. The nonlinear effects, which appear in the governing equations as quadratic and higher products of the state variables, are an order of magnitude smaller than the linear terms. Consequently, one encounters the linearized problem in the first stage of a perturbation analysis.

An important physical aspect of acoustic waves in an ideal fluid is the absence of dispersion, which results in a continual exchange of energy between various harmonics. Using the linear solution for a plane wave to approximate the nonlinear

J. H. Ginsberg
G. W. Woodruff School of Mechanical Engineering, Georgia Institute of Technology, Atlanta, GA, USA

M. F. Hamilton, D. T. Blackstock (eds.), *Nonlinear Acoustics*,
https://doi.org/10.1007/978-3-031-58963-8_10

terms in the governing equations has the effect of driving the differential equations by a field of sources that propagates through the fluid at the same speed as a small-signal wave. This represents a resonant field excitation, which results in cumulative growth of the waveform distortion. From a mathematical and physical viewpoint, describing this growth effect requires recognition that there are at least two relevant length scales. One is the short scale of wavelength, over which the propagation of the wave is sufficiently close to the small-signal limit to permit application of a straightforward perturbation technique. The long scale for the wave is characterized by the shock formation distance. In propagation over distances of that magnitude, the buildup of nonlinear effects often results in a wave that differs substantially from the one predicted by linear theory.

The primary limitation on the use of perturbation methods is reliance on the linear solution as the starting point. Most analyses have treated situations in which the wave fronts and rays of the corresponding small-signal wave coincide with a coordinate system in which the wave equation is separable, such as rectangular Cartesian and spherical coordinates. The presence of strong diffraction effects, notably in sound beams, leads to substantial complications. That this should be so is not surprising when one notes that no closed-form *linear* solution has been found for the entire acoustic field of a sound beam.

Many applications of perturbation techniques to nonlinear acoustics have been based on the exact nonlinear wave equation (Goldstein, 1960) governing the velocity potential for an ideal gas. The derivation of this equation, which may be found in Sect. 3.3.2, involves combining the mass and momentum conservation equations with the isentropic equation of state, such that the sound pressure p, particle velocity $\mathbf{u}$, and excess density ρ' are eliminated in favor of the velocity potential function ϕ. The result is

$$\mathbf{u} = \nabla\phi, \tag{10.1}$$

$$\frac{\rho'}{\rho_0} + 1 = \left(\frac{p}{P_0} + 1\right)^{1/\gamma}, \tag{10.2}$$

$$\int_0^p \frac{dp}{\rho} = \frac{c_0^2}{\gamma - 1}\left[\left(\frac{p}{P_0} + 1\right)^{(\gamma-1)/\gamma} - 1\right] = -\frac{\partial\phi}{\partial t} - \tfrac{1}{2}\nabla\phi \cdot \nabla\phi, \tag{10.3}$$

$$\frac{\partial^2\phi}{\partial t^2} - c_0^2\nabla^2\phi = -\left[2\nabla\frac{\partial\phi}{\partial t} + \tfrac{1}{2}\nabla(\nabla\phi \cdot \nabla\phi)\right] \cdot \nabla\phi$$
$$- (\gamma - 1)\left(\frac{\partial\phi}{\partial t} + \tfrac{1}{2}\nabla\phi \cdot \nabla\phi\right)\nabla^2\phi, \tag{10.4}$$

where $c_0 = (\gamma P_0/\rho_0)^{1/2}$ is the small-signal speed of sound, i.e., the phase speed of a plane wave of infinitesimal amplitude, P_0 and ρ_0 are the ambient values of the pressure and density, respectively, and γ is the ratio of specific heats.

10.2   Regular Perturbation Technique

As noted earlier, the acoustic state variables for pressure, particle velocity components, and density change are all $O(\varepsilon)$ or smaller. [When a term is identified as being $O(\varepsilon^n)$, its magnitude for very small ε is $\alpha\varepsilon^n$, where α is a finite positive number.] It is reasonable to expect that the potential function depends analytically on ε, in which case ϕ may be expanded in a power series relative to ε. In view of the assumption that the acoustic signal is $O(\varepsilon)$, such an expansion may be written as

$$\phi(\mathbf{x}, t, \varepsilon) = \varepsilon\phi_1(\mathbf{x}, t) + \varepsilon^2\phi_2(\mathbf{x}, t) + O(\varepsilon^3), \tag{10.5}$$

where $\mathbf{x} = (x, y, z)$. The term $O(\varepsilon^3)$ represents the order of magnitude of effects that are not included in the two-term expansion.

The basic idea of a perturbation analysis is that the expansion should be valid for any value of ε that is sufficiently small. Thus, when Eq. (10.5) is substituted into any of the governing equations, the coefficients associated with each order of magnitude should match. The first-order terms that result from following such a procedure for Eq. (10.4) are

$$\frac{\partial^2\phi_1}{\partial t^2} - c_0^2\nabla^2\phi_1 = 0. \tag{10.6}$$

Using this expression to eliminate the Laplacian of ϕ_1 leads to a slight simplification in the corresponding second-order equation, with the result that

$$\frac{\partial^2\phi_2}{\partial t^2} - c_0^2\nabla^2\phi_2 = -\frac{\partial}{\partial t}\left[(\nabla\phi_1 \cdot \nabla\phi_1) + \frac{\beta - 1}{c_0^2}\left(\frac{\partial\phi_1}{\partial t}\right)^2\right], \tag{10.7}$$

where the coefficient of nonlinearity $\beta = (\gamma + 1)/2$ has been introduced in order to accommodate liquids, for which $\beta = 1 + B/2A$ [see Eq. (2.16)]. (Such a replacement is permissible only if one does not progress to the third-order approximation.)

Note that the manner in which the second-order equation, Eq. (10.7), has been written is suggestive of the solution sequence, in that the first-order equation is solved for ϕ_1, and that solution then becomes a source term driving the second-order equation. Once the two perturbation equations have been solved and the potential ϕ formed according to Eq. (10.5), the corresponding particle velocity may be determined directly from Eq. (10.1), while evaluation of pressure involves solving Eq. (10.3) for p. Using the binomial theorem in conjunction with the fact that ϕ is $O(\varepsilon)$ leads to

$$p = -\rho_0\left[\frac{\partial\phi}{\partial t} + \tfrac{1}{2}\nabla\phi \cdot \nabla\phi - \frac{1}{2c_0^2}\left(\frac{\partial\phi}{\partial t}\right)^2\right]. \tag{10.8}$$

The difference between Eq. (10.5) and a conventional power series is that rather than seeking a convergent infinite series in ε, one expects that there is a sufficiently small ε such that the two-term representation of ϕ is as accurate as one wishes. If the magnitude of the $O(\varepsilon^2)$ term is independent of the location $\mathbf{x}$ and time t, the asymptotic series is said to be uniformly valid (or, synonymously, uniformly accurate). Because of the smallness of the Mach number ε, if Eq. (10.5) is uniformly valid, the second-order term ϕ_2 represents a very small effect, and there is no reason to proceed further. By definition, the linear solution corresponds to the limit as $\varepsilon \to 0$. Hence, $\varepsilon\phi_1$ represents the linear approximation, and Eq. (10.5) is referred to as the *quasilinear solution*. The first application of a regular perturbation series in nonlinear acoustics was an investigation of plane waves performed by Airy (1849); see Sect. 1.2.3 for historical context.

Because of the nondispersive nature of many types of waves in linear theory, a regular perturbation series often leads to a nonuniformly accurate solution in which the second-order term grows. The occurrence of nonuniform terms means that the effect grows in importance; it therefore must be described more carefully.

A simple demonstration of the failure of the quasilinear solution may be obtained by comparing it to the Poisson solution for a plane wave (Lamb, 1945) [see also Eq. (1.30)]. Consider a plane wave generated by harmonic excitation, $u_x = \varepsilon c_0 \sin \omega t$ at $x = 0$, where u_x denotes the x component of the particle velocity vector. Replacing u_x by $\partial\phi/\partial x$ and matching like orders of ε leads to the perturbation boundary conditions

$$\partial\phi_1/\partial x = c_0 \sin \omega t \quad \text{and} \quad \partial\phi_2/\partial x = 0 \quad \text{at} \quad x = 0. \tag{10.9}$$

The task now is to satisfy the perturbation differential equations, Eqs. (10.6) and (10.7), subject to these boundary conditions and the radiation condition, which requires that the solution represent a wave propagating in the positive x direction.

The first-order solution is readily identified as a harmonic wave,

$$\phi_1 = (c_0/k) \cos(\omega t - kx), \quad k = \omega/c_0. \tag{10.10}$$

The result of substituting this expression into the second-order differential equation, Eq. (10.7), is

$$c_0^2 \frac{\partial^2\phi_2}{\partial x^2} - \frac{\partial^2\phi_2}{\partial t^2} = \beta \frac{\omega^3}{k^2} \sin(2\omega t - 2kx). \tag{10.11}$$

If the phase of the inhomogeneous term were not proportional to $t - x/c_0$, the particular solution for ϕ_2 would have the form of this term. This is not possible here, because the nondispersive nature of the linear approximation leads to generation of a second harmonic that is always in phase with the first-order solution, and therefore continuously receives energy. A particular solution may be obtained by the method of variation of parameters,

$$\phi_2 = A(x)\sin(2\omega t - 2kx) + B(x)\cos(2\omega t - 2kx), \tag{10.12}$$

which when substituted into Eq. (10.11) leads to a pair of equations obtained by matching like coefficients of the sine and cosine terms. The solution of the resulting equations is

$$A = C_A, \quad B = \tfrac{1}{4}\beta c_0 x + C_B, \tag{10.13}$$

where C_A and C_B are constants corresponding to the complementary solution of Eq. (10.11). It is possible to determine these constants by satisfying the second-order boundary condition in Eqs. (10.9), but it will be shown that these values are not needed.

The growth of the amplitude $B(x)$ makes the corresponding solution for ϕ_2 the wave analog of the resonant response of an undamped one-degree-of-freedom oscillator. Growth of ϕ_2 with increasing x implies that the solution is not uniformly valid. However, it is premature to reach such a conclusion because the potential function is not a physical variable representing the state of the fluid. For example, if the particle velocity had an $O(\varepsilon^2)$ term that were constant, the corresponding potential would be proportional to x. It is therefore necessary to consider the particle velocity and pressure derived from this expression for ϕ. Substituting $\varepsilon\phi_1 + \varepsilon^2\phi_2$ into Eqs. (10.1) and (10.8) leads to

$$u_x = \varepsilon c_0 \sin(\omega t - kx) + \varepsilon^2[(2kC_B + \tfrac{1}{2}\beta\omega x)\sin(2\omega t - 2kx)$$
$$+ (\tfrac{1}{4}\beta\omega - 2kC_A)\cos(2\omega t - 2kx)] + O(\varepsilon^3), \tag{10.14}$$

$$p = \rho_0 c_0^2 \varepsilon \sin(\omega t - kx) + \rho_0 \varepsilon^2[(\tfrac{1}{2}\beta\omega c_0 x + 2\omega C_A)\sin(2\omega t - 2kx)$$
$$- 2\omega C_A \cos(2\omega t - 2kx)] + O(\varepsilon^3). \tag{10.15}$$

These expressions constitute the quasilinear solution. The presence of $\varepsilon^2 x$ terms in both expressions means that they are not uniformly valid. In comparison, the coefficients C_A and C_B, which result from the second-order boundary condition, represent contributions that are $O(\varepsilon^2)$ everywhere. Setting $C_A = C_B = 0$ simplifies the analysis without significantly affecting its accuracy. Indeed, the following sections will show that any term in p or $\mathbf{u}$ that remains $O(\varepsilon^2)$ independently of $\mathbf{x}$ may be ignored, regardless of whether it stems from the complementary or the particular solution for ϕ_2.

Because the second-order terms in Eqs. (10.14) and (10.15) grow in magnitude without bound relative to the respective first-order terms as x is increased, the regular perturbation solution is said to have a singularity at $x \to \infty$. Nevertheless, the quasilinear solution might contain useful features. For example, consider the Fubini solution (Fubini-Ghiron, 1935) for the pressure at a fixed location x corresponding to the excitation in Eq. (10.9) [see Eq. (4.49)],

$$p(x) = \sum_{n=1}^{\infty} P_n(x) \sin[n(\omega t - kx)],$$

$$P_n(x) = \varepsilon \rho_0 c_0^2 \frac{2J_n(\varepsilon n \beta kx)}{\varepsilon n \beta kx}. \tag{10.16}$$

The leading term in an expansion of the amplitudes P_n in powers of ε is

$$P_n = \varepsilon \rho_0 c_0^2 \frac{1}{n!} (\tfrac{1}{2} \varepsilon n \beta kx)^{n-1} + O(\varepsilon^{n+1}). \tag{10.17}$$

The coefficients of the fundamental and second harmonics in Eq. (10.15) (with $C_A = C_B = 0$) match the foregoing for $n = 1$ and $n = 2$, respectively.

If εx is sufficiently small, the difference between the harmonic amplitudes in the quasilinear solution and the exact solution is negligible. This is the notion underlying many analyses of sound beams and parametric arrays (see Sect. 8.3). Furthermore, a quasilinear solution that is not uniformly valid suggests the manner in which cumulative distortion effects behave. Singular perturbation techniques build on such information to obtain a uniformly valid solution.

10.3   Method of Multiple Scales

Nonuniform validity of the regular perturbation solution for a plane wave is a consequence of failure to recognize that a length scale other than the wavelength must be described. A comparable situation arises in nonlinear vibration, where the frequency is weakly dependent on amplitude. Let $\varepsilon \Omega_1$ denote this effect, and consider the following Taylor series expansion:

$$\sin[(\omega + \varepsilon \Omega_1)t] = \sin \omega t + \varepsilon \Omega_1 t \cos \omega t + O(\varepsilon^2). \tag{10.18}$$

Here, truncation of the series at the second term results in a nonuniformly valid expression having a singularity at $t \to \infty$, even though the full Taylor series is convergent for any t. In the same manner, nonuniform validity of the regular perturbation solution for the plane wave may be interpreted as being a consequence of an inadequate truncation of a full Taylor series for the nonlinear wave.

The term in Eq. (10.18) whose coefficient contains t is referred to as a secular term, because it is not a periodic response. Powerful techniques have been developed to remove secular terms, and thereby infer a uniformly valid solution. An early application of one such approach to nonlinear acoustics, the method of multiple scales, is described in a paper by Nayfeh and Kluwick (1976). Consider generalization of the boundary condition in Eq. (10.9) to the case of an arbitrary excitation,

$$\left.\frac{\partial \phi}{\partial x}\right|_{x=0} = \varepsilon c_0 f(t). \tag{10.19}$$

Because the magnitude of the troublesome second term in the regular perturbation solution is proportional to εx, whose value changes much more slowly than the length scale x associated with the linear solution, the potential is assumed to be an explicit function of two length variables,

$$x_0 = x, \quad x_1 = \varepsilon x. \tag{10.20}$$

Correspondingly, the perturbation expansion of the potential is taken as

$$\phi = \varepsilon \phi_1(x_0, x_1, t, \varepsilon) + \varepsilon^2 \phi_2(x_0, x_1, t, \varepsilon) + O(\varepsilon^3). \tag{10.21}$$

The derivatives of ϕ become

$$\frac{\partial \phi}{\partial x} = \varepsilon \frac{\partial \phi_1}{\partial x_0} + \varepsilon^2 \left(\frac{\partial \phi_1}{\partial x_1} + \frac{\partial \phi_2}{\partial x_0} \right),$$

$$\frac{\partial^2 \phi}{\partial x^2} = \varepsilon \frac{\partial^2 \phi_1}{\partial x_0^2} + \varepsilon^2 \left(2 \frac{\partial^2 \phi_1}{\partial x_0 \partial x_1} + \frac{\partial^2 \phi_2}{\partial x_0^2} \right),$$

$$\frac{\partial \phi}{\partial t} = \varepsilon \frac{\partial \phi_1}{\partial t} + \varepsilon^2 \frac{\partial \phi_2}{\partial t},$$

$$\frac{\partial^2 \phi}{\partial t^2} = \varepsilon \frac{\partial^2 \phi_1}{\partial t^2} + \varepsilon^2 \frac{\partial^2 \phi_2}{\partial t^2}. \tag{10.22}$$

These expressions are substituted into the nonlinear wave equation, Eq. (10.4), and the boundary condition, Eq. (10.19), and terms corresponding to like powers of ε in each are matched, which leads to the following set of perturbation equations:

$$O(\varepsilon): \quad \frac{\partial^2 \phi_1}{\partial t^2} - c_0^2 \frac{\partial^2 \phi_1}{\partial x_0^2} = 0, \quad \left.\frac{\partial \phi_1}{\partial x_0}\right|_{x_0=x_1=0} = c_0 f(t), \tag{10.23}$$

$$O(\varepsilon^2): \quad \frac{\partial^2 \phi_2}{\partial t^2} - c_0^2 \frac{\partial^2 \phi_2}{\partial x_0^2} = 2c_0^2 \frac{\partial^2 \phi_1}{\partial x_0 \partial x_1} - 2 \frac{\partial \phi_1}{\partial x_0} \frac{\partial^2 \phi_1}{\partial x_0 \partial t} - 2(\beta - 1) \frac{\partial \phi_1}{\partial t} \frac{\partial^2 \phi_1}{\partial x_0^2},$$

$$\left.\frac{\partial \phi_2}{\partial x_0}\right|_{x_0=x_1=0} = - \left.\frac{\partial \phi_1}{\partial x_1}\right|_{x_0=x_1=0}. \tag{10.24}$$

The general solution of the first-order differential equation having the form of a wave propagating in the positive x direction is

$$\phi_1 = \phi_1(\tau, x_1), \quad \tau = t - x_0/c_0. \tag{10.25}$$

Substitution of ϕ_1 into the second-order differential equation yields

$$\frac{\partial^2 \phi_2}{\partial t^2} - c_0^2 \frac{\partial^2 \phi_2}{\partial x_0^2} = -2c_0 \frac{\partial^2 \phi_1}{\partial \tau \partial x_1} - \frac{2\beta}{c_0^2} \frac{\partial \phi_1}{\partial \tau} \frac{\partial^2 \phi_1}{\partial \tau^2}. \tag{10.26}$$

Clearly, the term on the right-hand side of this equation is a function of τ, but any function of τ is also a homogeneous solution of the equation. As a result, if the right-hand side does not vanish, ϕ_2 will contain a term that is secular. Preventing the appearance of such a term yields the criterion determining the dependence of ϕ_1 on x_1,

$$\frac{\partial^2 \phi_1}{\partial \tau \partial x_1} + \frac{\beta}{c_0^3} \frac{\partial \phi_1}{\partial \tau} \frac{\partial^2 \phi_1}{\partial \tau^2} = 0. \tag{10.27}$$

In addition, Eq. (10.25) converts the boundary condition for ϕ_1 in Eq. (10.23) to

$$\left. \frac{\partial \phi_1}{\partial \tau} \right|_{x_0 = x_1 = 0} = -c_0^2 f(\tau). \tag{10.28}$$

The solution of Eq. (10.27) satisfying the boundary condition was shown by Nayfeh and Kluwick (1976) to be

$$\phi_1 = -c_0^2 g(\zeta) + \frac{\beta}{2} c_0 x_1 f^2(\zeta), \quad g'(\zeta) = f(\zeta),$$

$$\tau \equiv t - x_0/c_0 = \zeta - \frac{\beta}{c_0} x_1 f(\zeta). \tag{10.29}$$

Furthermore, because the occurrence of secular terms has been avoided, ϕ_2 contains only terms that are uniformly $O(\varepsilon^2)$. The smallness of ε means that there is no need to actually determine ϕ_2.

The last step in evaluating the signal is to determine the particle velocity and pressure corresponding to $\phi = \varepsilon \phi_1 + O(\varepsilon^2)$. Because of the dependence of ϕ_1 on ζ, application of Eqs. (10.1) and (10.8) requires use of the chain rule. Implicit differentiation of the second of Eqs. (10.29), using $x_0 = x$ and $x_1 = \varepsilon x$, leads to

$$\frac{\partial \zeta}{\partial t} = \left[1 - \frac{\beta}{c_0} x_1 f'(\zeta) \right]^{-1},$$

$$\frac{\partial \zeta}{\partial x} = -\frac{1}{c_0} [1 - \beta \varepsilon f(\zeta)] \left[1 - \frac{\beta}{c_0} x_1 f'(\zeta) \right]^{-1}, \tag{10.30}$$

where $f'(\zeta) \equiv df/d\zeta$. The corresponding particle velocity and pressure are

$$u = \varepsilon \frac{\partial \phi_1}{\partial \zeta} \frac{\partial \zeta}{\partial x} + \varepsilon \frac{\partial \phi_1}{\partial x_1} \frac{\partial x_1}{\partial x} + O(\varepsilon^2) = \varepsilon c_0 f(\zeta) + O(\varepsilon^2), \tag{10.31}$$

$$p = -\rho_0 \frac{\partial \phi_1}{\partial \zeta} \frac{\partial \zeta}{\partial t} + O(\varepsilon^2) = \varepsilon \rho_0 c_0^2 f(\zeta) + O(\varepsilon^2), \tag{10.32}$$

where $O(\varepsilon^2)$ denotes terms having the same magnitude for all x. That these results are consistent with the Poisson solution [Eq. (1.30)] is readily verified by substituting $f(\zeta) = u/\varepsilon c_0$ from Eq. (10.31) into Eqs. (10.29), solving the result for ζ, then substituting ζ back into Eq. (10.31), which leads to

$$u = \varepsilon c_0 f(t - x/\hat{c}), \quad \hat{c} = c_0 (1 - \beta u/c_0)^{-1}. \tag{10.33}$$

Because u is $O(\varepsilon)$, the phase speed $\hat{c}$ given above is the same to $O(\varepsilon^2)$ as the phase speed in the Poisson solution [see Eq. (3.19)].

An overview of the multiple-scale solution shows that although the secular terms exciting ϕ_2 were eliminated, the ensuing first-order potential ϕ_1 contains a term that grows with increasing x. This is consistent with the earlier observation that one cannot infer *a priori* from the potential function which terms are associated with nonuniform validity. The need to prevent the occurrence of secular terms in ϕ_2 stems from Eqs. (10.22) for the derivatives of ϕ, which show that such terms lead to comparable terms in the derivatives.

In general, the primary difficulty arising in the multiple-scale method is the fact that avoiding secular terms leads to one or more partial differential equations in the scaled independent variables. In the present case of a plane wave, it is possible that Nayfeh and Kluwick (1976) were assisted in finding the solution by the similarity of that equation to the one solved by Poisson (1808). The likelihood of obtaining an analytical solution of the secularity equations decreases substantially with increasing complexity of the basic problem—for example, waves propagating in more than one direction. However, even if one cannot solve the multiple-scale equations exactly, the method can still be useful, because the multiple-scale equation(s) for eliminating secular terms are of lower order than the full nonlinear wave equation. Hence, they might be more amenable to numerical analysis, which is a feature that has been exploited to study the KZK equation for sound beams (see Sect. 11.3.1). Also, the multiple-scale equations offer a different perspective. For example, Eq. (10.27) may be integrated with respect to time to obtain the lossless form of the Burgers equation (see Sect. 3.7 for an alternative derivation of the Burgers equation).

10.4   Method of Renormalization

Use of the method of multiple scales requires recognition that nonlinearity introduces an additional reference frame from which the wave propagation may be viewed. Lighthill's method of strained coordinates (1949), which preceded development of the method of multiple scales, approaches the problem from a different perspective. It considers the primary effect of nonlinearity to be a distortion of the space–time grid, such that the linear solution is obtained at a location and time instant that differ from the prediction of linear theory. This aspect of the plane-wave problem is apparent from the multiple-scale solution in the preceding section. For example, the expression for p in Eq. (10.32) has the form of the linear solution for a one-dimensional wave, except that ζ replaces $t - x/c_0$. The linear result is recovered when $\varepsilon \to 0$.

In Lighthill's approach, one introduces a perturbation series for independent, as well as dependent, variables. The series expansions are used to change variables in the governing differential equations. One then selects the various terms in the expansions to obtain a solution that, in addition to satisfying the differential equation and the boundary and initial conditions, contains only terms that are uniformly valid.

A difficulty is encountered when this approach is applied to problems in nonlinear acoustics. As shown by Eqs. (10.29), the correct potential function for the wave contains terms that grow with increasing distance. Hence, it is not obvious which terms in the differential equation resulting from the Lighthill transformation of coordinates lead to terms that are not uniformly valid. The method of renormalization, which was developed by Pritulo (1962), offers an alternative to the steps entailed in the Lighthill procedure. Rather than eliminating troublesome terms from the differential equation, its approach is to defer changing the independent variable until nonuniformly valid solutions for the relevant variables have been obtained. For nonlinear acoustics, this means correcting expressions for the particle velocity components and pressure, rather than the potential. This is the approach introduced by Ginsberg (1975a, 1975b) and later refined by Nayfeh and Kluwick (1976).

To see how the method of renormalization works, consider the regular perturbation solution in Sect. 10.2. According to Eqs. (10.14) and (10.15), the second-order terms lose validity with increasing values of εx. For this reason, a new coordinate variable ξ that resembles x is introduced. The transformation is assumed to have the form

$$x = \xi + \varepsilon x F(\xi, t), \qquad (10.34)$$

where $F(\xi, t)$ is a function to be determined. This function is referred to as a coordinate straining because the term $\varepsilon x F$ represents a small deviation from ξ being identical to x, as though ξ represents the markings on a ruler that is stretched and x measures the distance from the origin to these markings.

The task now is to identify the function F. Toward that end, Eq. (10.34) is substituted into Eq. (10.14), and the result is expanded in a Taylor series in ascending powers of ε. Because only $O(\varepsilon)$ and $O(\varepsilon^2)$ terms have been retained in Eq. (10.14), the Taylor series is truncated at $O(\varepsilon^2)$:

$$u = \varepsilon c_0 \{\sin(\omega t - k\xi) - \varepsilon k x F \cos(\omega t - k\xi)\}$$

$$+ \tfrac{1}{4}\varepsilon^2 \beta c_0 \{2kx \sin(2\omega t - 2k\xi) + \cos(2\omega t - 2k\xi)\} + O(\varepsilon^3). \qquad (10.35)$$

The presence of terms that are uniformly $O(\varepsilon^2)$ is acceptable, so F is selected to cancel only the ε^2 term whose magnitude is proportional to x. The condition $-F \cos(\omega t - k\xi) + \tfrac{1}{2}\beta \sin(2\omega t - 2k\xi) = 0$ is readily solved for F with the aid of a trigonometric identity. The $O(\varepsilon^2)$ term remaining in Eq. (10.35) has that magnitude uniformly, so it is of little importance. Consequently, the expression for particle velocity reduces to

$$u = \varepsilon c_0 \sin(\omega t - k\xi). \qquad (10.36)$$

Using this expression to remove the sine function appearing in F yields

$$x = \xi + \varepsilon \beta x \sin(\omega t - k\xi) = \xi + \beta x (u/c_0). \qquad (10.37)$$

This solution is identical to the multiple-scale result, Eq. (10.31), in the case where the excitation function is a sine.

A subtlety that proves to be relevant for multidimensional waves is the omission of pressure from the renormalization analysis. This was permissible because the property $p = \rho_0 c_0 u$ encountered in linear theory is also true for the regular perturbation solution in Eqs. (10.14) and (10.15), provided one ignores the terms that are uniformly $O(\varepsilon^2)$. Thus, any change of variables that renormalizes u also has the same effect on p. (For the same reason, any uniformly valid expression for p also yields an expression for the excess density ρ' having that property.)

In general, the particle velocity components and pressure obtained in the regular perturbation solution are not algebraically related. The coordinate straining transformation must be sufficiently general to provide the flexibility to renormalize any state variable that is not algebraically related to the others. This means that one might have to consider straining transformations of each position variable and, in some situations, time.

The ability to accommodate generalizations such as the foregoing is a direct result of the relative simplicity of the method of renormalization in comparison to the method of multiple scales. The fact that the procedure eliminates secular terms, and thereby yields a uniformly valid solution, without requiring that differential equations be solved has made it the method of choice for analyses of multidimensional problems. Balancing this is the fact that the method of multiple scales offers greater generality in the problems to which it may be applied. For example, it can accommodate the effects of dissipation in finite-amplitude plane waves by including an attenuation scale that matches the scale x_1 over which nonlinear distortion occurs (Nayfeh, 1981).

When the nonlinear wave equation, Eq. (10.4), is solved by a perturbation method, the solution is multivalued in regions where a shock exists. Most perturbation analyses have been limited to preshock conditions. However, it is possible to remove this restriction by supplementing the analysis with weak shock theory. Such an approach was employed by Coulouvrat (1991) to study finite-amplitude Gaussian sound beams. His analysis used the method of renormalization to solve the lossless form of the KZK equation [Eq. (3.65) with $\delta = 0$]. Although the signal in that situation is multidimensional, the Gaussian nature of the variation transverse to the propagation direction leads to a quasi-planar problem. In contrast, the remainder of the present chapter is devoted to the extension of perturbation techniques to treat phenomena that are inherently multidimensional.

10.5   Radiation from a Vibrating Plate

The mathematical understanding gained from the early perturbation solutions was crucial to addressing more complicated situations. The first multidimensional problem to be considered was two-dimensional waves radiating into an infinite half-space above a vibrating plate that is periodically supported at intervals L (Ginsberg, 1978a, 1978b; Nayfeh, 1981). This problem served initially as a framework for extending the analytical methods, but further investigation led to several generalizations, as well as new insights into the underlying physics of nonlinear acoustic waves.

The plate vibration was taken to be harmonic at angular frequency ω, and the periodic supports imposed a sinusoidal spatial variation in the transverse plate displacement w. For a coordinate origin placed at one of the supports, the displacement is described by

$$w = -\varepsilon(c_0/\omega) \cos \omega t \sin\left(\frac{n\pi x}{L}\right), \tag{10.38}$$

where n is a positive integer and ε is the acoustic Mach number for the normal velocity of the plate.

Continuity of velocity between the inviscid fluid and the plate leads to a requirement that the fluid's velocity component normal to the plate's surface at the deformed position of the plate match the plate's velocity component in that direction. The corresponding boundary condition for the velocity potential is

$$\nabla\phi \cdot \mathbf{e}_n \Big|_{z=w} = \frac{\partial w}{\partial t}, \tag{10.39}$$

where $\mathbf{e}_n$ is the unit normal to the deformed surface; the angle of $\mathbf{e}_n$ relative to the z axis is $-\arctan(\partial w/\partial x)$. The spatial periodicity requirement parallel to the plate replaces boundary conditions in the x direction, while the Sommerfeld radiation condition supplies the other boundary condition in the z direction.

A solution by the method of renormalization begins by obtaining the straightforward perturbation solution, in which the potential is represented by Eq. (10.5).

The first- and second-order differential equations are given by Eqs. (10.6) and (10.7) with the gradients expressed in terms of x, z coordinates. The nonlinear boundary condition in Eq. (10.39) may be simplified by using the fact that w is $O(\varepsilon)$ to replace it by a condition at $z = 0$. This is achieved by invoking the Taylor series $f(w) = f(0) + wf'(0) + O(\varepsilon^2)$. After substitution of the specified displacement w, the first-order boundary condition becomes

$$\frac{\partial \phi_1}{\partial z}\bigg|_{z=0} = \frac{1}{\varepsilon}\frac{\partial w}{\partial t} = c_0 \sin \omega t \sin\left(\frac{n\pi x}{L}\right). \tag{10.40}$$

The corresponding first-order solution is identical to the result of linear theory. Waves propagate away from the plate for $\omega L/c_0 > n\pi$, in which case

$$\phi_1 = \frac{c_0}{k_z} \cos(\omega t - k_z z) \sin\left(\frac{n\pi x}{L}\right), \tag{10.41}$$

$$k_z = (k^2 - n^2\pi^2/L^2)^{1/2}, \quad k = \omega/c_0. \tag{10.42}$$

When Eq. (10.41) is used to represent the inhomogeneous terms in Eq. (10.7), the differential equation for ϕ_2 that results contains squares of $\sin(n\pi x/L)$ and $\cos(n\pi x/L)$. Only the secular terms in ϕ_2 need be identified in the renormalization method. Using the identity for the square of a sine or cosine reveals that the right-hand side of Eq. (10.7) contains second harmonics of ϕ_1. These represent secular effects because they propagate in the same manner as a free wave. All excitation terms not leading to a secular solution for ϕ_2 are denoted as NST and not considered further. The simplified form of the differential equation for ϕ_2 is

$$c_0^2\left(\frac{\partial^2\phi_2}{\partial x^2} + \frac{\partial^2\phi_2}{\partial z^2}\right) - \frac{\partial^2\phi_2}{\partial t^2}$$

$$= -\frac{\beta\omega^3}{2k_z^2} \sin(2\omega t - 2k_z z) \cos\left(\frac{2n\pi x}{L}\right) + \text{NST}. \tag{10.43}$$

An additional shortcut follows as a consequence of ignoring nonsecular terms. The role of the homogeneous portion of the ϕ_2 solution is to satisfy the second-order boundary condition, which is the $O(\varepsilon^2)$ term in the Taylor series expansion of Eq. (10.39). Because such a solution represents a freely propagating wave arising from excitation on a boundary, and therefore cannot exhibit cumulative growth, there is no need to determine it. The particular solution for ϕ_2 may be determined by the variation of parameters approach used earlier. Because of the periodic nature of the dependence on x and t imposed by the plate, the amplitude of the trial function can depend only on the distance z from the boundary, so a suitable trial solution is

$$\phi_2 = A(z) \cos(2\omega t - 2k_z z) \cos\left(\frac{2n\pi x}{L}\right) + \text{NST}. \tag{10.44}$$

Selecting A such that the foregoing forms a particular solution of Eq. (10.43) yields

$$A = -c_0 Bz, \quad B = \frac{\beta}{8}\left(\frac{k}{k_z}\right)^3. \tag{10.45}$$

Note that a constant could be added to the above expression for A. It has been set to zero here because the above term grows with increasing distance from the plate, while the additive constant would not lead to such growth.

Because of the two-dimensional nature of the problem, it is necessary to evaluate the x and z particle velocity components and the pressure corresponding to Eqs. (10.41) and (10.44). The gradient components of $\varepsilon\phi_1 + \varepsilon^2\phi_2$ are

$$u_x = \varepsilon c_0 \frac{n\pi}{k_z L}\left[\cos(\omega t - k_z z)\cos\left(\frac{n\pi x}{L}\right)\right.$$

$$\left. + \varepsilon \frac{\beta}{4}\frac{k^3}{k_z^2} z \cos(2\omega t - 2k_z z)\sin\left(\frac{2n\pi x}{L}\right)\right] + \text{NST}, \tag{10.46}$$

$$u_z = \varepsilon c_0 \left[\sin(\omega t - k_z z)\sin\left(\frac{n\pi x}{L}\right)\right.$$

$$\left. - \varepsilon \frac{\beta}{4}\frac{k^3}{k_z^2} z \sin(2\omega t - 2k_z z)\cos\left(\frac{2n\pi x}{L}\right)\right] + \text{NST}. \tag{10.47}$$

The evaluation of p according to Eq. (10.8) leads to another shortcut associated with the omission of nonsecular effects. The quadratic terms appearing in Eq. (10.8) are a consequence of solving the pressure-density relation up to second order. Because quadratic terms are estimated by ϕ_1, which is bounded, these terms cannot be associated with secular effects. As a consequence, it is permissible to use $p = -\rho_0 \partial\phi/\partial t$, which leads to

$$p = (\rho_0 \omega / k_z) u_z. \tag{10.48}$$

Because the pressure is proportional to u_z, only the velocity components given by Eqs. (10.46) and (10.47) need to be considered. Each contains a second harmonic that grows relative to the first-order (linear) approximation, but the functional dependence of these secular terms is quite different. The corresponding coordinate straining transformation allows for alteration of both spatial coordinates from the physical variables (x, z) to a strained set (η, ξ). The growth factor for the secular term is z, so the coordinate transformation is

$$x = \eta + \varepsilon z F_x(\eta, \xi, t), \quad z = \xi + \varepsilon z F_z(\eta, \xi, t), \tag{10.49}$$

where F_x and F_z are to be determined. Substituting these expressions into the phase terms of the nonuniformly valid expressions for the velocity components, and expanding the result into Taylor series in ε, yields

$$u_x = \varepsilon c_0 \frac{n\pi}{k_z L}\left[\cos(\omega t - k_z\xi)\cos\left(\frac{n\pi\eta}{L}\right)\right.$$

$$- \varepsilon \frac{n\pi}{L} z F_x \cos(\omega t - k_z\xi)\sin\left(\frac{n\pi\eta}{L}\right)$$

$$+ \varepsilon k_z z F_z \sin(\omega t - k_z\xi)\cos\left(\frac{n\pi\eta}{L}\right)$$

$$\left.+ \varepsilon \frac{\beta}{4}\frac{k^3}{k_z^2} z \cos(2\omega t - 2k_z\xi)\sin\left(\frac{2n\pi\eta}{L}\right)\right] + \text{NST}, \tag{10.50}$$

$$u_z = \varepsilon c_0 \left[\sin(\omega t - k_z\xi)\sin\left(\frac{n\pi\eta}{L}\right)\right.$$

$$+ \varepsilon \frac{n\pi}{L} z F_x \sin(\omega t - k_z\xi)\cos\left(\frac{n\pi\eta}{L}\right)$$

$$- \varepsilon k_z z F_z \cos(\omega t - k_z\xi)\sin\left(\frac{n\pi\eta}{L}\right)$$

$$\left.- \varepsilon \frac{\beta}{4}\frac{k^3}{k_z^2} z \sin(2\omega t - 2k_z\xi)\cos\left(\frac{2n\pi\eta}{L}\right)\right] + \text{NST}. \tag{10.51}$$

The task of identifying the functions F_x and F_z that result in cancellation of all secular terms is achieved with the aid of trigonometric identities. After cancellation of the secular terms, the terms remaining in the particle velocity components are

$$u_x = \varepsilon c_0 \frac{n\pi}{k_z L}\cos(\omega t - k_z\xi)\cos\left(\frac{n\pi\eta}{L}\right),$$

$$u_z = (k_z/\rho_0\omega)p = \varepsilon c_0 \sin(\omega t - k_z\xi)\sin\left(\frac{n\pi\eta}{L}\right). \tag{10.52}$$

The coordinate transformation resulting from substituting the functions F_x and F_z back into Eqs. (10.49) is

$$x = \eta + \tfrac{1}{2}\varepsilon\beta\frac{k^3 L}{n\pi k_z^2} z \cos(\omega t - k_z\xi)\cos\left(\frac{n\pi\eta}{L}\right),$$

$$z = \xi + \tfrac{1}{2}\varepsilon\beta\frac{k^3}{k_z^3} z \sin(\omega t - k_z\xi)\sin\left(\frac{n\pi\eta}{L}\right). \tag{10.53}$$

A simpler set of expressions for the coordinate transformation is obtained by using the expressions for u_x and u_z to eliminate the trigonometric terms, which leads to

$$x = \eta + \tfrac{1}{2}\beta\frac{k^3 L^2}{n^2\pi^2 k_z} z \frac{u_x}{c_0}, \qquad z = \xi + \tfrac{1}{2}\beta\frac{k^3}{k_z^3} z \frac{u_z}{c_0}. \tag{10.54}$$

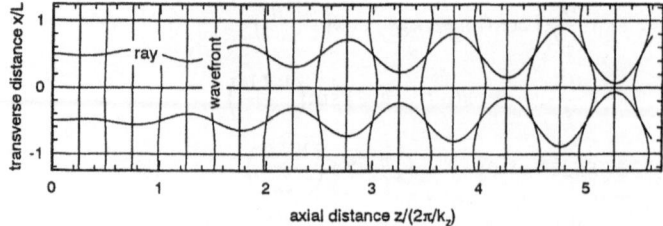

Fig. 10.1 Wave fronts and rays for radiation from a periodically supported plate: $\varepsilon = 0.005$, $\beta = 1.2, kL = 10, k_z L = 5, t = 0, n = 1$.

Fig. 10.2 Waveforms for radiation from a periodically supported plate: $\varepsilon = 0.005$, $\beta = 1.2, kL = 10, k_z L = 5$, $z/L = 1.76\pi, n = 1$.

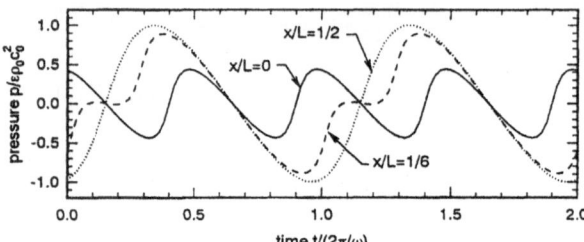

These expressions are particularly useful for understanding the effect of non-linearity. According to Eqs. (10.52), lines in (x, z) along which η is constant correspond to signals that departed from the boundary at the same time. These lines represent wave fronts of constant phase. Similarly, lines of constant ξ correspond to signals that departed from the same point on the plate, and therefore represent rays. A typical set of these lines is shown in Fig. 10.1.

In comparison to the linear approximation, in which $\eta = x$ and $\xi = z$, the effect of nonlinearity is to cause specific values of u_x and u_z associated with specific values of η and ξ to occur at a displaced position (x, z) obtained from Eqs. (10.54). When u_z is positive, the value of z is larger than ξ, which means that the signal has propagated farther than the signal predicted by linear theory. Similarly, signals corresponding to negative u_z propagate less far. This effect is like that for a plane wave, except that sinusoidal variation of u_z results in alternating advancement and retardation along the wave front specified by ξ.

The curvature of the rays (constant ξ) represents a phenomenon not encountered in plane waves. When u_x is positive, the value of x at which the signal is encountered is greater than the value $x = \eta$ at which the signal is predicted to occur by linear theory. Thus a ray is bent in the direction of the particle velocity transverse to it. This is the process of self-refraction. A corollary of this process is that the rays and wave fronts do not necessarily form an orthogonal mesh.

If one monitors the pressure at a fixed location as a function of time, the signal fluctuates as a result of the fluctuations in the rays due to the changing phase of u_z, as well as the changing wave fronts that arrive at that location. This can result in waveform distortion that differs substantially from that encountered in plane waves. Figure 10.2 shows typical waveforms at three locations equidistant from the plate.

Only the one for $\eta = L/2$, which is a straight ray ($x = \eta$) because $u_x = 0$ along it, displays the simple tendency to form a sawtooth shape having the fundamental period that is exhibited by plane waves. In contrast, the waveform on $x = L/2$ has the period of the second harmonic.

An important aspect of nonlinear acoustic waves is their tendency to form a shock. In general, shock formation in a solution that has been determined by the method of renormalization may be identified by seeking locations at which the mapping from the strained coordinates to the physical coordinates is not unique. In such cases, a set of values of the strained coordinates corresponds to two or more physical locations. The shock formation distance corresponds to the nearest location at which such a situation occurs, in which case the Jacobian of the transformation vanishes. Because the plate problem involves transformation from (η, ξ) to (x, z), the shock formation condition is attained when

$$\begin{vmatrix} \partial x/\partial\eta & \partial x/\partial\xi \\ \partial z/\partial\eta & \partial z/\partial\xi \end{vmatrix} = 0. \tag{10.55}$$

For the coordinate straining transformation in Eqs. (10.53), this leads to

$$\beta\frac{k^2}{2k_z^2}\sin\left[-\frac{n\pi\eta}{L} \pm (\omega t - k_z\xi)\right] = 1. \tag{10.56}$$

This relation can be used to determine the value of ξ corresponding to specified values of η and t at which the transformation is singular. These values may then be substituted into Eqs. (10.53) to determine the corresponding (x, z) position at which the condition occurs. Allowing η to vary with t held constant yields a locus of points. The locus of shock formation distances is the minimum value of z for a given x and all t at which the transformation is singular.

The locus of singular transformation points has an interesting property that was disclosed by Ginsberg (1978a). The slope of these loci may be obtained by considering the combination of Eqs. (10.53) and (10.56) to be a parametric description of x and z for fixed t, with the former set of relations represented as $x = f_1(\eta, \xi)$, $z = f_2(\eta, \xi)$, and solution of the latter yielding $\xi = F(\eta)$. Then the slope of the loci may be determined according to

$$\frac{dz}{dx}\bigg|_{\text{fixed } t} = \left(\frac{\partial f_2}{\partial\eta} + \frac{\partial f_2}{\partial\xi}\frac{\partial F}{\partial\eta}\right)\bigg/\left(\frac{\partial f_1}{\partial\eta} + \frac{\partial f_1}{\partial\xi}\frac{\partial F}{\partial\eta}\right). \tag{10.57}$$

Evaluation of the various derivatives leads to

$$\frac{dz}{dx}\bigg|_{\text{fixed } t} = \pm\frac{n\pi}{k_z L}. \tag{10.58}$$

In other words, the shock loci are straight lines. This feature is evident in the rays and wave fronts in Fig. 10.1. In the vicinity of $z = 0$, the wave fronts are approximately parallel to the plate and the rays are perpendicular to it. With increasing z, the curvature of both wave fronts and rays becomes more pronounced. When the shock condition is attained, a ray and a wave front become tangent, with the slope of that tangency being given by Eq. (10.58). This tangency is an alternative way of interpreting Eq. (10.55).

The observation that Eq. (10.58) is the shock condition motivates an algebraic decomposition of the acoustic field given by Eqs. (10.52) and (10.54) into a pair of plane waves propagating symmetrically with respect to $\mathbf{e}_z$, the z direction (Ginsberg, 1978b). As in the linear case, the trace velocity of each oblique plane wave tangential to the surface matches the phase speed of the respective oppositely directed waves forming the standing vibration pattern in Eq. (10.38). The propagation directions of the pair of plane waves are $\mathbf{n}_1$ and $\mathbf{n}_2$, at angle $\psi = \arctan(j\pi/k_z L)$ to the right and left, respectively, of the z axis. When $\mathbf{u}_j$ denotes the particle velocity associated with the plane wave propagating in direction $\mathbf{n}_j$ ($j = 1, 2$), Eqs. (10.52) and (10.54) may be rewritten as

$$\mathbf{u} = \mathbf{u}_1 + \mathbf{u}_2, \quad p = (\rho_0 \omega / k_z)\mathbf{u} \cdot \mathbf{e}_z, \tag{10.59}$$

$$\mathbf{u}_j = \frac{k}{2k_z} \varepsilon \mathbf{n}_j \cos(\omega t - k\xi_j). \tag{10.60}$$

The variable ξ_j is a straining of the propagation distance for wave j,

$$\mathbf{n}_j \cdot \mathbf{r} = \xi_j + \beta \frac{k}{k_z} z \frac{\mathbf{u}_j \cdot \mathbf{n}_j}{c_0}. \tag{10.61}$$

Nayfeh (1981) used the method of renormalization to prove the generality of Eq. (10.61) when the plate motion consists of a pair of arbitrary waves propagating in opposite directions along the plate. He also employed the method of multiple scales to extend the analysis to include dissipation, with the result that the individual plane waves were shown to satisfy the Burgers equation.

The similarity of Eq. (10.61) to the coordinate straining for a simple plane wave, Eq. (10.37), can lead to the erroneous interpretation that the individual plane waves propagate without interaction. The interaction phenomena here cause the growth factor in Eq. (10.61) to depend on the (normal) distance z from the field point to the plate. In contrast, if there were no interaction, the growth factor for each wave would depend on the (oblique) propagation distance $\mathbf{n}_j \cdot \mathbf{r} = z\cos\psi \pm x\sin\psi$ for that wave. Thus, although the wave fronts of each plane wave j are advanced or retarded relative to the corresponding small-signal location by an amount that depends on $\mathbf{u}_j$, the presence of the other wave causes a change in the distance on which the distortion depends.

The reason for the shock loci being straight is now apparent. The respective propagation directions $\mathbf{n}_j$ are perpendicular to the loci described by Eq. (10.58). The process of shock formation in the overall signal thus corresponds to situations in which either of the plane waves forms a shock.

An important observation follows from the foregoing analysis. It was possible to avoid considering the second-order boundary conditions because it was necessary only to identify those $O(\varepsilon^2)$ terms in the velocity potential that grow. However, bounded second-order terms can produce secular terms at the third order. Thus, if one were to continue to a higher level of approximation, all boundary conditions would need to be satisfied. Doing so in some cases is not possible, because the surface motion is not known to the necessary level of precision. (For example, this is the case for a projector whose face is fabricated from an array of piezoceramic elements.) To ignore these effects by linearizing the boundary condition, so that the only cubic terms retained are those appearing in the wave equation, Eq. (10.4), would lead to an inconsistent formulation.

10.6 Curvilinear Coordinates

Thus far in this chapter, nonlinear waves have been described in terms of Cartesian coordinates. The renormalization method has also been applied to cylindrically spreading waves (Ginsberg, 1978c, 1978d; Kelly and Nayfeh, 1981b) and spherically spreading waves (Kelly and Nayfeh, 1980) generated by a harmonic excitation whose amplitude varies spatially. As before, the analyses addressed problems whose linear analog can be solved by conventional separation of variables techniques. This section will treat situations in which the variation along a wavefront is described by a single angular wave function for the associated geometry. More general excitations are discussed in Sect. 10.7.

The primary innovation required to analyze nonlinear effects in curvilinear coordinates is the notion that these effects should be examined asymptotically for large propagation distances. Certain aspects of the overall procedure are apparent from the earlier developments. One forms a regular perturbation series for the velocity potential, whose first-order solution corresponds to linear theory. Depending on the type of wave, cylindrical or spherical Bessel functions describe the radial dependence. However, using such a solution to generate the nonlinear source distribution driving the second-order perturbation equations leads to a difficulty: One must generate a particular solution of the wave equation corresponding to products of the special functions appearing in the first-order solution.

The manner in which the difficulty is circumvented is a variant on the technique of matched asymptotic expansions. The procedure uses the simpler far-field representation to identify the secular terms in the perturbation series. The renormalization is also performed in the far field, but additional criteria are imposed to ensure that the solution behaves properly in the vicinity of the excitation. It is possible to consider the general procedure in outline form without tying it to a specific case.

Let (r, θ, χ) denote the curvilinear coordinates, with θ representing the azimuthal angle and r measuring the radial distance from the center of the exciting surface at $r = a$. For cylindrical waves, χ is the axial position while, for spherical waves, χ is the polar angle. The excitation consists of a harmonic particle velocity imposed on the surface defined by $r = a$. An excitation whose spatial dependence matches an angular wave function may be described by

$$u_r = \frac{\partial \phi}{\partial r}\bigg|_{r=a} = \tfrac{1}{2}\varepsilon c_0 f(\chi)\cos(n\theta)\exp(-i\omega t) + \text{c.c.}, \tag{10.62}$$

where c.c. will henceforth denote the complex conjugate of all terms preceding it. (This notation and the choice of a complex exponential to describe the time dependence are devices that expedite operations involving products of the first-order solution.) For the case of a vibrating infinite cylinder, $f(\chi)$ is a sinusoidal function, with $f(\chi) = 1$ for the two-dimensional case, while $f(\chi)$ is a Legendre function for the case of waves radiating from a vibrating sphere. It should be noted that the boundary condition, Eq. (10.62), has been linearized because only the growing part of the second-order solution is to be determined.

Let $h(\mu r)$ denote the radial part of the solution of the Helmholtz equation satisfying the Sommerfeld radiation condition at large distances. Thus, h is either the first cylindrical or the first spherical Hankel function of order n, and μ is the radial wave number. The corresponding first-order potential satisfying the boundary condition is

$$\phi_1 = \tfrac{1}{2}c_0 \frac{h(\mu r)}{h'(\mu a)} f(\chi)\cos(n\theta)\exp(-i\omega t) + \text{c.c.} \tag{10.63}$$

When this expression is used to evaluate the right-hand side of Eq. (10.7), terms containing products of $h(\mu r)$ are obtained. This presents a difficulty, because there are no known analytical solutions corresponding to such terms. However, if ε is sufficiently small, the wave will take on far-field properties before gradual growth of nonlinear distortion becomes significant. The far-field representation of either Hankel function is proportional to $r^{-\nu}\exp(i\mu r)$ ($\nu = \tfrac{1}{2}$ or 1). Because products of exponentials are exponentials, it is much simpler to develop the required particular solution using the far-field representation.

Let (^) denote the principal asymptotic form of a term for $r \to \infty$, and let $N(\phi)$ denote the nonlinear terms in Eq. (10.7). The next step in the perturbation analysis is to determine the far-field form $\hat{\phi}_2$ of the second-order potential. This entails obtaining a particular solution of

$$c_0^2 \nabla^2 \hat{\phi}_2 - \frac{\partial^2 \hat{\phi}_2}{\partial t^2} = N(\hat{\phi}_1). \tag{10.64}$$

Consistent with the far-field representation, it is permissible to simplify this step by ignoring terms that become unimportant at large r. For example, in cylindrical coordinates, the term $(1/r)\partial \phi_1/\partial r$ in $\nabla^2 \phi_1$ may be ignored.

Only the portion of the solution of Eq. (10.64) that represents cumulative growth needs to be determined. Because the first-order solution decays with increasing r, the secular terms in ϕ_2 are those that grow with respect to ϕ_1,

$$\lim_{r \to \infty} (|\hat{\phi}_2|/|\hat{\phi}_1|) = \infty. \tag{10.65}$$

The simplifications associated with a far-field analysis make it a straightforward matter to determine the portion of the particular solution of Eq. (10.64) that behaves according to Eq. (10.65). The corresponding far-field expressions for particle velocity and pressure, which are not uniformly valid, are determined from

$$\hat{\mathbf{u}} = \nabla(\varepsilon\hat{\phi}_1 + \varepsilon^2\hat{\phi}_2), \quad \hat{p} = -\rho_0\frac{\partial}{\partial t}(\varepsilon\hat{\phi}_1 + \varepsilon^2\hat{\phi}_2). \tag{10.66}$$

The last step in the far-field analysis is to use the renormalization procedure to remove the secular terms from the velocity components and pressure. Because the far-field pressure is in phase with the radial velocity, there is no need to treat p as a separate variable to be renormalized. Furthermore, several analyses have shown that secular terms can be removed without introducing a straining transformation for the azimuthal angle θ. (The physical reason for this will be discussed later.) Consequently, the coordinate straining is taken to be

$$r = \xi + \varepsilon g(r)F_r(\xi, \eta, t), \quad \chi = \eta + \varepsilon g(r)F_\chi(\xi, \eta, t), \tag{10.67}$$

where $g(r)$ represents the relative growth factor of the second-order terms, specifically, $|\hat{\phi}_2|/|\hat{\phi}_1|$. The process of selecting the straining functions to cancel all secular terms leads to the result that F_r and F_ξ are proportional to $\hat{u}_r$ and $\hat{u}_\chi$, respectively. (Except for the case of cylindrical three-dimensional waves, F_χ vanishes; the specific results will be listed below.)

At this juncture, the far-field representation of the wave has been determined. The last step is to use that result to obtain a description that is valid throughout the domain. This is done by imposing four criteria (Ginsberg, 1978a):

1. In the near field, where r is $O(1)$, the nonlinear solutions for pressure and particle velocity components for infinitesimal ε should match the corresponding linear solutions.
2. Asymptotic expansions for large r of the desired nonlinear solutions for pressure and particle velocity components should match the corresponding far-field solutions.
3. The growth factor in each coordinate straining transformation should match for large r the function $g(r)$ identified in the far-field analysis, but it should be modified such that each strained coordinate is identical to the corresponding physical coordinate at the surface from which the signal radiates.
4. The dependence of the coordinate straining transformation on a particle velocity component identified in the far-field analysis should also apply to the transformation covering the entire domain.

These criteria have not been proven mathematically. They are statements of two plausible physical arguments. The first, which is consistent with the notion of how nonlinear effects develop, is an assumption that these effects are minor in the vicinity of the vibrating surface. The second, which is somewhat more speculative, is an assumption that phenomena observed in the far field occur everywhere. Since coordinate strainings that depend on the velocity components represent self-refraction, the second assumption follows if self-refractive processes behave in the same manner everywhere.

After the secular terms are canceled, the far-field expressions for velocity and pressure are the same as the comparable forms of the linear solution, except that the strained coordinates replace the corresponding physical ones. Thus, the first and second criteria are satisfied when the strained coordinates replace the physical ones in the near-field form of the linear solution. The third and fourth criteria are addressed by using the coordinate straining form in which the dependence on the particle velocity components is explicit. If necessary, adjustment of the growth factor to make the strained and physical coordinates match at the boundary is achieved by shifting the origin for the dependence of the growth factor.

Ginsberg (1978b) considered the case where an infinite cylinder having radius a oscillates with an axial wavelength $2L$ and circumferential harmonic n. Two strained coordinates are needed to remove secular terms, ξ for the radial distance and η for the axial position. The pressure and particle velocity components were found to be

$$p = -\frac{1}{2}\varepsilon\rho_0 c_0 \frac{\omega H_n(\mu\xi)}{\mu H_n'(\mu a)} \cos(n\theta)\cos\left(\frac{\pi\eta}{L}\right)e^{-i\omega t} + \text{c.c.} + O(\varepsilon^2),$$

$$u_r = \frac{i}{2}\varepsilon c_0 \frac{H_n'(\mu\xi)}{H_n'(\mu a)} \cos(n\theta)\cos\left(\frac{\pi\eta}{L}\right)e^{-i\omega t} + \text{c.c.} + O(\varepsilon^2),$$

$$u_\theta = -\frac{i}{2}\varepsilon c_0 \frac{n H_n(\mu\xi)}{\mu r H_n'(\mu a)} \sin(n\theta)\cos\left(\frac{\pi\eta}{L}\right)e^{-i\omega t} + \text{c.c.} + O(\varepsilon^2),$$

$$u_x = -\frac{i}{2}\varepsilon c_0 \frac{\pi H_n(\mu\xi)}{\mu L H_n'(\mu a)} \sin(n\theta)\cos\left(\frac{\pi\eta}{L}\right)e^{-i\omega t} + \text{c.c} + O(\varepsilon^2).$$

$$(10.68)$$

In the foregoing, $H_n(\ )$ denotes the first Hankel function of order n, and a prime denotes differentiation with respect to the argument. The radial wave number is $\mu = (k^2 - \pi^2/L^2)^{1/2}$, and $k > \pi/L$ is assumed in order that the cylindrical waves be propagative rather than evanescent. The strained coordinates are given by

$$r = \xi + \beta(k/\mu)^3(r-a)(u_r/c_0),$$
$$x = \eta + \beta(k^3L^2/\pi^2\mu)(r-a)(u_x/c_0).$$
$$(10.69)$$

Ginsberg (1978a) performed a separate analysis of the special case where there is no axial variation in the oscillation of the cylinder. Most aspects of the results may be obtained from the foregoing by letting $L \to \infty$, $\mu \to k$, and $\cos(\pi\eta/L) = 1$, $\sin(\pi\eta/L) = 0$. The exception to this limit pertains to the insertion of the factor 2 preceding β in the transformation between r and ξ. This change will be explained below.

In addition to their studies of cylinders, Kelly and Nayfeh (1980) investigated spherical waves. Their analysis was complicated by the boundary motion, which required a series of spherical harmonics to represent the dependence on the polar angle ψ. For simplicity, the following results describe the case where the surface motion corresponds to only one spherical harmonic; the more general excitation is discussed in Sect. 10.7.

$$p = -\frac{1}{2}\varepsilon\rho_0 c_0^2 \frac{h_m(k\xi)}{h_m'(ka)} \cos(n\theta) P_m^n(\cos\psi) e^{-i\omega t} + \text{c.c.} + O(\varepsilon^2),$$

$$u_r = \frac{i}{2}\varepsilon c_0 \frac{h_m'(k\xi)}{h_m'(ka)} \cos(n\theta) P_m^n(\cos\psi) e^{-i\omega t} + \text{c.c.} + O(\varepsilon^2), \tag{10.70}$$

where $h_m(\ )$ denotes the spherical Hankel functions of the first kind and $P_m^n(\cos\psi)$ are associated Legendre functions. The corresponding expressions for u_ψ and u_θ, which are not presented for the sake of brevity, indicate that the magnitude of those velocity components decreases in the radial direction as $1/r$ relative to the magnitude of u_r. The coordinate straining transformation is

$$r = \xi + \beta r \ln(r/a)(u_r/c_0). \tag{10.71}$$

Much physical insight may be derived from an overview of these various results. Each wave propagates in the radial direction, with an amplitude that varies along the wave front. Unlike the situation for waves in a plate, it appears that in no case is the pressure proportional to the radial velocity. However, in the far field, where $kr \gg 1$ (in which case $k\xi \gg 1$ also), asymptotic expansions of the Hankel functions show that the pressure and radial particle velocity are in phase. This is consistent with the difference between near-field and far-field behavior of the corresponding linear solutions.

In the far field, the magnitude of the cylindrical Hankel functions is proportional to $1/\sqrt{\xi}$, while that of the spherical Hankel functions is proportional to $1/\xi$. Because ξ is, to a first approximation, the same as r, the foregoing represents the rate at which the radial velocity decays. Furthermore, with increasing radial distance, the particle velocity components in the angular directions (u_θ for cylindrical waves, and u_θ and u_ϕ for spherical waves) decay at a rate that is r times faster than the radial velocity. Because these velocity components become relatively weak, they do not play a role in self-refraction. Thus, in the far field, two-dimensional cylindrical waves and spherical waves have a quasi-one-dimensional behavior. Their particle velocity is approximately in the direction of propagation, and their distortion depends on the distance through which the wave has propagated.

The effect of spreading loss on the distortion process may be seen by considering the respective coordinate strainings. In the case of Eq. (10.71), the magnitude of ru_r in the far field is $O(\sqrt{r})$. Hence, the distortion effects, as measured by the difference between r and ξ, grow as $\sqrt{r}$ with increasing propagation distance. This is substantially more gradual than the effect for plane waves. For spherical waves in the far field, Eq. (10.71), the difference between r and ξ grows as $\ln r$. This growth factor is more gradual than for cylindrical waves because the wave spreads in two angular directions.

The difference between two-dimensional and three-dimensional curvilinear waves is comparable to the difference between plane waves and the two-dimensional waves generated by a vibrating plate. In each geometry there is a particle velocity component perpendicular to the propagation direction whose magnitude is comparable to the particle velocity in the direction of propagation. This leads to a pair of strained coordinates. The radial variable ξ describes the position of wave fronts of constant phase, while the variable η describes the rays of like phase emanating from a specified point on the boundary. The difference between η and the physical position x is proportional to the particle velocity in the x direction. Once again, this is the phenomenon of self-refraction. In both geometries, the absence of variation transverse to the propagation direction (that is, the limiting cases of plane waves and two-dimensional cylindrical waves) doubles the growth factor in the transformation for the propagation coordinate as a result of coalescence of effects.

10.7   Wave Groups

The small-signal analog of each of the waves considered thus far consists of a single wave that propagates without dispersion. This section describes the manner in which the preceding developments can be extended to handle more general excitations that generate several such waves. In multi-dimensional waves, the phase speed is dependent on the wavelength in the propagation direction, as well as on the frequency, so these waves disperse relative to one another. Perturbation techniques have yielded a general description of these waves based on collecting the various waves into groups having like phase velocities. The following discussion first addresses propagation in waveguides formed by parallel rigid walls, in which case the problem is formulated in terms of Cartesian coordinates (see also Sect. 5.3.1). Then the modifications of the analysis required to treat cylindrical and spherical waves propagating in open domains will be discussed.

Unless the cross section of a waveguide is rectangular, the various small-signal propagation modes have distinct phase speeds. Consequently, the effect of nonlinearity is confined to dependence of the phase speed on the acoustic Mach number (Keller and Millman, 1970; Nayfeh, 1975). This effect occurs at the third order, and therefore is quite weak. In contrast, in a hard-walled rectangular waveguide, linear theory indicates that numerous modes have the same phase speed. In nonlinear theory, the modes in such situations interact strongly. An initial investigation of propagation in a rectangular waveguide was performed by Vaidya and Wang (1977) using the method of multiple scales. However, the partial differential equations required to remove secular terms could be solved only numerically.

When one reconsiders the problem in Sect. 10.5 of nonlinear waves radiating from a vibrating plate, it is apparent that the rays that are straight correspond to parallel rigid walls, because the transverse velocity along both vanishes. This led Ginsberg (1979) to extend the renormalization method to the waveguide problem addressed previously by Vaidya and Wang (1977). The greater simplicity of the renormalization analysis yielded new insights into nonlinear modal interaction, which is the phenomenon to be considered in the present section.

In order to avoid unnecessary complications, only a two-dimensional waveguide formed by infinite rigid walls at $x = 0$ and $x = L$ shall be addressed. The excitation at $z = 0$ is an arbitrary periodic function of time having arbitrary spatial dependence on x. Linear theory predicts that the corresponding signal consists of an infinite sequence of propagative modes plus an evanescent part associated with frequencies that are below the cutoff for each transverse wavelength. The decay of the latter prevents buildup of the cumulative distortion process associated with nonlinearity, so only the propagating modes are retained. The primary new feature of the perturbation analysis is the manner in which these modes are rearranged into groups, according to their small-signal phase speed. Let $r_m > 0$ and $s_m \geq 0$ be prime numbers representing the lowest harmonic number and longest transverse wavelength, respectively, for the modes forming group m. (The plane-wave modes correspond to $m = 0$ and $s_m = 0$.) For an infinitesimal wave, the potential is a sum of the groups corresponding to the various prime integer pairs. The potential for each group is obtained by summing terms having the same harmonic number in time and both spatial directions,

$$(\phi_m)_{\text{lin}} = \frac{\varepsilon c_0^2}{2\omega} \sum_{n=1}^{\infty} A_{mn} \exp[inr_m\omega(t - z/c_m)] \cos(ns_m\pi x/L) + \text{c.c.}, \qquad (10.72)$$

where the coefficients A_{mn} are determined by satisfying the boundary condition at $z = 0$. The group's phase speed c_m satisfies the dispersion relation derived by requiring that Eq. (10.72) satisfy the linear wave equation, Eq. (10.6):

$$(c_0/c_m)^2 + (s_m\pi/r_m kL)^2 = 1, \quad k = \omega/c_0. \qquad (10.73)$$

The initial investigation by Ginsberg (1979) was restricted to relatively low frequencies, corresponding to $kL \ll \pi s_m/r_m$, so that the phase speed of each group m is distinct from the plane-wave group. In that case, each nonplanar group propagates without significant interaction with the plane-wave group. The high-frequency case is discussed at the end of this section. Only the highlights of the renormalization analysis shall be addressed here. The first-order potential matches the linear solution, Eq. (10.72). The sum of all propagative modes is used to evaluate the right side of the second-order perturbation equation, Eq. (10.7). Consideration of the corresponding particular solution reveals that all interactions of the harmonics forming a single group lead to secular terms that grow in proportion to z. In contrast, all interactions between harmonics in different groups represent source terms whose frequencies and wave numbers do not satisfy the dispersion relation for the linear wave equation, so they do not lead to secular terms. The second-order potential corresponding to the secular terms therefore consists of a sum over nondispersive groups that do not interact. The terms for each group correspond to sum- and difference-frequency generation by all modes in that group. The velocity components and pressure obtained by differentiating the potential exhibit similar behavior. As was true for the plate problem, both the transverse distance x and the axial distance z must be transformed in the coordinate straining. A different set of strained coordinates ξ_m and η_m applies to each group m. The straining transformation is quite similar to the result in Eqs. (10.53), except that all harmonics within the group are involved.

As is usual for a renormalization analysis, the expressions for the velocity components and pressure match the linear solution, except that for each group, x is replaced by η_m and z is replaced by ξ_m. The contribution of each group to the overall signal, to an accuracy of $O(\varepsilon^2)$, is

$$u_{xm} = \frac{\varepsilon \pi s_m c_0}{2kL} \sum_{n=1}^{\infty} n A_{mn} \exp[inr_m \omega(t - \xi_m/c_m)] \sin(n s_m \pi \eta_m/L) + \text{c.c.},$$

$$p_m = \rho_0 c_m u_{zm} \tag{10.74}$$

$$= \frac{1}{2i} \varepsilon \rho_0 c_0^2 r_m \sum_{n=1}^{\infty} n A_{mn} \exp[inr_m \omega(t - \xi_m/c_m)] \cos(n s_m \pi \eta_m/L) + \text{c.c.}$$

The functions and coefficients appearing in the coordinate straining are determined as part of the Taylor series expansion process. Ginsberg (1979) showed that the resulting terms are the same as in Eqs. (10.53), which describe the waves radiating from an infinite flat plate under harmonic excitation. In terms of the variables for the waveguide, the result is

$$x = \eta_m + \tfrac{1}{2}\beta z \left(\frac{r_m k L}{s_m \pi}\right)^2 \left(\frac{c_m u_{xm}}{c_0^2}\right), \qquad z = \xi_m + \tfrac{1}{2}\beta z \left(\frac{c_m^3 u_{zm}}{c_0^4}\right). \tag{10.75}$$

According to Eqs. (10.74), lines of constant ξ_m represent constant phase at a specific instant. Hence, Eqs. (10.75) indicate that the modal harmonics forming individual groups undergo waveform distortion that is determined by the total axial particle velocity in that group. This is comparable to the behavior of plane waves generated by an arbitrary periodic excitation. The multi-dimensional aspects are manifested by variation along the wave front, which is determined by the values of η_m. Since constant values of these quantities trace the phase back to the boundary, they represent rays. In the waveguide, the self-refraction of the rays for each group is determined by the transverse velocity associated with all modal harmonics in that group.

If one desires to evaluate waveforms, it is necessary to solve the coordinate straining transformation as a pair of coupled transcendental equations (Ginsberg, 1979). Once these equations have been solved, the corresponding contribution of that group to the signal may be evaluated. This process must be repeated for each group that is excited. Note that a shock, which corresponds to a vanishing value of the Jacobian of the straining transformation, may occur independently in any group.

High-frequency excitations were shown by Miao and Ginsberg (1986) to complicate the general theory. They studied the specific case where only the $(2, 0)$ mode in a hard-walled waveguide is excited, but the conclusion is general. The complication arises because each nondispersive group is essentially a sum of obliquely propagating plane waves, comparable to the representation of interacting plane waves in Eqs. (10.59)–(10.61). When the frequency is sufficiently high, the oblique directions may be very close to the propagation axis, in which case the corresponding axial phase velocities are very close to the velocity of the plane wave. This leads to coupling between the oblique and plane-wave groups, in which beating phenomena are encountered as a result of the slight mismatch in the axial phase speeds (see Fig. 5.5). Miao and Ginsberg (1986) did not provide an estimate of the maximum value of kL for which this type of interaction could be ignored, but Hamilton and TenCate (1988) found in their experiment that the effect was important for $kL = \sqrt{2}\pi s_m/r_m$, corresponding to a $45°$ propagation angle for the oblique plane waves forming the non-planar mode.

Analyses collecting effects into nondispersive groups have been performed for cylindrical and spherical waves (Kelly and Nayfeh, 1981a, 1981b). Like the procedure developed in Sect. 10.6, treatment of the curvilinear aspects entails study of the far-field behavior, after which the solution is projected back to the near field. Certain general conclusions are observed. A different set of strained coordinates is associated with each group. The wave fronts for each group are advanced or retarded in proportion to the group's total contribution to the radial particle velocity. Self-refraction of the rays for a group, in which the ray is deflected in the direction of the transverse velocity component of that group, is associated only with the axial velocity component in cylindrical waves. Neither the azimuthal velocity in cylindrical coordinates nor the azimuthal and polar velocity components in spherical coordinates lead to self-refraction, because spreading loss makes their influence negligible in the far field.

The phenomena discussed in this chapter arose as a result of boundary excitations whose spatial distribution could be represented by a series. Such a representation is not possible for many important problems, notably sound beams. For that reason, the renormalization method has been extended to treat signals whose linearized version is represented by integral transforms (Ginsberg, 1981, 1984; Coulouvrat, 1991; Frøysa and Coulouvrat, 1996), but the analyses are too complicated to discuss here.

References

Airy, G. B. (1849). The astronomer royal on a difficulty in the problem of sound. *Phil. Mag.* ser. 3, **34**, 401–405.
Coulouvrat, F. Y. (1991). An analytical approximation of strong nonlinear effects in bounded beams. *J. Acoust. Soc. Am.* **90**, 1592–1600.

Frøysa, K.-E., and Coulouvrat, F. (1996). A renormalization method for nonlinear pulsed sound beams. *J. Acoust. Soc. Am.* **99**, 3319–3328.

Fubini-Ghiron, E. (1935). Anomalies in the propagation of an acoustic wave of large amplitude. *Alta Frequenza* **4**, 530–581.

Ginsberg, J. H. (1975a). Multi-dimensional non-linear acoustic wave propagation. Part I: An alternative to the method of characteristics. *J. Sound Vib.* **40**, 351–358.

Ginsberg, J. H. (1975b). Multi-dimensional non-linear acoustic wave propagation. Part II: The non-linear interaction of an acoustic fluid and plate under harmonic excitation. *J. Sound Vib.* **40**, 359–379.

Ginsberg, J. H. (1978a). A re-examination of the non-linear interaction between an acoustic fluid and a flat plate undergoing harmonic excitation. *J. Sound Vib.* **60**, 449–458.

Ginsberg, J. H. (1978b). A new viewpoint for the two-dimensional non-linear acoustic wave radiating from a harmonically vibrating flat plate. *J. Sound Vib.* **63**, 151–154.

Ginsberg, J. H. (1978c). Propagation of nonlinear acoustic waves induced by a vibrating cylinder. Part I: The two-dimensional case. *J. Acoust. Soc. Am.* **64**, 1671–1678.

Ginsberg, J. H. (1978d). Propagation of nonlinear acoustic waves induced by a vibrating cylinder. Part II: The three-dimensional case. *J. Acoust. Soc. Am.* **64**, 1679–1687.

Ginsberg, J. H. (1979). Finite amplitude two-dimensional waves in a rectangular duct induced by arbitrary periodic excitation. *J. Acoust. Soc. Am.* **65**, 1127–1133.

Ginsberg, J. H. (1981). A uniformly accurate description of finite amplitude sound radiation from a harmonically vibrating planar boundary. *J. Acoust. Soc. Am.* **69**, 929–936.

Ginsberg, J. H. (1984). Nonlinear King integral for arbitrary axisymmetric sound beams at finite amplitudes. II: Derivation of uniformly accurate expression. *J. Acoust. Soc. Am.* **76**, 1208–1214.

Goldstein, S. (1960). *Lectures in Fluid Mechanics* (Wiley-Interscience, New York), Chapter 4.

Hamilton, M. F., and TenCate, J. A. (1988). Finite amplitude sound near cutoff in higher-order modes of a rectangular duct. *J. Acoust. Soc. Am.* **84**, 327–334.

Keller, J. B., and Millman, M. H. (1970). Finite-amplitude sound-wave propagation in a waveguide. *J. Acoust. Soc. Am.* **49**, 329–333.

Kelly, S. G., and Nayfeh, A. H. (1980). Non-linear propagation of directional spherical waves. *J. Sound Vib.* **72**, 25–37.

Kelly, S. G., and Nayfeh, A. H. (1981a). Non-linear propagation of general directional spherical waves. *J. Sound Vib.* **79**, 145–156.

Kelly, S. G., and Nayfeh, A. H. (1981b). Non-linear propagation of directional cylindrical waves. *J. Sound Vib.* **79**, 415–428.

Lamb, H. (1945). *Hydrodynamics*, 6th ed. (Dover, New York), pp. 483–484.

Lighthill, M. J. (1949). A technique for rendering approximate solutions to physical problems uniformly valid. *Phil. Mag.* **40**, 1179–1201.

Miao, H. C., and Ginsberg, J. H. (1986). Finite amplitude distortion and dispersion of a nonplanar mode in a waveguide. *J. Acoust. Soc. Am.* **80**, 911–920.

Nayfeh, A. H. (1973). *Perturbation Methods* (Wiley-Interscience, New York).

Nayfeh, A. H. (1975). Nonlinear propagation of a wave packet in a hard-walled circular duct. *J. Acoust. Soc. Am.* **57**, 803–809.

Nayfeh, A. H. (1981). Non-linear propagation of waves induced by general vibrations of plates. *J. Sound Vib.* **79**, 429–437.

Nayfeh, A. H., and Kluwick, A. (1976). A comparison of three perturbation methods for non-linear hyperbolic waves. *J. Sound Vib.* **48**, 293–299.

Poisson, S. D. (1808). Mémoire sur la théorie du son. *J. l'école polytech.* (Paris) **7**, 319–392.

Pritulo, M. F. (1962). On the determination of uniformly accurate solutions of differential equations by the method of perturbation of characteristics. *J. Appl. Math. Mech.* **26**, 661–667.

Vaidya, P. G., and Wang, K. S. (1977). Non-linear propagation of complex sound fields in rectangular ducts. Part I: The self-excitation phenomenon. *J. Sound Vib.* **50**, 29–42.

Chapter 11
Computational Methods

Jerry H. Ginsberg and Mark F. Hamilton

Contents

11.1 Introduction

The need to account simultaneously for the combined effects of nonlinearity in the continuity, momentum, and state equations, along with absorption originating from viscosity, heat conduction, and relaxation, creates an imposing analytical task for one-dimensional waves. With the advent of digital computers, it was natural for some researchers to pursue numerical solutions. This early work was founded on the recognition that the wide range of spatial and temporal scales involved in the various processes would cause straightforward numerical approximations of the basic equations to exceed the capacity of the available computers. As a result, the simulations relied on blending numerical techniques with physical approximations comparable to those used in parallel analytic endeavors.

J. H. Ginsberg
G. W. Woodruff School of Mechanical Engineering, Georgia Institute of Technology, Atlanta, GA, USA

M. F. Hamilton
Department of Mechanical Engineering, The University of Texas at Austin, Austin, TX, USA

© The Author(s) 2024
M. F. Hamilton, D. T. Blackstock (eds.), *Nonlinear Acoustics*,
https://doi.org/10.1007/978-3-031-58963-8_11

As understanding of the basic phenomena increased, interest turned toward multidimensional propagation problems involving sound beams and waveguides. The effects of diffraction and mode interactions in waveguides substantially complicate the methods required for numerical solutions. Indeed, even the linearized problem might not have an analytical solution. The general philosophy, as well as the specific techniques, of the numerical work in this area built on the developments for plane waves by selecting model equations and numerical techniques appropriate to the physical situation. Recent work has taken a different approach by bringing to nonlinear acoustics some of the general computational fluid dynamics techniques now in use to solve the full Navier–Stokes equations.

An overview of the computational investigations reveals that they have employed one of three philosophical approaches. Frequency-domain formulations rely on the fact that the periodicity of a steady-state signal radiated by a periodic source must remain the same. Consequently, the time-dependent part of the signal may be represented by a Fourier series whose position-dependent coefficients are required to satisfy a field equation appropriate to the particular problem. Direct time-domain simulations, which are suitable for signals having an arbitrary waveform, employ finite-difference techniques to approximate all aspects of the governing equations. Phenomenological approaches regard the various physical processes as being independent, and therefore superposable, when signals propagate over small distances. These approaches lead to hybrid algorithms in which calculations are performed in both the time and frequency domains.

In view of these three philosophical approaches, we have organized the present chapter as follows: Sect. 11.2 describes three algorithms for modeling one-dimensional (plane, cylindrical, and spherical) wave propagation in real fluids. The first is a frequency-domain approach, the second a time-domain approach, and the third a hybrid phenomenological approach. Each is described in sufficient detail that, with the assistance of a standard text on numerical methods (e.g., Ames, 1992), the reader should be adequately equipped to write simple, computationally efficient programs that can model a wide variety of real propagation problems in nonlinear acoustics. In Sect. 11.3, the topic of multidimensional wave fields is covered, with emphasis primarily on directional sound beams and, to a lesser extent, on guided waves. The algorithms in Sect. 11.3 employ the techniques introduced in Sect. 11.2 for describing nonlinearity and absorption. We therefore concentrate more on the techniques used to account for diffraction. Finally, in Sect. 11.4, a very general but computationally intensive approach to modeling nonlinear sound waves is briefly described.

11.2 One-Dimensional Waves

Although techniques employed by many researchers have contributed to the forms of the algorithms presented below, for historical perspective we call attention here only to the benchmark method developed by Cook (1962) to model progressive finite-amplitude plane waves in thermoviscous fluids. His was a phenomenological model based on the following assumptions: "(1) The distortion mechanism can be described by a change in phase velocity directly proportional to the particle velocity. (2) The absorption mechanism can be described by assuming that the rate of the absorption of each harmonic is proportional to the amount of the harmonic present and to the square of the frequency of the harmonic." This same approach, involving calculations in both the time and frequency domains, resurfaces in algorithms developed by others. Moreover, results from Cook's algorithm were often used as the basis of comparison for numerical procedures developed later.

Cook separates the effects of nonlinearity and absorption and superposes these effects over the same incremental propagation step in space. First, the waveform is distorted in the time domain according to the lossless theory described by Eq. (4.2) and illustrated in Fig. 4.1. Second, the Fourier integral of this implicit solution is evaluated numerically to obtain the harmonic amplitudes in the waveform. Third, the harmonic amplitudes are corrected for thermoviscous absorption over the same incremental step. Finally, the distorted and attenuated waveform is reconstructed by summing up the Fourier series, and the process is repeated over the next step in space. The algorithm described by Cook corresponds to a digital computer implementation of the graphical procedure used earlier by Fox and Wallace (1954). Cook retained up to 16 harmonics and used step sizes of up to 5% of the shock formation distance in his calculations.

The present development describes three simple, yet efficient, algorithms (frequency domain, time domain, and hybrid time-frequency domain) for modeling the propagation of one-dimensional progressive waves in homogeneous, thermoviscous fluids. These three algorithms incorporate basic approaches that are found in many methods currently encountered in the literature. For ease of comparison, we shall formulate each as a solution of the Burgers equation, Eq. (3.54):

$$\frac{\partial p}{\partial x} = \frac{\beta p}{\rho_0 c_0^3} \frac{\partial p}{\partial \tau} + \frac{\delta}{2c_0^3} \frac{\partial^2 p}{\partial \tau^2}, \tag{11.1}$$

where p is the sound pressure, $\tau = t - x/c_0$ is a retarded time associated with propagation in the $+x$ direction at the small-signal sound speed c_0, ρ_0 is the ambient density, β is the coefficient of nonlinearity, and δ is the sound diffusivity for a thermoviscous fluid. In practice, the algorithm described by Cook (1962) also corresponds to solving Eq. (11.1).[1] Consideration of spherical and cylindrical spreading, and inhomogeneity of the medium, is postponed to Sect. 11.2.4.

---

[1] In fact, Cook (1962) introduced a minor error by including the second term in his Eq. (11). This term was intended to provide an additional correction for finite-amplitude losses. However, only without this term does his algorithm reduce to the Burgers equation in the limit of zero step size. Fortuitously, the additional term is of negligible significance in the numerical calculations (Van Buren and Breazeale, 1968), and for practical purposes Cook's algorithm approximates the solution of the Burgers equation.

Algorithms for implementing numerical solutions are most appropriately expressed in dimensionless variables. We therefore let p_0 characterize the sound pressure amplitude and ω_0 the angular frequency of the signal at a source located at $x = 0$. For a pulse, ω_0 may correspond to an effective repetition frequency or to the dominant spectral component in the source waveform. We thus introduce the following dimensionless notation:

$$P = p/p_0, \quad \sigma = x/\bar{x}, \quad \theta = \omega_0\tau, \quad A = \alpha_0\bar{x}, \tag{11.2}$$

where $\bar{x} = \rho_0 c_0^3/\beta p_0 \omega_0$ is the lossless plane-wave shock formation distance for a signal with source condition $p = p_0 \sin \omega_0 t$, and $\alpha_0 = \delta\omega_0^2/2c_0^3$ is the thermoviscous attenuation coefficient at frequency ω_0. In terms of this notation, Eq. (11.1) becomes

$$\frac{\partial P}{\partial \sigma} = P\frac{\partial P}{\partial \theta} + A\frac{\partial^2 P}{\partial \theta^2}. \tag{11.3}$$

Equation (11.3) is the model equation considered in Sects. 11.2.1–11.2.3. With $P = \sin \theta$ at $\sigma = 0$, the shock formation distance in a lossless fluid ($A = 0$) is $\sigma = 1$, and determination of the incremental step size $\Delta\sigma$ in this case is independent of the physical source pressure p_0. The only adjustable coefficient is the absorption parameter A, which is the ratio of a shock formation distance to an absorption length (the Gol'dberg number $\Gamma = 1/A$ is often used for this ratio). For $A \gtrsim 1$ (see Sect. 4.5.4.1), nonlinear effects may be ignored in most situations of interest (self-demodulation of a pulse being one notable exception; see Sect. 8.3.5).

11.2.1 Frequency-Domain Algorithm

If attention is restricted to temporally periodic disturbances, the time dependence in Eq. (11.3) may be removed from explicit consideration by employment of a Fourier series representation. A number of authors have contributed to the development of frequency-domain algorithms. Fenlon (1971) was the first to derive coupled spectral equations equivalent to those given below, which he solved numerically for sound radiated from both monofrequency and bifrequency sources (including spherical and cylindrical spreading as described in Sect. 11.2.4). His article also contains an extensive review of relevant numerical and analytical solutions that appeared earlier. Subsequently, Korpel (1980) presented coupled spectral equations that account for arbitrary absorption and dispersion in progressive plane waves, the effects of which we include below.

Consider a trial solution of the form

$$P(\sigma, \theta) = \frac{1}{2} \sum_{n=-\infty}^{\infty} P_n(\sigma)e^{jn\theta}, \tag{11.4}$$

where $P_{-n}(\sigma) = P_n^*(\sigma)$ (P_n^* denotes the complex conjugate of P_n) because $P(\sigma, \theta)$ is a real function. Substitution of Eq. (11.4) into Eq. (11.3) yields, with linear terms on the left, nonlinear terms on the right, and $P(\partial P/\partial\theta) = \frac{1}{2}(\partial P^2/\partial\theta)$,

$$\sum_{n=-\infty}^{\infty} \left(\frac{dP_n}{d\sigma} + n^2 A P_n \right) e^{jn\theta} = \frac{1}{4} \frac{\partial}{\partial\theta} \sum_{l=-\infty}^{\infty} \sum_{m=-\infty}^{\infty} P_l P_m e^{j(l+m)\theta} \qquad (11.5)$$

$$= \sum_{n'=-\infty}^{\infty} \left(j\frac{n'}{4} \sum_{m=-\infty}^{\infty} P_m P_{n'-m} \right) e^{jn'\theta}. \qquad (11.6)$$

Equation (11.6) was obtained by letting $n' = l + m$ and taking the derivative with respect to θ. The left and right sides are equal for arbitrary θ only with $n' = n$ and with the terms in the parentheses equal to each other:

$$\frac{dP_n}{d\sigma} + n^2 A P_n = j\frac{n}{4} \sum_{m=-\infty}^{\infty} P_m P_{n-m}. \qquad (11.7)$$

Taken together, Eqs. (11.4) and (11.7) are an exact reformulation of Eq. (11.3) for periodic disturbances.

To perform the convolution on the right-hand side of Eq. (11.7), it is convenient to rewrite the summation as

$$\sum_{m=-\infty}^{\infty} P_m P_{n-m} = \sum_{m=1}^{n-1} P_m P_{n-m} + 2 \sum_{m=n+1}^{\infty} P_m P_{m-n}^*, \qquad (11.8)$$

where the relation $P_{-n} = P_n^*$ was taken into account.[2] In this way, the summations include only quantities P_n for which n is positive. Moreover, one may now associate the products $P_m P_{n-m}$ in the first summation with sum-frequency generation, and the products $P_m P_{m-n}^*$ in the second summation with difference-frequency generation. The main benefit of using Eq. (11.8) is associated with the required truncation of the summation in Eq. (11.4) to a finite number of harmonics for the purpose of numerical calculations. Specifically, let M be the number of harmonics retained in the computations and write

$$P(\sigma, \theta) = \frac{1}{2} \sum_{n=-M}^{M} P_n(\sigma) e^{jn\theta} = \frac{1}{2} \sum_{n=1}^{M} P_n(\sigma) e^{jn\theta} + \text{c.c.} \qquad (11.9)$$

---

[2] From Eq. (11.7), note that $P_0(\sigma) = P_0(0)$, and therefore, dc pressure generation ($n = 0$) is not described by the Burgers equation. We therefore take $P_0 = 0$.

in place of Eq. (11.4), where c.c. represents the complex conjugate of the preceding expression. Use of the finite Fourier series corresponds to replacing ∞ by M in the second summation on the right-hand side of Eq. (11.8), and Eq. (11.7) thus becomes

$$\frac{dP_n}{d\sigma} = -n^2 A P_n + j\frac{n}{4}\left(\sum_{m=1}^{n-1} P_m P_{n-m} + 2\sum_{m=n+1}^{M} P_m P_{m-n}^*\right). \tag{11.10}$$

Equation (11.10) is a coupled system of M ordinary differential equations in the M complex unknowns $P_1, P_2, \cdots, P_M$. Once obtained, the solutions P_n may be substituted into Eq. (11.9) to construct the waveform. Alternatively, with $P_n = a_n - jb_n$ (a_n and b_n both real), Eq. (11.9) becomes

$$P(\sigma, \theta) = \sum_{n=1}^{M}[a_n(\sigma)\cos n\theta + b_n(\sigma)\sin n\theta]. \tag{11.11}$$

The same substitution may be used to rewrite Eq. (11.10) directly in terms of a_n and b_n, which yields two sets of equations following separation of real and imaginary terms.

An efficient method for solving Eq. (11.10) is a standard Runge–Kutta routine that marches the solution forward over an incremental step $\Delta\sigma$. The initial values $P_n(0)$ for the source waveform $P(0, \theta)$ are given by

$$P_n(0) = \frac{1}{\pi}\int_{-\pi}^{\pi} P(0, \theta)e^{-jn\theta}\, d\theta. \tag{11.12}$$

For example, with $P(0, \theta) = \sin\theta$, Eq. (11.12) yields $P_1(0) = -j$, with $P_n(0) = 0$ for $n > 1$. When calculations are to be extended into a domain where shocks develop in the waveform [e.g., for $\sigma > 1$ with $P(0, \theta) = \sin\theta$], a finite value for the absorption coefficient A is required for stability. The smaller the value of A, the larger must be the number M of harmonics to reduce the effect of Gibbs oscillations in the neighborhoods of shocks. Owing to the spectral convolution that accounts for nonlinearity, the computation time increases as M^2. To accurately model the propagation of time waveforms with thin shocks, several hundred harmonics must be retained.

Many fewer harmonics are needed if only harmonic propagation curves are of interest. Trivett and Van Buren (1981) investigated errors associated with finite values of the step size $\Delta\sigma$ and harmonic number M on the spectral components calculated for small n. Their algorithm is based on coupled equations in terms of the coefficients a_n and b_n in Eq. (11.11). They report that for a sinusoidal source condition and with no absorption, numerical errors in the first through fifth harmonic amplitudes at $\sigma = 1$ are less than 1% for $\Delta\sigma = 0.05$ and $M = 15$. Much larger values of M are required as shocks form because of the increased flow of energy upward in the frequency spectrum, which results in excessive accumulation in the highest harmonics retained in the calculations ($n \sim M$). They suggested that a remedy for this would be to modify the algorithm to attenuate the highest several harmonics by limiting their amplitudes so that they do not exceed the amplitudes of the neighboring lower harmonics. However, their computations instead relied on truncation of the series at a sufficient number of terms to prevent the error from propagating down to the harmonics of interest.

Transient source conditions, namely pulses, can be modeled by choosing ω_0 to be a sufficiently low repetition frequency that adjacent pulses in the periodic sequence do not interact as a result of the potential increase in pulse duration due to nonlinearity and absorption. The penalty for performing calculations of transient waveforms in the frequency domain is the substantially larger number of harmonics required. For example, with a tone burst repeated at a rate equal to one-tenth its center frequency, the computations require approximately $10M$ harmonics if M are required to accurately describe cw radiation at the center frequency of the pulse. The computation time in this case increases by a factor of order 100.

Pishchal'nikov et al. (1996) proposed a novel method for substantially reducing the computation time required for evaluating the nonlinear distortion of waveforms containing shocks. The expansion in Eq. (11.4) is rewritten as two, one labeled $P^{(1)}$ for the summation over harmonics $1 \leq n \leq M$, as in Eq. (11.9), and the other labeled $P^{(2)}$ for the summation over harmonics $M + 1 \leq n \leq \infty$. The M low-frequency components $P_n^{(1)}$ are calculated numerically using Eq. (11.10) (but with all $1 \leq n \leq \infty$ harmonics retained in the summations), and the $M + 1 \leq n \leq \infty$ high-frequency components $P_n^{(2)}$ are determined analytically subject to an appropriate matching condition. It is assumed that $P^{(1)}$ adequately describes the continuous waveform segments between the shocks, and that $P^{(2)}$ accounts for the shocks themselves. On the basis of weak shock theory, the high-frequency components are expressed in the form $P_n^{(2)} = (P_s/n)e^{jn\theta_s}$ [see Eq. (4.179)], where P_s and θ_s are constants. This approximation of $P_n^{(2)}$ permits analytical expressions to be obtained for the summations over harmonics $M + 1 \leq n \leq \infty$ in both Eqs. (11.4) and (11.10). The values of P_s and θ_s are calculated with a matching condition that is a function of the highest two frequency components determined numerically, $P_M^{(1)}$ and $P_{M-1}^{(1)}$. Calculated waveforms presented by these authors reveal well-defined shocks with no Gibbs oscillations for $M \sim 10$ to 30, and they are in good agreement with the results obtained via straightforward numerical solution of Eq. (11.10), i.e., with no analytical approximations but with considerably larger values of M. The algorithm was demonstrated for waveforms containing a single shock (or one per cycle for sinusoidal sources), and the authors do not discuss how the algorithm should be modified to accommodate arbitrary numbers of shocks.

Equation (11.10) may be augmented in a straightforward way to account for arbitrary (not just thermoviscous) absorption and dispersion by making the substitution

$$n^2 A \rightarrow A_n + jD_n, \tag{11.13}$$

where A_n and D_n are the absorption and dispersion coefficients, respectively. Here $D_n = n\omega_0\bar{x}(c_n^{-1} - c_0^{-1})$ takes into account the deviation of the phase speed c_n of the nth harmonic component from the reference sound speed c_0. The effect of the dispersion terms in the coupled spectral equations is discussed in detail by Korpel (1980). Figure 5.3 provides an example of calculations based on Eq. (11.10), taking into account absorption and dispersion due to relaxation, in addition to classical thermoviscous absorption.

Finally, reflection from a surface at $\sigma = \sigma_r$ having arbitrary complex reflection coefficient R_n at frequency $n\omega_0$ is easily taken into account in the frequency domain. The main assumption is that the mutual nonlinear interaction of the incident and reflected waves is negligible in comparison with the nonlinear distortion of these waves independently of one another. Such an assumption is particularly reasonable for pulses, for which the overlap occurs only within distances on the order of one pulse length away from the reflector. Given the solutions $P_n^i(\sigma_r)$ for the incident wave at the location of the reflecting plane, one sets $P_n^r(\sigma_r) = R_n P_n^i(\sigma_r)$ to initialize the spectral amplitudes for the reflected wave, and the propagation algorithm continues as for the incident wave. An early application of this approach, based on an algorithm similar to Cook's (1962), was used by Van Buren and Breazeale (1968) to study reflection, primarily in relation to total reflection ($|R_n| = 1$) with phase shifts of 90° and 180°. For the case of a pressure release surface (180°), when reflection takes place in the preshock region and absorption is not taken into account, the waveform that is inverted upon reflection undistorts back to the original source waveform.

11.2.2   Time-Domain Algorithm

The time-domain algorithm presented here is described in detail by Cleveland et al. (1996b). Although our discussion is restricted to the solution of Eq. (11.3), these authors also included effects of relaxation, and inhomogeneity of the medium in the geometrical acoustics approximation. Their algorithm is a modification of the time-domain procedure developed previously by Lee and Hamilton (1995) for modeling pulsed sound beams in thermoviscous fluids.

Begin by separating Eq. (11.3) into the two equations

$$\frac{\partial P}{\partial \sigma} = P\frac{\partial P}{\partial \theta}, \tag{11.14}$$

$$\frac{\partial P}{\partial \sigma} = A\frac{\partial^2 P}{\partial \theta^2}. \tag{11.15}$$

Equations (11.14) and (11.15) are solved independently over each incremental step $\Delta\sigma$. Separation of Eq. (11.3) as in Eqs. (11.14) and (11.15) is referred to as operator splitting, and solving such equations independently to approximate the solution of the original equation is referred to as the method of fractional steps (Ames, 1992).

Time-domain solutions require discretization of the waveform. We shall assume that the waveform samples are separated initially by uniform time intervals $\Delta\theta$ and label each with index i, where $i = 1, 2, \cdots, M$. The samples at times θ_i^k are labeled P_i^k, where the index k designates the current value of σ, with $k+1$ corresponding to the solution following propagation over the incremental spatial step $\Delta\sigma$ out to the next location, $\sigma + \Delta\sigma$. The coordinate pairs (θ_i^k, P_i^k) and $(\theta_i^{k+1}, P_i^{k+1})$ thus define the discretized waveforms at steps k and $k + 1$, respectively.

Consider first Eq. (11.14), for which an analytic solution is available. The general solution is Eq. (4.29), which in the dimensionless notation used here becomes $P(\sigma, \theta) = f(\theta + \sigma P)$, where $f(\theta)$ is an arbitrary function of time prescribed at the source, $\sigma = 0$. To determine the waveform at step $\sigma + \Delta\sigma$ given a waveform $P(\sigma, \theta)$ at the previous step σ, the solution may be recast as

$$P(\sigma + \Delta\sigma, \theta) = P(\sigma, \theta + P\Delta\sigma). \tag{11.16}$$

The discretized solution of Eq. (11.14) on the basis of Eq. (11.16) is obtained by introducing the following time-base transformation:

$$\theta_i^{k+1} = \theta_i^k - P_i^k \Delta\sigma. \tag{11.17}$$

The waveform samples at the new times θ_i^{k+1} are labeled P_i^{k+1}. Note that the values of P_i^k and P_i^{k+1} are equivalent; these points are merely translated along the time axis. Thus, for $P_i^k > 0$, wavelets are shifted earlier in the retarded time frame, corresponding to the fact that their local propagation speeds are greater than c_0. For $P_i^k < 0$, the wavelets are delayed. A restriction must be imposed on the application of Eq. (11.17) to ensure that a multivalued waveform is not predicted, i.e., that the sequential ordering of the time samples $\theta_1^{k+1} < \theta_2^{k+1} < \cdots < \theta_M^{k+1}$ is maintained. Single-valued solutions of Eq. (11.16) are ensured for $\Delta\sigma < [\max(\partial P/\partial\theta)]^{-1}$, and therefore the sequential ordering of the time samples is preserved for

$$\Delta\sigma < \frac{\Delta\theta}{\max|P_i^k - P_{i-1}^k|}. \tag{11.18}$$

Consequently, waveforms with steeper slopes (e.g., shocks with shorter rise times) require smaller spatial steps. Finally, note that following application of Eq. (11.17), the waveform samples are no longer spaced uniformly in time. Linear interpolation is used to resample the waveform to reestablish the uniform intervals $\Delta\theta$ in preparation for solving Eq. (11.15).

To include absorption over same propagation step from σ to $\sigma+\Delta\sigma$ for which nonlinear distortion was introduced in the previous paragraph, take the solutions $(\theta_i^{k+1}, P_i^{k+1})$ just obtained and relabel them (θ_i^k, P_i^k). That is, the output from the nonlinear distortion algorithm becomes the input for the absorption algorithm over the same incremental step (this is the essence of the method of fractional steps). The desired solution now is that of Eq. (11.15), which is the diffusion equation. As described in standard texts, a stable solution algorithm for the diffusion equation is provided by an implicit Crank–Nicolson method with forward-space, centered-time finite differences. When the finite differences prescribed by the Crank–Nicolson method are introduced in Eq. (11.15), the resulting system of equations may be expressed as

$$\mathbf{A} \cdot \mathbf{P}^{k+1} = \mathbf{B} \cdot \mathbf{P}^k. \tag{11.19}$$

Here $\mathbf{P}^k = (P_1^k, P_2^k, \cdots, P_M^k)^\mathrm{T}$ is the known, sampled waveform vector at step k, and $\mathbf{A}$ and $\mathbf{B}$ are coefficient matrices. Both $\mathbf{A}$ and $\mathbf{B}$ are tridiagonal, and therefore the Thomas algorithm can be used to solve for the solution vector $\mathbf{P}^{k+1}$ explicitly at step $k + 1$ (Ames, 1992).

The algorithm just described, based on application of Eqs. (11.17) and (11.19) independently over the same (sufficiently small) propagation step $\Delta\sigma$, generates a solution that includes the combined effects of nonlinearity and absorption in accordance with the Burgers equation, Eq. (11.3). The computation time for both the nonlinear distortion algorithm and the absorption algorithm (as well as linear interpolation) is proportional to the number of time samples M. Recall that the computation time for the frequency-domain algorithm described in Sect. 11.2.1 is proportional to M^2, where M in that context is the number of frequency components. Although the number of either time samples or frequency components required to accurately represent a waveform with a given resolution corresponds nominally to the same value M, the time-domain algorithm is not necessarily more efficient than the frequency-domain algorithm. For monofrequency sources, or when only harmonic propagation curves rather than time waveforms are of interest, the frequency-domain algorithm is usually more efficient. The time-domain algorithm is usually preferable for describing pulses, random waveforms, and step shocks.

The time-domain algorithm is restricted in regard to the types of absorption and dispersion laws that can be taken into account, whereas the frequency-domain algorithm can accommodate arbitrary absorption and dispersion via Eq. (11.13). However, a method for including the effect of multiple relaxation processes in the time-domain algorithm is available (Cleveland et al., 1996b). The nonlinear distortion routine is unaltered, and the algorithm produces a tridiagonal matrix system in the form of Eq. (11.19). Consequently, the effect of relaxation is included in the algorithm in the same way as the effect of thermoviscous absorption.

11.2.3   Combined Time-Frequency Domain Algorithm

A third approach is to include nonlinear distortion in the time domain, and absorption and dispersion in the frequency domain. One such method, which has come to be known as the Pestorius algorithm (Pestorius, 1973; Pestorius and Blackstock, 1974), is described here. The time-domain portion of the Pestorius algorithm is based on weak shock theory (Sect. 4.4.1). This approach proves advantageous when nonlinear effects are strong in comparison with small-signal absorption, because weak shock theory includes losses at the shocks and dispenses with the need to take a specific absorption mechanism into account; in this case, all computations may be performed in the time domain. However, weak shock theory is based on the assumption of perfectly thin shocks, and therefore rise times and other features related to shock structure are not described. A listing of the code with detailed explanation is provided by Pestorius (1973).

To perform the time-domain operations for the nonlinear distortion algorithm, two arrays are now used to define the waveform. As in Sect. 11.2.2, the coordinate pairs (θ_i^k, P_i^k) are the samples along all continuous segments of the waveform between any shocks that may exist. A second set of coordinates $(\hat{\theta}_i^k, \hat{P}_{i,a}^k, \hat{P}_{i,b}^k)$ is associated with the shocks. Here, $\hat{\theta}_i^k$ identifies the position of the ith shock within the waveform, with $\hat{P}_{i,a}^k$ and $\hat{P}_{i,b}^k$ the corresponding pressures immediately ahead of and behind that shock, respectively. In this way, the shock locations in time are defined with greater precision than would correspond to the spacing between the points θ_i^k associated with the more smoothly varying, continuous segments of the waveform. Whereas calculation of $(\theta_i^{k+1}, P_i^{k+1})$ for propagation over step $\Delta\sigma$ is relatively straightforward, as in Sect. 11.2.2, calculation of $(\hat{\theta}_i^{k+1}, \hat{P}_{i,a}^{k+1}, \hat{P}_{i,b}^{k+1})$ requires more attention to the structure of the waveform under consideration. Our intent below is to describe only the general framework of the weak shock propagation algorithm.

Thus, consider a waveform in which shocks already exist, but for simplicity assume that over the propagation step $\Delta\sigma$ under consideration, no new shocks are formed and no shock overtakes (i.e., merges with) another. The nonlinear distortion algorithm then consists of two steps. The first step is to use Eq. (11.17) to distort the continuous segments of the waveform between the shocks, which gives $(\theta_i^{k+1}, P_i^{k+1})$. The second step is to calculate $(\hat{\theta}_i^{k+1}, \hat{P}_{i,a}^{k+1}, \hat{P}_{i,b}^{k+1})$ as follows. From weak shock theory, Eq. (4.163) provides an expression for movement of the shock in the retarded time frame, which in the dimensionless notation used here becomes, for the ith shock,

$$\frac{d\hat{\theta}_i}{d\sigma} = -\tfrac{1}{2}(\hat{P}_{i,a} + \hat{P}_{i,b}). \tag{11.20}$$

The discretized form of this relation is

$$\hat{\theta}_i^{k+1} = \hat{\theta}_i^k - \tfrac{1}{2}(\hat{P}_{i,a}^k + \hat{P}_{i,b}^k)\Delta\sigma. \tag{11.21}$$

It remains to determine $\hat{P}_{i,a}^{k+1}$ and $\hat{P}_{i,b}^{k+1}$. For this purpose, define $\theta_{i,a}^k$ to be the time sample of the continuous segment of the waveform ahead of the shock that is closest to $\hat{\theta}_{i,a}^k$ prior to application of the nonlinear distortion algorithm. Similarly, define $\theta_{i,b}^k$ to be the time sample of the waveform behind the shock that is closest to $\hat{\theta}_{i,b}^k$. We thus have $\theta_{i,a}^k < \hat{\theta}_i^k < \theta_{i,b}^k$. For $\theta_{i,a}^{k+1} < \hat{\theta}_i^{k+1} < \theta_{i,b}^{k+1}$ following application of Eqs. (11.17) and (11.21), the shock remains positioned between the same (although now displaced) time samples of the continuous waveform, and the shock pressures are unchanged. Thus, set $\hat{P}_{i,a}^{k+1} = \hat{P}_{i,a}^k$ and $\hat{P}_{i,b}^{k+1} = \hat{P}_{i,b}^k$ in this case. However, if the shock overtakes the point ahead ($\theta_{i,a}^{k+1} > \hat{\theta}_i^{k+1}$), or if the point behind overtakes the shock ($\hat{\theta}_i^{k+1} > \theta_{i,b}^{k+1}$), linear interpolation of the solution for the now multivalued continuous waveform, (θ_i^{k+1}, P_i^{k+1}), determines the value of $\hat{P}_{i,a}^{k+1}$ or $\hat{P}_{i,b}^{k+1}$, respectively, at time $\hat{\theta}_i^{k+1}$. It is this interpolation procedure that accounts for losses at the shocks.

Although the nonlinear distortion algorithm just described is straightforward to implement, further consideration must be given to circumstances related to the birth and merger of shocks, and to determination of when absorption reduces the strength of a shock so much that it is no longer described accurately by weak shock theory. When the effects of absorption and dispersion on the continuous segments of the waveform are negligible in comparison with nonlinearity, the distortion algorithm may be applied repeatedly over subsequent steps $\Delta\sigma$ up to the desired output location, with energy loss at the shocks taken into account automatically by weak shock theory. Under these conditions, the present algorithm is substantially more efficient for modeling propagation of waveforms with shocks than those described in Sects. 11.2.1 and 11.2.2.

When effects of absorption and dispersion on the continuous segments of the waveform cannot be ignored, they are taken into account in the frequency domain. Let C_n^k be the nth spectral coefficient in a Fourier decomposition of the waveform at step k. Following application of the nonlinear distortion algorithm over step $\Delta\sigma$, consider the resulting solution to be the input to the algorithm for absorption and dispersion over the same step and apply Eq. (11.13) as follows:

$$C_n^{k+1} = C_n^k \exp[-(A_n + jD_n)\Delta\sigma]. \tag{11.22}$$

The resulting frequency spectrum is transformed back into the time domain for application of the nonlinear distortion algorithm over the next incremental step. When absorption and dispersion are weak over step $\Delta\sigma$ but not negligible over the desired total propagation distance, it is sufficient to apply Eq. (11.22) after, say, every m propagation steps taken with the distortion algorithm. In this case, the correction factor $\exp[-(A_n + jD_n)m\Delta\sigma]$ is applied each time in the frequency domain. Judicious use of the frequency-domain corrections reduces errors introduced by repeated application of the fast Fourier transform. The overall computation time for the algorithm is proportional to $M \log M$ as a result of the FFT, because the computation times associated with the nonlinearity and absorption algorithms are each proportional to M.

A word of caution is offered if absorption is applied in the frequency domain when shocks exist in the waveforms. As noted above, energy loss at the shocks is taken into account automatically in the time domain via weak shock theory. The potential therefore exists for slightly overestimating absorption at the shocks through Eq. (11.22). This concern is absent from a subsequent modification of the algorithm by Anderson (1974), who avoided use of the weak shock relation in Eq. (11.21) and relied instead on thermoviscous absorption through Eq. (11.22) to stabilize the waveform steepening.

The algorithm described above was developed originally to model the propagation of arbitrary waveforms, and in particular finite-amplitude noise, for which excellent agreement was obtained with the measurements shown in Fig. 13.4a. In this case, distortion of the continuous waveform segments and the associated shock formation shifted energy upward in the frequency spectrum, while the coalescence of shocks shifted energy back down. Figure 13.6 shows results from the Pestorius algorithm for the interaction of a finite-amplitude tone with narrowband noise (Webster and Blackstock, 1978a). The same code was used to study acoustical saturation (Webster and Blackstock, 1977).

11.2.4   Spreading

We now discuss methods for including effects due to spherical and cylindrical spreading in the algorithms described in Sects. 11.2.1–11.2.3 on the basis of the generalized Burgers equation, Eq. (3.58):[3]

$$\frac{\partial p}{\partial r} + \frac{m}{r} p = \frac{\beta p}{\rho_0 c_0^3} \frac{\partial p}{\partial \tau} + \frac{\delta}{2c_0^3} \frac{\partial^2 p}{\partial \tau^2}, \tag{11.23}$$

where $m = 1$ for spherical waves, $m = \frac{1}{2}$ for cylindrical waves. Here r is the radial coordinate, and the retarded time is now defined by $\tau = t - (r - r_0)/c_0$, where r_0 is taken to be the source radius. The constant phase term r_0/c_0 is included for convenience so that the value of the pressure $p(r, \tau)$ at the source is given by $p(r_0, t)$. For plane waves set $m = 0$, in which case the form of Eq. (11.1) is recovered. In terms of the dimensionless variables defined in Eqs. (11.2) (but here with $\sigma = r/\bar{x}$, where $\bar{x}$ is again the plane-wave shock formation distance when the signal is sinusoidal at the source), Eq. (11.23) becomes

$$\frac{\partial P}{\partial \sigma} + \frac{m}{\sigma} P = P \frac{\partial P}{\partial \theta} + A \frac{\partial^2 P}{\partial \theta^2}. \tag{11.24}$$

The source condition is taken now at σ equal to $\sigma_0 = r_0/\bar{x}$, rather than at $\sigma = 0$ as for plane waves.

---

[3] We consider here only diverging (outgoing) waves. For converging (incoming) waves, modifications associated with the sign changes discussed in Sect. 3.8 must be taken into account.

Modification of the frequency-domain algorithm in Sect. 11.2.1 is trivial; simply introduce the term $-mP_n/\sigma$ on the right-hand side of Eq. (11.10) and solve as before. Trivett and Van Buren (1981) used a substitution of the form $P_n = \sigma^{-m}e^{-n^2A\sigma}Q_n$ to factor out effects due to both spreading and absorption, the result of which eliminates the spreading and absorption terms in Eq. (11.10), but yields a modified form of the nonlinear terms in the resulting equations for Q_n. The time-domain algorithm in Sect. 11.2.2 is augmented by including the following relation in the set of Eqs. (11.14) and (11.15):

$$\frac{\partial P}{\partial \sigma} = -\frac{m}{\sigma}P. \tag{11.25}$$

Given $P(\sigma, \theta)$, the analytic solution of Eq. (11.25) at $\sigma + \Delta\sigma$ is

$$P(\sigma + \Delta\sigma, \theta) = (1 + \Delta\sigma/\sigma)^{-m}P(\sigma, \theta). \tag{11.26}$$

This solution is incorporated in the split step algorithm in the same way as the solutions for effects due to nonlinearity and absorption. The same modification can be introduced in the combined time-frequency domain algorithm described in Sect. 11.2.3.

Alternatively, as done by Fenlon (1971) to include effects of spreading in the frequency-domain algorithm, one may define the new variables (Blackstock, 1964; see also Sect. 4.6.1)

$$Q = (\sigma/\sigma_0)^m P, \quad \zeta = \int_{\sigma_0}^{\sigma} (\sigma_0/\sigma)^m \, d\sigma, \tag{11.27}$$

the substitution of which in Eq. (11.24) yields

$$\frac{\partial Q}{\partial \zeta} = Q\frac{\partial Q}{\partial \theta} + \tilde{A}(\zeta)\frac{\partial^2 Q}{\partial \theta^2}. \tag{11.28}$$

The source location $r = r_0$ (i.e., $\sigma = \sigma_0$) corresponds to $\zeta = 0$, with $Q(0, \theta) = P(\sigma_0, \theta)$. Equation (11.28) differs formally from Eq. (11.3) only via the spatially varying absorption parameter $\tilde{A}(\zeta)$, which is taken into account in the algorithms in Sects. 11.2.1–11.2.3 simply via direct substitution of $\tilde{A}(\zeta)$ everywhere for A. For $m = 1$, one obtains $\zeta = \sigma_0 \ln(\sigma/\sigma_0)$ and $\tilde{A}(\zeta) = Ae^{\zeta/\sigma_0}$, and for $m = \frac{1}{2}$, we have $\zeta = 2\sigma_0[(\sigma/\sigma_0)^{1/2} - 1]$ and $\tilde{A}(\zeta) = A(1 + \zeta/2\sigma_0)$. In this stretched coordinate system, nonlinear distortion occurs in the same way as for plane waves, whereas the effect of absorption increases exponentially with distance for spherical waves and linearly with distance for cylindrical waves. Webster and Blackstock (1978b) used the transformation in Eqs. (11.27) to include the effect of spherical spreading in the nonlinear distortion routine described in Sect. 11.2.3.

The algorithms in Sects. 11.2.1–11.2.3 may be further augmented to model propagation in inhomogeneous media by introducing a general ray-tube spreading term [see, e.g., Eqs. (3.60) and (12.58)] and position-dependent medium parameters (Cleveland et al., 1996b). In this case, a separate algorithm is required to determine the ray path and thus the values of the medium parameters along that path (σ in this case measures distance along the ray path). Cleveland et al. (1996a) made these modifications in the algorithms described in Sects. 11.2.2 and 11.2.3 to model sonic boom propagation through the atmosphere, and they compared the results obtained from these computer codes when they were run for identical conditions. Despite the different procedures used in the two codes, the outputs are shown to be in excellent agreement. When real atmospheric absorption and dispersion were taken into account, the time-domain code (Sect. 11.2.2) ran the fastest. When an ideal lossless medium was assumed, the weak shock algorithm (Sect. 11.2.3) was most efficient.

11.3 Directional Three-Dimensional Waves

Here we describe algorithms that are primarily intended to model the propagation of directional sound beams, although application of these same algorithms to guided waves having frequencies well above cutoff (so that the waves may be described as quasi-plane) is also discussed. In each of these algorithms, nonlinearity is taken into account by the nonlinear term in the Burgers equation, i.e., to within the accuracy required for describing nonlinear effects consistently for quasi-plane waves. In Sect. 11.3.1, three algorithms based on the KZK (Khokhlov–Zabolotskaya–Kuznetsov) nonlinear parabolic wave equation are described. As in Sect. 11.2, these three algorithms illustrate approaches relying to different extents on time- and frequency-domain calculations. In Sect. 11.3.2, a time-domain algorithm based on another nonlinear parabolic wave equation, referred to as the NPE, is described. In Sect. 11.3.3, a frequency-domain algorithm is discussed that avoids the parabolic approximation inherent in the diffraction terms in the KZK and NPE models.

11.3.1 Solutions of the KZK Equation

The three algorithms described in the following subsections are based on the KZK equation, Eq. (3.65). We write the KZK equation here in a form that is obtained following integration with respect to time:

$$\frac{\partial p}{\partial z} = \frac{\beta p}{\rho_0 c_0^3} \frac{\partial p}{\partial \tau} + \frac{\delta}{2c_0^3} \frac{\partial^2 p}{\partial \tau^2} + \frac{c_0}{2} \int_{-\infty}^{\tau} (\nabla_{\perp}^2 p) \, d\tau, \tag{11.29}$$

where $\nabla_{\perp}^2 = \partial^2/\partial x^2 + \partial^2/\partial y^2$ is a Laplacian that operates in the plane perpendicular to the axis of the beam. Here z is taken as the direction of propagation, with $\tau =$

$t - z/c_0$. Otherwise, Eq. (11.29) differs from Eq. (11.1) only by the term containing $\nabla_\perp^2 p$, which accounts for diffraction effects in the paraxial approximation of the region around the beam axis. Only axisymmetric radiation shall be considered, for which $\nabla_\perp^2 = \partial^2/\partial r^2 + r^{-1}(\partial/\partial r)$, where $r^2 = x^2 + y^2$. As shown in Sect. 3.9, it is consistent in the parabolic approximation to use the linear plane-wave impedance relation and write $p = \rho_0 c_0 u_z$, where u_z is the z component of the particle velocity vector. This relation is significant because many problems involving sound beams are defined with normal velocity source conditions—for example, baffled piston radiation.

11.3.1.1 Combined Time-Frequency Domain Algorithm

The pioneering numerical investigations of nonlinear effects in directional sound beams were performed by Bakhvalov et al. (1976, 1978a, 1978b, 1979a, 1979b, 1979c, 1980) in a series of articles that are summarized in a book published later (Bakhvalov et al., 1987). The appendix of their book contains a detailed description of the numerical algorithm, and a very brief discussion appears in the first of their articles. Their calculations are based on a dimensionless version of Eq. (11.29) expressed in terms of the following quantities (the notation used here differs in minor ways from theirs, but it is helpful for discussions in Sects. 11.3.1.2 and 11.3.1.3):

$$P = \frac{p}{p_0}, \quad R = \frac{r/a}{1 + z/z_1}, \quad Z = \frac{z}{z_0}, \quad T = \omega_0 \tau, \tag{11.30}$$

where p_0 and ω_0 are the characteristic pressure amplitude and angular frequency at the source, respectively, $z_0 = \frac{1}{2} k_0 a^2$ is the corresponding Rayleigh distance (i.e., diffraction length) in terms of a characteristic source radius a and wave number $k_0 = \omega_0/c_0$, and z_1 is an adjustable length scale. Equation (11.29) thus becomes

$$\frac{\partial P}{\partial Z} = N P \frac{\partial P}{\partial T} + A \frac{\partial^2 P}{\partial T^2} + \frac{1}{4(1 + CZ)^2} \int_{-\infty}^{T} (\nabla_R^2 P) \, dT, \tag{11.31}$$

where $\nabla_R^2 = \partial^2/\partial R^2 + R^{-1}(\partial/\partial R)$ and $C = z_0/z_1$. In contrast with the scaling used in Eqs. (11.2) and (11.3), here the Rayleigh distance z_0 rather than the plane-wave shock formation distance $\bar{x}$ defines the characteristic length scale in the direction of propagation, with $A = \alpha_0 z_0$ and $N = z_0/\bar{x}$. The algorithm described by Bakhvalov et al. (1976, 1987) computes the effects of absorption and diffraction in the frequency domain, and it calculates the nonlinear distortion in the time domain using the method of Godunov (1959) (which yields solutions consistent with algorithms based on weak shock theory, such as described in Sect. 11.2.3).

We concentrate on the frequency-domain calculations, and in particular the diffraction algorithm. The purpose of scaling the radial coordinate r as in Eq. (11.30) is that for $CZ \ll 1$ (which corresponds to the near field for $C \sim 1$), we have $R \simeq$

r/a, which is an appropriate cylindrical coordinate for calculations where the beam is well collimated. For $CZ \gg 1$, we have $R \simeq (z_1/a)\tan\theta$ (throughout Sect. 11.3, θ is the angle with respect to the axis of the beam, defined by $\tan\theta = r/z$), and $R = $ const thus follows the spherical divergence of the beam in the far field, with the expansion rate of the spatial grid determined by z_1 (or equivalently, C). The value of C was chosen empirically.

An expansion as in Eq. (11.4) was employed, with P_n depending also on R, and substitution into Eq. (11.31) leads to a system of equations similar to Eq. (11.10) but with the addition of a diffraction term. Ignoring the nonlinear terms, the effect of which is taken into account in the time domain (the FFT is used to alternate between the time and frequency domains), we write here only the frequency-domain equation that accounts for diffraction (i.e., ignoring absorption as well):

$$\frac{\partial P_n}{\partial Z} = \frac{\nabla_R^2 P_n}{j4n(1 + CZ)^2}. \tag{11.32}$$

Because Eq. (11.32) is parabolic, corresponding to a diffusion-like process, it may be integrated forward in the $+Z$ direction without concern for backward-propagating waves, as is the case with Eq. (11.10). Despite the three-dimensional nature of the problem, there is no need for a boundary condition at $Z = \infty$. The field is represented by a uniformly spaced mesh of points extending in the axial and radial directions, the latter covering the bounded region from $R = 0$ to $R = R_{max}$. Standard backward finite-difference approximations of the Z and R derivatives were used (further discussion of appropriate algorithms is postponed to Sect. 11.3.1.2). The condition of axisymmetry requires $\partial P_n/\partial R = 0$ at $R = 0$, whereas R_{max} is selected to be sufficiently large that the pressure is negligibly small at the largest radial distances (information relevant to selection of R_{max} corresponds to linear diffraction effects). It follows that the condition appropriate to $R = R_{max}$ is $P_n = 0$.

The scope of the problems investigated by Bakhvalov et al. is revealed by the titles of their articles. Source amplitude distributions at $Z = 0$ were described by either the polynomial function $(1 - R^2)^2$ for $R < 1$ and zero for $R > 1$, or the exponential $\exp(-R^{2m})$, where m is a positive integer (e.g., $m = 1$ for Gaussian sources, $m = \infty$ for pistons). Radiation from circular pistons was approximated by letting $m = 8$, thus ensuring a smooth function that may have been helpful for providing greater numerical stability. Attention was devoted primarily to the near field, and the maximum distance at which numerical results were presented was $Z = 4$. The algorithm was used also by Zhileikin et al. (1980) to investigate the propagation of high-frequency sound in a rigid cylindrical tube. They set $z_1 = \infty$ ($C = 0$) to obtain a cylindrical grid that matches the waveguide geometry, and the boundary condition $\partial P_n/\partial R = 0$ was applied at the wall.

Alternative algorithms employing frequency domain calculations to account for diffraction and time-domain calculations to account for nonlinearity were developed by McKendree (1981) and Frøysa et al. (1993).

11.3.1.2   Frequency-Domain Algorithm

The algorithm described here, which is based exclusively on frequency-domain calculations, has been the most widely used to analyze periodic signals radiated by circular pistons, both unfocused and focused. The first entirely frequency-domain algorithm was developed by Aanonsen et al. (1984) for the near field. Like the algorithm developed by Bakhvalov et al. (1987), theirs considered radiation from axisymmetric projectors. An expansion as in Eq. (11.4) was again employed, the substitution of which into the KZK equation leads to a system of equations similar to Eq. (11.10) but with an additional term given by the right side of Eq. (11.32) with $C = 0$. [More precisely, Aanonsen et al. used an expansion in the form of Eq. (11.11), and therefore two coupled sets of equations were obtained for the Fourier coefficients a_n and b_n.]

Aanonsen et al. employed a simple implicit backward difference scheme to integrate the resulting system of equations numerically. Consideration of boundary conditions is much the same as described in Sect. 11.3.1.1. However, since an expanding coordinate grid was not used in this work, the computation time for far-field calculations was high. Calculations for radiation from pistons were confined to the near field ($Z \leq 1$), although calculations for Gaussian beams were performed out to $Z \simeq 7$ (because the more slowly varying field structure of a Gaussian beam permits use of larger spatial steps). The effect of varying the number M of harmonics retained was investigated, with a maximum value $M = 30$ used in that work.

To improve the efficiency of the algorithm for calculations of the far field, Hamilton et al. (1985) introduced a modified form of the transformation in Eqs. (11.30):

$$
P = \left(1 + \frac{z}{z_0}\right)\frac{p}{p_0}, \quad R = \frac{r/a}{1 + z/z_0}, \quad Z = \frac{z}{z_0}, \quad T = \omega_0\tau - \frac{r^2/a^2}{1 + z/z_0},
$$

$$
\tag{11.33}
$$

the substitution of which into Eq. (11.29) yields

$$
\frac{\partial P}{\partial Z} = \frac{NP}{1 + Z}\frac{\partial P}{\partial T} + A\frac{\partial^2 P}{\partial T^2} + \frac{1}{4(1 + Z)^2}\int_{-\infty}^{T} (\nabla_R^2 P)\, dT, \tag{11.34}
$$

where the definitions $A = \alpha_0 z_0$ and $N = z_0/\bar{x}$ remain the same as in Eq. (11.31). Here, the radial coordinate assumes the value $R \simeq \frac{1}{2}k_0 a \tan\theta$ for $Z \gg 1$. We also have $p/p_0 \simeq P/Z$ and $\omega_0\tau \simeq T + k_0 r^2/2z$ for $Z \gg 1$, indicating that amplitude decay and wave-front curvature associated with spherical spreading have been factored out of the transformed variables. Now substitute the expansion

$$
P(R, Z, T) = \frac{1}{2}\sum_{n=1}^{M} P_n(R, Z)e^{jnT} + \text{c.c.} \tag{11.35}
$$

in Eq. (11.34) to obtain the coupled spectral equations

$$\frac{\partial P_n}{\partial Z} = \frac{\nabla_R^2 P_n}{j4n(1+Z)^2} - n^2 A P_n$$

$$+ \frac{jnN}{4(1+Z)} \left(\sum_{m=1}^{n-1} P_m P_{n-m} + 2 \sum_{m=n+1}^{M} P_m P_{m-n}^* \right). \tag{11.36}$$

Given, for example, the monofrequency source condition $p = p_0 f(r) \sin \omega_0 t$ at $z = 0$, where $f(r)$ is an arbitrary amplitude distribution, the corresponding source condition for Eq. (11.36) is $P_1(R, 0) = -jf(R) \exp(jR^2)$, with $P_n(R, 0) = 0$ for $n > 1$. For a piston, $f(R)$ is unity for $R < 1$ and zero elsewhere. As in Sect. 11.2.1, the computation time is associated mainly with the nonlinear terms and is therefore proportional to M^2.

Two finite-difference methods are used to compute the diffraction term (Berntsen, 1990). Very near the source (e.g., for $Z \lesssim 0.1$), where the wave field can vary rapidly (particularly for the case of radiation from a piston), a fully implicit backward-difference method is used. A Crank–Nicolson method is used throughout the remainder of the field. Typical step sizes for piston radiation are $\Delta R \simeq 0.03$ (with $R_{max} \simeq 8$) and $\Delta Z = (1 + Z)^2 (\Delta Z)_0$, where $(\Delta Z)_0 \simeq 0.003$ is the axial step size at the source. Increasing ΔZ in proportion to $(1 + Z)^2$ corresponds to the presence of this factor in the diffraction term, and it reduces the number of calculations in the more slowly varying far-field region of the beam. The transformed coordinates permitted calculations to be made out to $Z = 30$, and simulations to be run for earlier experiments demonstrating the effects of acoustical saturation on far field beam patterns (Hamilton et al., 1985).

A modification of the coordinate transformation in Eqs. (11.33) is available for focused beams (Hart and Hamilton, 1988). With d taken to be the focal length and with the source plane now at $z = -d$ (and thus the focal plane at $z = 0$), the transformation is

$$P = (z/d \pm \delta) \frac{p}{p_0}, \quad R = \frac{r/a}{z/d \pm \delta}, \quad Z = \frac{z}{d}, \quad T = \omega_0 \tau - \frac{Gr^2/a^2}{z/d \pm \delta}, \tag{11.37}$$

where $G = k_0 a^2 / 2d$ is a characteristic linear focusing gain (see Sect. 8.3.3), and δ here is a small positive quantity that governs the rate at which the transformed spatial grid converges in the vicinity of the focus. Once again, the result of substituting the transformation and a complex Fourier series in the form of Eq. (11.35) into the KZK equation is a set of coupled spectral equations like those in Eq. (11.36), which permits use of the same finite-difference algorithm that is used for unfocused beams. For the region between the source and focal planes, the minus signs in Eqs. (11.37) are used to produce a converging grid. Computations beyond the focal plane use the plus signs, which yield a diverging grid. Boundary conditions for the latter are

obtained from a matching procedure based on the computed harmonics at $z = 0$
obtained from the solution in the prefocal region.

Other applications of the frequency-domain algorithm have included the study
of radiation from bifrequency (Kamakura et al., 1989; Naze Tjøtta et al., 1990;
Naze Tjøtta et al., 1991), pulsed (Baker and Humphrey, 1992), and rectangular
(Kamakura et al., 1992; Baker et al., 1995) sources. The harmonic amplitudes shown
in Figs. 8.5, 8.6, and 8.8 were computed with the present algorithm.

11.3.1.3 Time-Domain Algorithm

The time-domain algorithm developed by Lee and Hamilton (1995) solves
Eq. (11.34) directly with finite differences. As in Sect. 11.2.2, the method of
fractional steps is used to solve the equation term by term. The set of equations
here is

$$\frac{\partial P}{\partial Z} = \frac{NP}{1+Z} \frac{\partial P}{\partial T}, \tag{11.38}$$

$$\frac{\partial P}{\partial Z} = A \frac{\partial^2 P}{\partial T^2}, \tag{11.39}$$

$$\frac{\partial P}{\partial Z} = \frac{1}{4(1+Z)^2} \int_{-\infty}^{T} (\nabla_R^2 P) \, dT. \tag{11.40}$$

Note the resemblance of Eqs. (11.38) and (11.39), respectively, to Eqs. (11.14) and
(11.15).

The nonlinear distortion algorithm that replaces Eq. (11.17) is

$$T_i^{k+1} = T_i^k - N P_i^k \ln\left(1 + \frac{\Delta Z}{1+Z}\right), \tag{11.41}$$

for which the step size restriction in Eq. (11.18) is replaced by

$$\ln\left(1 + \frac{\Delta Z}{1+Z}\right) < \frac{\Delta T}{N \max |P_i^k - P_{i-1}^k|}. \tag{11.42}$$

Resampling is required before proceeding to the absorption and diffraction algorithms. Although Lee and Hamilton describe a nonlinearity algorithm in which
the distortion and resampling procedures are combined in a single operation, the
step size restriction is more stringent than that in Eq. (11.42). Since Eq. (11.39)
is formally equivalent to Eq. (11.15), its solution is obtained from Eq. (11.19) as
before.

We consider now the diffraction term. Some care must be given to choosing
an appropriate time window, defined by $T_{min} \leq T \leq T_{max}$, to follow the
waveform throughout the field. An advantage of the transformed retarded time T

in Eqs. (11.33) is that it follows the spherical curvature of the wave front, and therefore with T_{min} and T_{max} held constant, the pulse does not shift significantly within the time window as distance from the beam axis is increased. Without this transformation, in the original retarded time frame τ, it would become necessary for τ_{min} and τ_{max} to be functions of distance from the beam axis. For a tone burst, it is usually sufficient to take T_{min} to be the time just a few cycles ahead of the source waveform, where the field is always approximately zero in the retarded time frame. The value of T_{max} must be considerably farther behind the trailing edge of the pulse, normally about ten cycles, to account for the arrival of the diffracted edge wave in the near field.

The integral in Eq. (11.40) is evaluated with a trapezoidal rule, with the lower limit replaced by T_{min}. Otherwise, the diffraction routine is similar to that used in the frequency domain. In the near field, a fully implicit backward-difference method is used, and beyond this the Crank–Nicolson method is used, with the same spatial grid parameters as cited in Sect. 11.3.1.2. Both methods yield tridiagonal matrices that can be solved explicitly with the Thomas algorithm (Ames, 1992), and the computation time of each is proportional to the number M of time samples in each waveform across the beam. Given a source condition $p = p_0 f(r, t)$ at $z = 0$, the boundary condition for the algorithm is $P = f(R, T + R^2)$ at $Z = 0$. Using comparisons with analytic solutions for pulses along the axis of a circular piston, Lee and Hamilton (1995) discuss numerical considerations for the diffraction algorithm related to proper description of the edge wave and sharp transient effects. The computation time for the entire algorithm is proportional to M.

Problems that have been modeled with the algorithm have focused on piston radiation and include shock formation in tone bursts and random noise waveforms that are propagated out to $Z = 100$ (Lee and Hamilton, 1995), self-demodulation in fluids with strong absorption (Averkiou et al., 1993), and the effect of relaxation (Cleveland et al., 1996b). See also Fig. 8.7.

11.3.2 Time-Domain Algorithm Based on the NPE

An alternative to the time-domain algorithm described in Sect. 11.3.1.3 is based on the nonlinear progressive-wave equation referred to as the NPE (McDonald and Kuperman, 1987), which we write here in the form

$$\frac{Dp}{Dt} = -\frac{\beta p}{\rho_0 c_0} \frac{\partial p}{\partial z} - \frac{c_0}{2} \int_{-\infty}^{z} (\nabla_\perp^2 p) \, dz, \qquad (11.43)$$

where here $D/Dt = \partial/\partial t + c_0(\partial/\partial z)$ corresponds to differentiation in the retarded time frame $\tau = t - z/c_0$ used above. Comparison of Eq. (11.43) with Eq. (11.29) shows that the NPE may be characterized as a lossless KZK equation with the roles of propagation distance and time reversed. The NPE appears frequently in the literature with an additional term that accounts for small variations in the ambient sound speed, which is included below. Absorption may also be included in the NPE in a manner similar to the way the corresponding term appears in the KZK equation.

In acoustics, one is normally interested in radiation from a source at a fixed location, where the waveform is prescribed as a function of time. A boundary condition posed in this way is appropriate for the Burgers and KZK equations. However, the NPE requires a space waveform at a given instant in time. A matching routine—based on linear theory, for example—is therefore needed to initialize the NPE algorithm when the source condition is posed as a boundary-value problem, after which the forward integration in time may proceed. When implementing a numerical solution of Eq. (11.43), one must recognize that z is position relative to the moving reference frame, so the z domain to be discretized represents a moving window that must capture the entire pulse. To produce a time waveform at a fixed location in space, it is necessary to select $\Delta t = \Delta z/c_0$ so that each trailing mesh point in the moving window arrives at the position of interest at the successive time steps.

As in Sect. 11.3.1.3, each step in the integration process uses a splitting method in which different numerical approximation techniques are applied to the various terms in order to ensure the stability and accuracy of the approximation. Diffraction is taken into account by approximating the integral with a trapezoidal rule and applying finite differences according to a Crank–Nicolson scheme. The principal distinctive feature of the NPE algorithm is the use of a second-order upwind flux-corrected transport scheme (McDonald and Ambrosiano, 1984; McDonald and Kuperman, 1985) to include the nonlinear effects in Eq. (11.43) (as well as the refraction effects introduced below, if included). This nonlinearity algorithm accounts for not only waveform distortion but also shock propagation without requiring an absorption term in Eq. (11.43). The nonlinearity algorithm produces results consistent with those obtained on the basis of the weak shock algorithm described in Sect. 11.3.1.3.

The first applications of the NPE were to nonlinear propagation in ocean acoustic waveguides (McDonald and Kuperman, 1987; Ambrosiano et al., 1990). In these cases, the following form of the NPE was used:

$$\frac{Dp}{Dt} = -\frac{c_0}{2r}p - \frac{\partial}{\partial r}\left(c_1 p + \frac{\beta p^2}{2\rho_0 c_0}\right) - \frac{c_0}{2}\int_{-\infty}^{r}\frac{\partial^2 p}{\partial z^2}\,dr, \tag{11.44}$$

where we now have $D/Dt = \partial/\partial t + c_0(\partial/\partial r)$ in a cylindrical coordinate system where z is ocean depth and r is radial distance in the horizontal plane. The term $-c_0 p/2r$ accounts for cylindrical spreading. Refraction is taken into account through the new quantity c_1, which is a small correction to the local ambient small-signal sound speed, given now by $c_0 + c_1(r, z)$, with $|c_1| \ll c_0$.

McDonald and Kuperman (1987) tested their code with two problems, propagation of a plane nonlinear N wave and linear propagation in a waveguide. The former verified that their nonlinearity algorithm produced results in agreement with weak shock theory. For the latter, they considered propagation of a Gaussian waveform in a homogeneous fluid bounded below by a rigid surface (at which $\partial p/\partial z = 0$)

and above by a free surface (at which $p = 0$). Comparison of the numerical results with alternative solutions verified the diffraction algorithm, principally its ability to account for the dispersion introduced by waveguides. The main nonlinear problem considered by McDonald and Kuperman was propagation of a weak shock through a caustic in an inhomogeneous ocean waveguide. The incident wave was taken to originate from an explosive source in a region of constant sound speed below a depth z_d, while above z_d the sound speed increases linearly with z. Rays arriving from below z_d form a caustic at some depth above z_d. Another application of the NPE to ocean acoustics may be found in the paper by Ambrosiano et al. (1990), which presents simulations of propagation in a homogeneous fluid ($c_1 = 0$) bounded above by a pressure release surface and below by a penetrable fluid bottom. Attention was focused on the dependence of bottom penetration on wave amplitude.

Too and Ginsberg (1992a) modified the NPE algorithm to model axisymmetric sound beams on the basis of Eq. (11.43), with $\nabla_\perp^2 = \partial^2/\partial r^2 + r^{-1}(\partial/\partial r)$. Their axial step size Δz was 5% of the wavelength at the source frequency, the radial step size was $\Delta r = 5\Delta z$, and the grid width was $r_{max} = 3a$ for calculations within the Rayleigh distance z_0, where a is again the source radius. The radial boundary conditions are the same as those used in the KZK algorithms.

To test the accuracy of the diffraction algorithm, Too and Ginsberg considered linear radiation from a piston projector in an infinite baffle. The waveform window close to the source was initialized with the Rayleigh diffraction integral, and numerical results at different distances were compared with results obtained from direct numerical integration of the Rayleigh integral. Sufficiently long tone bursts were used to simulate continuous radiation. Harmonic beam profiles calculated numerically are shown to be extremely close to the prediction from the Rayleigh integral, even at near-field locations where the paraxial approximations inherent in both the NPE and KZK equations are normally assumed to be unsuitable. The authors speculate that the good agreement is due to initialization of the routine with the Rayleigh integral, rather than the more common approach of using the plane-wave impedance relation to calculate the pressure in the plane of the baffled piston.

As discussed in Sect. 11.3.1.2, it becomes necessary to expand the transverse grid when one is interested in propagation of a sound beam well beyond the near field. Too and Ginsberg (1992b) thus developed a spherical coordinate version of the NPE computer code. The dependent variable for this formulation is $q = Rp$, based on R being the (spherical) radial distance from the source. The moving window was propagated in the radial direction, so the transverse Laplacian $\nabla_\perp^2$ contained θ derivatives. The main features of the resulting equation are like those of Eq. (11.43), although with space-dependent coefficients in the diffraction term. Both unfocused and focused beams were investigated using this approach. To model focused beams, Too and Ginsberg placed the origin of the spherical coordinate at the focal distance of a concave baffled projector.

11.3.3 Frequency-Domain Algorithm Without the Parabolic Approximation

Christopher and Parker (1991a) removed the effect of the parabolic approximation inherent in the KZK and NPE algorithms by using a phenomenological approach. Like other algorithms described in this chapter, theirs uses a split step superposition procedure to describe different physical effects. The key advantage of their method is the inclusion of diffraction according to an exact formulation based on the Kirchhoff–Helmholtz integral, with the principal source of error in the diffraction subroutine arising only through the use of discrete Hankel transforms to perform the required convolutions. Otherwise, their methods for including effects of absorption and nonlinearity are essentially the same as those in the frequency domain approach described in Sect. 11.2.1.

The focus here is therefore on the diffraction algorithm. Consider a time harmonic pressure field $p = \frac{1}{2} p_n(x, y, z)e^{jn\omega_0 t}$ that is known in plane z. In the absence of nonlinearity and absorption, the following form of the Kirchhoff–Helmholtz integral determines the pressure in plane $z + \Delta z$ for arbitrarily large Δz:

$$p_n(x, y, z + \Delta z) = \iint_{-\infty}^{\infty} p_n(x', y', z) h_n(x - x', y - y') \, dx' dy', \qquad (11.45)$$

where

$$h_n(x, y) = (1 + jnk_0 R)\frac{e^{-jnk_0 R}}{2\pi R^3} \Delta z, \quad R = [x^2 + y^2 + (\Delta z)^2]^{1/2}, \qquad (11.46)$$

and $k_0 = \omega_0/c_0$. Equation (11.45) thus provides the pressure in terms of the two-dimensional spatial convolution of $p_n(x, y, z)$ and $h_n(x, y)$ over the xy plane. The convolution may be performed by multiplying the 2-D spatial Fourier transforms of $p_n(x, y, z)$ and $h_n(x, y)$ and then taking the inverse transform of the resulting product to obtain $p_n(x, y, z + \Delta z)$. This general approach is used widely for modeling diffraction phenomena (Goodman, 1968).

Christopher and Parker were interested mainly in axisymmetric sound fields, for which the following expansion of the pressure is employed:

$$p(r, z, t) = \frac{1}{2} \sum_{n=1}^{M} p_n(r, z)e^{jn\omega_0 t} + \text{c.c.} \qquad (11.47)$$

The axisymmetric propagation function $h_n(r)$ is given by Eqs. (11.46) with $R = [r^2 + (\Delta z)^2]^{1/2}$. The diffraction substep is performed using an efficient discrete Hankel transform (DHT) algorithm developed by Johnson (1987), which relates the values of a radial function $f(r)$ at a discrete set of points r_l to the values of a transformed function $\tilde{f}(\kappa)$ at a discrete set of wave numbers κ_m. With W denoting the width of a radial window containing $N - 1$ discrete samples, the transform and its inverse are approximated by

$$\tilde{f}(\kappa_m) = \frac{2\pi W^2}{j_N^2} \sum_{l=1}^{N-1} Y_{ml} f(r_l), \quad f(r_l) = \frac{1}{2\pi W^2} \sum_{m=1}^{N-1} Y_{lm} \tilde{f}(\kappa_m), \qquad (11.48)$$

where $r_l = j_l W/j_N$, $\kappa_m = j_m/2\pi W$, $Y_{lm} = 2J_0(j_l j_m/j_N)/J_1^2(j_m)$, and j_m is the mth zero of the zeroth-order Bessel function, i.e., $J_0(j_m) = 0$. Note that because the values of j_m do not occur at constant increments, this transform pair corresponds to nonuniformly spaced radial locations and wave numbers.

In order to carry out the convolution, the functions p_n and h_n are sampled at the locations r_l to obtain transformed functions $\tilde{p}_n(\kappa_m, z)$ and $\tilde{h}_n(\kappa_m)$. The DHT of the propagated signal is then, from Fourier transform theory, $\tilde{p}_n(\kappa_m, z + \Delta z) = \tilde{p}_n(\kappa_m, z)\tilde{h}_n(\kappa_m)$.[4] The corresponding radial distribution $p_n(r_l, z + \Delta z)$ is obtained by applying the inverse DHT in Eqs. (11.48), which yields

$$p_n(r_l, z + \Delta z) = \frac{1}{2\pi W^2} \sum_{m=1}^{N-1} Y_{lm} \tilde{p}_n(\kappa_m, z)\tilde{h}_n(\kappa_m). \qquad (11.49)$$

Finally, this result for $p_n(r_l, z + \Delta z)$ from the diffraction subroutine is taken to be the input $p_n(r_l, z)$ back at plane z for subroutines that account for absorption, dispersion, and nonlinearity. Since the calculations are performed in the frequency domain, absorption and dispersion can be included as in Eq. (11.22). Nonlinear distortion over step Δz is taken into account as for plane waves, i.e., according to the solution of Eq. (11.10) with $A = 0$ [alternatively, absorption and dispersion may be included in this same step via Eq. (11.13)]. Christopher and Parker also introduced small correction factors to account approximately for wave-front curvature and the fact that the signal does not propagate precisely in the z direction.

Like other authors, Christopher and Parker addressed the accumulation of energy in the highest harmonics close to the truncation number M. They estimated that their computation time increases in proportion to M^5 [M^2 from nonlinearity in the frequency domain and M^2 from the DHT (the authors also discussed the alternative of using a 2-D FFT), and they increased spatial resolution roughly in proportion to M], so they employed a scheme for artificially increasing the attenuation of the higher harmonics in order to help reduce the value of M. The selections of Δz and Δr are based on resolution requirements, while their suggestion for the radial mesh size is based on the Nyquist radial sampling rate associated with the fundamental frequency component.

---

[4] In a separate treatment of linear propagation, Christopher and Parker (1991b) discussed the alternative of using the analytical Hankel transform of $h_n(r)$, rather than obtaining $\tilde{h}_n(\kappa_m)$ numerically, which would enhance numerical efficiency. However, they showed that to do so can lead to increased error, such as aliasing.

A notable application of this algorithm has been the modeling of shock wave propagation produced by spark source lithotripters (see Sect. 15.6), in which both diffraction and nonlinear effects are very strong. Such lithotripters employ a spark source at one focus of an axisymmetric ellipsoidal reflector of finite depth, the firing of which produces an outgoing spherical wave. The reflected wave converges in the neighborhood of the second ellipsoidal focus. The diffraction algorithm was found to be very accurate for this challenging geometry (Christopher, 1994), as was verified by comparison with a linear analytical solution based on the Kirchhoff–Helmholtz integral.

11.4   General Time-Domain Algorithm

Each of the computational algorithms discussed thus far applies to situations in which the propagation is predominantly in a single coordinate direction. Multidimensional features are included in some formulations by allowing for variation of the signal transverse to the propagation direction. When the scale of the transverse variation is comparable to the wavelength scale associated with propagation, the fundamental assumptions on which these formulations are based cease to be valid. In contrast, computational fluid dynamics (CFD) programs are used regularly to study inherently multidimensional problems. The difficulty with employing such programs to address nonlinear acoustics resides in the smallness of the acoustic Mach number, even for the highest amplitudes of usual interest. Application of a standard CFD code in such situations can result in the nonlinear effects being lost in the numerical approximation errors. One recent line of study, by Inoue and Yano (1993, 1996), provides numerical solutions of the exact equations for a perfect gas, with dissipation taken into account only at the shocks. They have used their results to illustrate differences with predictions based on weak shock theory when acoustic Mach numbers approach unity (Inoue and Yano, 1993), and also to model streaming in sound beams at similar amplitudes (Yano and Inoue, 1996).

In the present section we discuss another recent line of study, by Sparrow and Raspet (1991), who developed a finite-difference approximation of the Navier–Stokes–Fourier equations and thus take dissipation into account explicitly. Specifically, the foundation for their treatment is the set of continuity, momentum, heat transfer, energy, and constitutive field equations. They assume that all relaxation effects, including bulk viscosity, are negligible. Following expansion of all field variables about their equilibrium states and discarding of high-order terms, their resulting basic equations, which contain lossless linear and quadratic terms, as well as lossy linear terms, correspond to Eqs. (3.30), (3.31), (3.33), and (3.34), plus an auxiliary state equation. Sparrow and Raspet used first-order substitutions to rewrite several second-order terms, and assumed an axisymmetric cylindrical geometry [(r, z) coordinates, with u_r and u_z the corresponding components of the particle velocity vector]. This led to a system of equations in the following form:

$$\frac{\partial \mathbf{w}}{\partial t} + \frac{\partial \mathbf{F}_1}{\partial r} + \frac{\partial \mathbf{G}_1}{\partial z} = \mathbf{F}_2 + \mathbf{G}_2, \tag{11.50}$$

where $\mathbf{w} = (\rho', u_r, u_z, s)^{\mathrm{T}}$, $\mathbf{F}_1 = [(\rho_0 + \rho')u_r, p/\rho_0, 0, 0]^{\mathrm{T}}$, $\mathbf{G}_1 = [(\rho_0 + \rho')u_z, 0, p/\rho_0, 0]^{\mathrm{T}}$, ρ' is the density perturbation, and s is the specific entropy. The elements of $\mathbf{F}_2$ and $\mathbf{G}_2$ are lengthy expressions containing derivatives with respect to r and z, but not with respect to t. The equations were thus manipulated such that, for the purpose of numerical solution, the only time derivative is that which appears explicitly in Eqs. (11.50).

Initial conditions are given for $\mathbf{w}$ over all space, and Eqs. (11.50) are solved numerically by stepping forward in time using the splitting method described by Maestrello et al. (1981). A uniform discretization in space and time is employed in which any variable $\psi(r, z, t)$ is approximated by $\psi(i\Delta r, j\Delta z, n\Delta t)$. Equations (11.50) are separated into

$$\frac{\partial \mathbf{w}}{\partial t} + \frac{\partial \mathbf{F}_1}{\partial r} = \mathbf{F}_2, \quad \frac{\partial \mathbf{w}}{\partial t} + \frac{\partial \mathbf{G}_1}{\partial z} = \mathbf{G}_2. \tag{11.51}$$

The basic idea is to integrate one of Eqs. (11.51), in the process ignoring the dependence of $\mathbf{w}$ on the spatial variable that does not appear explicitly as a derivative in that equation. This produces updated values that are used to integrate the other equation. In the next time step, the integration sequence is reversed, with the second equation in the sequence for the prior time step now solved first.

A symbolic representation of these operations involves defining solution operators $f_r(\Delta t)$ and $f_z(\Delta t)$ for each equation, such that solution of each of Eqs. (11.51) after one time step may be written as $\mathbf{w}(t + \Delta t) = f_r(\Delta t)\mathbf{w}(t)$ and $\mathbf{w}(t + \Delta t) = f_z(\Delta t)\mathbf{w}(t)$, respectively. After two time steps, the solution that is thereby obtained is

$$\mathbf{w}(t + 2\Delta t) = f_r(\Delta t)f_z(\Delta t)f_z(\Delta t)f_r(\Delta t)\mathbf{w}(t). \tag{11.52}$$

The symmetry of the operations between the two time steps is required for the approximation to be accurate to second order in time. Expressions for $f_r(\Delta t)$ and $f_z(\Delta t)$ that are accurate to fourth order in space and second order in time are provided by Sparrow and Raspet. At the boundaries of the integration region, where the spatial mesh does not provide sufficient points to evaluate the operators, Sparrow and Raspet use second-order symbolic operators. In order to maintain acceptable accuracy, they found it necessary to set $c_0\Delta t/\Delta x \simeq \frac{1}{4}$.

In a finite-difference representation of a wave equation, discontinuities can lead to local oscillations that eventually disperse throughout the discretized domain. Sparrow and Raspet addressed this phenomenon by using an artificial viscosity term that adds a dissipation term to the kinematic wave equation. The artificial viscosity is introduced in such a way that in the absence of absorption and nonlinearity, each of Eqs. (11.51) assumes the form $\partial \psi/\partial t + c_0 \partial \psi/\partial x = -\mu \partial^4 \psi/\partial x^4$. The corresponding numerical approximation entails adding a correction term to the value of the variable obtained by stepping forward in the absence of artificial viscosity. With ψ_j^n denoting the value of a variable at time step n and spatial grid point $j\,\Delta x$ [x is either r for the first of Eqs. (11.51), or z for the second], the artificial viscosity correction used by Sparrow and Raspet is

$$\psi_j^{n+1}|_{\text{new}} = \psi_j^{n+1}|_{\text{old}} - \nu(\psi_{j+2}^n - 4\psi_{j+1}^n + 6\psi_j^n - 4\psi_{j-1}^n + \psi_{j-2}^n), \qquad (11.53)$$

which is accurate to fourth order. The viscosity coefficient is therefore given by $\mu = \nu(\Delta x)^4/\Delta t$, and the corresponding attenuation increases as the fourth power of the frequency. Hence, the artificial viscosity has the effect of attenuating the high-frequency components, while leaving the lower-frequency components relatively unaffected. (The value of ν must be chosen carefully, because too large a value may influence the relevant features of the signal, and there will be no beneficial effect if the coefficient is too small.) Sparrow and Raspet incorporated the viscosity correction into the operators $f_r(\Delta t)$ and $f_z(\Delta t)$ for ρ', the $f_r(\Delta t)$ operator for u_r, and the $f_z(\Delta t)$ operator for u_z.

Any finite-difference treatment of an open domain must truncate that domain. Sparrow and Raspet (1990) previously had developed an absorbing boundary condition for linear wave propagation in order to prevent reflection of signals at the artificial boundary. The specific approximation was a variant of the Bayliss–Turkel condition, which they determined on the basis of quantitative examples to obtain the most accurate approximation of transmission into an open domain. Sparrow and Raspet (1991) used the same boundary condition in their subsequent nonlinear simulation, even though it is a linear relation. They did so on the basis of the assumption that nonlinearity and absorption should not be significant effects over the scale of a few grid intervals.

The examples considered by Sparrow and Raspet (1991) treated various aspects of the propagation of a spark pulse in air. The source waveform is given in terms of the sound pressure, and the required initial spatial waveform for the pressure, as well as the initial spatial waveforms required for all other field variables, is calculated using lossless linear theory. Computations were performed for three problems: free-field propagation of a spherical pulse, normal incidence of a planar pulse on a rigid surface, and oblique incidence of a spherical pulse on a hard surface. Waveforms calculated for the first problem were found to be in good agreement with those obtained with the Pestorius algorithm (Sect. 11.2.3), apart from some numerical oscillations just ahead of the shock. For the second problem, good agreement was obtained with an analytic expression for the pressure at the wall. No alternative solutions were available for the third problem. The authors commented on the extreme computational effort of their algorithm. Even for the first problem, in which a moving time window was used to improve the efficiency, long execution time was required on a supercomputer.

Sparrow and Raspet (1991) closed their paper by calling for additional work to establish the limits of their formulation, as well as to develop improved algorithms that would decrease computation times and increase accuracy. It is likely that several numerical techniques in general use for computational fluid dynamics can be adapted successfully to handle acoustics problems. Such tools are essential, for there are many situations in which the propagation is inherently a multidimensional phenomenon that is beyond the reach of our present analytical capabilities.

References

Aanonsen, S. I., Barkve, T., Naze Tjøtta, J., and Tjøtta, S. (1984). Distortion and harmonic generation in the nearfield of a finite amplitude sound beam. *J. Acoust. Soc. Am.* **75**, 749–768.

Ambrosiano, J. J., Plante, D. R., McDonald, B. E., and Kuperman, W. A. (1990). Nonlinear propagation in an ocean acoustic waveguide. *J. Acoust. Soc. Am.* **87**, 1473–1481.

Ames, W. F. (1992). *Numerical Methods for Partial Differential Equations*, 3rd ed. (Academic Press, New York).

Anderson, M. O. (1974). The propagation of a spherical N wave in an absorbing medium and its diffraction by a circular aperture. Technical Report ARL-TR-74-25 (Applied Research Laboratories, The University of Texas at Austin, AD 787 878).

Averkiou, M. A., Lee, Y.-S., and Hamilton, M. F. (1993). Self-demodulation of amplitude and frequency modulated pulses in a thermoviscous fluid. *J. Acoust. Soc. Am.* **94**, 2876–2883.

Baker, A. C., Berg, A. M., Sahin, A., and Naze Tjøtta, J. (1995). The nonlinear pressure field of plane, rectangular apertures: Experimental and theoretical results. *J. Acoust. Soc. Am.* **97**, 3510–3517.

Baker, A. C., and Humphrey, V. F. (1992). Distortion and high frequency generation due to nonlinear propagation of short ultrasonic pulses from a plane circular piston. *J. Acoust. Soc. Am.* **92**, 1699–1705.

Bakhvalov, N. S., Zhileikin, Ya. M., Zabolotskaya, E. A., and Khokhlov, R. V. (1976). Nonlinear propagation of a sound beam in a nondissipative medium. *Sov. Phys. Acoust.* **22**, 272–274.

Bakhvalov, N. S., Zhileikin, Ya. M., Zabolotskaya, E. A., and Khokhlov, R. V. (1978a). Focused high-amplitude sound beams. *Sov. Phys. Acoust.* **24**, 10–15.

Bakhvalov, N. S., Zhileikin, Ya. M., Zabolotskaya, E. A., and Khokhlov, R. V. (1978b). Propagation of finite-amplitude sound beams in a dissipative medium. *Sov. Phys. Acoust.* **24**, 271–275.

Bakhvalov, N. S., Zhileikin, Ya. M., Zabolotskaya, E. A., and Khokhlov, R. V. (1979a). Harmonic generation in sound beams. *Sov. Phys. Acoust.* **25**, 101–106.

Bakhvalov, N. S., Zhileikin, Ya. M., and Zabolotskaya, E. A. (1979b). Parametric interaction of sound beams. *Sov. Phys. Acoust.* **25**, 280–283.

Bakhvalov, N. S., Zhileikin, Ya. M., and Zabolotskaya, E. A. (1979c). Nonlinear propagation of Gaussian beams. *Sov. Phys. Acoust.* **25**, 458–460.

Bakhvalov, N. S., Zhileikin, Ya. M., and Zabolotskaya, E. A. (1980). Nonlinear propagation of sound beams with a uniform amplitude distribution. *Sov. Phys. Acoust.* **26**, 95–100.

Bakhvalov, N. S., Zhileikin, Ya. M., and Zabolotskaya, E. A. (1987). *Nonlinear Theory of Sound Beams* (American Institute of Physics, New York).

Berntsen, J. (1990). Numerical calculations of finite amplitude sound beams. In *Frontiers of Nonlinear Acoustics*, M. F. Hamilton and D. T. Blackstock, eds. (Elsevier Applied Science, London), pp. 191–196.

Blackstock, D. T. (1964). On plane, spherical, and cylindrical sound waves of finite amplitude in lossless fluids. *J. Acoust. Soc. Am.* **36**, 217–219.

Christopher, P. T. (1994). Modeling the Dornier HM3 lithotripter. *J. Acoust. Soc. Am.* **96**, 3088–3095.

Christopher, P. T., and Parker, K. J. (1991a). New approaches to nonlinear diffractive field propagation. *J. Acoust. Soc. Am.* **90**, 488–499.

Christopher, P. T., and Parker, K. J. (1991b). New approaches to the linear propagation of acoustic fields. *J. Acoust. Soc. Am.* **90**, 507–521.

Cleveland, R. O., Chambers, J. P., Bass, H. E., Raspet, R., Blackstock, D. T., and Hamilton, M. F. (1996a). Comparison of computer codes for the propagation of sonic boom waveforms through isothermal atmospheres. *J. Acoust. Soc. Am.* **100**, 3017–3027.

Cleveland, R. O., Hamilton, M. F., and Blackstock, D. T. (1996b). Time-domain modeling of finite-amplitude sound in relaxing fluids. *J. Acoust. Soc. Am.* **99**, 3312–3318.

Cook, B. D. (1962). New procedure for computing finite-amplitude distortion. *J. Acoust. Soc. Am.* **34**, 941–946.

Fenlon, F. H. (1971). A recursive procedure for computing the nonlinear spectral interactions of progressive finite-amplitude waves in nondispersive fluids. *J. Acoust. Soc. Am.* **50**, 1299–1312.

Fox, F. E., and Wallace, W. A. (1954). Absorption of finite amplitude sound waves. *J. Acoust. Soc. Am.* **26**, 994–1006.

Frøysa, K.-E., Naze Tjøtta, J., and Berntsen, J. (1993). Finite amplitude effects in sound beams. Pure tone and pulsed excitation. In *Advances in Nonlinear Acoustics*, H. Hobæk, ed. (World Scientific, Singapore), pp. 233–238.

Godunov, S. K. (1959). Difference method for the numerical calculation of discontinuous solutions of hydrodynamical equations. *Mat. Sb.* **47**, 271–306 (in Russian).

Goodman, J. W. (1968). *Introduction to Fourier Optics* (McGraw-Hill, New York).

Hamilton, M. F., Naze Tjøtta, J., and Tjøtta, S. (1985). Nonlinear effects in the farfield of a directive sound source. *J. Acoust. Soc. Am.* **78**, 202–216.

Hart, T. S., and Hamilton, M. F. (1988). Nonlinear effects in focused sound beams. *J. Acoust. Soc. Am.* **84**, 1488–1496.

Inoue, Y., and Yano, T. (1993). Propagation of strongly nonlinear plane waves. *J. Acoust. Soc. Am.* **94**, 1632–1642.

Johnson, H. F. (1987). An improved method for computing a discrete Hankel transform. *Comp. Phys. Comm.* **43**, 181–202.

Kamakura, T., Hamada, N., Aoki, K., and Kumamoto, Y. (1989). Nonlinearly generated spectral components in the nearfield of a directive sound source. *J. Acoust. Soc. Am.* **85**, 2331–2337.

Kamakura, T., Tani, M., Kumamoto, Y., and Ueda, K. (1992). Harmonic generation in finite amplitude sound beams from a rectangular aperture source. *J. Acoust. Soc. Am.* **91**, 3144–3151.

Korpel, A. (1980). Frequency approach to nonlinear dispersive waves. *J. Acoust. Soc. Am.* **67**, 1954–1958.

Lee, Y.-S., and Hamilton, M. F. (1995). Time-domain modeling of pulsed finite-amplitude sound beams. *J. Acoust. Soc. Am.* **97**, 906–917.

Maestrello, L., Bayliss, A., and Turkel, E. (1981). On the interaction of a sound pulse with the shear layer of an axisymmetric jet. *J. Sound Vib.* **74**, 281–301.

McDonald, B. E., and Ambrosiano, J. J. (1984). High order upwind flux methods for scalar hyperbolic equations. *J. Comp. Phys.* **56**, 448–460.

McDonald, B. E., and Kuperman, W. A. (1985). Time domain solution of the parabolic equation including nonlinearity. *Comp. and Math. with Appl.* **11**, 843–851.

McDonald, B. E., and Kuperman, W. A. (1987). Time domain formulation for pulse propagation including nonlinear behavior at a caustic. *J. Acoust. Soc. Am.* **81**, 1406–1417.

McKendree, F. S. (1981). A numerical solution of the second-order nonlinear acoustic wave equation in one and three dimensions, Ph.D. dissertation, The Pennsylvania State University.

Naze Tjøtta, J., Tjøtta, S., and Vefring, E. H. (1990). Propagation and interaction of two collimated finite amplitude sound beams. *J. Acoust. Soc. Am.* **88**, 2859–2870.

Naze Tjøtta, J., Tjøtta, S., and Vefring, E. H. (1991). Effects of focusing on the nonlinear interaction between two collinear finite amplitude sound beams. *J. Acoust. Soc. Am.* **89**, 1017–1027.

Pestorius, F. M. (1973). Propagation of plane acoustic noise of finite amplitude. Technical Report ARL-TR-73-23 (Applied Research Laboratories, The University of Texas at Austin, AD 778 868). This report is an adaptation of the Ph.D. dissertation by the same author (The University of Texas at Austin, 1973).

Pestorius, F. M., and Blackstock, D. T. (1974). Propagation of finite-amplitude noise. In *Finite-Amplitude Wave Effects in Fluids*, L. Bjørnø, ed. (IPC Science and Technology Press, Guildford, Surrey, England), pp. 24–29.

Pishchal'nikov, Yu. A., Sapozhnikov, O. A., and Khokhlova, V. A. (1996). A modification of the spectral description of nonlinear acoustic waves with discontinuities. *Acoust. Phys.* **42**, 362–367.

Sparrow, V. W., and Raspet, R. (1990). Absorbing boundary conditions for a spherical monopole in a set of two-dimensional acoustics equations. *J. Acoust. Soc. Am.* **87**, 2422–2427.

Sparrow, V. W., and Raspet, R. (1991). A numerical method for general finite amplitude wave propagation in two dimensions and its application to spark pulses. *J. Acoust. Soc. Am.* **90**, 2683–2691. See also Sparrow, V. W. (1993). Time domain computations in nonlinear acoustics without one-way assumptions. In *Computational Acoustics*, Vol. 1, R. L. Lau, D. Lee, and A. R. Robinson, eds. (Elsevier Science, London), pp. 359–369.

Too, G. P. J., and Ginsberg, J. H. (1992a). Nonlinear progressive wave equation model for transient and steady-state sound beams. *J. Acoust. Soc. Am.* **92**, 59–68.

Too, G. P. J., and Ginsberg, J. H. (1992b). Cylindrical and spherical coordinate versions of NPE for transient and steady-state sound beams. *J. Vib. Acoust.* **114**, 420–424.

Trivett, D. H., and Van Buren, A. L. (1981). Propagation of plane, cylindrical, and spherical finite amplitude waves. *J. Acoust. Soc. Am.* **69**, 943–949.

Van Buren, A. L., and Breazeale, M. A. (1968). Reflection of finite-amplitude ultrasonic waves. II. Propagation. *J. Acoust. Soc. Am.* **44**, 1021–1027.

Webster, D. A., and Blackstock, D. T. (1977). Finite-amplitude saturation of plane sound waves in air. *J. Acoust. Soc. Am.* **62**, 518–523.

Webster, D. A., and Blackstock, D. T. (1978a). Collinear interaction of noise with a finite-amplitude tone. *J. Acoust. Soc. Am.* **63**, 687–693.

Webster, D. A., and Blackstock, D. T. (1978b). Experimental investigation of outdoor propagation of finite-amplitude noise. NASA Contractor Report 2992 (Langley Research Center).

Yano, T., and Inoue, Y. (1996). Strongly nonlinear waves and streaming in the near field of a circular piston. *J. Acoust. Soc. Am.* **99**, 3353–3372.

Zhileikin, Ya. M., Zhuravleva, T. M., and Rudenko, O. V. (1980). Nonlinear effects in the propagation of high-frequency sound waves in tubes. *Sov. Phys. Acoust.* **26**, 32–34.

Chapter 12
Propagation in Inhomogeneous Media (Ray Theory)

Christopher L. Morfey and Frederick D. Cotaras

Contents

C. L. Morfey
Institute of Sound and Vibration Research, University of Southampton, Southampton, UK

F. D. Cotaras
Defense Research Establishment Atlantic, Dartmouth, NS, Canada

© The Author(s) 2024   337
M. F. Hamilton, D. T. Blackstock (eds.), *Nonlinear Acoustics*,
https://doi.org/10.1007/978-3-031-58963-8_12

12.1   Ray Theory and Its Extension to Finite Amplitude Propagation

12.1.1   Introduction

We begin this chapter by showing how geometrical acoustics, or ray theory, may be extended to allow for nonlinear waveform distortion, in the relatively simple case where the medium is stationary. Examples principally related to long-range ocean propagation are presented in Sect. 12.2. In Sect. 12.3 the theory is generalized to deal with a moving medium; it then becomes necessary to distinguish between ray and wavenormal directions. Further examples, related to aeroacoustics and sonic boom propagation, follow in Sect. 12.4.

The theoretical development involves a number of approximations, which are reviewed in Sect. 12.5. Essentially they restrict the application of the results to weakly nonlinear acoustic waves, propagating in a slowly varying medium that is not strongly dispersive. Finally, Sect. 12.6 discusses the relevant acoustical properties of water and seawater.

12.1.2   Definitions and Approximations—Linear Ray Theory

An inhomogeneous moving medium, in its undisturbed state, has a density $\rho_0(\mathbf{x})$, sound speed $c_0(\mathbf{x})$ and velocity $\mathbf{w}(\mathbf{x})$ which are functions of vector position $\mathbf{x}$.[1] A *slowly varying medium* is one in which these quantities change by only a small amount over distances comparable with a typical acoustic wavelength λ.

---

[1] Throughout this chapter, the undisturbed state of the medium is regarded as steady. Extension to a time-dependent medium is fairly straightforward provided the time dependence is slow enough, although it is more complicated to implement numerically. Specifically, the inequalities in (12.1) must also hold over time intervals comparable with the acoustic period.

Specifically,

$$\frac{\Delta\rho_0}{\rho_0}, \frac{\Delta c_0}{c_0}, \frac{|\Delta\mathbf{w}|}{c_0} \ll 1 \quad \text{for} \quad |\Delta\mathbf{x}| \approx \lambda. \tag{12.1}$$

Because of restriction (12.1), it is possible to regard both the properties and the motion of the medium as locally uniform on a wavelength scale. Ray acoustics is the approximate description, valid in the far field of the sound source, that results from:

- Treating the sound field locally as a set of plane progressive wave fronts, propagating in a single direction, in a uniform medium
- Allowing the wave-front directions to change according to the laws of refraction
- Imposing energy conservation to determine the local field strength

A concise account may be found in Pierce (1989).

12.1.3 Illustration—Rays in a Horizontally Stratified Medium

Figure 12.1 shows sound waves propagating from a source at A to a receiver at B; the intervening medium is stationary, but has a sound speed $c_0(z)$ that varies with the downward coordinate z.[2] Such a model, with horizontally stratified properties, provides a simplified representation of the Earth's oceans or atmosphere; in the latter case, the wind would normally be included as a horizontal vector $\mathbf{w}(z)$, and the complications that this introduces are addressed in Sects. 12.3 and 12.4.

A ray is the path traced by wavelets (such as P in Fig. 12.1a) that propagate, relative to the local medium, at the local sound speed in the wavenormal direction. The ray AB in Fig. 12.1a is curved, but its projection in the horizontal plane is a straight line (see Fig. 12.1b), as long as the ambient medium is stationary and the sound speed depends only on the coordinate z.

In a stationary fluid[3] medium, the unit vector $\mathbf{n}$ in Fig. 12.1a, normal to the local wave fronts, coincides with the ray propagation direction. When we deal with moving media in Sect. 12.3, it will be necessary to distinguish the wavenormal direction $\mathbf{n}$ from the ray direction $\mathbf{m}$; for stratified moving-medium problems, where the wavenormal direction is described by a grazing angle χ, we shall denote the corresponding ray angle by θ.

The process of refraction in the stationary-medium example of Fig. 12.1 is described by

---

[2] Defining z downward is conventional in ocean acoustics, whereas in atmospheric acoustics z is usually defined upward (as in Fig. 12.4).

[3] Or solid; Sects. 12.1 and 12.2 still apply, provided the waves are compressional and the medium is isotropic.

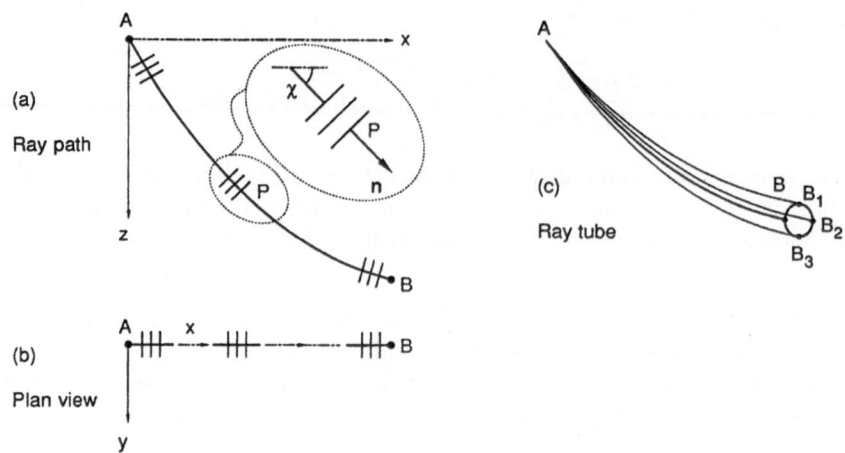

Fig. 12.1 (**a, b**) Geometry of a typical ray path in a stratified medium at rest, showing the wavenormal grazing angle χ at a typical point P on the ray path. (**c**) Ray tube generated by a set of adjacent rays, all originating at A.

$$\frac{c(z)}{\cos \chi(z)} = \frac{c_+}{\cos \chi_+} = \text{const}, \tag{12.2}$$

which is Snell's law. Here and in the remainder of the chapter, we omit the zero subscript denoting undisturbed values of the sound speed, except where it is essential. The $+$ subscript notation introduced in Eq. (12.2) will be used frequently; it denotes the value at a reference point on the ray, usually the starting point.

The local wave-front orientation, described by Eq. (12.2), determines the ray geometry. In Fig. 12.1c, rays are drawn from A to a set of neighboring points B_1, B_2, etc., as well as to B. The bundle of rays defined in this way forms a ray tube; its cross-sectional area at any point is denoted by $A(s)$, with coordinate s measured along the ray path.

12.1.4 Signal Variation Along a Ray—Linear Theory

In a slowly varying medium that is lossless, a simple time-domain equation connects the signal at any point along a ray to its earlier value at a point nearer the source. By "signal" we mean any of the linear variables (pressure, particle velocity, etc.) that characterize the locally plane progressive wave; the acoustic pressure p is used throughout this chapter. The lossless-medium linear relation is

$$\frac{p(s, \tau)}{p(s_+, \tau)} = B(s, s_+), \tag{12.3}$$

where τ is the shifted, or retarded, time variable defined by

$$\tau = t - \int_{s_+}^{s} \frac{ds'}{c(s')}. \tag{12.4}$$

The amplitude factor B is determined by the ray geometry and the local value of ρc:

$$B = \left(\frac{\rho c}{\rho_+ c_+} \right)^{1/2} \left(\frac{A}{A_+} \right)^{-1/2}. \tag{12.5}$$

An equivalent linear-theory statement is given in Sect. 8-5 of Pierce (1989), along with a derivation.

Equations (12.3)–(12.5) have a simple physical interpretation that is helpful when we come to derive a nonlinear version. The integral in Eq. (12.4) is the time taken for small-amplitude acoustic waves to travel from s_+ to s along the ray. The amplitude factor accounts for energy conservation along a ray tube; the instantaneous energy flow rate is the product of the local area A and the local intensity I ($= p^2/\rho c$), and its value is required to remain constant as the signal is tracked along the ray path.

Finally, we note the following important consequence of Eqs. (12.3)–(12.5): The quantity $B^{-1} p(s, \tau)$ is *invariant along a ray*, according to linear geometrical acoustics, for any given value of the shifted time variable τ. This statement is expressed in mathematical form by

$$B^{-1} p = \tilde{p}(s, \tau) = p_+(\tau). \tag{12.6}$$

In Sects. 12.1.6–12.1.10 we show how Eq. (12.6) may be generalized to finite amplitudes.

12.1.5 Propagation in a Lossy Medium

Before moving to finite amplitudes, we note for future reference how Eq. (12.3) may be generalized to describe signal propagation along rays in a lossy inhomogeneous medium. This is accomplished by transforming the equation into the frequency domain, and then applying factors to account for attenuation and dispersion. In the frequency domain, the lossless-medium equation (12.3) is equivalent to a transfer function $T(\omega, s, s_+) = B \exp(j\omega \Delta t)$, where Δt is the time delay $(t - \tau)$ given by Eq. (12.4). The correction factor to be applied to T to account for attenuation and dispersion is

$$F(\omega, s, s_+) = \exp \left\{ - \int_{s_+}^{s} \tilde{\alpha}(\omega, s') \, ds' \right\}, \tag{12.7}$$

where $\tilde{\alpha}$ is the complex attenuation coefficient [see Eq. (5.5)]. Multiplying T by F is equivalent to local substitution of the lossless propagation coefficient $j\omega/c$ by $(\tilde{\alpha} + j\omega/c)$ at each point along the ray path.

12.1.6   Nonlinear Ray Theory—Additional Assumptions Needed

The theory of small-amplitude lossless propagation along a ray, as given in Sect. 12.1.4, is extended in Sects. 12.1.7–12.1.10 to weakly nonlinear signals. The phenomenon of waveform distortion—described for plane progressive waves in Chap. 4—is shown to follow analogous rules in a nonuniform medium, insofar as ray theory is valid.

Two key assumptions are involved in this extension:

1. Every element of the starting waveform (at $s = s_+$) may be regarded as following a common ray path, despite the fact that different elements travel at different propagation speeds;
2. The effects of attenuation and dispersion may be incorporated into the lossless analysis retrospectively (Pestorius, 1973) in the manner of Sect. 12.1.5, although corrections will need to be made at intervals along the propagation path, rather than in one step at the end.

Assumption 2 is not self-evident, because rays in a nonuniform dispersive medium follow frequency-dependent paths: the rainbow or optical prism provides a vivid analogy. However, we shall assume the effects of dispersion to be weak enough that rays joining any two given points remain sharply defined along their whole length, and do not suffer "chromatic aberration."

12.1.7   Wavelet Propagation Speed and Ray Velocity

The following terminology is useful in discussing waveform distortion at finite amplitudes. A small segment of the signal time history is called a *wavelet*; an example is PP' in Fig. 12.2. Each wavelet is in fact associated with a spatially extended wave front, which is regarded as locally plane, and propagates in direction **n** relative to the undisturbed fluid. The *wavelet propagation speed*, again relative to the undisturbed fluid, is given to first order by

$$v_w \simeq c_0 + \beta u \simeq c_0 + (\beta/\rho c)p, \tag{12.8}$$

where u is the acoustic particle velocity and β is the coefficient of nonlinearity.[4] In the small-amplitude limit, v_w reduces to c_0.

---

[4] See Eqs. (2.16). Note that in the last term of Eqs. (12.8), ρ and c should strictly be interpreted as the undisturbed values $\rho_0(\mathbf{x})$ and $c_0(\mathbf{x})$; the difference this makes to v_w is of second order, however. We shall therefore drop the zero subscript wherever there is no risk of confusion, while retaining it in the zero-order c_0 term.

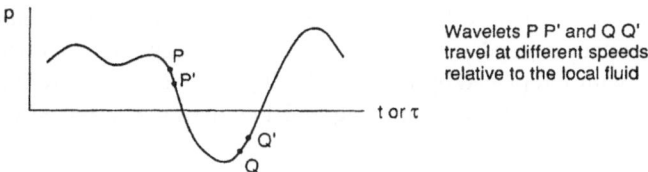

Fig. 12.2 Pressure waveform as a function of time. Adjacent points P, P' or Q, Q' on the waveform define a wavelet.

Our concern is to obtain a nonlinear version of the arrival-time equation (12.4) for any given ray path. We do this by introducing the *nonlinear ray velocity*, defined in general (for a moving medium) by

$$\mathbf{v}_g = v_w \mathbf{n} + \mathbf{w}; \qquad\qquad (12.9)$$

its direction is, by assumption 1 above, along the original linear ray, and its magnitude is the rate at which a wavelet advances along the ray path. The notation $\mathbf{v}_g$ is used because the ray velocity is also the group velocity. Equations (12.8) and (12.9) imply that, in Fig. 12.2 for example, $\mathbf{v}_g(P) \neq \mathbf{v}_g(Q)$, and therefore the wavelets at P and Q take different times to travel the same distance along the ray path.

An explicit expression is developed below for the amplitude-dependent wavelet travel time for the special case of a stationary medium, in which $\mathbf{w}$ is zero. The corresponding analysis for a moving medium is deferred to Sect. 12.3.

12.1.8   Signal Travel Times in a Nonmoving Medium

It is convenient, from this point on, to characterize or label each wavelet by the time τ_w at which it passes the reference point $s = s_+$. The time at which it reaches point s is denoted by $t(s, \tau_w)$, and the travel time between s_+ and s is given by

$$\Delta t = t(s, \tau_w) - \tau_w = \int_{s_+}^{s} \frac{ds'}{v_g(s', \tau_w)}. \qquad (12.10)$$

Because $\mathbf{w}$ is zero, Eqs. (12.8) and (12.9) give

$$\frac{1}{v_g} = \frac{1}{v_w} \simeq \left(c_0 + \frac{\beta}{\rho c} p\right)^{-1} \simeq \frac{1}{c_0} - \frac{\beta}{\rho c^3} p(s, \tau_w). \qquad (12.11)$$

An important consequence of Eqs. (12.10) and (12.11) is that the wavelet travel time differs from $\int_{s_+}^{s} ds'/c_0(s')$, which means that the wavelet does not arrive at a retarded time $\tau = \tau_w$ (as it would according to linear theory). To evaluate the integral in Eq. (12.10) for a finite-amplitude wavelet, we assume that Eqs. (12.3) and (12.6) remain a good approximation if they are applied *following the local wavelet*, i.e., with τ replaced by τ_w. Substituting $p = Bp_+(\tau_w)$, from the modified Eq. (12.6), into the nonlinear term of Eq. (12.11) gives the corrected wavelet travel time as[5]

$$\Delta t \simeq \int_{s_+}^s \frac{ds'}{c_0(s')} - p_+(\tau_w) \int_{s_+}^s \frac{\beta}{\rho c^3} B(s', s_+)\, ds'. \qquad (12.12)$$

Note that $p_+(\tau_w)$ is the time history of the pressure at the starting position ($s = s_+$). Equation (12.12) provides a nonlinear estimate of the wavelet travel time, with the first term on the right corresponding to the linear result, Eq. (12.4). The second term represents a nonlinear correction; positive pressures take less time to arrive than negative pressures.

12.1.9 Terminology—Arrival Phase and Characteristic Phase

The shifted, or retarded, time $\tau = t - (\Delta t)_{\text{lin}}$, where $(\Delta t)_{\text{lin}}$ is the linear signal travel time from some reference point on the ray path, is called by some authors the time phase variable or simply the phase (Hayes and Runyan, 1972). For nonlinear propagation, Eq. (12.12) gives the phase variable following a wavelet as

$$\tau(s, \tau_w) = \tau_w - p_+(\tau_w) \int_{s_+}^s \frac{\beta}{\rho c^3} B\, ds'. \qquad (12.13)$$

Equation (12.13) determines the value of τ for the arriving signal as a function of τ_w and s; we shall call this quantity $\tau(s, \tau_w)$ the *arrival phase* of the signal, and τ_w the *characteristic phase*. Hayes and Runyan[6] call τ_w the *linear phase*, since the retarded time of arrival of a *small-amplitude* wavelet at any point on the ray is given by $\tau = \tau_w$.

---

[5] The integral factor in the second term is called the *age variable* by Pierce (1989; see Sect. 11-8). The same name is used for a different but related quantity by Hayes and Runyan (1972) in their important paper on sonic boom propagation. In neither case does the age variable have the dimensions of time! In Sect. 12.1.10 the "age variable" terminology will be avoided and Eq. (12.12) will be written in an alternative form, analogous to the plane-wave result in Eq. (4.26).

[6] Hawkings (1974) adopts an apparently similar but incompatible terminology in which τ_w is called the phase, and τ is then (logically) called the linear phase!

12.1.10 Summary of Results for a Lossless Nonmoving Medium

The essential feature of nonlinear geometrical acoustics in a lossless medium, as explained in Sect. 12.1.8, is that $\tilde{p}(s|\tau_w = \text{const})$ is invariant along each ray path, rather than $\tilde{p}(s|\tau = \text{const})$. Here $\tilde{p}$ is the modified pressure p/B introduced in Eq. (12.6). Thus the arriving wavelet amplitude is given by

$$\tilde{p}(s, \tau_w) = p_+(\tau_w), \qquad (12.14)$$

while the actual arrival phase $\tau(s, \tau_w)$ is related to τ_w by Eq. (12.13). The point of Eq. (12.14) is that the wavelet energy conservation argument that led to Eq. (12.6) carries over unaltered to weakly nonlinear waves (Whitham, 1974, Chap. 9). Some insight into this conclusion may be gained by noting that when a plane wave front develops into a shock, the fraction of the forward-traveling energy that is scattered into backward-traveling waves is of order ε^4, where ε $(= \Delta p/\rho c^2)$ is the acoustic Mach number of the initial wave front (Morfey and Sparrow, 1993).

The arrival phase equation (12.13) may be written in the equivalent form

$$\tau(s, \tau_w) = \tau_w - \left(\frac{\beta}{\rho c^3}\right)_+ \tilde{x} p_+(\tau_w); \quad \tilde{x} = \int_{s_+}^{s} \frac{\Lambda}{\Lambda_+} \left(\frac{A}{A_+}\right)^{-1/2} ds', \qquad (12.15)$$

where $\Lambda = \beta \rho^{-1/2} c^{-5/2}$. It may already have occurred to the reader that Eqs. (12.14) and (12.15) are identical in form to corresponding plane-wave results given by Eqs. (3.61), although the nonlinear propagation being described here takes place along curved rays of varying ray-tube area. The quantity $\tilde{x}$ is a generalized version of the plane-wave propagation distance, and will be referred to as the *reduced path length*; the modified pressure $\tilde{p}$ will be called the *reduced pressure*. Examples illustrating the calculation of $\tilde{x}$ are presented in the next section. We note that, as discussed in Sect. 4.2, the nonlinearly distorted waveform may become multivalued: Several wavelets, with different values of τ_w, may share the same arrival phase τ. This is an indication that losses need to be taken into account, either via weak shock theory (the simplest option) or—where that is not appropriate—via the stepwise calculation procedure outlined in Sect. 12.3.5 below.

12.2 Examples of Nonlinear Propagation in a Stationary Medium

12.2.1 Introduction

The results derived in Sect. 12.1 show that extending linear ray theory to weakly nonlinear signals requires evaluation of a reduced path length $\tilde{x}$, which is used to calculate a nonlinear correction to the signal phase variable at any given point on a ray.

The evaluation of $\tilde{x}$ from Eqs. (12.15) involves the ray geometry, through the ray-tube area $A(s)$, and a particular combination of the medium properties (β, ρ, c), defined following Eqs. (12.15) and denoted by $\Lambda(s)$. Several examples that illustrate the calculation of $\tilde{x}$ are presented in this section, beginning with one-dimensional cases and ending with numerical calculations for a stratified model ocean. Other examples of ray theory applied to nonlinear waves propagating through nonuniform media may be found in an early review by Ostrovsky (1976).

12.2.2 Waveguide or Ray Tube with Power-Law Area Variation

Figure 12.3a shows a narrow waveguide with rigid walls, containing a uniform fluid at rest. The cross-sectional area is

$$A(x) = A_+\left(\frac{x}{x_+}\right)^{2n};\qquad(12.16)$$

n may be either negative (as in Fig. 12.3a) or positive. Plane acoustic waves propagate axially along the waveguide, in the direction of increasing x. Note that with $n = \frac{1}{2}$ or 1, Eq. (12.16) could also represent a ray tube in a cylindrically or spherically spreading sound field, generated by a source at $x = 0$.

Propagation from x_+ to x along the waveguide, without passing through $x = 0$, corresponds to a reduced path length [see also Eqs. (4.291)]

$$\tilde{x}(x, x_+) = \int_{x_+}^{x} \left(\frac{x'}{x_+}\right)^{-n} dx';\qquad(12.17)$$

thus

$$\tilde{x} = \frac{x_+}{1-n}\left[\left(\frac{x}{x_+}\right)^{1-n} - 1\right] \quad \text{(for } n \neq 1\text{)},\qquad(12.18)$$

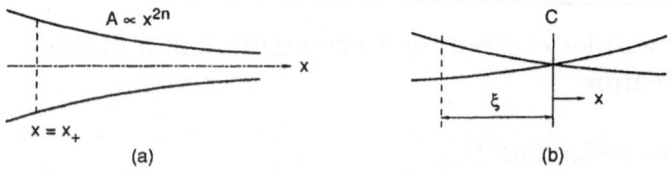

Fig. 12.3 (a) Waveguide (or ray tube) with power-law area variation. (b) Ray tube in the neighborhood of a caustic. The ray tube passes through a line focus at C (cylindrical focusing).

and when n is zero, we recover the plane-wave result $\tilde{x} = x - x_+$. For the special case $n = 1$ [see also Eqs. (4.281)],

$$\tilde{x} = x_+ \ln \frac{x}{x_+} \quad \text{(spherical spreading).} \tag{12.19}$$

Now suppose both x_+ and x are positive. Equations (12.18) and (12.19) show that for all values of n less than or equal to 1, $\tilde{x}$ increases indefinitely in the propagation direction; but if n exceeds 1 (faster-than-spherical spreading), $\tilde{x}$ approaches an asymptotic upper limit. This latter situation has been called "waveform freezing," since it implies that the difference between τ and τ_w never grows beyond a certain value, however far the wavelet propagates. A similar situation can arise when the properties of the medium vary along the ray path, as will be shown in Sect. 12.2.4 below.

12.2.3   Rays Approaching a Caustic or Focus

When neighboring rays pass through a caustic or focus, as in Fig. 12.3b, the ray-tube area vanishes and the ray approximation breaks down locally. However, it is instructive to examine the behavior of $\tilde{x}$ as the focus is approached from a short distance away.

In Fig. 12.3b, C is the point of zero area, and ξ is the distance (terminating at C) over which the reduced path length will be evaluated. The same ray-tube area variation applies here as in the previous example, namely Eq. (12.16), which allows for C being a caustic point ($n = \frac{1}{2}$) or a three-dimensional focus ($n = 1$). Note that x is measured from C in the propagation direction, so we are now working in the region of negative x.

Applying Eqs. (12.18) and (12.19) gives $\tilde{x}$ for a signal traveling from $x = -\xi$ to $x = 0$:

$$\tilde{x} = \frac{\xi}{1 - n} \quad \text{(for } n < 1\text{);} \qquad \tilde{x} \to \infty \quad \text{(for } n \geq 1\text{).} \tag{12.20}$$

Thus cylindrical focusing ($n = \frac{1}{2}$), as at a caustic, leads to the definite value $\tilde{x} = 2\xi$. Likewise, propagation beyond the caustic out to $x = \xi'$ adds a further amount $2\xi'$ to $\tilde{x}$.

There is a matching problem at the caustic itself, which for sufficiently weak waves may be overcome by using the linear matching procedure: A $\pi/2$ phase advance (time factor $e^{j\omega t}$ replaced by $e^{j(\omega t + \pi/2)}$ for positive ω, and by $e^{j(\omega t - \pi/2)}$ for negative ω) is applied to all frequency components, as explained in Sect. 9-4 of Pierce (1989). The local use of linear theory to traverse the caustic is given a formal justification by Hunter and Keller (1987); it was first proposed by Obermeier (1974) and by Ostrovsky et al. (1976).

The reduced path length becomes singular as a spherical focus is approached ($n = 1$), which points to the need for a more accurate nonlinear-propagation model in such situations. Specifically, the assumption that rays follow the paths given by linear theory may have to be abandoned (Whitham, 1974, Sects. 8.3 and 8.8; Sturtevant and Kulkarny, 1976; Fridman, 1982; see also Sects. 12.5.3 and 12.5.4), and dissipative processes and shock formation will need to be included (see Sects. 12.1.5 and 12.1.6).

12.2.4 Ray Propagation in an Isothermal Still Atmosphere

The sound speed is uniform in this example, so the rays are straight, but the density varies with the height z. It is instructive to compare $\tilde{x}$ results for plane, cylindrically-spreading and spherically-spreading waves. The ray-tube area in all three cases follows the power law $A(s) \propto s^{2n}$ (with s measured along the ray from the source), with $n = 0$, $\frac{1}{2}$, and 1, respectively.

The vertical variation of β and ρ, in an isothermal atmosphere consisting of an ideal gas at constant absolute temperature T, is given by

$$\beta(z) = \text{const} = \beta_+; \quad \rho(z) = \rho_+ e^{-z/H} \quad (H = RT/g, \text{ the scale height}).$$
$$(12.21)$$

Here z is measured upward from a convenient reference height, and the subscript $+$ labels properties of the medium at that height; R is the specific gas constant, and g is the acceleration due to gravity.

The reduced path length will be calculated for propagation from A (at $z = z_+ = 0$) to B, along the ray tube shown in Fig. 12.4; i.e., from a vertical distance h above the source to a vertical distance $h + \xi$. The ray path is straight, but it slopes at an arbitrary angle. From Eqs. (12.15) and (12.21), Λ varies along the propagation path as follows:

$$\Lambda = \Lambda_+ e^{\xi/2H}.$$
$$(12.22)$$

The ray-tube area ratio A/A_+ depends on the source distance h below the reference plane, the upward propagation distance ξ, and the exponent n. Provided h and $\xi + h$ have the same sign,[7]

$$A = A_+ \left(\frac{\xi + h}{h}\right)^{2n}.$$
$$(12.23)$$

---

[7] Not necessarily positive; the present description covers downward propagation if ξ and h are negative.

Fig. 12.4 Definition sketch for upward propagation of finite-amplitude sound in an isothermal atmosphere.

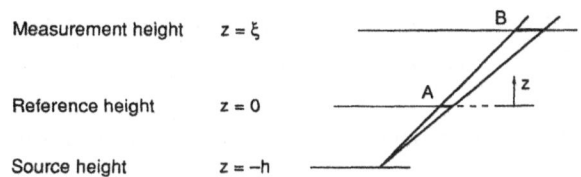

Measurement height	$z = \xi$
Reference height	$z = 0$
Source height	$z = -h$

Substituting Eqs. (12.22) and (12.23) in the second of Eqs. (12.15) gives

$$\tilde{x} = \int_0^\xi e^{\xi'/2H} \left(1 + \frac{\xi'}{h} \right)^{-n} d\xi'. \tag{12.24}$$

The presence of the exponential factor in the integrand above has a powerful effect once $|\xi|$ becomes comparable with the scale height or greater. Upward-directed rays then suffer *enhanced* nonlinear distortion (greatly increased $\tilde{x}$), compared with propagation over the same path in the reference medium whose density is ρ_+. Downward-directed rays (ξ negative) suffer *reduced* distortion on account of the exponential factor: However large the path length, the value of $\tilde{x}$ never exceeds an asymptotic upper limit, in contrast to the uniform-medium case (Sect. 12.2.2), where $\tilde{x}$ increases indefinitely. The former situation was discussed by Ostrovsky and Fridman (1985), who showed that upward-propagating sound waves from a source near the ground are strongly dissipated on reaching the ionosphere, on account of shock formation in this low-density region.

For downward-propagating rays, Hayes (1968, 1969) had already noted the existence of an upper limit on $\tilde{x}$, and had coined the phrase "waveform freezing" for this situation (cf. Sect. 12.2.2). Hayes was concerned with the nonlinear evolution of sonic boom signatures from high-flying aircraft, for which an appropriate ray-tube model is Eq. (12.23) with $n = \frac{1}{2}$, rather than $n = 1$ as used by Ostrovsky and Fridman (1985). It is interesting to compare the asymptotic $\tilde{x}$ approached by downward-propagating rays in the limit $|\xi| \gg H$ for all three values of the spreading index ($n = 0, \frac{1}{2}, 1$), i.e., for plane, cylindrical, and spherical waves. These asymptotic limiting values correspond to propagation along the same ray tube in the reference medium, but over a "frozen" vertical distance ξ_∞:

$$|\xi_\infty| = H F(n, h/H). \tag{12.25}$$

In the limit $h \ll H$, the factor $F = |\xi_\infty|/H$ is given by

$$F(0, 0) = 2/e^\gamma \doteq 1.1229; \quad F(\tfrac{1}{2}, 0) = \pi/2 \doteq 1.5708; \quad F(1, 0) = 2, \tag{12.26}$$

while for $h \gg H$, the effect of wave-front spreading becomes negligible, and the plane-wave result ($F = 2$) applies regardless of n.

For the next two examples we introduce a modification factor G, applied to the actual path length, as a convenient way to quantify $\tilde{x}$ in more general situations.

12.2.5   Long-Range Propagation in a Depth-Dependent Ocean: Definition of the Factor G

Results for $\tilde{x}$ obtained in Sects. 12.2.6 and 12.2.7 will be presented in terms of the path-length modification factor G, defined for rays spreading from a point source by the relation

$$\tilde{x}(s, s_+) = s_+ \ln\left[1 + \left(\frac{s - s_+}{s_+}\right)G\right]. \tag{12.27}$$

Implicit in the choice of Eq. (12.27) to represent $\tilde{x}$ is the assumption that, from the source out to the reference position $s = s_+$, the inhomogeneity of the medium may be neglected. This means that we are assuming $s_+ \ll H$, where H is the vertical inhomogeneity scale.

The factor G in Eq. (12.27) may then be interpreted as follows. Suppose that the medium beyond $s = s_+$ is replaced by a reference medium, with uniform properties matching those at s_+. For this reference case, Eq. (12.19) gives

$$\tilde{x}(R, s_+) = s_+ \ln\frac{R}{s_+} \quad \text{(at spherical radius } R\text{).} \tag{12.28}$$

The two values of $\tilde{x}$ in Eqs. (12.27) and (12.28) will be equal if

$$R - s_+ = (s - s_+)G. \tag{12.29}$$

Thus G is the factor by which the actual curved path length must be multiplied in order to produce the same $\tilde{x}$ (and hence the same nonlinear distortion, for any given starting amplitude) under ideal spherical spreading conditions in the reference medium, starting from the same radius s_+.

Finally, we note that Eq. (12.27) may be simplified in the limit of *long-range* propagation, defined by $s \gg s_+$; the long-range version, which we shall use in the examples that follow, is

$$\tilde{x} \simeq s_+ \ln\left(\frac{s}{s_+}G\right). \tag{12.30}$$

12.2.6   Analytical Results: Ocean with Linear Sound-Speed Profile

In accord with convention, the vertical coordinate z is measured downward from the ocean surface. If the sound speed varies linearly with depth, the ray paths are circular arcs. The ray-tube area ratio A/A_+ is given in terms of the current ray angle θ and the ray launch angle λ by

$$\frac{A}{A_+} \approx \frac{rd}{r_+^2} \cos\theta \cos\lambda \quad \left(d = \frac{\partial z}{\partial\lambda}\bigg|_r\right), \tag{12.31}$$

with both angles measured downward from the horizontal. In this expression, r denotes horizontal range measured from the source; its value at the reference point near the source is r_+. Out to $r = r_+$ the rays are approximated by straight lines, as noted in Sect. 12.2.5.

Combining Eq. (12.31) with standard ray path expressions (Tolstoy and Clay, 1987; Frisk, 1994) gives

$$\frac{A}{A_+} = \left(\frac{s}{s_+}\right)^2 \left(\frac{\sin\theta - \sin\lambda}{\theta - \lambda}\right)^2 \sec^2\lambda. \tag{12.32}$$

Note that in this example $s/(\theta - \lambda)$ remains constant along a given ray; its value is determined by the (constant) sound-speed gradient. By introducing the gradient, either of (s, θ) may be eliminated in favor of the other, but the form of Eq. (12.32) is particularly convenient.

We now introduce the assumption that Λ and c are related, along each ray, by the power law $\Lambda c^n = \text{const}$. This provides a first approximation to typical deep-ocean profiles (Morfey, 1984; see Appendix C). Then

$$\frac{\Lambda}{\Lambda_+} = \left(\frac{c}{c_+}\right)^{-n} = \left(\frac{\cos\theta}{\cos\theta_+}\right)^{-n} \simeq \left(\frac{\cos\lambda}{\cos\theta}\right)^n. \tag{12.33}$$

Inserting Eqs. (12.32) and (12.33) in Eqs. (12.15) gives $\tilde{x}$ as an integral along the ray path. We use the relation $ds/s = d\theta/(\theta - \lambda)$ to change the integration variable from s to ray angle θ:

$$\tilde{x} \simeq s_+ \cos\lambda \int_{\theta_+}^{\theta} \frac{d\theta}{\sin\theta - \sin\lambda} \left(\frac{\cos\lambda}{\cos\theta}\right)^n. \tag{12.34}$$

The integral is to be evaluated for $s_+ \ll s$, which means that θ_+ approaches λ. Although the limiting $\tilde{x}$ is singular, the factor G defined in Eq. (12.30) remains finite. The limiting value of G can be expressed analytically for $n = 0$ and $n = 1$ (Gradshteyn and Ryzhik, 1980). The case $n = 1$ yields

$$G(\theta, \lambda) = \frac{\sin\theta - \sin\lambda}{(\theta - \lambda)\cos\theta} \left[\tan\left(\frac{\pi}{4} + \frac{\theta}{2}\right) \Big/ \tan\left(\frac{\pi}{4} + \frac{\lambda}{2}\right)\right]^{\sin\lambda}, \tag{12.35}$$

and is of practical interest since in the deep ocean, the product Λc remains roughly constant below the sound channel axis.

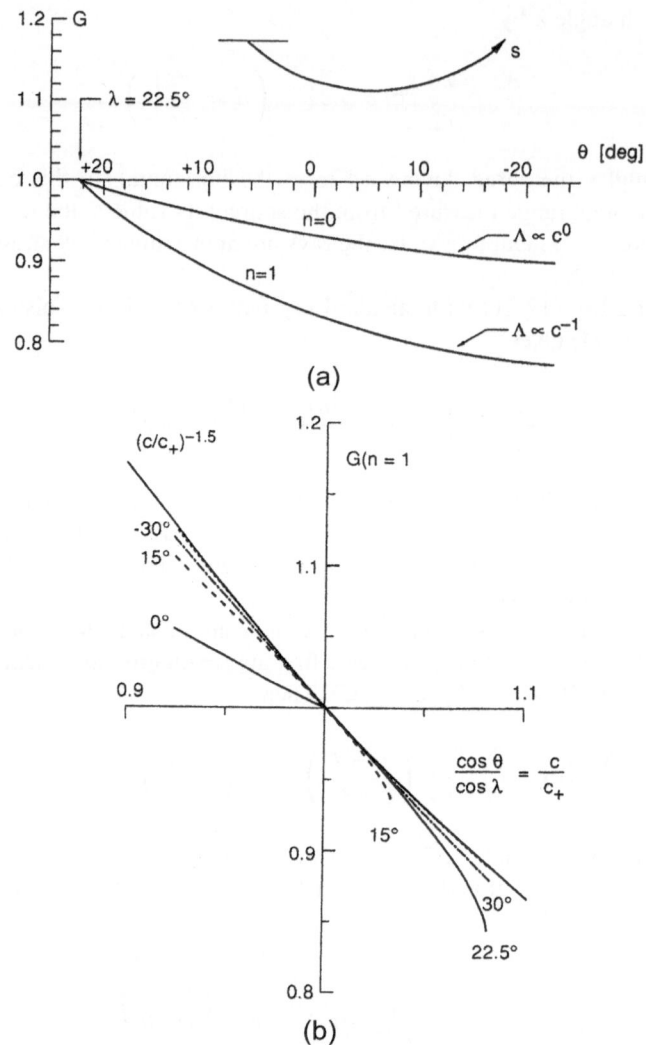

Fig. 12.5 (a) Values of the path-length modification factor G calculated for a stratified medium with a linear sound-speed profile, using two different models for Λ as a function of sound speed. The ray is launched in the direction of increasing sound speed at an angle of 22.5° to the horizontal, and G is shown as a function of the subsequent ray angle θ. (b) Further results for the situation described in (a), extended to a range of launch angles to either side of the horizontal. Plotting G versus $\cos\theta/\cos\lambda$ permits comparison with the asymptotic theory of Sect. 12.2.7.

Note that the sound-speed gradient does not appear in Eq. (12.35), which predicts G in terms of n and the initial and final ray angles. As an illustration of the way G varies along a typical ray, Fig. 12.5a shows G plotted against θ for a ray launched downward at $\lambda = 22.5°$. The ray-tube area effect alone ($\Lambda = \text{const}$) produces the curve labeled $n = 0$; G is less than 1 for this particular ray path because the area $A(s)$ increases more rapidly than s^2. Adding a semirealistic $\Lambda(s)$ variation accentuates the effect ($n = 1$ curve), since the ray starts out downward and Λ decreases with depth.

A wider range of results for the same sound-speed profile and $n = 1$ is shown in Fig. 12.5b, where the horizontal axis is the sound-speed ratio c/c_+. The effective path length for nonlinear distortion is increased when rays are launched upward and decreased when rays are launched downward. Plotting G versus c/c_+ reveals the approximate power-law relation $G \approx (c/c_+)^{-1.5}$, which is given theoretical support by the analysis described in Sect. 12.2.7.

12.2.7   Sound-Speed Profiles with Curvature—Asymptotic Results

If we remove the limitation to linear $c(z)$ profiles, it is more difficult to find analytical expressions for $\tilde{x}$. Some progress can be made by asymptotic methods, with the relative variation in sound speed treated as a small quantity: results obtained in this way are summarized below. Otherwise, $\tilde{x}$ may be evaluated numerically; examples are presented in Sect. 12.2.8. The results in this chapter may be compared with the earlier work of Pelinovsky et al. (1979).

It is convenient for asymptotic analysis to define the sound-speed profile in inverse form, with z given as a function of the quantity

$$\sigma = c(z)/c(0) - 1. \tag{12.36}$$

The following two-parameter profile family includes the linear profile of Sect. 12.2.6 as a special case:

$$\frac{z}{a} = \sigma - \tfrac{1}{2}b\sigma^2 \quad (a, b \text{ constants}). \tag{12.37}$$

It leads to a particularly simple prediction for the factor G in Eq. (12.30)—and hence for $\tilde{x}$—when terms of relative order σ^2 are discarded. The final result is

$$G \simeq \left(\frac{c}{c_+}\right)^{-(n+1/2)}, \tag{12.38}$$

where n is the index relating Λ and c, introduced in Eq. (12.33). Details of the analysis are given elsewhere (Morfey, 1984, Appendix D).

Besides the restriction on profile shape imposed by Eq. (12.37), which can represent the gradient and curvature of $c(z)$ only in an average sense, two further limitations apply to the asymptotic Eq. (12.38):

1. dc/ds must not change sign along the ray path; rays which have passed through a vertex, or crossed the sound channel axis, are not validly described by Eq. (12.38) beyond that point.
2. The small-σ limit strictly requires $\sigma \ll \sin^2\lambda$, so the asymptotic theory does not apply to horizontally launched rays.

Since the linear sound-speed profile belongs to the profile family defined above, the exact expression for G already found as Eq. (12.35) (for $n = 1$) provides a test of Eq. (12.38). The appropriate asymptotic prediction is $G \simeq (c/c_+)^{-1.5}$, and this line is plotted in Fig. 12.5b alongside the exact results. The agreement improves as $|\lambda|$ increases, which is consistent with limitation (2) above.

12.2.8   Realistic Ocean Profiles—Numerical Results

Figure 12.6 shows values of G calculated numerically from Eqs. (12.15) and (12.30) for a depth-dependent ocean typical of the North Pacific (Morfey, 1984). The two source depths in Fig. 12.6a and b are above and below the sound channel axis,

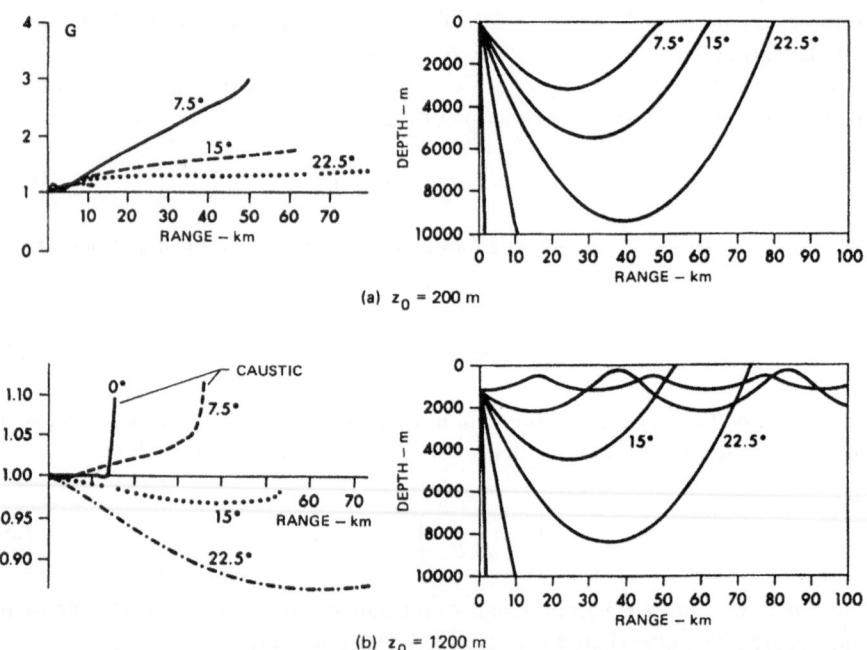

Fig. 12.6 Numerical calculations of the path-length modification factor G, for two source depths in a typical North Pacific ocean profile with the sound channel axis at a depth of 600 m. The source in (**a**) is above the sound channel axis at 200 m depth, and the source in (**b**) is below the axis at 1200 m depth.

respectively. Noteworthy features are, in Fig. 12.6a, the quite large G (of order 4) attained by near-horizontal rays, and, in Fig. 12.6b, the rapid increase in G as the caustic surface is approached. (The numerical calculations were performed in discrete steps of horizontal range, and therefore stop short of the actual caustic.)

When these and similar numerical examples are compared with the analytical approximation of G in Eq. (12.38), using $n = 1$, it appears that nearly all the realistic cases that fall within restriction 1 of Sect. 12.2.7 are quite well described by the asymptotic prediction. The agreement between asymptotic and numerically calculated values of $(G - 1)$ is typically within a factor of 1.4. However, restriction 1 imposes a severe limitation, and the high G values seen in Fig. 12.6a are outside the scope of Eq. (12.38): The rays involved have all passed through a sound-speed minimum.

12.3   Finite Amplitude Ray Propagation in Moving Media

12.3.1   Introduction

The linear theory of geometrical acoustics in moving media was established by Blokhintsev (1946a, 1946b); a concise account, with additional references, appears in the book by Pierce (1989). Our aim in this section is to show how the linear theory may be corrected to allow for first-order effects of nonlinearity. The procedure parallels that given in Sect. 12.1, and we shall not repeat the detail of the earlier discussion, but rather focus on the differences caused by a mean ambient flow. Practical illustrations are deferred to Sect. 12.4.

12.3.2   Signal Variation Along A Ray—Linear Theory

For small-amplitude signals Eq. (12.3) still applies, namely

$$\frac{p(s, \tau)}{p(s_+, \tau)} = B(s, s_+), \tag{12.39}$$

but Eqs. (12.4) and (12.5) are modified to allow for the mean flow:

$$\tau = t - \int_{s_+}^{s} \frac{ds'}{v_g(s')}, \tag{12.40}$$

$$B = \frac{c\rho^{1/2}(ADv_g)^{-1/2}}{c_+\rho_+^{1/2}(ADv_g)_+^{-1/2}}. \tag{12.41}$$

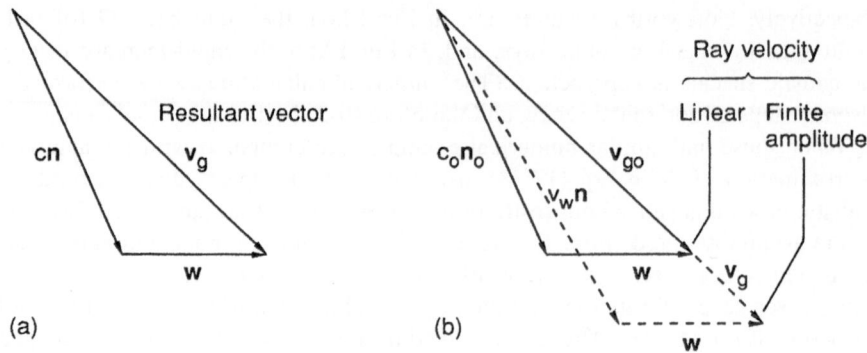

Fig. 12.7 (a) Definition of the ray velocity in a moving medium; $\mathbf{w}$ is the mean flow velocity vector, $\mathbf{n}$ is the unit wavenormal vector, and $\mathbf{v}_g$ is the ray velocity vector. (b) Vector triangles as in (a), compared for linear and finite-amplitude wave fronts traveling in the same ray direction.

Here v_g is the magnitude of the ray velocity as defined in Eq. (12.9) and Fig. 12.7, and D is a Doppler factor determined by the component $w_n = \mathbf{w} \cdot \mathbf{n}$ of the mean velocity field:

$$D = 1 + w_n/c = (1 + w_n/Dc)^{-1}. \tag{12.42}$$

The factor D is the ratio of the frequency f heard by a fixed observer to the frequency f_m heard by an observer moving with the local flow.

12.3.3 Signal Travel Times in a Moving Medium

We now modify the results above to allow for nonlinear propagation. The discussion parallels that in Sects. 12.1.7 and 12.1.8, except that for a fixed ray direction, the wavenormal direction in a moving medium is amplitude-dependent. This is illustrated in Fig. 12.7b, which compares the ray vector triangles for a small-amplitude and a finite-amplitude wavelet.

It follows that when the travel time integral in Eq. (12.40) is evaluated along the original ray path, and we wish to take account of v_w variations between wavelets, the appropriate first-order approximation is

$$v_g \simeq v_{g0} + \left.\frac{\partial v_g}{\partial v_w}\right|_{m,w} (v_w - c_0). \tag{12.43}$$

The wavenormal direction $\mathbf{n}$ is allowed to vary in evaluating the partial derivative, while the ray direction $\mathbf{m}$ is held fixed. This corresponds to using the linear ray path for the nonlinear travel time estimate. The resulting error in Δt is of second order in $(v_w - c_0)$, because small variations in the ray path produce no first-order change in travel time, by definition.

The partial derivative in Eq. (12.43) is obtained by differentiating the relation

$$v_w^2 = w^2 + v_g^2 - 2wv_g\cos\theta, \tag{12.44}$$

where θ is the angle between the vectors $\mathbf{v}_g$ and $\mathbf{w}$. Holding $\mathbf{m}$ constant is equivalent to holding θ constant. Thus

$$\left.\frac{\partial v_w}{\partial v_g}\right|_{\mathbf{m},\mathbf{w}} = \frac{v_g - v_w\cos\theta}{v_w} = \mathbf{n}\cdot\mathbf{m} \quad \text{(from Fig. 12.7b).} \tag{12.45}$$

Inversion gives

$$\left.\frac{\partial v_g}{\partial v_w}\right|_{\mathbf{m},\mathbf{w}} = (\mathbf{n}\cdot\mathbf{m})^{-1} = \frac{v_g}{v_w + w_n}, \tag{12.46}$$

after making the substitutions $\mathbf{n}\cdot\mathbf{m} = \mathbf{n}\cdot\mathbf{v}_g/v_g$ and $\mathbf{v}_g = v_w\mathbf{n} + \mathbf{w}$. Finally, when Eq. (12.46) is substituted into Eq. (12.43) along with Eq. (12.8) for v_w, we get

$$v_g \simeq v_{g0} + \frac{v_g}{Dc}\beta u, \tag{12.47}$$

which is accurate to first order in the velocity perturbation u. Note that Eq. (12.42) has been used to rewrite $v_w + w_n$ in terms of D.

The nonlinear travel time of a given wavelet may now be estimated to the same accuracy, as an integral over the linear ray path:

$$\Delta t = \int_{s_+}^{s}\frac{ds'}{v_g(s',\tau_w)} \simeq \int_{s_+}^{s}\frac{ds'}{v_{g0}} - p_+(\tau_w)\int_{s_+}^{s}\frac{\beta}{\rho c^2}\frac{1}{Dv_g}B(s',s_+)\,ds'. \tag{12.48}$$

This result generalizes Eq. (12.12), derived for a nonmoving medium in Sect. 12.1. As before, $p_+(\tau_w)$ is the time history of the pressure at the starting position ($s = s_+$).

As a final comment, we note that if the partial derivative in Eq. (12.43) had been evaluated with $\mathbf{n}$ rather than $\mathbf{m}$ held constant, then the coefficient v_g/Dc in Eq. (12.47) would have been inverted. The difference becomes significant as the mean flow Mach number increases. The inverted result was obtained in an early paper by Gubkin (1958) that is of historical interest. The correct expressions, Eqs. (12.47) and (12.48), are used—although in a different form—by Hayes and Runyan (1972), and we shall demonstrate agreement by recovering their result in Sect. 12.4.3. For a recent derivation by a different method, the reader is referred to Robinson (1991).

12.3.4 Summary of Results for a Lossless Moving Medium

The wavelet invariance relation remains as in Eq. (12.14), namely

$$\tilde{p}(s, \tau_w) = p_+(\tau_w), \tag{12.49}$$

but B is now given by Eq. (12.41). The arrival phase $\tau(s, \tau_w)$ follows from Eq. (12.48) as

$$\tau(s, \tau_w) = \tau_w - p_+(\tau_w) \int_{s_+}^{s} \frac{\beta}{\rho c^2} \frac{1}{D v_g} B \, ds'. \tag{12.50}$$

This result generalizes Eq. (12.13) of Sect. 12.1. It may be written in the equivalent form

$$\tau(s, \tau_w) = \tau_w - \left(\frac{\beta}{\rho c^2}\right)_+ \left(\frac{1}{D v_g}\right)_+ \tilde{x} p_+(\tau_w),$$

$$\tilde{x} = \int_{s_+}^{s} \frac{\beta \rho^{-1/2} c^{-1}}{\beta_+ \rho_+^{-1/2} c_+^{-1}} \left(\frac{D v_g}{D_+ v_{g+}}\right)^{-3/2} \left(\frac{A}{A_+}\right)^{-1/2} ds', \tag{12.51}$$

which corresponds to Eqs. (12.15) of Sect. 12.1.

12.3.5 Extension to a Lossy Moving Medium

We have already indicated in Sects. 12.1.5 and 12.1.6 how the effects of attenuation and dispersion may be allowed for in a stepwise calculation scheme (Pestorius, 1973). The ray path is subdivided into suitably small segments, and over each successive segment the propagation process is simulated in two stages. First the time-domain nonlinear distortion algorithm is applied as if the medium were lossless, and then a frequency-domain correction factor is applied to the output to account for linear attenuation and dispersion over the ray segment.

The frequency-domain correction factor given in Sect. 12.1.5 must be modified because of the mean flow. Two physical effects are involved:

1. In a reference frame moving with the local mean flow, the apparent frequency is ω/D rather than ω.
2. The distance by which a wave front advances relative to the fluid in time δt is $v_w \delta t$, which is different from the distance $v_g \delta t$ traveled along the ray path.

Between any two points (s_1, s_2), the linear frequency-domain correction factor is therefore

$$F(\omega, s_2, s_1) = \exp\left\{-\int_{s_1}^{s_2} \tilde{\alpha}(\omega/D, s)(c/v_g) \, ds\right\}, \tag{12.52}$$

where $\tilde{\alpha}$ is the complex attenuation coefficient in the linear plane progressive wave relation $p = p_0 \exp\{j\omega(t - x/c) - \tilde{\alpha}x\}$ [see Eq. (5.5)].

The segment length $s_2 - s_1$ is determined in practice as a compromise between computation speed, which requires large segments, and accuracy, for which the ray path segments should be small, but not so small as to introduce cumulative errors due to repeated resampling. Further discussion of step-size criteria is given in Chap. 4 of Robinson (1991).

12.3.6 Alternative Numerical Solution Techniques

The attraction of the Pestorius approach, mentioned in Sects. 12.1.5 and 12.3.5 (see also Sect. 11.2.3), is that each of the two physical phenomena of (1) nonlinear distortion and (2) waveform diffusion and dispersion is handled in the most directly appropriate manner. The disadvantage is the need to transform repeatedly between the time and frequency domains, which can introduce inaccuracy (because of the need to resample) and slows the calculation.

An alternative procedure, developed by Lee and Hamilton (1991, 1995) and used by Cleveland et al. (1996), is based entirely within the time domain and represents the linear attenuation and dispersion steps by means of tridiagonal matrix operators (see Sect. 11.2.2). No FFTs are involved, and the calculation is therefore faster. A further advantage is the ease with which transient pulses or single shocks can be represented.

12.3.7 Differential Equations for the Pressure Waveform

The results given in Sects. 12.3.4 and 12.3.5 can be expressed in differential equation form. This leads in the most general case—for a lossy medium—to a generalized Burgers equation.

We begin with the lossless case, for which Eq. (12.49) gives $\partial \tilde{p}/\partial s|_{\tau_w} = 0$, and rewrite this result using the identity

$$\left.\frac{\partial \tilde{p}}{\partial s}\right|_{\tau} = \left.\frac{\partial \tilde{p}}{\partial s}\right|_{\tau_w} - \left.\frac{\partial \tilde{p}}{\partial \tau}\right|_s \left.\frac{\partial \tau}{\partial s}\right|_{\tau_w}. \tag{12.53}$$

The first term on the right is zero, and the second term is evaluated using Eq. (12.50), which yields

$$\left.\frac{\partial \tau}{\partial s}\right|_{\tau_w} = -\frac{\beta}{\rho c^2 D v_g} B p_+(\tau_w). \tag{12.54}$$

The resulting differential equation for the pressure $p(s, \tau) = B\tilde{p}$ is

$$\frac{\partial p}{\partial s}\bigg|_{\tau} - \frac{B'}{B}p - \frac{\beta}{\rho c^2 D v_g}Bp_+(\tau_w)\frac{\partial p}{\partial \tau}\bigg|_{s} = 0, \tag{12.55}$$

which to the same order of accuracy may be written as

$$\frac{\partial p}{\partial s}\bigg|_{\tau} - \frac{B'}{B}p - \frac{\beta}{2\rho c^2 D v_g}\frac{\partial p^2}{\partial \tau}\bigg|_{s} = 0, \tag{12.56}$$

since $p(s, \tau)$ is equal to $Bp_+(\tau_w)$ according to linear theory. Here B' denotes the derivative of $B(s)$.

On the other hand, linear ray theory for a lossy medium gives

$$\frac{\partial p}{\partial s}\bigg|_{\tau} - \frac{B'}{B}p = L(p), \tag{12.57}$$

where L is a linear differential operator, equivalent in the frequency domain to the complex factor F of Eq. (12.52) [see also Eq. (5.1)]. Combining the effects of the lossy medium and nonlinear distortion into a single differential equation finally yields

$$\frac{\partial p}{\partial s}\bigg|_{\tau} - \frac{B'}{B}p - \frac{\beta}{2\rho c^2 D v_g}\frac{\partial p^2}{\partial \tau}\bigg|_{s} = L(p). \tag{12.58}$$

This is a generalized, or augmented, version of the Burgers equation that includes mean flow [cf. Eq. (3.58)]. In its original form, the Burgers equation is limited to plane waves in a uniform medium at rest, with losses restricted to the classical frequency-squared dependence of attenuation; whereas Eq. (12.58) describes propagation along a ray tube, in a moving medium, with no restriction on the frequency dependence of the attenuation. The basis of the augmented equation is the assumed smallness of both the nonlinear and dissipative terms relative to the first two terms, which describe lossless linear propagation.

12.4   Examples of Nonlinear Propagation in a Moving Medium

12.4.1   Introduction

Mean flow effects on sound propagation are a function of Mach number; they are therefore important in aeroacoustics, but hardly ever so in ocean acoustics, where typical Mach numbers are of order 10^{-3}. In this section we shall examine two nonlinear propagation problems from aeroacoustics, focusing exclusively on the lossless-medium aspects: sonic boom propagation in Sect. 12.4.3, and shock-wave radiation from a supersonic ducted fan rotor in Sect. 12.4.4. The flow-field model in both these examples is two-dimensional and stratified, with the stratification coordinate perpendicular to the flow in Sect. 12.4.3 and parallel to the flow in Sect. 12.4.4:

$$
\begin{aligned}
\mathbf{w} &= \{w_x(z), w_y(z), 0\} \quad \text{(sonic boom)}; \\
\mathbf{w} &= \{w_x(x), 0, 0\} \qquad \text{(supersonic fan)}.
\end{aligned}
\tag{12.59}
$$

First, however, we consider the simplest possible situation of one-dimensional propagation in collinear flow.

12.4.2   One-Dimensional Propagation in a Variable-Area Flow Duct

Let the duct have rigid walls and carry a slowly varying mean flow in the x direction, with local Mach number $M(x) = w/c$. Superimposed on the flow is a finite-amplitude sound field, which consists of plane waves propagating in the x direction. The waveform evolution along the duct is described by the equations summarized in Sect. 12.3.4, with $v_{g0} = c_0(1 + M)$ [from Eq. (12.9)] and $D = 1 + M$ [from Eq. (12.42)].

It can be seen from Eqs. (12.51) that the nonlinear shift in signal arrival time is proportional to $\tilde{x}$, and the contribution of any short duct segment Δx to the $\tilde{x}$ integral is proportional to $\Delta x (D v_g)^{-3/2}$ or $\Delta x (1 + M)^{-3}$. Thus sound waves traveling against a near-sonic flow, with M close to -1, will experience greatly increased nonlinear distortion on account of the mean flow. Moreover, if M passes through -1, with dM/dx remaining finite, the $\tilde{x}$ integral becomes singular. This situation arises at the throat of a smoothly contoured convergent-divergent nozzle when the nozzle is choked. Sound waves attempting to travel upstream from the throat (where the flow is sonic) suffer unlimited nonlinear distortion and dissipation, associated with the infinite time they take to escape from the throat convergence.

12.4.3   Propagation in a Stratified Atmosphere with Horizontal Wind

Hayes and Runyan (1972) used a nonlinear modification of geometrical acoustics, as described in Sect. 12.3, to model sonic-boom propagation from a supersonic aircraft to the ground. They represented the atmosphere as a horizontally stratified lossless fluid with a horizontal mean wind velocity $\mathbf{w}(z)$. Our aim in this section is to recover their main result using Eqs. (12.49)–(12.51).

The coordinate system to be used is that shown in Fig. 12.1. We choose the z axis downward, against the normal convention for atmospheric acoustics, in order to avoid negative values of the grazing angle χ. The x axis is defined as perpendicular to the wave fronts in plan view (Fig. 12.1b). The situation differs from the zero-wind case described in Sect. 12.1.3, however, because the transverse wind produces a ray velocity component $v_{gy} = w_y(z)$ in the y direction. Furthermore, the ray grazing angle θ in the xz plane is no longer the same as the wavenormal angle χ, but is given by

$$\tan \theta = \frac{v_{gz}}{v_{gx}} = \frac{c \sin \chi}{w_x + c \cos \chi} \neq \tan \chi. \tag{12.60}$$

An important invariant in the present problem is the horizontal phase speed $v = c(z)/\cos \chi(z) + w_x(z) = \text{const}$, which follows from the stratified-atmosphere assumption. Since the wind velocity component w_n in the wavenormal direction is given by $w_x \cos \chi$, the invariance relation is equivalent to

$$v = (c + w_n)/\cos \chi = cD/\cos \chi = \text{const}, \tag{12.61}$$

where D is the Doppler factor defined in Eq. (12.42). We can use these results to express the nonlinear time shift $\tau - \tau_w$ as an integral involving χ along the ray path. There is some advantage, as long as the ray does not pass through a vertex, in using z rather than s as the integration variable, and working with the horizontal ray-tube area A_z—this is the area cut by a horizontal plane, related to A by $A_z v_{gz} = A v_g$. Since the vector $\mathbf{v}_g$ defines the ray direction, increments of z and s are related by

$$\left.\frac{dz}{ds}\right|_{\text{ray}} = \frac{v_{gz}}{v_g} = \frac{A}{A_z}. \tag{12.62}$$

The reduced distance integral [$\tilde{x}$ in Eqs. (12.51)] simplifies in terms of the new variables, since the integrand contains the group of terms $(ADv_g)^{-3/2} A \, ds' = (A_z Dv_{gz})^{-3/2} A_z \, dz'$, and we can use Eqs. (12.60) and (12.61) to write

$$A_z Dv_{gz} = A_z Dc \sin \chi = A_z v \cos \chi \sin \chi. \tag{12.63}$$

Substitution in Eqs. (12.51) leads after some algebra to the following result for the nonlinear time shift:

$$\tau - \tau_w = -\left(\frac{\beta}{\rho c^2}\right)_+ \frac{1}{(\cos \chi \sin \chi)_+} \frac{p_+(\tau_w)}{v}$$

$$\times \int_{z+}^{z} \frac{\beta \rho^{-1/2} c^{-1}}{\beta_+ \rho_+^{-1/2} c_+^{-1}} \left(\frac{\cos \chi \sin \chi}{\cos \chi_+ \sin \chi_+}\right)^{-3/2} \left(\frac{A_z}{A_{z+}}\right)^{-1/2} dz', \tag{12.64}$$

which is exactly equivalent to Eqs. (11) and (14) of Hayes and Runyan (1972). It applies equally to upward or downward propagation (with χ negative in the former case), but at a vertex ($\chi = 0$) the integrand is singular.

To complete the description, note that the wavelet amplitude variation along the ray path is determined by Eqs. (12.39) and (12.41). Use of Eq. (12.63) gives the amplitude factor $B(s, s_+)$ as

$$B = \left(\frac{\rho c^2}{\rho_+ c_+^2} \right)^{1/2} \left(\frac{A_z \cos \chi \sin \chi}{A_{z+} \cos \chi_+ \sin \chi_+} \right)^{-1/2}, \tag{12.65}$$

which also agrees with Hayes and Runyan (1972).

12.4.4   Upstream Shock Radiation from a Supersonic Ducted Fan

When a ducted axial-fan rotor runs at supersonic tip speeds, so that the relative Mach number exceeds 1 over the outer portion of each blade, the upstream pressure field rotating with the fan blades develops into a shock pattern that spirals up the inlet duct. The periodic blade-to-blade pressure waveform becomes saturated within a short distance upstream of the rotor face, and may be estimated using weak shock theory. The first published analyses (Morfey and Fisher, 1970; Hawkings, 1971) modeled the fan rotor as a two-dimensional cascade with uniform flow upstream. The extension to two-dimensional inlet ducts of variable area, with axially varying Mach number M_x, density ρ, and sound speed c, was worked out by Hawkings (1974), and we shall show in this section how Hawkings's result may be recovered by an approach similar to that used by Hayes and Runyan for the sonic-boom problem.

Figure 12.8a shows a two-dimensional representation of the fan and its upstream-propagating pressure field. The mean flow approaching the rotor is axial, with Mach number M_x. In the rotor reference frame, the relative mean flow is nonaxial, with components as shown in Fig. 12.8a. The angle μ between the relative velocity vector and the forward-propagating wave fronts is the Mach angle, defined by $\sin \mu = 1/M_{\text{rel}}$. Much of the analysis of Sect. 12.4.3 carries over to the supersonic-rotor problem, with the following changes to accommodate the different geometry: The coordinate z is measured upstream from the rotor face; the tangential phase speed is $v = c M_t$ in the y direction, and the ray velocity component in the upstream direction is $v_{gz} = c(\sin \chi - M_x)$. Equations (12.61) and (12.62) apply unaltered, and the horizontal phase speed invariant is replaced by

$$v = \frac{c - w_x \sin \chi}{\cos \chi} = \text{const}, \tag{12.66}$$

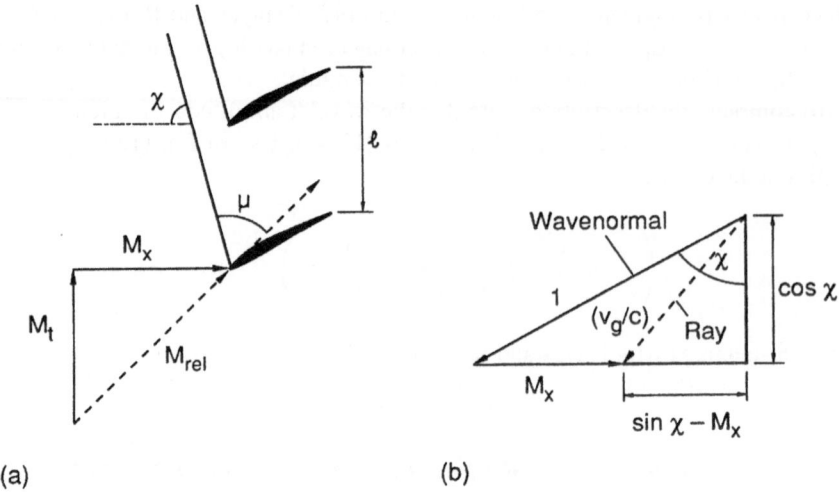

(a)   (b)

Fig. 12.8 (a) Wave-front geometry for shock waves propagating upstream of a supersonic fan with axial inflow. (b) Relation between ray and wavenormal vectors.

where $-w_x \sin \chi$ is the mean flow velocity in the wavenormal direction. Also, corresponding to Eq. (12.63), $A_z D v_{gz} = A_z v \cos \chi (\sin \chi - M_x)$. Here D has been replaced by the first of the equivalent expressions $D = M_t \cos \chi = 1 - M_x \sin \chi$; note that the wavenormal angle χ, which appears in all these relations, is expressible as a function of the relative Mach number components (M_t, M_x), in view of the last equality. Explicit expressions are given by Morfey and Fisher (1970); see their Eq. (13).

Finally, the ray-tube area A_z transverse to the duct axis is proportional to H, the duct height in the radial direction, since the present 2-D model permits no spreading in the tangential direction. The nonlinear shift in wavelet arrival time is therefore given by an equation similar to Eq. (12.64), but with the substitutions $\sin \chi \rightarrow (\sin \chi - M_x)$ and $A_z \rightarrow H$. An alternative version of the arrival-time result uses the equivalence $\sin \chi - M_x = q^2 \cos \chi$, with $q = (M_{\text{rel}}^2 - 1)^{1/4}$, where M_{rel} is the relative Mach number $(M_t^2 + M_x^2)^{1/2}$; this follows from the $\chi(M_t, M_x)$ relationship mentioned in the previous paragraph. In nondimensional form, therefore,

$$\frac{\tau - \tau_w}{T} = -\left(\frac{\beta}{\rho c^2}\right)_+ \frac{p_+(\tau_w)}{[\cos \chi (\sin \chi - M_x)]_+}$$

$$\times \int_{s_+}^{s} \frac{\beta}{\beta_+} \left(\frac{c}{c_+}\right)^{-1} \left(\frac{\rho H}{\rho_+ H_+}\right)^{-1/2} \left(\frac{q \cos \chi}{q_+ \cos \chi_+}\right)^{-3} d(z/l), \quad (12.67)$$

where l is the blade spacing (Fig. 12.8a) and l/v defines the blade-passing period T.

Equation (12.67) is equivalent to Hawkings's result, apart from an algebraic error (involving powers of ρ and c) in his factor B [Hawkings, 1974; see Eqs. (2.6) and (2.12)]. A crucial role is played by the quantity

$$q \cos \chi = [\cos \chi (\sin \chi - M_x)]^{1/2} = f(M_{rel}, M_x), \tag{12.68}$$

which appears in the integrand and is analogous to $(1 + M)$ in the 1-D example of Sect. 12.4.2. As M_x approaches 1, $f(M_{rel}, M_x)$ becomes small; its asymptotic dependence on $1 - M_x$ is in fact $f(M_{rel}, M_x) \sim (1 - M_x)$, which emphasizes the qualitative similarity with Sect. 12.4.2. Small values of $q \cos \chi$ imply large contributions to the integral in Eq. (12.67), and therefore rapid nonlinear distortion.

12.5 Review of Approximations

12.5.1 Introduction

The theoretical framework set out in Sects. 12.1 and 12.3, and the examples of Sects. 12.2 and 12.4, are based on two key approximations, which were introduced in Sect. 12.1.6. Our aim in what follows is to indicate the physical origin of the errors involved, and to deduce magnitude estimates. The issue of amplitude-dependent ray paths is covered in Sects. 12.5.2–12.5.4, and the issue of frequency-dependent ray paths in Sect. 12.5.5.

12.5.2 The Use of Small-Signal Ray Paths to Describe Finite-Amplitude Propagation

The approach used throughout this chapter has been to start from the results of linear geometrical acoustics, and to seek first-order corrections for propagation of finite-amplitude signals. For describing propagation between any two points (e.g., A and B in Fig. 12.1a), the theory uses the small-signal connecting ray path (L_0) in two ways: First, the path L_0 is used as the integration path in the travel time equations (12.12) and (12.48); second, adjacent paths L_0 are used to calculate the ray-tube area $A(s)$, which appears in the Blokhintsev weighting factor $B(s, s_+)$ of Eqs. (12.5) and (12.41). However, finite-amplitude wavelets[8] launched from A with different values of p or u will in general get to B by slightly different ray paths L_w, with initial wavenormal directions $\mathbf{n}(A) = \mathbf{n}_+$ that depend on wavelet amplitude (compare Sect. 12.3.3). The physical reasons are explained in Sect. 12.5.3 below; they relate to the difference between the finite-amplitude wavelet propagation speed and the local sound speed in the ambient medium, i.e., $\delta v_w = v_w - c_0$.

---

[8] We exclude shocks at this stage; they are considered in Sect. 12.5.4.

The perturbation in ray path geometry between any two points A and B is first-order in δv_w, and implies a first-order perturbation in the wavenormal launch direction $\mathbf{n}_+$ if the wavelets traveling their different ray paths are to reassemble at B. The same first-order perturbation also applies to the integrands in the travel time equations (12.12) and (12.48), so one might expect a similar perturbation in the wavelet travel time $(\Delta t)_w$. However, the signal travel time along any ray path is, by definition, stationary with respect to small perturbations of the path. It follows that *errors* in $(\Delta t)_w$ incurred by using L_0 as the path of integration are of second order in δv_w, whereas the nonlinear correction to $(\Delta t)_w$ given in Eqs. (12.12) and (12.48) is of first order. The use of L_0 is therefore justified, to the accuracy sought in the present chapter.

A numerical calculation scheme based on this approach, called by Ostrovsky (1976) the "linear-ray" approximation, was developed by Warshaw (1980) for predicting blast pressure waveforms in the upper atmosphere due to ground-level explosions. Some success was achieved in matching experimental data (Warshaw and Dubois, 1981). It is interesting to note that propagation in the opposite (downward) direction, as with sonic booms, does not provide a good source of experimental comparison for a stratified-medium model because of the turbulent boundary layer in the lowest few hundred meters of the atmosphere.

12.5.3 Physics of Ray Path Dependence on Wavelet Amplitude

A finite-amplitude wavelet surface will generally propagate with a slightly different ray direction, and also a slightly different ray curvature, compared with a linear wave front. The direction effect occurs only in a moving medium; the curvature effect occurs in any nonuniform medium. We consider the two phenomena in turn. First, in a moving medium, Fig. 12.7 shows that the wavenormal and ray directions are different. The angle between them is determined by the scalar product $\mathbf{n} \cdot \mathbf{m} = \mathbf{n} \cdot (v_w \mathbf{n} + \mathbf{w})/v_g = (v_w + w_n)/v_g$. Thus, with zero wind, the angle is zero; but if $\mathbf{w}$ has a component orthogonal to $\mathbf{n}$, there is a finite angular separation, which varies with v_w and thus with the wavelet amplitude. It follows that different wavelets with the same value of $\mathbf{n}$ will propagate in different directions; the shift in ray direction relative to the small-signal value is obtained from Fig. 12.7b as

$$\delta\mathbf{m} \simeq \mathbf{m} \times (\mathbf{n} \times \mathbf{w})\delta v_w/v_w^2; \quad \text{i.e.,} \quad |\delta\mathbf{m}| \simeq \frac{w_{\text{trans}}}{c^2}|\delta v_w|, \tag{12.69}$$

which depends on the wind speed $w_{\text{trans}} = |\mathbf{n} \times \mathbf{w}|$ parallel to the wave front surface.

Second, in a nonuniform medium—which we shall take to be at rest in order to simplify the discussion—the curvature of rays is controlled by the ratio $(\mathbf{n} \times \nabla v_w)/v_w$ (Pierce, 1989; see Sect. 8-3). The numerator equals $\mathbf{n} \times \nabla c$, but the denominator is amplitude-dependent; it follows that different wavelets follow differently curved ray paths. If the increment in ray curvature relative to the small-signal value is denoted by $\delta\kappa$, then $\delta\kappa/\kappa_0 \simeq \delta v_w/c$.

It is clear from the discussion above that amplitude-dependent changes in ray direction and curvature are of first order in δv_w, and therefore proportional to the wavelet amplitude. This fact has already been used in Sect. 12.5.2.

12.5.4   Shock Dynamics

With one important exception, the propagation of weak shock fronts falls under the preceding discussion, provided we replace the wavelet propagation speed v_w with the *shock propagation speed* (Sect. 4.4.1):

$$v_s \simeq \tfrac{1}{2}(v_{wa} + v_{wb}) + O(|\delta v_w|^2). \tag{12.70}$$

Here the subscripts a and b denote values ahead of the shock and behind the shock. The exception arises because the strength of a shock—and therefore v_s—can vary *over the shock surface*, even when the shock front propagates into a uniform medium at rest. The strength of a wavelet, on the other hand, is by definition constant over a wavelet surface, within the ray acoustics approximation.

Variation of v_s over the shock surface implies $\mathbf{n} \times \nabla v_s \neq 0$. The shock ray path curvature is related to this quantity (see Sect. 12.5.3); a nonzero value implies that elements of the shock surface propagate along curved ray paths, even though the medium ahead of the shock may be uniform. The result is that spatially nonuniform curved shocks exhibit *defocusing*.

A situation where shock defocusing has practical significance arises in shock-wave lithotripsy. Müller (1987) has shown experimentally how an initially spherical but nonuniform converging shock front, produced by reflecting a uniform spherical outgoing shock from an ellipsoidal surface, fails to produce a sharp focus. Instead, the strongest region of the shock front travels faster and the shock front becomes flattened as the focus is approached. Henshaw et al. (1986) and Schwendeman (1988, 1993) have demonstrated similar phenomena numerically.

As with all the phenomena discussed so far in Sect. 12.5, the consequent deviations from the approximate model of nonlinear propagation (Sects. 12.1 and 12.3) are of second order in the field strength. We next examine a different source of error, related to dispersion in the acoustic medium.

12.5.5   The Assumption That Ray Paths Are Independent of Frequency

The theoretical development of Sects. 12.1 and 12.3 is based on the assumption that ray paths do not vary either with amplitude or with frequency. Frequency dependence is certainly absent when the medium is uniform, since the rays are straight lines.[9] Likewise, there is no frequency dependence if the medium is nondispersive. However, a dispersive nonuniform medium will produce frequency-dependent refraction, and signal components at different frequencies will propagate along different rays.

The practical justification for the model described in this chapter is that the atmosphere and ocean exhibit extremely weak acoustic dispersion up to ultrasonic frequencies. Order-of-magnitude estimates for typical ocean ray paths indicate a spread of launch angles λ, for propagation between two fixed points A and B, of $\delta\lambda \approx 10^{-5}|\theta - \lambda|$ over the range 0–10 kHz. Here $|\theta - \lambda|$ is the ray turning angle between A and B. It seems reasonable to assume that outgoing wave fronts separated by so small a launch angle should be identical.

There remains the possibility that significant arrival-time differences might accumulate, over a long propagation path, between wavelets emitted at these slightly different launch angles. The difference due to basing travel times on the true (dispersive) ray path, rather than a standard (e.g., zero-frequency) ray path, is of order $(\delta c/c)^2$. Here δc is the frequency-dependent variation in sound speed. The absence of a first-order error term is due to the stationary-phase principle; it means that frequency-dependent travel times are given to good accuracy by integration along a standard ray path, as in Eqs. (12.12) and (12.48). However, the effect of the second-order correction remains unexplored in the literature.

12.6   Acoustical Properties of Water and Seawater

Numerical implementation of the equations derived in Sects. 12.1–12.4 requires a knowledge of the acoustical properties ρ, c, β and Λ [defined following Eqs. (12.15)]. Values of these quantities for water and seawater are documented in a paper by Cotaras and Morfey (1993); in particular, polynomial expressions are provided for β and Λ as functions of temperature, pressure, and seawater salinity.

Acknowledgments The authors thank Steve Warshaw, Leick Robinson, and Lev Ostrovsky for useful discussions and comments during the preparation of this chapter.

---

[9] Here and throughout Sect. 12.5.5, linear-signal or small-amplitude rays are understood.

References

Blokhintsev, D. I. (1946a). *Acoustics of a Nonhomogeneous Moving Medium* (Gostekhizdat, Moscow). Available in English translation as *National Advisory Committee for Aeronautics Technical Memorandum* No. 1399, Washington, D.C. (1956).

Blokhintsev, D. I. (1946b). The propagation of sound in an inhomogeneous and moving medium I. *J. Acoust. Soc. Am.* **18**, 322–328. See Eq. (28).

Cleveland, R. O., Hamilton, M. F., and Blackstock, D. T. (1996). Time-domain modeling of finite-amplitude sound in relaxing fluids. *J. Acoust. Soc. Am.* **99**, 3312–3318.

Cotaras, F. D., and Morfey, C. L. (1993). Polynomial expressions for the coefficient of nonlinearity β and $\beta/(\rho c^5)^{1/2}$ for fresh water and seawater. *J. Acoust. Soc. Am.* **94**, 585–588.

Fridman, V. E. (1982). Self-refraction of small amplitude shock waves. *Wave Motion* **4**, 151–161.

Frisk, G. V. (1994). *Ocean and Seabed Acoustics: A Theory of Wave Propagation* (Prentice-Hall, Englewood Cliffs, N.J.), pp. 218–219.

Gradshteyn, I. S., and Ryzhik, I. M. (1980). *Tables of Integrals, Series, and Products* (Academic Press, San Diego).

Gubkin, K. E. (1958). Propagation of discontinuities in sound waves. *J. Appl. Math. Mech.* **22**, 787–793.

Hawkings, D. L. (1971). Multiple tone generation by transonic compressors. *J. Sound Vib.* **17**, 241–250.

Hawkings, D. L. (1974). The effects of inlet conditions on supersonic cascade noise. *J. Sound Vib.* **33**, 353–368.

Hayes, W. D. (1968). State of the art of sonic boom theory. *NASA SP-180: Second Conference on Sonic Boom Research*, I. R. Schwartz, ed. (National Aeronautics and Space Administration, Washington, D.C.), pp. 181–182.

Hayes, W. D. (1969). Review of sonic boom theory. *Aerodynamic Noise: Proceedings of AFOSR–UTIAS Symposium, Toronto* (20–21 May 1968) (University of Toronto Press, Toronto), pp. 387–396.

Hayes, W. D., and Runyan, H. L., Jr. (1972). Sonic-boom propagation through a stratified atmosphere. *J. Acoust. Soc. Am.* **51**, 695–701.

Henshaw, W. D., Smyth, N. F., and Schwendeman, D. W. (1986). Numerical shock propagation using geometrical shock dynamics. *J. Fluid Mech.* **171**, 519–545.

Hunter, J. K., and Keller, J. B. (1987). Caustics of nonlinear waves. *Wave Motion* **9**, 429–443.

Lee, Y.-S., and Hamilton, M. F. (1991). Nonlinear effects in pulsed sound beams. *Ultrasonics International 91 Conference Proceedings*, D. E. Gray, ed. (Butterworth-Heinemann, Oxford), 177–180.

Lee, Y.-S., and Hamilton, M. F. (1995). Time-domain modeling of pulsed finite-amplitude sound beams. *J. Acoust. Soc. Am.* **97**, 906–917.

Morfey, C. L. (1984). Nonlinear propagation in a depth-dependent ocean. Technical Report ARL-TR-84-11 (Applied Research Laboratories, The University of Texas at Austin).

Morfey, C. L., and Fisher, M. J. (1970). Shock-wave radiation from a supersonic ducted rotor. *Aeronautical Journal of the Royal Aeronautical Society* **74**, 579–585.

Morfey, C. L., and Sparrow, V. W. (1993). Plane compression front steepening in nonlinear media forms both a shock and a reflected wave. *J. Acoust. Soc. Am.* **93**, 3085–3088.

Müller, M. (1987). Experimental investigations on focusing of weak spherical shock waves in water by shallow ellipsoidal reflectors. *Acustica* **64**, 85–93.

Obermeier, F. (1974). Sonic boom behaviour near a caustic. *Noise Mechanisms, AGARD Conference Proceedings 131*, Paper 17.

Ostrovsky, L. A. (1976). Short-wave asymptotics for weak shock waves and solitons in mechanics. *Int. J. Non-Linear Mechanics* **11**, 401–416.

Ostrovsky, L. A., and Fridman, V. E. (1985). Dissipation of high-intensity sound in an isothermal atmosphere. *Sov. Phys. Acoust.* **31**, 374–375.

Ostrovsky, L. A., Pelinovsky, E. N., and Fridman, V. E. (1976). Propagation of finite-amplitude sound waves in an inhomogeneous medium with caustics. *Sov. Phys. Acoust.* **22**, 516–520.

Pelinovsky, E. N., Petukhov, Yu. V., and Fridman, V. E. (1979). Approximate equations for the propagation of strong acoustic signals in the ocean. *Izv. Acad. Sci. USSR, Atmos. Oceanic Phys.* **15**, 299–304.

Pestorius, F. M. (1973). Propagation of plane acoustic noise of finite amplitude. Technical Report ARL-TR-73-23 (Applied Research Laboratories, The University of Texas at Austin).

Pierce, A. D. (1989). *Acoustics: An Introduction to Its Physical Principles and Applications* (Acoustical Society of America, New York).

Robinson, L. D. (1991). Sonic boom propagation through an inhomogeneous, windy atmosphere. Ph.D. dissertation, The University of Texas at Austin.

Schwendeman, D. W. (1988). Numerical shock propagation in nonuniform media. *J. Fluid Mech.* **188**, 383–410.

Schwendeman, D. W. (1993). A new numerical method for shock wave propagation based on geometrical shock dynamics. *Proc. R. Soc. Lond.* **A441**, 331–341.

Sturtevant, B., and Kulkarny, V. A. (1976). The focusing of weak shock waves. *J. Fluid Mech.* **73**, 651–671.

Tolstoy, I., and Clay, C. S. (1987). *Ocean Acoustics* (Acoustical Society of America, New York), p. 54.

Warshaw, S. I. (1980). On a finite amplitude extension of geometric acoustics in a moving, inhomogeneous atmosphere. Report UCRL-53055 (Lawrence Livermore Natl. Lab., Livermore, Calif.).

Warshaw, S. I., and Dubois, P. F. (1981). Preliminary theoretical acoustic and RF sounding calculations for MILL RACE. Report UCID-19231 (Lawrence Livermore Natl. Lab., Livermore, Calif.).

Whitham, G. B. (1974). *Linear and Nonlinear Waves* (John Wiley, New York), Chapter 9.

Chapter 13
Statistical Phenomena

Sergey N. Gurbatov and Oleg V. Rudenko

Contents

13.1 Introduction

Statistical nonlinear acoustics concerns phenomena associated with the propagation of intense acoustic noise in homogeneous and randomly inhomogeneous media. Investigations of these phenomena are of considerable practical interest because of the existence of both natural and man-made sources of intense acoustic noise. Blast waves in the atmosphere and the ocean, jet engine noise, and intense fluctuating sonar signals are examples of low-frequency disturbances for which nonlinear effects can be substantial. The ultrasonic frequency range encompasses noise spectra produced by cavitation and acoustic emission. Finally, thunder and seismic waves are examples of intense noise produced by natural processes.

Monographs by Rudenko and Soluyan (1977), Gurbatov et al. (1991), and Naugol'nykh and Ostrovsky (1997), and reviews by Webster and Blackstock (1978a, 1978b), Gurbatov et al. (1983), and Rudenko (1986) describe the advances in statistical nonlinear acoustics during the 1970s and 1980s. Extensive reference to other work may be found in these publications. In this short chapter, however,

S. N. Gurbatov
State University of Nizhni Novgorod, Department of Radiophysics, Nizhni Novgorod, Russia

O. V. Rudenko
Moscow State University, Department of Physics, Moscow, Russia

© The Author(s) 2024
M. F. Hamilton, D. T. Blackstock (eds.), *Nonlinear Acoustics*,
https://doi.org/10.1007/978-3-031-58963-8_13

we cannot describe all experimental results and theoretical methods. Instead, we confine our review to discussion of the more important statistical phenomena that are observed, and to qualitative explanations based on the simplest theoretical models.

13.2 Evolution Equation and Statistical Functions

The general analysis of plane wave propagation proceeds as follows. Let the finite amplitude propagation be described by an evolution equation, for example, the Burgers equation [Eq. (3.54)]:

$$\frac{\partial p}{\partial x} - \frac{\beta p}{\rho_0 c_0^3} \frac{\partial p}{\partial \tau} = \frac{\delta}{2c_0^3} \frac{\partial^2 p}{\partial \tau^2}. \tag{13.1}$$

Here p is the acoustic pressure, x is the coordinate along which the plane wave propagates, $\tau = t - x/c_0$ is retarded time, c_0 and ρ_0 are the equilibrium sound speed and medium density, respectively, β is the coefficient of nonlinearity, and δ is the sound diffusivity. A source located at $x = 0$ produces the signal

$$p(x = 0, t) = \zeta(t). \tag{13.2}$$

If $\zeta(t)$ is a regular function, for example, a time harmonic or pulsed signal, Eq. (13.1) can be solved subject to the boundary condition in Eq. (13.2). The dynamic problem is thus reduced to calculation of the waveform $p(x, \tau)$ at an arbitrary point in the medium. The waveform determines all other properties of the signal, for example, its frequency spectrum, peak pressure, intensity, and so on.

If, however, the input signal $\zeta(t)$ is a random process, the received signal $p(x, \tau)$ will be an irregular function of time. A series of measurements $(p_1, p_2, \ldots, p_N)$ at distance x yields N different realizations, each one a random variation of the pressure with time. Averaging various characteristics of the wave over an ensemble of realizations provides a specific statistical description of the random field. In the analysis of noise, one encounters properties such as one-dimensional and two-dimensional probability density functions [$W(p)$ and $W(p_1, p_2)$, respectively], the cross-correlation function $R(x, \tau_1, \tau_2) = \langle p(x, \tau_1) p(x, \tau_2) \rangle$ (where $\langle \cdot \rangle$ represents the expected value of the quantity inside), the intensity spectrum $S(x, \omega)$, average pressure $\langle p \rangle$, and others. When these properties are known, one can answer questions such as: In what manner does the noise wave energy decrease with x? How is the energy redistributed in the frequency domain? What is the probability of large outliers, i.e., large pressure excursions?

We now introduce the basic formulas for calculating general properties of the noise. The ensemble average is defined as follows:

$$\langle f \rangle = \frac{1}{N} \lim_{N \to \infty} \sum_{n=1}^{N} f(p_n) = \int_{-\infty}^{\infty} f(p) W(x, p) \, dp, \tag{13.3}$$

where $f(p)$ is an arbitrary function. In particular, with $f = p$, we obtain the average pressure, whereas for $f = p^2$, we obtain p_{rms}^2, where p_{rms} is the root-mean-square, or effective, pressure of the wave (note also that p_{rms}^2 is proportional to intensity). The variance σ_p (units of pressure) is defined by the relation $\sigma_p^2 = \langle(p - \langle p \rangle)^2\rangle$, and thus for $\langle p \rangle = 0$ we have $\sigma_p = p_{\text{rms}}$. The cross-correlation function is defined in terms of the two-dimensional density function:

$$R(x, \tau_1, \tau_2) = \langle p(x, \tau_1) p(x, \tau_2) \rangle = \iint_{-\infty}^{\infty} p_1 p_2 W(x, \tau_1, \tau_2, p_1, p_2) \, dp_1 dp_2.$$

(13.4)

If the process is stationary, then $R = R(x, \tau)$, where here $\tau = \tau_1 - \tau_2$, and for $\langle p \rangle = 0$ we have $\sigma_p^2 = R(x, 0)$. The intensity spectrum is given by the Fourier transform of the correlation function:

$$S(x, \omega) = \frac{1}{2\pi} \int_{-\infty}^{\infty} R(x, \tau) e^{-j\omega\tau} d\tau = \frac{1}{\pi} \int_0^{\infty} R(x, \tau) \cos \omega\tau \, d\tau. \qquad (13.5)$$

In practice, instead of averaging Eq. (13.3) over an ensemble of realizations (which is referred to as statistical averaging), one normally averages over time:

$$\bar{f} = \lim_{T \to \infty} \frac{1}{T} \int_{t_0}^{t_0+T} f[p(x, t)] \, dt. \qquad (13.6)$$

For ergodic processes, statistical averaging and time averaging produce the same result. See Papoulis (1965) for extended discussions of the above definitions.

13.3   Basic Phenomena in Nonlinear Noise Fields

The phase speed of sound waves in homogeneous fluids exhibits a very weak dependence on frequency, and therefore the spectral components interact efficiently. For progressive plane waves, the condition for wave resonance (Vinogradova et al., 1990), $k_3 = k_1 + k_2$, is satisfied for any frequencies that are related according to $\omega_3 = \omega_1 + \omega_2$, where the wave numbers are defined by $k_n = \omega_n/c_0$ and the phase speed c_0 does not depend on frequency. Resonant interactions result in an avalanche-type generation of spectral lines, and therefore in pronounced spectral broadening. In the time domain, this process corresponds to formation of steep waveform profiles, i.e., shock fronts. Following shock formation, the behavior of the signal changes as nonlinear attenuation and acoustical saturation become important (Sect. 4.4.3.4).

We now discuss specific features related to the behavior of a random acoustic perturbation, confining our attention for the time being to lossless progressive plane waves. We thus set $\delta = 0$ in Eq. (13.1), in which case a solution that satisfies

Eq. (13.2) may be written in the form [Eq. (4.29)]

$$p = \zeta\left(\tau + \frac{\beta}{\rho_0 c_0^3} x p\right).$$ (13.7)

For distances x that are small in comparison with the nonlinear length scale $\bar{x} = \rho_0 c_0^3/\beta\omega_0 A$ (where ω_0 and A are a characteristic frequency and pressure amplitude, respectively, of the signal), one can expand Eq. (13.7) in terms of the small parameter $x/\bar{x}$:

$$p \simeq \zeta(\tau) + \frac{\beta}{\rho_0 c_0^3} x \zeta(\tau)\zeta'(\tau),$$ (13.8)

where $\zeta' = d\zeta/d\tau$. For a quasi-harmonic input signal $\zeta(t) = A(t)\cos[\omega_0 t + \phi(t)]$, Eq. (13.8) acquires the form

$$p \simeq A(\tau)\cos[\omega_0\tau + \phi(\tau)] + \frac{\beta}{4\rho_0 c_0^3} x \frac{d}{d\tau} A^2(\tau)\{1 + \cos[2\omega_0\tau + 2\phi(\tau)]\}.$$ (13.9)

If the amplitude A and phase ϕ do not depend on time, we obtain a regular harmonic signal. In accordance with Eq. (13.9), nonlinearity then leads to generation of a second-harmonic component at frequency $2\omega_0$, the amplitude of which is proportional to A^2 and increases linearly with propagation distance [compare Eq. (4.54)]:

$$A_2 = (\beta\omega_0/2\rho_0 c_0^3) x A^2.$$ (13.10)

In what follows it is assumed that a quasi-harmonic signal with random amplitude and phase modulation is specified at the source. When the modulation is slowly varying, A and ϕ can be considered constant with respect to the time derivative in Eq. (13.9), and the noise signal will simply generate a second harmonic with random amplitude A_2. The average intensity of the second-harmonic component, $\langle A_2^2 \rangle$, is greater than in the case of a regular harmonic source excitation, that is, in the absence of the random modulation. This result follows from the inequality $\langle A^4 \rangle \geq \langle A^2 \rangle^2$. In particular, if an input signal has Gaussian statistics, the amplitude distribution will follow the Rayleigh distribution $W(A) = (A/\sigma_p^2)\exp(-A^2/2\sigma_p^2)$, with $\langle A^4 \rangle = 2\langle A^2 \rangle^2$. In this case, it follows from Eq. (13.10) that the average intensity of the second harmonic within the noise field is twice as large as that of the second harmonic within the field of a regular signal of the same intensity. By analogy, it is observed that for Gaussian noise, generation of the nth-harmonic component, with amplitude $A_n \sim A^n$, is $n!$ times more efficient than harmonic generation by a regular signal.

   The efficiency of spectral interactions within a noise field, and therefore the rate at which harmonic amplitudes increase with distance, is thus associated with the general properties of nonlinear wave propagation, which become more pronounced with increasing signal amplitude. In a noise field one can identify outliers, i.e., values that are anomalously large or small in comparison with the average value of the property under consideration (for example, extreme values of the instantaneous signal amplitude). Inasmuch as nonlinearity emphasizes large outliers in a random process, the nonlinear processes are more efficient for noise than for regular signals.

   The previous observations are now illustrated with a simple example. Consider three realizations of a random source waveform, formed with portions of three sinusoids: one with average amplitude $\langle A \rangle$, the second an outlier with twice the average amplitude, $2\langle A \rangle$, and the third a signal with half the amplitude of the first, $\frac{1}{2}\langle A \rangle$. Let the three realizations occur with equal probability. A nonlinear transformation of the type in Eq. (13.10), $a = \alpha A^2$, yields for each realization the second-harmonic amplitudes $a_1 = \alpha \langle A \rangle^2$, $a_2 = 4\alpha \langle A \rangle^2$, and $a_3 = \frac{1}{4}\alpha \langle A \rangle^2$. The average intensity is thus $\langle a^2 \rangle = \frac{1}{3}(a_1^2 + a_2^2 + a_3^2) \simeq 6\alpha^2 \langle A \rangle^4$. Therefore, the contribution of the large outlier with amplitude $2\langle A \rangle$ results in $\langle a^2 \rangle$ being increased by approximately a factor of 6, as compared with the value of $a^2 = \alpha^2 \langle A \rangle^4$ observed for the field with amplitude $\langle A \rangle$.

   The influence of large outliers, which was discussed in the previous example of second-harmonic generation, proves also to be important in all random nonlinear processes, for example, generation of combination frequencies, parametric interactions, and self-interactions of randomly modulated waves.

   Let us illustrate another important property of finite-amplitude noise. Consider two quasi-harmonic reference signals, one with amplitude modulation and the other with phase modulation, but both having similar correlation functions and spectra at the source. We now avoid the quasistatic approximation. Equation (13.9) implies that an amplitude-modulated signal will generate not only a second-harmonic component, but also low- (i.e., difference-) frequency components in the spectrum: $p_- \sim dA^2/d\tau$. In the quasilinear approximation [e.g., Eq. (13.9)], this effect does not occur for a signal with only phase fluctuations. Thus, in nonlinear problems, the magnitude of the input spectrum of a wave does not define its evolution process uniquely.

   A random signal can be characterized completely with a reference spectrum or correlation function (second-order moments) only if it possesses Gaussian statistics, for which all higher-order moments can be calculated from the second-order moments. In general, however, additional information is required on the statistics of the input signal, i.e., on its higher-order moments or cumulant functions.

   For stationary signals with Gaussian statistics at the source, Eq. (13.9) yields for the correlation function

$$R(x, \tau) = R_0(\tau) - \left(\frac{\beta x}{2\rho_0 c_0^3} \right)^2 \frac{d^2 R_0^2}{d\tau^2}, \tag{13.11}$$

where $R_0(\tau) = R(0, \tau)$. Since convolution of the spectrum corresponds to squaring the correlation function [recall Eq. (13.5)], Eq. (13.11) yields for the evolution of the noise spectrum associated with a progressive plane wave in a lossless fluid

$$S(x, \omega) = S_0(\omega) + \left(\frac{\beta x}{2\rho_0 c_0^3}\right)^2 \omega^2 \int_{-\infty}^{\infty} S_0(\omega - \Omega) S_0(\Omega) \, d\Omega, \qquad (13.12)$$

where $S_0(\omega) = S(0, \omega)$. The second term in Eq. (13.12) describes the generation of new spectral components, at frequencies $\omega = \omega_1 + \omega_2$, whose intensities are proportional to an integral of the spectral density product $S_0(\omega_1) S_0(\omega_2)$. Here, because of the factor ω^2 that multiplies the integral in Eq. (13.12), harmonic generation proceeds more effectively at higher than at lower frequencies. In particular, Eq. (13.12) reveals that the spectral component at zero frequency is unchanged: $S(x, 0) = S_0(0)$.

In a medium without dispersion, multiple interactions take place among the components of the spectrum, and therefore solutions obtained via the method of successive approximations yield only a qualitative description of how the noise spectrum is transformed due to nonlinearity. For quantitative analysis (Rudenko and Soluyan, 1977; Gurbatov et al., 1983; Gurbatov et al., 1991; Rudenko, 1986), one must resort to exact or asymptotic solutions of evolution equations such as Eq. (13.1).

13.4   Evolution of Quasi-Monochromatic Signals

The evolution of a tonal signal of amplitude A and frequency ω_0 can be described by Eq. (13.1). In this case, the nonlinear length scale is the shock formation distance in an ideal lossless fluid ($\delta = 0$), $\bar{x} = \rho_0 c_0^3 / \beta \omega_0 A$. For $x < \bar{x}$, waveform distortion occurs, but the energy in the wave remains constant. Shocks are formed for $x > \bar{x}$, and energy dissipation increases with the amplitudes of the shocks. For $x \gg \bar{x}$, the wave is transformed into a succession of triangular sawtooth profiles having the same slope, $\partial p / \partial \tau = -\rho_0 c_0^3 / \beta x$. In this advanced stage, the shock amplitudes and wave energy no longer depend on the source amplitude. Small-signal attenuation ($\delta \neq 0$) ultimately reduces nonlinear effects, makes the shock fronts smoother, and at finite distances (of the order of the dissipation length $x_d = 1/\alpha_0$, where $\alpha_0 = \delta \omega_0^2 / 2 c_0^3$ is the small-signal attenuation coefficient at frequency ω_0) completely suppresses the effect of nonlinearity.

For a signal with random modulation, the manifestation of nonlinear effects is different in each characteristic period of the waveform—the higher the local amplitude, the more pronounced are nonlinear effects at a given distance. Since nonlinearity emphasizes large outliers, the distortion of a randomly modulated signal is, on average, stronger than that of a tonal signal. The phase modulation also becomes stronger when higher harmonics are generated. Near the source the random phase of the nth harmonic is equal to $n\phi$. As a result, the spectral line widths increase with n.

During the initial stages of the distortion process, when one can neglect dissipation and shock formation, the evolution of the wave is described by Eq. (13.1) with $\delta = 0$, and exact expressions for the correlation function and the spectrum can be obtained (Rudenko and Soluyan, 1977; Gurbatov et al., 1991). Here, however, we shall follow a more simple approach, based on the quasi-static approximation, to obtain the expression for the correlation function. For a source excitation $A\sin(\omega_0 t + \phi)$, the radiated field can be represented by the Fubini solution [Eq. (4.49)]

$$p = A\sum_{n=1}^{\infty} \frac{2J_n(nx/\bar{x})}{(nx/\bar{x})}\sin(n\omega_0\tau + n\phi),\tag{13.13}$$

the nth term of which corresponds to the nth harmonic component, where J_n is the Bessel function of the first kind. It is assumed in the quasi-static approximation that Eq. (13.13) is also valid for slow random variations of the amplitude A and phase ϕ. Neglecting the influence of shocks in the waveform, we obtain the correlation function by averaging Eq. (13.13) according to Eq. (13.4). For a Gaussian input signal with correlation function $R_0(\tau) = \sigma_p^2 r(\tau)\cos\omega_0\tau$, where $r(\tau)$ $[r(0) = 1]$ is a slowly varying envelope responsible for the finite width of the spectral line ($\Delta\omega \ll \omega_0$), and using a two-point probability distribution for A and ϕ, we obtain

$$R(x,\tau) = \sigma_p^2\sum_{n=1}^{\infty} \frac{2\exp[-(nx/\bar{x})^2]}{(nx/\bar{x})^2}I_n[(nx/\bar{x})^2 r(\tau)]\cos n\omega_0\tau,\tag{13.14}$$

where I_n is the modified Bessel function, and $\sigma_p = p_{\text{rms}}$, with σ_p^2 characterizing the reference noise intensity. The characteristic distortion distance for the random wave is defined as $\bar{x} = \rho_0 c_0^3/\beta\omega_0\sigma_p$. Comparison of Eqs. (13.13) and (13.14) shows that for the same source intensities, harmonic generation (as well as depletion of energy from the fundamental component) proceeds more rapidly in the case of noise. At small distances ($x \ll \bar{x}$), the ratio of the intensity I_n^N of harmonics generated by a random wave to the intensity I_n^S of harmonics generated by a regular harmonic signal is $I_n^N/I_n^S = n!$.

Calculations of the harmonic intensities, with dissipation taken into account, are presented in Fig. 13.1a. The effect of weak attenuation is included in Eq. (13.14) by multiplying the arguments of the exponential and Bessel functions by $e^{-2\alpha_0 x} = \exp[-(2/\Gamma)(x/\bar{x})]$, where α_0 is the amplitude attenuation coefficient at frequency ω_0 and $\Gamma = 2\beta c_0\sigma_p/\delta\omega_0$ is the effective Gol'dberg number (Rudenko, 1986). Figure 13.1a compares the spatial behavior of the normalized second ($n = 2$) and third ($n = 3$) harmonic intensities in the field of a tonal signal (I_n^S, dashed curves) and narrowband noise [solid curves, calculated with the nth term, $I_n^N(x) \equiv \sigma_p^{-2}R_n(x,0)$, in the modified form of Eq. (13.14)]. The curves have been

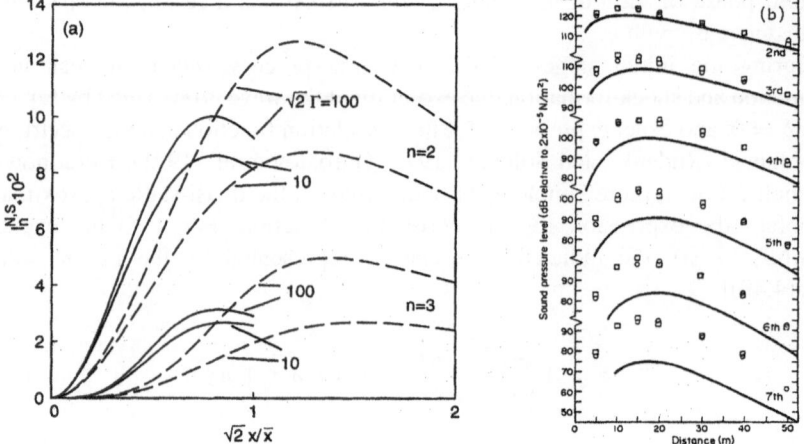

Fig. 13.1 (a) Comparison between the spatial behavior of the second- ($n = 2$) and third-($n = 3$) harmonic intensities generated by a tonal signal (I_n^S, dashed curves) and narrowband noise (I_n^N, solid curves) (Rudenko, 1986). (b) Measurements of harmonic levels ($n = 2-7$) for a monochromatic signal (curves) and noise (o, narrowband noise; □, third-octave-band noise) (Pernet and Payne, 1971).

constructed for two values of the Gol'dberg number. The reference intensities of the noise and the tonal signal were assumed equal, and the center frequency of the noise spectrum coincides with the frequency of the tone. Decreasing Γ, which decreases the role of nonlinearity because of the increased relative importance of absorption, has a more pronounced influence on the tonal components. For $\sqrt{2}\Gamma < 10$ the solid curves lie above the dashed curves for all x, because for small Γ, generation of tonal harmonics is suppressed by dissipation, whereas the noise contains outliers that are subject to a relatively weaker effect of dissipation. The harmonics in the random wave are generated mainly by the large-amplitude outliers, and therefore the generation process is, on average, more efficient than that in the regular wave.

The phenomena predicted in Fig. 13.1a were studied experimentally by Pernet and Payne (1971). High-intensity noise, with sound pressure levels between 120 dB[1] and 140 dB at the source, was transmitted in a polyethylene tube of length 75 m and diameter 4.9 cm. Measurements were made for signals having two bandwidths, 6% ("narrowband") and 23% (third-octave band) of the center frequencies at 0.5, 1, 2, and 3.2 kHz. The tube walls were 0.56 cm thick, and the tube was buried in sand to suppress flexural modes. A reasonable approximation of plane wave propagation was thus achieved. Although boundary-layer losses were observed, the experiment provided a good approximation of propagation in free space.

Typical measurements of the second- through seventh-harmonic sound pressure levels are shown in Fig. 13.1b for a monochromatic source (solid curves) and noise

---

[1] The reference value for all sound pressure levels reported in this chapter is $p_{ref} = 20\,\mu$Pa.

(circles for narrowband noise, squares for third-octave-band noise). Uncertainties in the measurements were reported to be ± 1 dB for the lower harmonics, whereas spectral overlap introduced additional errors in measurements of the higher harmonics. The increased efficiency of harmonic generation in the noise wave is nevertheless demonstrated very clearly. Estimates for the ratio I_n^N/I_n^S were obtained by extrapolation to small distances, which yielded 2.2 ($n = 2$), 5.6 (3), 16 (4), 45 (5), 200 (6), and 500 (7). For $n = 2$ and $n = 3$, there is fair agreement with the theoretical value of $n!$. Disagreement for $n \geq 4$, apart from the previously mentioned spectral overlap, may be due to measurement difficulties in the near field, and to effects of dissipation.

Along with the increase in efficiency of harmonic generation in random waveforms, Eq. (13.14) implies that the bandwidths $(\Delta\omega)_n$ of the spectral lines increase with n. In particular, if the spectral line at the fundamental frequency is characterized by a Lorentz function, we have $(\Delta\omega)_n = n(\Delta\omega)$ in the initial stage of propagation, and we have $(\Delta\omega)_n = \sqrt{n}(\Delta\omega)$ if it is characterized by a Gaussian function. For sufficiently large n, at which $(\Delta\omega)_n$ is of order ω_0, the "wings" of the frequency bands overlap and a monotonically decreasing high-frequency spectrum is formed. Behavior of the high-frequency spectrum is determined by the thin structures of the shock fronts.

If the shock fronts may be considered to be perfect discontinuities of zero duration, the spectrum falls off according to the power law ω^{-2}. In a dissipative medium at sufficiently high frequencies, this law is replaced with the exponential dependence $e^{-\omega t_r}$, where t_r characterizes the shock rise time.

In a medium with weak attenuation and dispersion (e.g., sound waves in pipes, where these effects are caused by the thermoviscous boundary layer along the walls), discontinuities in the slope of the waveform occur in the vicinity of the shock front. The high-frequency spectrum falls off more rapidly in this case, as ω^{-4}. Spectral broadening and generation of a high-frequency wing in the spectrum were studied experimentally by Bjørnø and Gurbatov (1985). Figure 13.2a shows

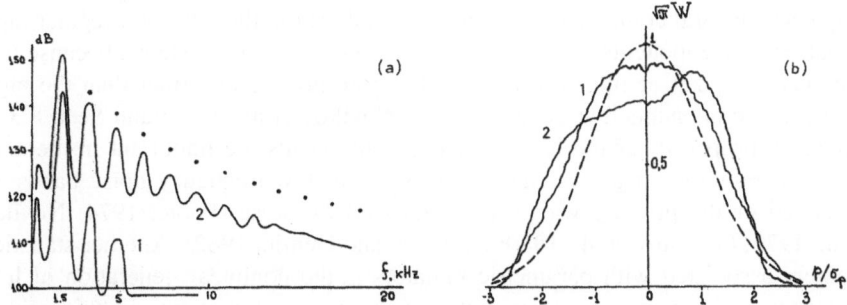

Fig. 13.2 (a) Measurements of narrowband noise spectra near the source (curve 1) and 10 m away (curve 2, shown with 20 dB increase for clarity). Dots are the harmonic amplitudes at 10 m for a tonal signal with the same intensity (Bjørnø and Gurbatov, 1985). (b) Measured probability distributions for narrowband noise at 7.5 m (curve 1) and 17.5 m (curve 2) away from the source (Sakagami et al., 1982). Dashed curve depicts initial Gaussian distribution.

the spectra measured near the source ($x = 0.35\,\text{m}$, curve 1) and at distance $x = 10\,\text{m}$ (curve 2). The noise level at the source was 150 dB, with 1.5 kHz center frequency and 316 Hz bandwidth. The dots are the harmonic intensities at $x = 10\,\text{m}$ for a monofrequency source. The difference in levels between the results for the monofrequency source and for the noise source increases with n. When broadening of the spectral lines is taken into account, it can be shown that the total intensity of each harmonic is approximately the same as that for the tonal signal. Deviations from this law are due to the fact that the measurements were taken at distances $x \simeq 2\bar{x}$, where shocks had already formed. Since for $x > \bar{x}$ the slope of the linear portions of the sawtooth waveform, and the corresponding shock amplitudes, depends only weakly on the source amplitude, the total harmonic intensities for the noise and the regular signal are approximately the same. Because of spectral broadening in the noise signal, a universal asymptotic behavior is always obtained at high frequencies.

After shocks are formed in the waveform, the probability density function changes drastically. Prior to shock formation, the one-point distribution in Eq. (13.3) is maintained, $W(x, p) = W(0, p)$ (Webster and Blackstock, 1979; Gurbatov et al., 1991). After shock formation, however, nonlinear attenuation at the shocks suppresses large outliers. In this case the probability of outliers $|p|$ exceeding $p_{\max} = 2\pi\rho_0 c_0^3/\beta\omega_0 x$ vanishes identically (i.e., all pressures are within the limits $-p_{\max} < p < p_{\max}$), and the probability density function tends toward a uniform distribution that no longer varies with distance. The tendency of a Gaussian distribution to become uniform was observed experimentally by Sakagami et al. (1982). Figure 13.2b depicts the probability distribution for narrowband noise propagating in a pipe out to the distances 7.5 m (curve 1) and 17.5 m (curve 2). The sound pressure level at the source was 152 dB, and the center frequency was 1 kHz. It is evident that as the wave propagates, the probability distribution becomes less similar to a Gaussian distribution (dashed curve) and approaches a uniform distribution. Asymmetry of the spectra may be attributed to boundary-layer dispersion.

An important property of narrowband noise is its ability to generate low-frequency spectral components that are not produced in the field of a regular input signal. Such components are important for parametric sonar systems, because they experience weaker attenuation and can therefore propagate farther than the high-frequency components that generate them (Novikov et al., 1987; and Sect. 8.3.4). Statistical properties of the low-frequency components are important in the analysis of parametric sonar systems in which either low-frequency noise bands are generated or the primary waves are influenced by noise (Foote, 1974; Novikov et al., 1976; Gurbatov et al., 1980; Gurbatov and Demin, 1982). Another statistical problem associated with parametric radiators is the nonlinear generation of low-frequency components in a randomly inhomogeneous medium. In this case, a high-frequency wave propagating through random medium inhomogeneities is subject to fluctuations that result in the appearance of fluctuating low-frequency components and in a decrease in the efficiency of parametric sonar (Gurbatov and Pronchatov-Rubtsov, 1982; Zaitsev and Raevskii, 1990). The scattered low-frequency components carry information about the random inhomogeneities of the medium. Since it is easy to generate low frequency signals with relatively large bandwidths (because small relative changes in the frequencies of the primary waves produce large variations in the difference frequency), information can be obtained about the spatial spectrum of the inhomogeneities over a wide range of wave numbers. A more detailed discussion of these important problems is beyond the scope of this review.

We conclude this section by noting that one type of modulation can be transformed into another during propagation of a randomly modulated wave (Gurbatov and Shepelevich, 1978). For example, as an amplitude modulated wave propagates, the amplitude modulation is reduced and a low-frequency component is generated. This low-frequency component, through interaction with the high-frequency harmonics, produces phase modulation of the latter. As a result, additional broadening and merging of spectral lines occur, and asymptotic relations describe an ever-growing portion of the spectrum.

13.5 Evolution of Broadband Spectra: Acoustic Turbulence

We note that Eq. (13.1) was suggested by Burgers (1948) as a model describing two basic phenomena that are typical of hydrodynamic turbulence: the nonlinear evolution of the frequency spectrum, and damping of disturbances having small length scales (i.e., high frequencies); see Sect. 4.5.1 for historical context. The evolution of random acoustic waveforms is therefore referred to as acoustic turbulence. The apparent simplicity of the Burgers equation stimulated many investigations in which various statistical methods based on this equation were used to describe nonlinear random wave fields [see, for example, the references provided by Gurbatov et al. (1991)].

Interest in the nonlinear evolution of broadband acoustic spectra was motivated by measurements of intense aircraft noise (Howell and Morfey, 1981, 1987), which revealed anomalously low attenuation at high frequencies. Within the 5–10 kHz frequency range, the attenuation at 500 m was observed to be 10 dB less than expected. Effects of temperature variations in the atmosphere, humidity, and other properties of the medium were found to be insufficient to explain this effect. It was therefore assumed that the anomalously high levels of the high-frequency components are attained through energy transfer from the intense low-frequency components of the spectrum.

Morfey (1984) presented detailed measurements of parameters related to jet engine noise under natural conditions. Figure 13.3a shows the attenuation of noise in one-third octave bands radiated by an aircraft with four jet engines. The data in the upper and lower parts of the figure correspond, respectively, to propagation over distances that varied from $R_1 = 262$ m to $R_2 = 345$ m and from $R_2 = 345$ m to $R_3 = 501$ m. Note that the solid curves, based on nonlinear theory, are in much better agreement with the measurements (circles) than are the calculations based on

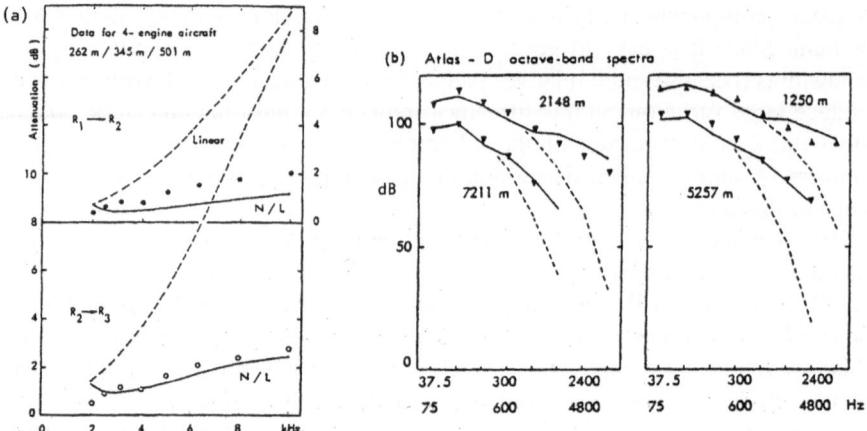

Fig. 13.3 (a) Measured attenuation of jet engine noise, in one-third-octave bands. (b) Measured octave-band spectra of rocket noise. Solid curves are nonlinear predictions, dashed curves are linear predictions (Morfey, 1984).

linear theory (dashed curves). If the transfer of energy upward in the spectrum is not taken into account, the attenuation is overestimated. Figure 13.3b shows octave-band measurements (triangles) of noise generated by an Atlas-D rocket engine. The measurements at frequencies greater than 1 kHz lie appreciably higher than the dashed curves (linear theory), and nonlinear theory is again in much better agreement with experiment.

A series of experiments with finite-amplitude noise has also been performed in the laboratory (Pestorius and Blackstock, 1974; Watanabe and Urabe, 1981; Bjørnø and Gurbatov, 1985; Robsman, 1991). The experiments performed by Pestorius and Blackstock (1974) were conducted with sound pressure levels up to 160 dB in an air-filled pipe of length 29.3 m. Figure 13.4a shows two measured waveforms, one 0.3 m (1 ft) away from the source and the other 26 m (85 ft) away. Two processes are clearly revealed: waveform steepening and an increase in the time scale of the oscillations. The first process leads to shock formation and energy transfer to higher frequencies, which is also observed in outdoor propagation of jet noise (Morfey, 1984). The second process is associated with the relative velocities of the shock fronts—in particular, their coalescence (Rudenko and Soluyan, 1977; Gurbatov et al., 1991), which transfers energy from the central part of the spectrum to the low-frequency range. Thus, the noise spectrum broadens because of energy transfer to both higher and lower frequencies. Figure 13.4b shows the noise spectra measured by Pestorius and Blackstock (1974) in 50 Hz bands at three distances from the source. Similar behavior of the spectrum was observed by Watanabe and Urabe (1981), who were interested in the statistical properties of the slopes of the waveform segments in between the shocks (as in the lower waveform in Fig. 13.4a).

Robsman (1991) observed similar spectral evolution in a solid (Fig. 13.5). Normally, intensities on the order of watts per centimeter squared at frequencies of tens

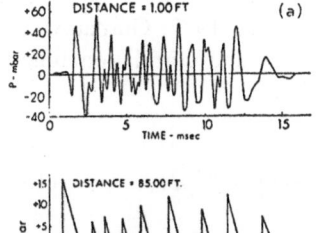

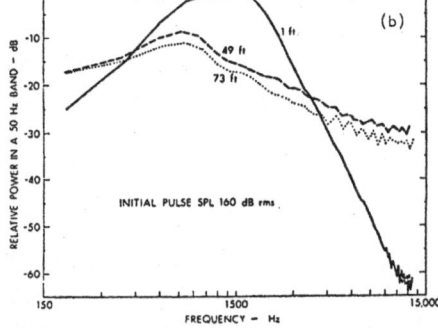

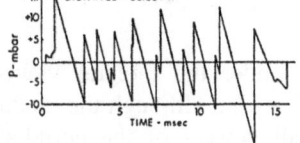

Fig. 13.4 (**a**) Comparison of measured noise waveforms near (at distance 1 ft = 0.3 m) and far away (85 ft = 29 m) from the source. (**b**) Related noise spectra measured at distances 1 ft (0.3 m), 49 ft (15 m), and 73 ft (27.3 m) (Pestorius and Blackstock, 1974).

Fig. 13.5 Measured evolution of a broadband noise spectrum in a solid containing cracks and micropores (○, input spectrum; △, spectrum at 0.5 m; ●, spectrum at 2.5 m) (Robsman, 1991).

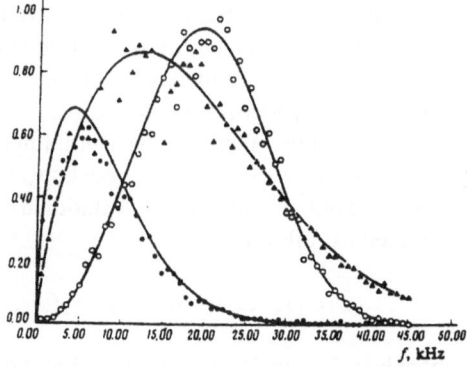

of kilohertz are insufficient for observation of nonlinear effects. However, Robsman used a 3-m-long concrete beam with cracks. Solid media with cracks and micropores can exhibit anomalously high nonlinearity (Ostrovsky, 1991; Naugol'nykh and Ostrovsky, 1997), and, indeed, the measured nonlinearity parameter was found to be 800 (instead of a value of order 10, which is typical for homogeneous solids). The source was attached to the end of the beam and radiated with an acoustic intensity of 5 W/cm^2. Figure 13.5 shows the source spectrum (open circles), together with measurements at 0.5 m (triangles) and 2.5 m (closed circles), away from the source. Since the measured nonlinear effects correlate with structural defects in the material, conclusions can be made regarding the strength of the material.

Theoretical descriptions of the evolution of broadband noise have been obtained for two limiting cases, at small distances $x \ll \bar{x}$, where the influence of the infrequently occurring shocks can be ignored, and for $x \gg \bar{x}$, where the shocks determine the statistical properties. It is reasonable to use the value $\bar{x} = \rho_0 c_0^3 \tau_0 / \beta \sigma_p$ for a nonlinear distortion length in the case of broadband noise, where τ_0 is the time scale (the inverse characteristic frequency) of the disturbance, and $\sigma_p = p_{rms}$ is again the characteristic pressure amplitude.

Near the source, where waveform distortion is described by Eq. (13.7), one can perform exact statistical averaging (Rudenko and Soluyan, 1977; Gurbatov et al., 1991). For a source waveform with Gaussian statistics and correlation function $R_0(\tau)$, the intensity spectrum is given by the expression

$$S(\sigma, \omega) = \frac{\sigma_p^2 e^{-(\omega\tau_0\sigma)^2}}{2\pi(\omega\tau_0\sigma)^2} \int_{-\infty}^{\infty} \left[e^{(\omega\tau_0\sigma)^2 R_0(\tau)/\sigma_p^2} - 1 \right] e^{-j\omega\tau} \, d\tau, \qquad (13.15)$$

where $\sigma = x/\bar{x}$. Expansion of the exponential under the integral yields a power series in terms of the initial correlation function R_0. Although the medium nonlinearity is quadratic, the expansion contains all powers of the correlation function of the source waveform [in contrast to Eq. (13.12)], which correspond to multiple interactions of the harmonic components. Equation (13.15) makes it possible to explain qualitatively the evolution of the frequency spectra observed in the experiments described previously. For example, it follows from Eq. (13.15) that nonlinearity leads to the transfer of wave energy upward in the spectrum. Also, if the source spectrum falls off at low frequencies as $S_0(\omega) \propto \omega^n$, $n \geq 2$, nonlinearity will cause the low-frequency spectrum to evolve into one following the universal asymptotic dependence $S(x, \omega) \propto \omega^2$.

Along with the intensity spectrum, important information in statistical nonlinear acoustics is provided by higher-order spectra and, in particular, by the bispectrum. If $C(\omega)$ is a Fourier transform of a stationary process, then the bispectrum $S_2(\omega_1, \omega_2)$ is defined as follows:

$$\langle C(\omega_1) C(\omega_2) C^*(\omega_3) \rangle = S_2(\omega_1, \omega_2) \delta(\omega_1 + \omega_2 - \omega_3), \qquad (13.16)$$

where here δ is the Dirac delta function. For an input signal with Gaussian statistics, $S_2 \equiv 0$. Comparison of Eq. (13.16) with the phase synchronism conditions shows that the bispectrum reflects the process of three-wave interaction in a quadratic medium, and, hence, the value of the bispectrum characterizes the extent of nonlinear effects. The bispectrum can be used to determine whether energy is shifted upward or downward in the frequency spectrum. Measurements of the bispectrum were reported by Watanabe and Urabe (1981), whose results demonstrate clearly the breakdown of Gaussian statistics as a result of nonlinear wave distortion.

At distances $x > \bar{x}$, the waveforms appear as a succession of randomly located shocks connected by straight line segments, i.e., random sawtooth waveforms (recall the second waveform in Fig. 13.4a). The shocks propagate at random speeds (because their amplitudes are random), which leads to coalescence of the shocks and therefore an increase in the time scale of the turbulence, $\tau_0(x)$. An analogy can be made between the behavior of the shocks and that of an ensemble of particles. Each shock corresponds to a particle whose velocity coincides with the propagation speed of the shock, and mass is proportional to shock amplitude. The instant when two shocks merge corresponds to a perfectly inelastic collision, with coalescence of the particles following the laws of conservation of mass and momentum. This analogy is used in kinetic equations such as the Boltzmann equation, and it allows one to obtain a statistical description of the random distribution of the shocks (Khokhlova et al., 1990).

The second method for analyzing acoustic turbulence is based on an asymptotic solution of the Burgers equation (Gurbatov et al., 1991). In the region where shocks are well developed ($x \gg \bar{x}$), one- and two-dimensional probability density functions for the pressure field can be derived, as can correlation functions, intensity spectra, and probability distributions for the amplitudes and velocities of the shocks. As a result of the multiple coalescence of shocks, information about the input signal structure is lost, and the statistical characteristics become self-similar. All statistical characteristics are defined by the time scale $\tau_0(x)$ of the turbulence, which increases according to the following law, in terms of $\sigma = x/\bar{x}$ (Gurbatov et al., 1991):

$$\tau_0(\sigma) = \tau_0 \sigma^{1/2} \ln^{-1/4}(\sigma), \tag{13.17}$$

where it is assumed that $S_0(\omega = 0) = 0$. Because of shock coalescence, the number of shocks per unit time decreases as $n = \tau_0^{-1}(\sigma) \sim \sigma^{-1/2}$, and the following self-similar noise spectrum is obtained:

$$S(\sigma, \omega) = \sigma^{-2} \tau_0^3(\sigma) S_1[\omega \tau_0(\sigma)]. \tag{13.18}$$

Equation (13.18) exhibits two asymptotic properties, the dependence $S \sim \sigma^{-3/2} \omega^{-2}$ at high frequencies, and $S \sim \sigma^{1/2} \omega^2$ at low frequencies. On the whole, however, the maximum of the energy spectrum is shifted to lower frequencies as $\tau_0^{-1}(\sigma)$. Shock coalescence, and the associated transfer of energy to lower frequencies, causes noise to be attenuated more slowly than a regular signal.

With finite but sufficiently small viscosity, a sawtooth waveform exists out to a distance where the effective Gol'dberg number becomes small. Beyond this distance, the high-frequency dependence $S \sim \omega^{-2}$ is replaced with the exponential law $S \sim \exp[-(\omega \tau_f)^\nu]$, $\nu < 1$, because of variations in the rise times of the shock fronts. Consequently, the spectrum falls off more slowly at high frequencies than for a quasi-monochromatic signal. Also, unlike that for a periodic signal, the attenuation of finite-amplitude noise cannot be predicted by linear theory until distances become very large (Gurbatov et al., 1991).

13.6   Interaction of a Regular Wave with Noise

Of all the problems encompassed by statistical nonlinear acoustics, those involving interactions between regular waves and noise are perhaps of greatest interest from the viewpoint of application. For example, an intense regular wave can be used to control the behavior of a random wave by increasing or decreasing the energy in the noise in a certain frequency range or by pumping the noise energy from one portion of the frequency spectrum to another. On the other hand, because of nonlinear coupling, a regular wave that propagates through a noise field attenuates more rapidly than when the same wave propagates in the absence of noise.

The nonlinear interaction of a regular signal with noise depends on many factors, including the characteristics of both the regular wave (periodic, pulsed, weak, or intense) and the noise (broadband, quasi-harmonic, Gaussian or non-Gaussian statistics), and the ratio of the characteristic frequencies of the interacting waves.

Cavitation noise spectra, acoustic radiation from jet engines, and other sources of intense disturbances consist of discrete spectral lines embedded in a background of broadband noise. The generation of various combination frequencies broadens the spectra associated with the discrete signal frequencies and their harmonics, and the continuous part of the spectrum grows rapidly with increasing propagation distance (Bechert and Pfizenmaier, 1975; Rennick and Scott, 1976). In particular, Bechert and Pfizenmaier (1975) measured an increase of 7 dB in the level of broadband noise produced by a jet stream when a high-intensity signal (130 dB) was introduced.

Numerical predictions obtained by Webster and Blackstock (1978a) for the interaction of noise with an intense tone are shown in Fig. 13.6. The spectra were calculated at increasing distances from the source using the algorithm described in Sect. 11.2.3. The results show clearly how the propagation of noise with an intense tone is accompanied by generation of harmonics of the tone that broaden and intensify the continuous part of the spectrum (the inset for $\sigma = 0$ is a sample of the input noise waveform). At 20 kHz, the noise level increases by 25 dB as the distance increases from $\sigma = 0.8$ to $\sigma = 1.2$ (not shown in Fig. 13.6). Webster and Blackstock (1978a) also performed an experimental investigation of the interaction of a tone

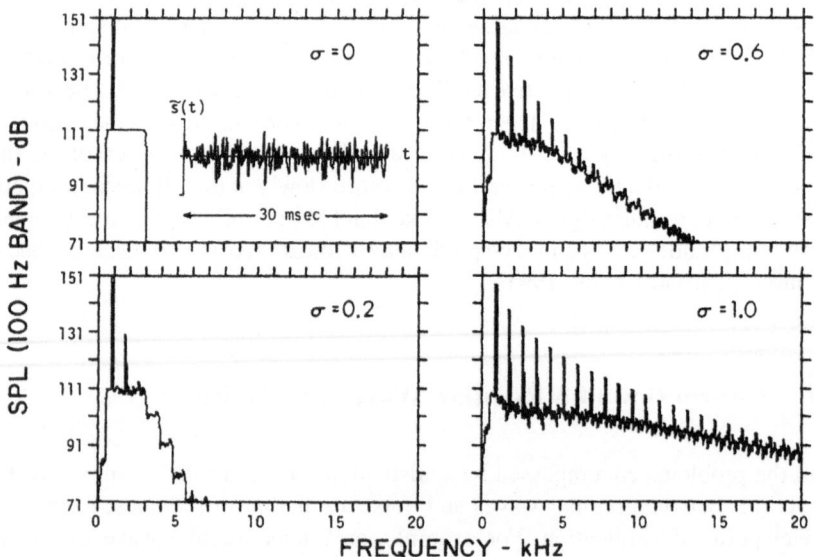

Fig. 13.6 Numerical simulation of the interaction of an intense tone with noise, as a function of $\sigma = x/\bar{x}$ (Webster and Blackstock, 1978a).

with noise in a plane-wave tube filled with air. In the experiment corresponding to Fig. 13.6, an 864 Hz tone with 151 dB source level was embedded in the lower end of a three-octave noise band with 128 dB overall source level. Spectra were measured next to the source and at two locations downstream, at distances 7.4 and 15 m, once with both the tone and the noise propagating together, and once with just the noise. It was observed that in the absence of the intense tone, the noise spectrum remained virtually unchanged and the noise thus propagated according to linear theory. In the presence of the tone, however, the noise spectrum was transformed as in Fig. 13.6, with energy extending up to at least 50 kHz. Similar results were obtained when the experiment was repeated with the tone embedded in the upper end of the noise band.

The anomalously large increase in level at high frequencies is most pronounced near the shock formation distance of the regular wave, and it can be observed for arbitrary ratios of the frequencies of the interacting waves (Gurbatov, 1981). The physical explanation is that noise introduces random fluctuations at the shocks in the intense regular wave, which in turn produce broadening of the spectral lines for the harmonics of the regular wave. The fine structure of the nonlinearly generated frequency bands depends on the frequency ratio of the interaction components, while the dependence of the noise level in the vicinity of the nth signal harmonic on distance and harmonic number is weak or completely absent in this case.

We demonstrate this effect for the case in which the characteristic frequency ω_N of the noise $\zeta(t)$ is much less than the frequency ω_0 of an intense tone. If $A_n(x)$ are the harmonic amplitudes of the regular signal, then for $\omega_N \ll \omega_0$, the field of the nth harmonic can be formulated as

$$p_n(x, \tau) = A_n(x) \cos\left[n\omega_0\left(\tau + \frac{\beta}{\rho_0 c_0^3} x \zeta(\tau)\right)\right]. \tag{13.19}$$

The interaction leads to random phase modulation of the high-frequency wave, and the phase shift $\Phi_n(\tau) = n\omega_0(\beta/\rho_0 c_0^3) x \zeta(\tau)$ increases with both harmonic number and distance. If the phase shift is small, we obtain for the "scattered" component, from Eq. (13.19),

$$p_n^{\text{scat}}(x, \tau) = A_n(x) \left[n\omega_0 \frac{\beta}{\rho_0 c_0^3} x \zeta(\tau) \right] \sin n\omega_0 \tau, \tag{13.20}$$

and consequently, near the nth harmonic, the spectrum of the scattered signal follows that of the low-frequency noise, $\zeta(t)$. In the sawtooth region, where the amplitudes of the regular signal harmonics vary as $A_n \sim 1/nx$, the scattered component has an amplitude independent of distance x and equal to that of the low-frequency noise up to very high frequencies. This effect is associated with the amplification and frequency transformation of the noise that was observed experimentally by Webster and Blackstock (1978a).

It should be noted that in addition to the possibility of increasing the level of the noise by introducing an intense tone, one can also use the tone to suppress the noise level in certain narrow frequency bands. To suppress the noise, the frequency of the intense tone must be chosen sufficiently high that neither it nor the spectral pedestal that is generated at neighboring frequencies during propagation overlaps with the input noise spectrum.

The suppression of low-frequency noise by an intense regular signal was observed by Zorin et al. (1975). The experimental arrangement consisted of a tube 11 cm in diameter, constructed with five detachable sections, each 1 m long. The noise source, an air jet, was placed on the axis at one end of the tube. Part of the air stream was directed into narrowband signal generators—ultrasonic whistles—located at the input end around the jet opening. To reduce the effect of the jet, the measurements were taken at a distance at least 1 m away, and at a 45° angle with respect to the axis of the tube. In the absence of the ultrasonic signal, the noise spectrum remained virtually unchanged at distances from 1 to 5 m. The presence of the regular signal caused the noise level in a portion of the frequency band to drop by 10–15 dB [see Fig. 9a of Rudenko (1986)]. Increasing either the amplitude of the signal or the distance of observation was associated with further suppression of the noise level in that frequency band.

We note that noise can also be suppressed by an intense regular signal of much lower frequency than the input noise spectrum [see Fig. 10 of Rudenko (1986)].

Let us now proceed to discussion of another important problem, increased attenuation of a weak signal as a result of its interaction with an intense noise wave. This problem is related to the attenuation of sound in solids as a result of interaction with thermal phonons. Different aspects of this problem have also been considered in the context of nonlinear acoustics (Westervelt, 1976; Rudenko and Soluyan, 1977). It has been shown that for collinear interaction of Gaussian noise with a weak harmonic wave, propagation of the latter is described by

$$p_n(x, \tau) = A \exp[-\alpha_0 x - (\beta \omega_0 \sigma_p / \rho_0 c_0^3)^2 x^2] \cos \omega_0 \tau. \qquad (13.21)$$

Here α_0 is the classical thermoviscous attenuation coefficient at frequency ω_0. The additional attenuation, associated with the second term in the exponential, is determined completely by the noise intensity σ_p^2 and does not depend on the frequencies of the noise and the tone.

Experimental investigations of the suppression of a regular signal by noise have been conducted by several researchers (Zorin et al., 1975; Moffett et al., 1978; Stanton and Beyer, 1978; Burov et al., 1978; Larraza et al., 1996). Burov et al. (1978) obtained measurements in a tank filled with water. A source was pulsed at 11.5 MHz with a pressure amplitude of 1.2 kPa. The noise spectrum was concentrated at 0.6–1.9 MHz, with intensity 0.3 W/cm^2, which corresponds to $\sigma_p = 69$ kPa. Observed attenuation of the 11.5 MHz signal was in reasonable agreement with Eq. (13.21).

The attenuation factor e^{-ax^2} in Eq. (13.21) is valid only for collinear wave interactions. When the noise field is isotropic, with components that propagate in all directions with equal probability, the attenuation with distance follows the standard exponential law e^{-bx}. The attenuation factor b for the latter case was calculated by Westervelt (1976). For noncollinear interaction where the regular signal and the noise propagate in distinct yet different directions, asynchronous nonlinear interactions occur, which are qualitatively different from those described above (Lind and Hamilton, 1991).

13.7   Conclusion

Statistical nonlinear acoustics today possesses all the attributes of a developed science, and one can refer to hundreds of published works concerning theory, experiment, and applications in this area. These problems are of interest not only to specialists in acoustics, but also to specialists in statistical radiophysics, the theory of nonlinear waves, mechanics of turbulence, and the physics of solids. Therefore, our investigation does not pretend to encompass entirely the current state of the art. In particular, it practically ignores the mathematical methods of statistical nonlinear acoustics, which are significantly different from the traditional methods used in nonlinear statistical optics and in the statistical theory of plasmas (Gurbatov et al., 1991). Our review is dedicated only to what, in our opinion, are some of the most important problems in this area, and it can serve as a brief introduction to this interesting and rapidly developing branch of science.

Acknowledgment   This work was supported in part by RFBR-INTAS Grant No. 95-723.

References

Bechert, D., and Pfizenmaier, E. J. (1975). On the amplification of broad band jet noise by a pure tone excitation. *J. Sound Vib.* **43**, 581–587.

Bjørnø, L., and Gurbatov, S. N. (1985). Evolution of universal high-frequency asymptotic forms of the spectrum in the propagation of high-intensity acoustic noise. *Sov. Phys. Acoust.* **31**, 179–182.

Burgers, J. M. (1948). A mathematical model illustrating the theory of turbulence. In *Advances in Applied Mechanics*, Vol. 1, R. von Mises and T. von Kármán, eds. (Academic Press, New York), pp. 171–199.

Burov, V. A., Krasil'nikov, V. A., and Tagunov, E. Ya. (1978). Experimental investigation of collinear interaction of a weak signal with low frequency noise waves. *Vestnik (Moscow State University Bulletin)* **19**, 53–58 (in Russian).

Foote, K. G. (1974). Wideband response of the parametric acoustic array. In *Finite-Amplitude Wave Effects in Fluids*, L. Bjørnø, ed. (IPC Science and Technology Press, England), pp. 145–150.

Gurbatov, S. N. (1981). Transformation of the statistical characteristics of noise in interaction with a strong regular wave. *Sov. Phys. Acoust.* **27**, 475–480.

Gurbatov, S. N., and Demin, I. Yu. (1982). Statistical analysis of the operation of a parametric source radiating nonlinearly confined spherical beams. *Sov. Phys. Acoust.* **28**, 19–22.

Gurbatov, S. N., Demin, I. Yu., and Malakhov, A. N. (1980). Influence of phase fluctuations on the characteristics of parametric arrays. *Sov. Phys. Acoust.* **26**, 217–220.

Gurbatov, S. N., Malakhov, A. N., and Saichev, A. I. (1991). *Nonlinear Random Waves and Turbulence in Nondispersive Media: Waves, Rays and Particles* (Manchester University Press, Manchester).

Gurbatov, S. N., and Pronchatov-Rubtsov, N. V. (1982). Influence of regular and random inhomogeneities of the medium on the characteristics of parametric radiators. *Sov. Phys. Acoust.* **28**, 456–459.

Gurbatov, S. N., Saichev, A. I., and Yakushkin, I. G. (1983). Nonlinear waves and one-dimensional turbulence in nondispersive media. *Sov. Phys. Usp.* **26**, 857–876.

Gurbatov, S. N., and Shepelevich, L. G. (1978). Spectra of discontinuous noise waves. *Radiophysics and Quantum Electronics* **21**, 1131–1138.

Howell, G. P., and Morfey, C. L. (1981). A parametric study of nonlinear effects in aircraft noise propagation. *9th International Symposium on Nonlinear Acoustics—Book of Abstracts* (University of Leeds, Leeds, U.K.), p. 37.

Howell, G. P., and Morfey, C. L. (1987). Non-linear propagation of broadband noise signals. *J. Sound Vib.* **114**, 189–201.

Khokhlova, V. A., Rudenko, O. V., and Sapozhnikov, O. A. (1990). Sawtooth waves: One dimensional statistical ensembles and thermal self-focusing of beams. In *Frontiers of Nonlinear Acoustics: 12th ISNA*, M. F. Hamilton and D. T. Blackstock, eds. (Elsevier, London), pp. 47–63.

Larraza, A., Denardo, B., and Atchley, A. (1996). Absorption of sound by noise in one dimension. *J. Acoust. Soc. Am.* **100**, 3554–3560.

Lind, S. J., and Hamilton, M. F. (1991). Noncollinear interaction of a tone with noise. *J. Acoust. Soc. Am.* **89**, 583–591.

Moffett, M. B., Konrad, W. L., and Carlton, L. F. (1978). Experimental demonstration of the absorption of sound by sound in water. *J. Acoust. Soc. Am.* **63**, 1048–1051.

Morfey, C. L. (1984). Aperiodic signal propagation at finite amplitudes: Some practical applications. In *Proceedings of 10th International Symposium on Nonlinear Acoustics*, A. Nakamura, ed. (Teikohsha Press, Kadoma, Japan), pp. 199–206.

Naugol'nykh, K. A., and Ostrovsky, L. A. (1997). *Nonlinear Wave Processes in Acoustics* (Cambridge University Press, Cambridge).

Novikov, B. K., Rudenko, O. V., and Chirkin, A. S. (1976). On the synthesis of characteristics of parametric radiating systems from a low frequency noise spectrum. *Nonlinear Hydrocoustics–76* (Taganrog, Russia), pp. 15–18 (in Russian).

Novikov, B. K., Rudenko, O. V., and Timoshenko, V. I. (1987). *Nonlinear Underwater Acoustics* (Acoustical Society of America, New York).

Ostrovsky, L. A. (1991). Wave processes in media with strong acoustic nonlinearity. *J. Acoust. Soc. Am.* **90**, 3332–3337.

Papoulis, A. (1965). *Probability, Random Variables, and Stochastic Processes* (McGraw-Hill, New York). See especially Chapters 9 and 10.

Pernet, D. F., and Payne, R. C. (1971). Non-linear propagation of signals in air. *J. Sound Vib.* **17**, 383–396.

Pestorius, F. M., and Blackstock, D. T. (1974). Propagation of finite-amplitude noise. In *Finite-Amplitude Wave Effects in Fluids*, L. Bjørnø, ed. (IPC Science and Technology Press, England), pp. 24–29.

Rennick, D. F., and Scott, D. S. (1976). Interactions between aerosols and single and multiple finite-amplitude acoustic fields. In *Proceedings of the 6th International Symposium on Nonlinear Acoustics*, Vol. 1, R. V. Khokhlov, ed. (Moscow State University Press, Moscow), pp. 220–240.

Robsman, V. A. (1991). Nonlinear transformation of noise spectra in the acoustic diagnostics of concrete structures. *Sov. Phys. Acoust.* **37**, 541–543.

Rudenko, O. V. (1986). Interactions of intense noise waves. *Sov. Phys. Usp.* **29**, 620–641.

Rudenko, O. V., and Soluyan, S. I. (1977). *Theoretical Foundations of Nonlinear Acoustics* (Plenum, New York). See especially Chapter 10.

Sakagami, K., Akoi, S., Chou, I. M., Kamakura, T., and Ikegaya, K. (1982). Statistical characteristics of finite amplitude acoustic noise propagating in a tube. *J. Acoust. Soc. Japan* **3**, 43–45.

Stanton, T. K., and Beyer, R. T. (1978). The interaction of sound with noise in water. *J. Acoust. Soc. Am.* **64**, 1667–1670.

Vinogradova, M. B., Rudenko, O. V., and Sukhorukov, A. P. (1990). *Theory of Waves* (Nauka, Moscow) (in Russian).

Watanabe, Y., and Urabe, Y. (1981). Changes of zero-crossing slopes of a finite-amplitude noise propagating in a tube. *Jap. J Appl. Phys.* **20**, Suppl. 20–3, 35–40.

Webster, D. A., and Blackstock, D. T. (1978a). Collinear interaction of noise with a finite-amplitude tone. *J. Acoust. Soc. Am.* **63**, 687–693.

Webster, D. A., and Blackstock, D. T. (1978b). Experimental investigation of outdoor propagation of finite-amplitude noise. NASA Contractor Report 2992, Langley Research Center.

Webster, D. A., and Blackstock, D. T. (1979). Amplitude density of a finite amplitude wave. *J. Acoust. Soc. Am.* **65**, 1053–1054.

Westervelt, P. J. (1976). Absorption of sound by sound. *J. Acoust. Soc. Am.* **59**, 760–764.

Zaitsev, V. Yu., and Raevskii, M. A. (1990). Parametric sound radiation in a medium with random inhomogeneities. *Sov. Phys. Acoust.* **36**, 156–160.

Zorin, V. A., Kolotilov, N. N., Rudenko, O. V., Cherepetskaya, E. B., and Chirkin, A. S. (1975). Transformation of a turbulent noise spectrum in the presence of an intense initial signal: Theoretical and experimental investigation. *Symposium on Acoustic-Hydrodynamic Phenomena* (Nauka, Moscow), pp. 264–268 (in Russian).

Chapter 14
Parametric Layers, Four-Wave Mixing, and Wave-Front Reversal

Harry J. Simpson and Philip L. Marston

Contents

14.1   Introduction

The present chapter concerns sound interactions in situations that differ from those analyzed in most of the other chapters of this book. The main processes examined in the present chapter involve either of the following: (1) situations where the coefficient of nonlinearity $\beta = 1 + B/2A$ is large in a flat layer, or (2) four-wave mixing processes where the perturbation of the local speed of sound in the medium (or other induced phase perturbations) varies in proportion to the square of the amplitude of an acoustic pump wave. Situation 1 and certain manifestations of situation 2 have the following common aspect: The response of heterogeneities within the fluid medium governs the resulting sound-wave production. In both cases the host fluid medium is commonly taken to be water. In situation 1 the relevant heterogeneities are usually taken to be bubbles of a gas, and the relevant response is the volume pulsation of the bubble. One manifestation of situation 2 examined here concerns an initially spatially uniform suspension of particles. The relevant

H. J. Simpson
Naval Research Laboratory, Washington, DC, USA

P. L. Marston
Department of Physics, Washington State University, Pullman, WA, USA

© The Author(s) 2024
M. F. Hamilton, D. T. Blackstock (eds.), *Nonlinear Acoustics*,
https://doi.org/10.1007/978-3-031-58963-8_14

response of the particles is the formation of layers of enhanced concentration induced by the radiation pressure of sound. In that case the layers of enhanced particle concentration result in the Bragg reflection of sound.

Both categories of nonlinear response mentioned have analogies in nonlinear optics. Similarities between nonlinear acoustical and optical phenomena have been reviewed by Bunkin et al. (1986). Some of the applications of the optical analogies include optical phase conjugation (or wave-front reversal) and optical probes of physical processes in heterogeneous media or in adverse environments. While there is insufficient space to examine the optical analogies, selected references will be given. As in the optical case, the acoustic response of the nonlinear medium to a diverging acoustic probe wave can be to produce a converging wave that in certain circumstances may be focused back toward the source of the probe wave. Reversal of acoustic wave fronts is not limited to processes depending on nonlinear acoustics. Recent advances in transducer array fabrication and the rapid electronic parallel storage and regeneration of array signals have facilitated the array synthesis of a type of time-reversed ultrasonic wave (Nikoonahad and Pusateri, 1989; Fink, 1993).

The analysis of the nonlinear acoustical processes given here is limited to quasilinear responses in a sense that will be evident from the formulations used. It is the intent of the authors that principal sections may be read independently.

14.2   Focused-Wave Production by Parametric Mixing in a Nonlinear Layer

14.2.1   Formulation and General Case of Parametric Focusing

In this section, the interaction of acoustic waves within a flat layer of nonlinear fluid is analyzed and is shown to produce an outgoing focused wave when certain conditions are met. It will be convenient to designate one of the waves incident on the layer as the acoustic pump wave and the other incident wave as the acoustic probe wave. The pump and probe waves have different frequencies. Nonlinear processes in the fluid outside the layer are neglected throughout the analysis so that linear wave equations can be used to describe the propagation toward and away from the layer. For the purpose of illustrating the production of focused waves by the nonlinear mixing processes, the external fluid is taken to be homogeneous and to have a small-signal sound speed c_0. The parametric wave produced by the mixing process has a frequency given by the difference in frequency of the pump and probe waves. A physical picture of the mixing process for the case of a spherical probe wave is that the response of the nonlinear layer is to act like a dynamic Fresnel zone reflection or transmission grating. The grating evolves in time in such a way as to focus a component of the scattered pump wave and to shift the frequency of the component. Because three waves are essential to the description of the process (the

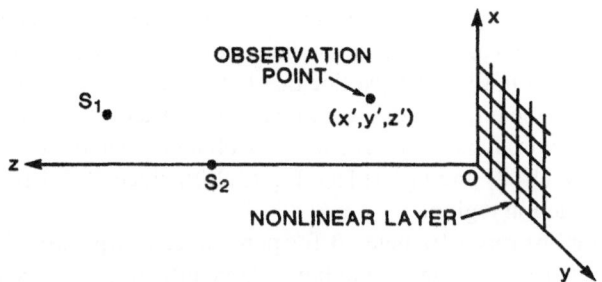

Fig. 14.1 Pump and probe waves spread out from point sources at S_1 and S_2, respectively, and interact in a nonlinear layer where $B/2A$ is enhanced. The enhancement is normally the result of a layer of microbubbles in water. The resulting sound radiated at the difference frequency has foci located symmetrically on both sides of the layer.

incoming pump and probe waves and the outgoing difference-frequency wave), the process is sometimes described as three-wave mixing.

The geometry under consideration is shown in Fig. 14.1. Let S_1 denote the location of the source of an acoustic pump wave of radian frequency $\omega_1 = 2\pi f_1$, and let S_2 denote the location of the source of an acoustic probe wave of frequency $\omega_2 = 2\pi f_2$. The case where $\omega_1 > \omega_2$ will be analyzed. The difference-frequency wave has a frequency $\omega_3 = \omega_1 - \omega_2$. The Cartesian coordinates of the mth source will be denoted by (x_m, y_m, z_m), $m = 1$ and 2, and we may take S_2 at $(0, 0, z_2)$ and S_1 at $(x_1, 0, z_1)$. The small-signal sound pressure due to these sources is

$$p_1(x, y, z, t) = \tfrac{1}{2}[P_1(\mathbf{r}) \exp(j\omega_1 t) + P_2(\mathbf{r}) \exp(j\omega_2 t)$$
$$+ P_1^*(\mathbf{r}) \exp(-j\omega_1 t) + P_2^*(\mathbf{r}) \exp(-j\omega_2 t)], \tag{14.1}$$

$$P_m(\mathbf{r}) = (A_m/R_m) \exp(-jk_m R_m), \quad m = 1, 2, \tag{14.2}$$

where P_m is the complex amplitude of the mth component at $\mathbf{r}$, asterisks denote complex conjugation, $k_m = \omega_m/c_0$, A_m is the source strength, and $R_m(x, y, z)$ is the distance at some field point $\mathbf{r} = (x, y, z)$ from S_m. In the parametric generation of sound the local instantaneous source strength density q is associated with the usual quadratic nonlinearity (Westervelt, 1963). When applied to the situation under consideration, the Westervelt parametric source equation [Eq. (3.46) with $\delta = 0$] becomes

$$\nabla^2 p_2 - \frac{1}{c_0^2}\frac{\partial^2 p_2}{\partial t^2} = -\rho_0 \frac{\partial q}{\partial t}, \tag{14.3}$$

$$q = \frac{\beta(z)}{(\rho_0 c_0^2)^2}\frac{\partial p_1^2}{\partial t}. \tag{14.4}$$

where ρ_0 denotes the ambient density, and the coefficient of nonlinearity β is taken to be a function of z so as to localize the interaction process at or near the plane of constant z, as shown in Fig. 14.1. Notice that in the quasilinear analysis given here, q depends only on the pressure of the pump and probe acoustic waves as given by a linear analysis. Hydrodynamic corrections for obliquely incident waves have been neglected in Eqs. (14.3) and (14.4) [see Eq. (5.26)], since $B/2A$ is assumed to be large in the interaction region.

Since the nonlinearity parameter β for pure water is typically about 3.5, some comments are appropriate concerning how a large effective enhancement of β may be achieved in a layer. The most common enhancement considered is the result of mixing stabilized microbubbles with water. While the full analysis is complicated by inertia and heat flow, one of the reasons for the enhancement of $B/2A$ and therefore β follows from consideration of the quasistatic response of a bubbly mixture. The quasistatic enhancement is a consequence of the enhancement of $B/2A = [\rho c(\partial c/\partial P)_s]_0$ in the equation of state, where ρ and c are evaluated for the mixture and the partial derivative is evaluated for adiabatic conditions (Everbach et al., 1991), and the subscript 0 indicates the equilibrium state; see also Eq. (2.7). For a mixture of gas bubbles in water, the sound velocity c is a strong function of bubble volume fraction μ when μ is less than about 10^{-3}. The velocity of sound increases with decreasing μ. [See the definition of c_{00} in Eq. (5.63), and measurements by Karplus shown by Duvall and Taylor (1971).] Because of the relatively high compressibility of the gas within the bubbles, an increase in pressure can result in a decrease in the effective μ, giving an enhanced pressure dependence for the effective sound speed c. Consequently $B/2A$ of the mixture is enhanced through the factor $(\partial c/\partial P)_s$. An explicit expression for B/A as a function of μ is given by Eq. (5.66).

The inclusion of the effects of inertia on the response of the bubble gives rise to resonance enhancement of the bubble's response associated with either the pump or probe frequency or the difference frequency. In resonance regions, damping becomes important and, depending on the bubble size, heat exchange between the bubble and the surrounding water may be relevant. A partial review of previous studies of the enhancement is given by Wu and Zhu (1991), with measurements of second-harmonic generation in the presence of microbubbles (stabilized by thin polycarbonate membranes) suggesting effective B/A of 10^4 to 10^5. Theoretical studies include those of Welsby and Safar (1970), Zabolotskaya and Soluyan (1973), and Nazarov and Sutin (1989); see also Sect. 5.3.2. Other experimental or combined experimental and theoretical studies include those of Kobelev and Sutin (1980), Kustov et al. (1985, 1986, 1987), and Asada and Watanabe (1990). While there is no doubt that stabilized microbubbles can give rise to local enhancements of β, the actual dynamics of the bubble's response may result in phase shifts of the difference-frequency sound produced that differ from those predicted by Eq. (14.4). For a spatially uniform bubble layer, such phase shifts would not depend on the spatial position. Their presence will not affect the analysis here of parametric focal properties, which follows Marston and Kargl (1990).

The spatial features of the parametrically generated wave field indicative of focusing may be seen from the following analysis of the situation illustrated in Fig. 14.1. Inspection of Eqs. (14.1)–(14.4) shows that components of the parametrically generated wave p_2 oscillate at $\omega_3 = \omega_1 - \omega_2$, with q being proportional to $P_2^*(x, y, z) P_1(x, y, z)$ and $P_2(x, y, z) P_1^*(x, y, z)$, where (x, y, z) are the coordinates at the site of the interaction. The contribution to p_2 that oscillates at ω_3 will be designated as $p_d(x', y', z', t)$. The standard solution of Eq. (14.3) based on superposition for a homogeneous fluid gives

$$p_d(x', y', z', t) = -\int \frac{S(\mathbf{r}, t - |\mathbf{r}' - \mathbf{r}|/c_0)}{4\pi |\mathbf{r}' - \mathbf{r}|} d^3\mathbf{r}, \tag{14.5}$$

$$S(\mathbf{r}, t) = \frac{\omega_3^2}{2\rho_0 c_0^4} [\beta(z) P_2^*(\mathbf{r}) P_1(\mathbf{r}) \exp(j\omega_3 t) + \beta(z) P_2(\mathbf{r}) P_1^*(\mathbf{r}) \exp(-j\omega_3 t)]. \tag{14.6}$$

Consider the case of a layer of thickness L that is sufficiently small that the phase shift for propagation through the layer, approximated as $c_0 L/\omega_m$, is much less than unity for $m = 1$, 2, and 3. Then for a layer centered in the plane $z = z_0$, the spatial dependence of β may be approximated as

$$\beta(z) \simeq \beta_e(\omega_1, \omega_2) L \delta(z - z_0), \tag{14.7}$$

where β_e on the right-hand side denotes the effective coefficient of nonlinearity within the layer, and δ is the Dirac delta function. For the case of a layer of microbubbles, β_e may depend on ω_1 and ω_2 because of resonances. To simplify the evaluation of Eq. (14.5), it is convenient to take $z_0 = 0$. The Fresnel approximation may be used to simplify the various propagation distances (see, e.g., Marston, 1992). Equations (14.6) and (14.7) give

$$p_d(x', y', z') \simeq -\frac{\omega_3^2 L}{8\pi \rho_0 c_0^4 |z'|} (J + J^*), \tag{14.8}$$

$$J = \beta_e P_2^*(0) P_1(0) \exp(j\omega_3 t - jk_3 r') \iint\limits_{-\infty}^{\infty} e^{-j(x^2 + y^2)F/2} e^{jxG} e^{jyH} dx dy, \tag{14.9}$$

$$F = \frac{k_3}{|z'|} + \frac{k_1}{z_1} - \frac{k_2}{z_2}, \quad G = \frac{k_1 x_1}{z_1} + \frac{k_3 x'}{|z'|}, \quad H = \frac{k_3 y'}{|z'|}, \tag{14.10}$$

where the axes have been chosen such that $x_2 = y_2 = 0$ as illustrated in Fig. 14.1, and r' designates the distance from O to the observation point (x', y', z'). The function in Eq. (14.8) may be written $J + J^* = 2\,\mathrm{Re}(J)$. The integral in Eq. (14.9) has been simplified by approximating the spreading factor by $|z'|^{-1}$, since it is the dependence on phase that usually determines p_d. Consequently, the incident-wave amplitudes P_m are evaluated only at the origin O, and aperture effects have been neglected. For observation points on the far side of the layer, z' is negative. Notice that the parametric field is symmetric in this level of approximation: $p_d(x', y', -z', t) = p_d(x', y', z', t)$.

A field point (x', y', z') giving an infinite J in Eq. (14.9) corresponds to a focal point of the parametrically generated wave and will be denoted by (x_f, y_f, z_f). The condition for divergence is $F = G = H = 0$. There are two foci, each with $y_f = 0$ and with

$$z_f = \frac{\pm(\omega_1 - \omega_2)}{(\omega_2/z_2) - (\omega_1/z_1)}, \quad x_f = -x\frac{\omega_1|z_f|}{\omega_3 z_1}, \tag{14.11}$$

where the $\pm$ sign implies that there are forward- and backward-directed foci that are symmetric with respect to the layer. Notice that except for special cases, (x_f, y_f, z_f) is shifted away from the probe source location (x_2, y_2, z_2). An important special case corresponds to phase conjugation (discussed in Sect. 14.2.2), where $(x_f, y_f, z_f) = (x_2, y_2, z_2)$ with $\omega_1 = 2\omega_2$, in which case the parametric-wave frequency ω_3 corresponds to the probe frequency ω_2. This focal condition is met by taking $x_1 = 0$ and z_1 infinite, so that the pump wave is a plane wave at normal incidence. This is consistent with symmetry and the related analysis in Sect. 14.4.

When the field point (x', y', z') is displaced from the focal point (x_f, y_f, z_f), J may be approximated using the two-dimensional stationary phase approximation (SPA) as reviewed by Marston (1992). The SPA may be used to illustrate the convergence of the parametrically generated wave front, as will now be illustrated. It is sufficient to require that $y' = 0$ and to restrict attention to $z_f > 0$. The phase of the integrand is stationary at $x = x_s$, where

$$x_s = (\Delta z)^{-1}(z_f x' - z' x_f), \quad \Delta z = z_f - z'. \tag{14.12}$$

The geometrical construction illustrated in Fig. 14.2 shows that a ray from the layer at $(x_s, 0, 0)$ drawn through the field point $(x', 0, z')$ also intersects the focus at $(x_f, 0, z_f)$. A paraxial assumption that such rays, as well as those from S_1 and S_2 to O, are nearly parallel to the z axis, and the SPA of Eq. (14.9), give

$$p_d(x', y', z', t) \simeq \mathrm{Re}\{\mathcal{G}(A_2^*/R)\exp[j\omega_3 t + jk_3 R\,\mathrm{sgn}(\Delta z) - j(\pi/2)\,\mathrm{sgn}(\Delta z)]\}, \tag{14.13}$$

$$\mathcal{G} = -\frac{P_1(0)}{2\rho_0 c_0^2}\beta_e k_3 L\left[\frac{z_f}{R_{02}}\exp(jk_2 R_{02} - jk_3 R_{0f})\right], \tag{14.14}$$

where R denotes the distance from (x', y', z') to (x_f, y_f, z_f), and R_{0f} and R_{02} are the distances from O to F and S_2, respectively. The function $\mathrm{sgn}(\Delta z)$ is ± 1 depending on the corresponding sign of Δz. Equation (14.13) describes the

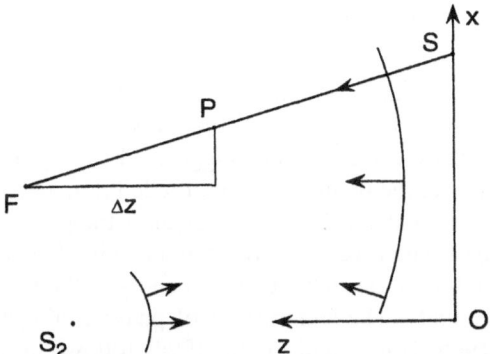

Fig. 14.2 F at $(x_f, 0, z_f)$ is a focal point, and P at $(x', 0, z')$ is some observation point displaced from it. The phase of the integral in Eq. (14.9) is stationary for $x = x_s$ and $y = 0$, which are the coordinates of S. The parametrically generated wave front converges on F, and the stationary phase contribution is associated with the ray to F that passes through P. In the case of phase conjugation, F and S_2 are the same point.

convergence of a spherical wave front toward (x_f, y_f, z_f). If the wave is allowed to pass through the focus unobstructed, it then spreads out, since the space-time dependence becomes $\exp[j(\omega_3 t - k_3 R)]$. Upon the wave's passing through the focus, there is a phase advance of the wave by π, as expected for a three-dimensional focus (Marston, 1992). In obtaining Eq. (14.13), use was made of the expression $P_2 = A_2/R_2$ for the probe wave that spreads out from S_2. From the form of Eq. (14.13), $\mathcal{G}$ is the complex "gain" due to parametric mixing in the layer.

As the focal point is approached, R vanishes, the SPA becomes invalid, and it is necessary to include effects of diffraction. Let w denote the smaller of the widths of the pump and probe beams at the nonlinear layer. By analogy with the focal properties of a converging lens of diameter w, the magnitude of the pressure at the focus is estimated to be

$$|p_d|_{\max} \simeq (k_3 w^2/2\pi z_f)|\mathcal{G} P_2(0)|, \qquad (14.15)$$

which apart from numerical factors close to unity agrees with an analysis of Kustov et al. (1986).

Equations (14.8)–(14.10) are symmetric with respect to moving either S_1 or S_2 (or both) to the other side of the screen so that z_m in Eq. (14.11) can be replaced by $|z_m|$. For example, the layer could function as part of a parametric receiving system by allowing the probe wave to be incident from the right-hand side of the layer. The analysis may be generalized for pump or probe waves that converge toward virtual source locations at S_1 or S_2 beyond the nonlinear layer.

14.2.2   Wave Front Reversal and Phase Conjugation

Consider now the special case in which $\omega_3 = \omega_2$ and the pump conditions are such that the parametric wave is focused back to the probe source at S_2. This corresponds to the conditions of phase conjugation as suggested by the proportionality of P_d to the complex conjugate of the pump-wave amplitude in Eq. (14.13). Since $\omega_3 = \omega_2$, the converging wave is associated with a reversal of the probe wave front incident on the nonlinear layer. The reversed wave may be shifted as a result of the phase of $\mathcal{G}$, though the conditions are such that the quantity in the brackets in Eq. (14.14) may be replaced by unity. The significance of phase conjugation (Pepper, 1982; Zel'dovich et al., 1985; Gower and Proch, 1994) follows by examining the case where the probe wave front incident on the parametric layer has been distorted by a phase perturbation due to an inhomogeneity within the medium. Suppose that a local portion of the probe wave front is delayed slightly; as a consequence of the complex conjugation of A_2 in Eq. (14.13), the corresponding portion of the outgoing wave front is advanced in space. The advancement is such that after propagation back through the same inhomogeneity, a spherical wave front is produced that converges on S_2. This property has motivated the development of optical phase conjugation (Pepper, 1982, 1986) and the analysis of acoustical analogs (Jackson and Dowling, 1991; Fink, 1993; Cassereau and Fink, 1993).

While some other nonlinear acoustical mechanisms for phase conjugation will be discussed later in this chapter, it is appropriate to review some of the experiments carried out with bubble layers. Kustov et al. (1985, 1986) demonstrated the parametric production of focused sound by a bubble layer by measuring the evolution of the amplitude and width of the parametrically generated beam. These were maximized and minimized, respectively, at the focus. (The reader is cautioned that experiments described as phase conjugation by those authors actually involve probe and parametrically generated waves that differ in frequency, since they have $f_2 = 60$ kHz and $f_3 = 40$ kHz.) One of the difficulties in making $|\mathcal{G}|$ in Eq. (14.14) sufficiently large for practical applications stems from the large magnitude of $\rho_0 c_0^2$ in the denominator. (For water, $\rho_0 c_0^2 \simeq 22{,}200$ atm.) Other complications include attenuation and [as discussed by Marston and Kargl (1990)] the dephasing of the pump and probe waves relative to the midplane of the bubble layer as a result of spreading or corrugations of the layer. Perhaps parametrically focused sound in water could be more thoroughly explored if other mechanisms for producing a large enhancement of the effective value of $B/2A$ in a layer could be developed. Other ways of demonstrating focusing include (1) measuring for the parametrically generated wave the phase dependence on the transverse coordinates x' and y' for comparison with Eq. (14.13) for values of z' displaced from the focus; and (2) for such (x', y', z') demonstrating the flow of energy along the ray shown in Fig. 14.2 by selected blocking of possible ray paths. Parametric demodulation and experiments with $\omega_1 - \omega_2 \ll \omega_1$ are described by Kustov et al. (1987) and Asada and Watanabe (1990). The latter report values of β_e as large as 78 in a layer of hydrogen bubbles in water produced by electrolysis. Zabolotskaya (1984) analyzed a system in which the $2\omega_2$ component of the interaction results from the response of bubbles to a pair of opposing pump waves having the same frequency as the probe wave.

14.3 Four-Wave Mixing Resulting from Responses to Radiation Pressure

14.3.1 Four-Wave Mixing Mediated by a Suspension of Particles

The processes examined in this section differ from those considered in Sect. 14.2, since here the interaction of three waves produces a fourth wave of interest. Furthermore, the processes depend on the slow response of the medium to ultrasonic radiation pressure of incident waves. Such processes have been explored in optics, where an aqueous suspension of dielectric spheres was used to produce what is an artificial Kerr medium (Smith et al., 1981, 1982). In optics, the terminology "Kerr medium" is used to describe a medium in which the complex optical refractive index changes in proportion to the local intensity (Pepper, 1982; Gower and Proch, 1994). An analogous change in the acoustical refractive index occurs when a suspension of particles or droplets that is initially spatially homogeneous is altered by the radiation pressure of sound. As a consequence of the response to radiation pressure, the sound speed or density of the effective medium varies spatially. The scattering of sound (typically that of a probe wave) by the variations results in the production of a fourth wave. The interaction can be used as a probe of suspension properties or (see Sect. 14.3.2) for phase conjugation. The process is efficient in the following sense: A pump wave having an amplitude of only 5000 Pa can result in appreciable levels of coherent reflection of a probe wave from a sparse suspension in an interaction volume of only a few cubic centimeters. A suspension of bubbles should also exhibit this interaction, though it differs from Zabolotskaya's dynamic four-wave mixing interaction mentioned at the end of Sect. 14.2.2.

The analysis and experimental results summarized here are those of Simpson (1992) and Simpson and Marston (1993, 1995). The interaction geometry is shown in Fig. 14.3. A standing pump wave is established in a suspension of particles by counterpropagating waves having wave vectors $\mathbf{k}_1$ and $\mathbf{k}_2$ that are equal in magnitude but opposite in direction. In response to the radiation pressure of the standing wave, particles within the suspension will migrate to form spatially periodic variations in the number n of particles per unit volume. For the measurements with solid particles to be described, as for the case of most solid particles in water, the particles are attracted to the pressure nodes (velocity antinodes) of the standing wave. The spacing between the enriched layers is $d = \lambda_1/2$, where $\lambda_1 = 2\pi/k_1$ is the wavelength of the pump wave. A probe wave having a wave number $k_3 = \omega_3/c_0 > 2k_1$ is Bragg reflected by the induced grating of particles when the probe-wave frequency ω_3 and angle of incidence θ meet the condition

Fig. 14.3 Acoustic
four-wave mixing mediated
by a suspension. The particle
suspension is spatially
modulated in response to
radiation pressure. Bragg
reflection from the resulting
grating generates the wave
directed along $\mathbf{k}_4$.

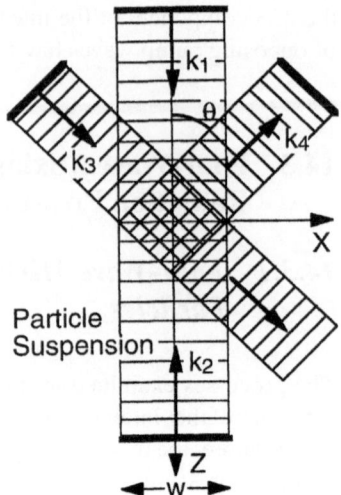

$$k_3 d \cos \theta = m\pi. \tag{14.16}$$

When this Bragg condition is met, the waves reflected with an angle of reflection
equal to the angle of incidence interfere constructively, and the reflected signal is
thus enhanced in proportion to the number of reflecting layers for the usual case of
a sparse grating.

The process of grating formation is important to the understanding of the
evolution and magnitude of the Bragg reflected signal. Consider first the equilibrium
properties of the grating that are reached a sufficiently long time after the standing
pump wave is established. The analysis is based on the radiation pressure on
small spherical objects (particles or droplets of oil) in a standing wave that are
sufficiently sparse that only the radiation pressure due to the incident standing wave
is significant. The approximation of the radiation pressure on a small spherical solid
particle is the same as that for a fluid particle of the same radius provided that the
adiabatic compressibility of the particle is related to the elastic properties of the solid
by $\kappa_p = \rho_p^{-1}(c_l^2 - \frac{4}{3}c_t^2)^{-1}$, where ρ_p is the density and c_l and c_t are the velocities
for longitudinal and transverse waves. Gor'kov (1962) expressed the radiation force
on a spherical particle in a standing wave as $-\nabla U$, where U is an effective potential
[see Eq. (6.96)]. For a standing-wave pressure given by $p = p_0 \sin k_1 z \sin \omega_1 t$, the
potential becomes

$$U = \frac{V_p \kappa_0 p_0^2}{8} (A - B \cos 2kz), \tag{14.17}$$

where $V_p = 4\pi a^3/3$ is the volume of the particle and here

$$A = -\left(\frac{\kappa_p}{\kappa_0} - \frac{4\rho_0 - \rho_p}{2\rho_p + \rho_0}\right), \quad B = -\left(\frac{\kappa_p}{\kappa_0} - \frac{5\rho_p - 2\rho_0}{2\rho_p + \rho_0}\right), \tag{14.18}$$

where $\kappa_0 = 1/\rho_0 c_0^2$ is the compressibility of the outer (or host) liquid. The resulting force is the same as the prediction by Yosioka and Kawasima (1955) calculated by a different method (see also Sect. 6.3.1.3). The viscous corrections have been shown to be weak when the thickness of the oscillating Stokes layer around the particle is small in comparison to the particle radius a (Doinikov, 1994). Gor'kov argued that for a suspension of weakly interacting particles, the particle number density $n_B(z)$ varies with position z according to a Boltzmann distribution $n_B(z) = C \exp(-U/k_B T_0)$, where T_0 denotes the ambient temperature and k_B denotes Boltzmann's constant. The constant C may be expressed in terms of the number N of particles per area contained in one layer, which is taken to be a constant, independent of standing-wave amplitude. (It is assumed that the migration of particles into the standing-wave region shown in Fig. 14.3 is inhibited.) The number density becomes (Simpson and Marston, 1995)

$$n_B(z) = \frac{k_1 N \exp(q \cos 2k_1 z)}{\pi I_0(q)}, \quad q = V_p B p_0^2 \beta_0 / 8 k_B T_0, \tag{14.19}$$

where I_0 is a modified Bessel function.

The widths and peak concentrations of the resulting bands of particles depend strongly on the amplitude of the standing wave. The solid curves in Fig. 14.4 show the predicted equilibrium values of $n_B(z)$ for different standing-wave amplitudes. The calculation is for SDVB (styrene-divinylbenzene) particles having radii $a = 12.5\,\mu m$ and compressibility $\kappa_p = 2.35 \times 10^{-10}\ kg^{-1}m^{-1}s^2$ in a density matched saline solution with $\rho_0 = \rho_p = 1.05$ g/cm^3 and $\kappa_0 = 3.86 \times 10^{-10}\ kg^{-1}m^{-1}s^2$. The calculation is shown for a pump-wave frequency $f_2 = 800$ kHz that has $d = 0.981$ mm. The number density in the absence of a pump wave is $2.00 \times 10^{10}\ m^{-3}$, which corresponds to a sparse volume fraction of only 1.64×10^{-4}. Inspection of Fig. 14.4 shows that substantial increases in the concentration are predicted.

The banding of particles or droplets due to ultrasonic standing waves in aqueous suspensions or emulsions is easily seen by eye. Figure 14.5 shows a photograph of the phenomenon for the case of an emulsion of dodecane oil drops in a 1.2 MHz ultrasonic standing wave in water. The drops were made neutrally buoyant by dissolving a small amount of CCl$_4$ in the oil. A small amount of surfactant was added to help stabilize the emulsion so that the drop radii were typically less than $50\,\mu m$.

Calculation of the Bragg reflected amplitude makes use of an effective media approximation where the local acoustic properties depend on $n_B(z)$. The experiment can be arranged so that the disturbance of the grating of particles by the radiation pressure of the probe wave is negligible. There are two approaches that have been found to give equivalent predictions for sparse gratings (Simpson and Marston, 1995). These approaches may be summarized as follows: (1) replace $n_B(z)$ by a

Fig. 14.4 The spatial
Bolztmann distribution of
particles in the radiation
pressure potential well is
shown for an example of
neutrally buoyant SDVB
(styrene-divinylbenzene)
particles in a saline solution.
The distribution (solid curve)
becomes narrower for
increasing pump pressure
amplitudes p_0. The dashed
curve is the local distribution
assumed in one of the Bragg
scattering models.

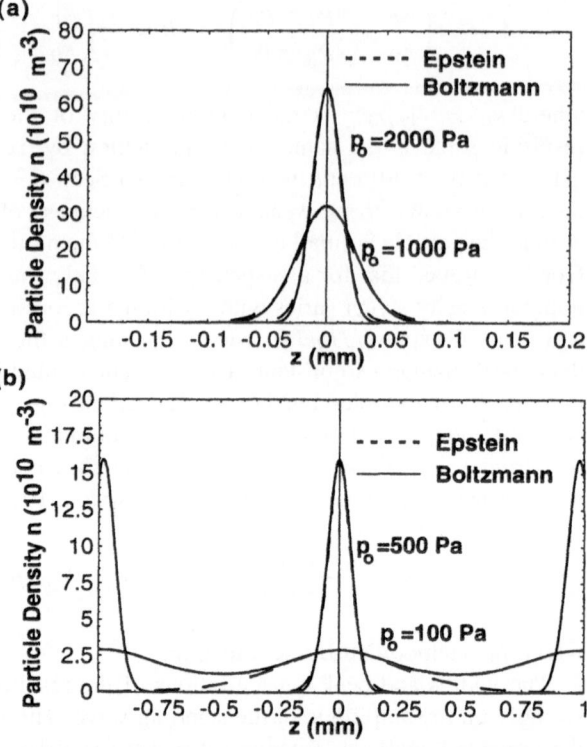

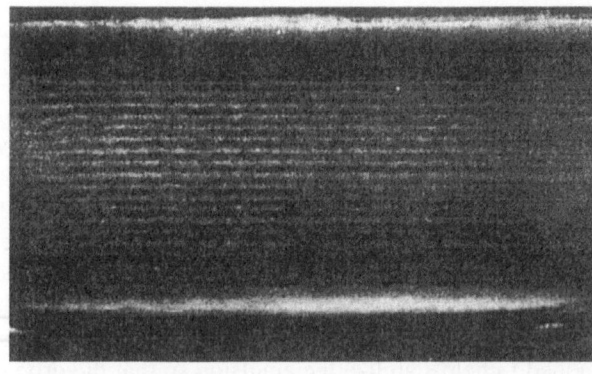

Fig. 14.5 Bands in an
emulsion of dodecane oil
droplets that have formed in
response to the radiation
pressure of an ultrasonic
standing wave. The distance
between bands is 0.7 mm,
which is half of the
wavelength.

nearly equivalent distribution for which the scattering problem by a single layer is known and calculate the many-layer case using a Born approximation that neglects the depletion of the probe wave; or (2) evaluate the spatial Fourier coefficients of the acoustical refractive index and express the scattering amplitude using a coupled-wave theory (developed for thick optical holograms) simplified with the aforementioned Born approximation. Approach 1 was facilitated by using the symmetric refractive index profile known as an Epstein layer, for which an exact expression for the reflectivity is known. The refractive index profile is related to a particle number density profile $n_B(z)$ for a single layer through an effective media approximation. The dashed curves in Fig. 14.4 show $n_B(z)$ in the vicinity of the velocity amplitude at $z = 0$ for different standing-wave pressure amplitudes p_0. Comparison with $n_B(z)$ confirms that the Epstein layer should be a good approximation for sufficiently large p_0. The limiting form of the reflectivity and the effective media approximation were also confirmed in the sparse particle limit by comparison with Rayleigh backscattering for a single layer of particles.

To approximate the reflection of the probe waves off of the system of layers diagrammed in Fig. 14.3, the complex amplitude reflected by each layer is superposed in the Born approximation. The resulting amplitude reflection coefficient for the array is

$$R_A(k_3, \theta) = \nu N_l R_E(k_3, \theta) \frac{\sin(N_l k_3 d \cos \theta)}{N_l \sin(k_3 d \cos \theta)}, \qquad (14.20)$$

where $R_E(k_3, \theta)$ is the reflection coefficient for a single Epstein layer for the indicated wavenumber and angle of incidence, ν is an aperture factor equal to unity for the case of an unbounded plane probe wave, and N_l is the total number of reflecting layers. A reference phase has been assumed, making it unnecessary to include a separate propagation phase shift in Eq. (14.16). Neglecting aperture restrictions gives $N_l \simeq w(d \tan \theta)^{-1}$ for a grating of width w. Notice that $|R_A|$ is maximized when the Bragg condition given by Eq. (14.16) is met. At the Bragg condition, one obtains $R_A = \nu N_l R_E$ because the reflections from the layers have the same phase. To model experiments, the factor ν becomes less than unity to account for the volume of overlap between a probe beam and the region viewed by the receiver (Simpson, 1992). For θ near $45°$, it is a good approximation to take $\nu_l \simeq \frac{1}{3}$, as was done for the plot of $|R_A(k_3, \theta)|$ shown in Fig. 14.6. The properties assumed for the suspension and the pump wave are as for Fig. 14.4. The reader is referred to Simpson and Marston (1995) for expressions relating R_E to the suspension properties and to the pump amplitude p_0. While the peak reflectivities may appear to be small, it should be remembered that they increase roughly in proportion to the initial volume fraction, which is only 1.64×10^{-4}.

In method 2 for calculating the scattering, the application of diffraction-grating coupled-wave theory (Gaylord and Moharam, 1982) makes it necessary to restrict θ and k_3 to values such that the Bragg condition is met. Our resulting expression for the reflection at the mth Bragg order is exact within the assumptions of the Born approximation and effective media theory for the situation where $\rho_p = \rho_0$

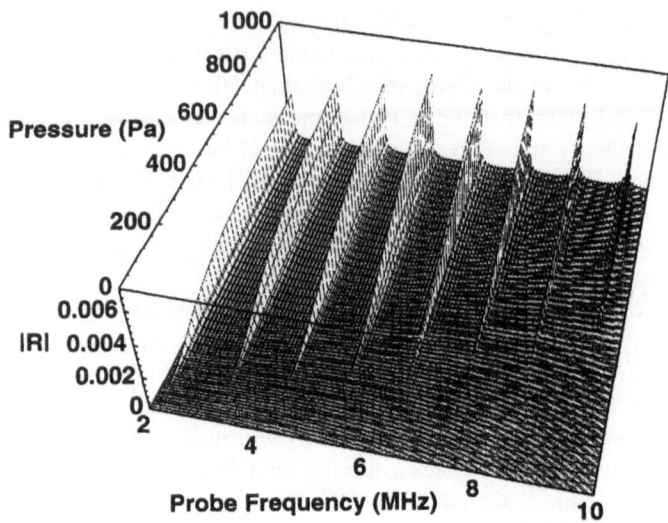

Fig. 14.6 Predicted equilibrium reflection coefficient as a function of pump amplitude p_0 for the SDVB particle suspension considered in Fig. 14.4. The ridges correspond to the Bragg condition. The angle of incidence is $\theta = 41°$.

and the particle number density is given by $n_B(z)$. The magnitude of the amplitude reflectivity at the mth Bragg order is

$$|R_m| = \nu C |1 - \kappa_p/\kappa_0| I_m(q)/I_0(q), \qquad (14.21)$$

where $C = (V_p N/d)(k_3 w/2 \sin \theta)$, and the aperture factor ν has been inserted as discussed following Eq. (14.20). For small values of q, the ratio $I_m(q)/I_0(q)$ becomes $q^m 2^m m!$ and $|R_m|$ is proportional to p_0^{2m}. Since q is proportional to the ratio of the depth of the potential energy well U to $k_B T$, large values of q correspond to highly localized layers of particles, as shown in Fig. 14.4 for the case $p_0 = 2000$ Pa. The resulting reflectivity saturates as $q \to \infty$ and $I_m/I_0 \to 1$, since all of the particles in the suspension scatter in the Bragg direction with the same phase. A direct comparison (Simpson and Marston, 1995) shows that the transition from the p_0^{2m} onset of the reflectivity to saturation in Eq. (14.21) is well approximated by $|R_A|$ from Eq. (14.20) at the Bragg condition. This transition corresponds to the ridges evident in Fig. 14.6.

The general behavior evident in Fig. 14.6 was observed in experiments (Simpson, 1992), with some differences that illustrate limitations of our analysis. Figure 14.7 shows the measured reflectivity magnitude as a function of pump pressure for a fixed angle of incidence $\theta = 40.3°$. The measurement of $|R|$ as a function of frequency made use of a calibration procedure that accounted for the frequency response of the probe-source/receiver system, and the probe tone bursts were sufficiently long to achieve steady-state reflection by the grating. (The number of cycles is

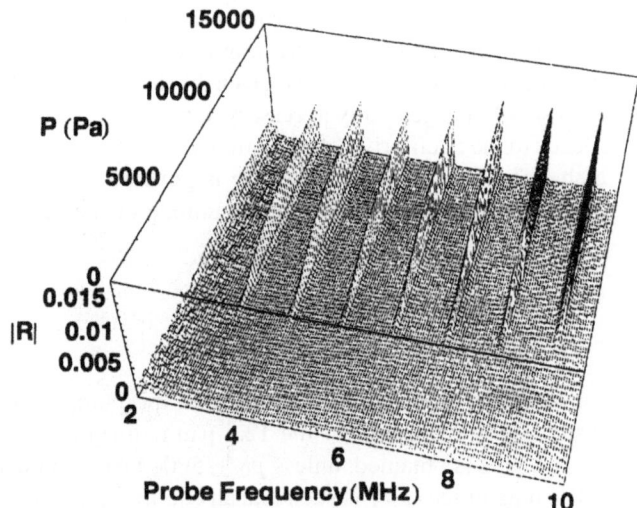

Fig. 14.7 Measured reflection coefficient for the sparse suspension of SDVB particles considered in Fig. 14.4. The buildup of the reflectivity with increasing pump pressure p_0 is at higher amplitudes than calculated in Fig. 14.6 because the approach to equilibrium is long.

Fig. 14.8 A slice through the measured response surface in Fig. 14.7 is shown in the saturated region ($p_0 = 20{,}000$ Pa) as the solid curve. The prediction from Eq. (14.23) as a function of pump frequency is shown as the dashed curve.

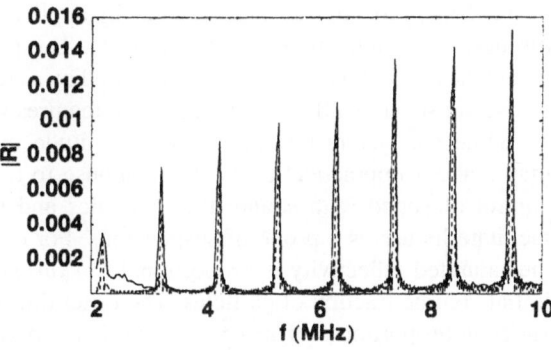

larger than the number N_l of layers within the grating.) The ridges evident in Fig. 14.7 correspond to the Bragg maxima in Fig. 14.6, and the measured $|R|$ in the saturation region is similar to (or slightly greater than) the predicted value as shown in Fig. 14.8. A significant difference between Figs. 14.6 and 14.7 is that the observed onset of the Bragg reflectivity ridge occurs for a larger apparent value of p_0 than predicted. The reason for this is that, for small p_0, the time for the induced grating to reach equilibrium becomes too large to be achievable. [See the discussion of Eq. (14.22) below.] The measurements in Fig. 14.7 (and several other scans not shown here) were carried out by increasing p_0 in small steps of about 100 Pa. After each new p_0 is established, there is a 1- to 3-minute waiting period before a probe frequency scan is recorded. The probe tone bursts had f_3 ranging from 2 to 10 MHz and f_3 was incremented in 50 kHz steps.

The dynamics of migrating suspended particles in an ultrasonic standing wave are analyzed by Higashitani et al. (1981). The drag force which opposes the motion in response to the radiation force $-\nabla U$ is given by the Stokes drag, since the Reynolds number is small. It is generally necessary to include the effect of Brownian diffusion, which leads to the equilibrium distribution $n_B(z)$. The early-time response of particles to a change in the radiation pressure is governed by the Stokes drag, which leads to the following estimate for the magnitude of the grating formation time:

$$\tau_f \simeq \frac{9\eta d^2}{2\pi\kappa_0 Ba^2 p_0^2},\tag{14.22}$$

which depends strongly on the particle radius a and the pump amplitude p_0 as well as the viscosity η. For example, for the 12.5 μm radius particles used in the experiment, $\tau_f \geq 2 \times 10^3$ s is obtained, unless $p_0 \geq 5000$ Pa for sudden application of a pump wave. The magnitude of τ_f is such that an equilibrium grating is expected only for the region where $|R|$ saturates in Fig. 14.7 when the suspension responds to the sequence of steps used in the experiments. Simpson (1992) analyzed the early-time behavior of an initially homogeneous suspension in response to sudden application of pump waves having amplitudes $p_0 \geq 7000$ Pa. The time evolution of the Bragg reflectivity was calculated by application of the coupled-wave method discussed in conjunction with Eq. (14.21). The predicted behavior gives a period where the reflectivity is negligible, followed by a steep rise. The onset delay and rate of rise are strongly affected by p_0. These features were confirmed by experiments.

While the factors that affect the magnitude of the grating formation time τ_f may make it impractical to use the response to radiation pressure to facilitate the control of sound with sound, the dynamics and magnitude of the response may facilitate its use as a probe of suspensions. For example, Eq. (14.21) predicts that the saturated reflectivity is proportional to $k_3 w/\sin\theta$ and to $V_p N/d$, which is the initial volume fraction of particles. The onset time for the growth of the reflectivity varies in proportion to $\eta/a^2 p_0^2$, where a is the particle radius. It is remarkable that changes in the equilibrium grating reflectivity in response to small changes in the pump wave intensity are predicted to be larger in the acoustic case than in the optical case investigated by Smith et al. (1981). The reason is that the potential well depth when expressed in terms of the pump-wave intensity and the corresponding value q in Eqs. (14.19) are proportional to the product of the particle volume V_p and reciprocal of the wave velocity. Since the speed of sound is much less than the speed of light, the required acoustic pump wave intensity is relatively small.

14.3.2 Phase Conjugation Involving Responses to Radiation Pressure

The response to the combined radiation pressure of pump and probe waves can be used to facilitate acoustical phase conjugation. Sato et al. (1990) demonstrated that a suspension of microspheres could be used to generate a conjugate wave. The interaction geometry may be described using Fig. 14.3. The system is configured with $\omega_1 = \omega_2 = \omega_3$. The wave $\mathbf{k}_1$ serves as a reference wave in the generation of a hologram that forms in response to the radiation pressure of the standing wave associated with the interference by the probe wave $\mathbf{k}_3$. The Bragg diffraction of a readout wave $\mathbf{k}_2$ by the induced grating produces a phase conjugated wave having $\mathbf{k}_4 = -\mathbf{k}_3$. For practical reasons, the waves $\mathbf{k}_1$ and $\mathbf{k}_3$ are turned off during the readout process, and the degradation of the grating by particle diffusion is inhibited by application of a higher-frequency standing wave.

An earlier observation of acoustical phase conjugation (Andreeva et al., 1982) and subsequent experiments (Bunkin et al., 1985) made use of the response of a free horizontal water surface to radiation pressure. A pump wave propagates vertically to reflect from the surface and produces a standing wave. A probe wave having the same frequency is incident on the surface at an angle. The wave may be diverging from a point source. The combined radiation pressure of the pump and probe waves produces a stationary pattern of ripples (elevations and depressions). The pattern serves as a phase grating that diffracts a portion of the incident pump wave to produce a conjugate probe wave. The situation is analyzed by Zel'dovich et al. (1985).

14.4 Kinematic Processes and Miscellaneous Phase Conjugation Processes

Conjugation processes described here as "kinematic" are closely related to those discussed in Sect. 14.2.2. Figure 14.9 shows the essential features of the interaction. A probe wave of frequency ω_2 diverges from a point source at S_2. Near the plane $z = 0$ is a region where the velocity of sound or attenuation may be temporally modulated with a frequency ω_1. To illustrate the focal properties of the transmitted wave, it is sufficient to restrict attention to phase modulation in a region of narrow width $L \ll |z_2|$. Let the phase delay introduced by paraxial propagation through that region be given by $(kL - 2\varepsilon \sin \omega_1 t)$, where ε is a small parameter and $k = \omega_2/c_0$. The Fresnel approximation of the complex amplitude of the outgoing wave at $z = L/2$ is

$$p \simeq \frac{A_2}{|z_2|} e^{-jk[|z_2|+(L/2)]} e^{-jk(x^2+y^2)/2|z_2|} e^{j(\omega_2 t+2\varepsilon \sin \omega_1 t)}. \tag{14.23}$$

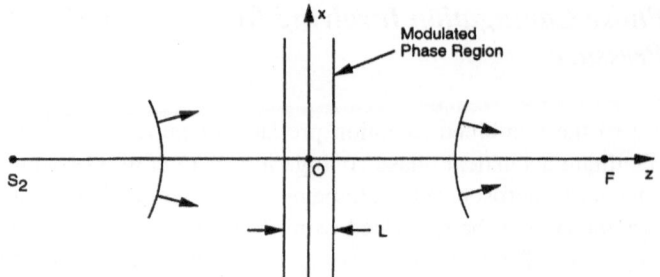

Fig. 14.9 Configuration for kinematic production of a focused wave by propagation through a thin region where the phase velocity is modulated at frequency ω_1. If the probe-wave frequency ω_2 equals $\omega_1/2$, the focus F is symmetric about the midplane with the source point S_2. The effect of modulation analyzed applies to various parametric phase conjugation methods.

Since ε is small, the approximation $\exp(j2\varepsilon \sin \omega_1 t) \simeq 1 + \varepsilon \exp(j\omega_1 t) - \varepsilon \exp(-j\omega_1 t)$ may be used. From the third term, there is a component of the transmitted wave that has an amplitude $\varepsilon A_2/|z_2|$ and a time dependence of $\exp[j(\omega_2 - \omega_1)t]$. Restrict attention to the case where $\omega_1 = 2\omega_2$. The time factor becomes $\exp(-j\omega_2 t)$, so that the dependence of phase on $x^2 + y^2$ is such that the wave converges toward point F at $z = |z_1|$. This wave has all the features of a conjugate wave except that it is forward-directed. (For example, the analysis may be used to describe the focus produced at the far side of the layer for the situation shown in Fig. 14.1. The oscillating phase shift is the result of the response of the nonlinear layer to the pump wave. Processes we describe here as kinematic are described as "parametric" by most other authors.) To produce a conjugate wave with the system shown in Fig. 14.9, a mirror is placed in the plane $z = L/2$. Double passage of the probe wave through the modulation region gives a conjugate outgoing wave of amplitude $2\varepsilon A_2/|z_2|$.

Several mechanisms for achieving the required time-dependent phase shift have been investigated, and only a summary is given here. Brysev et al. (1990) demonstrated the production of conjugate ultrasonic longitudinal waves in a solid where the photoelastic effect was used to make the conjugate wave visible. The phase modulation mechanism was an oscillating magnetic field applied to a propagation region of Ni-ferrite magnetoelastic solid. Ohno and Takagi (1992), and references cited therein, report conjugation of ultrasonic waves in water by application of an oscillating electric field to a LiNbO$_3$ crystal in contact with the water. The conjugator was used to reduce the aberrations of acoustical images of an object within submerged agar having a rippled surface. An appropriate kinematic phase shift is produced by replacing the modulated region in Fig. 14.9 by a (not necessarily flat) surface oscillating at frequency $2\omega_2$. While a full analysis would include the parametric interaction of the wave radiated at frequency $2\omega_2$ with the incident and reflected probe waves, inspection of the modified form of Eq. (14.23) reveals the presence of a kinematically generated conjugate wave. Brysev et al. (1982) summarize a related experiment, and Lyamshev and Sakov (1988) analyze phase conjugation due to scattering by a sphere pulsating at frequency $2\omega_2$. The production of focused difference-frequency sound due to reflection at a free water surface was demonstrated by Simpson (1992).

Several authors have investigated microwave acoustic phase conjugation in solids or solid powders. See Fossheim (1985) for a review.

Acknowledgment The authors' research related to topics in this chapter has been supported by the Office of Naval Research.

References

Andreeva, N. P., Bunkin, F. V., Vlasov, D. V., and Karshiev, K. (1982). Experimental observation of acoustic phase conjugation at a liquid surface. *Sov. Tech. Phys. Lett.* **8**, 45–46.

Asada, T., and Watanabe, Y. (1990). Experiments of parametric amplification using nonlinear vibration of bubble under water. In *Frontiers of Nonlinear Acoustics: Proceedings of 12th ISNA*, M. F. Hamilton and D. T. Blackstock, eds. (Elsevier Science Publishers, London), pp. 485–490.

Brysev, A. P., Bunkin, N. F., Vlasov, D. V., and Gervits, L. L. (1982). Plane parametric phase-conjugation mirror. *Sov. Tech. Phys. Lett.* **8**, 241–243.

Brysev, A. P., Bunkin, F. V., Vlasov, D. V., Krutyanskii, L. M., Preobrazhesnkii, V. L., Pyl'nov, Yu. V., Stakhovskii, A. D., and Ékonomov, N. A. (1990). Spatial distribution of the field of a longitudinal ultrasonic wave. *Sov. Phys. Acoust.* **36**, 91–92.

Bunkin, F. V., Kravtsov, Y. A., and Lyakhov, G. A. (1986). Acoustic analogues of nonlinear optics phenomena. *Sov. Phys. Usp.* **29**, 607–619.

Bunkin, F. V., Vlasov, D. V., Karshiev, K., and Stakhovskii, A. D. (1985). Experimental observation of the suppression of a wave field by means of a phase-conjugating (wavefront-reversing) mirror (PCM). *Sov. Phys. Acoust.* **31**, 80–81.

Cassereau, D., and Fink, M. (1993). Focusing with plane time-reversal mirrors: An efficient alternative to closed cavities. *J. Acoust. Soc. Am.* **94**, 2373–2386.

Doinikov, A. A. (1994). Acoustic radiation pressure on a compressible sphere in a viscous fluid. *J. Fluid Mech.* **267**, 1–21.

Duvall, G. E., and Taylor, Jr., S. M. (1971). Shock parameters in a two component mixture. *J. Composite Materials* **5**, 130–139.

Everbach, E. C., Zhu, Z., Jiang, P., Chu, B. T., and Apfel, R. E. (1991). A corrected mixture law for B/A. *J. Acoust. Soc. Am.* **89**, 446–447.

Fink, M. (1993). Time-reversal mirrors. *J. Phys. D: Appl. Phys.* **26**, 1333–1350.

Fossheim, K. (1985). Phonon echoes, polarizations echoes, and acoustic phase conjugation in solids. In *Nonequilibrium Phonon Dynamics*, W. E. Bron, ed. (Plenum Press, New York), pp. 277–312.

Gaylord, T. K., and Moharam, M. G. (1982). Planar dielectric grating diffraction theories. *Appl. Phys. B* **28**, 1–14.

Gor'kov, L. P. (1962). On the forces acting on a small particle in an acoustical field in an ideal fluid. *Sov. Phys. Dokl.* **6**, 773–775.

Gower, M., and Proch, M., eds. (1994). *Optical Phase Conjugation* (Springer-Verlag, Berlin).

Higashitani, K., Fukushima, M., and Matsuno, Y. (1981). Migration of suspended particles in plane stationary ultrasonic field. *Chem. Eng. Sci.* **36**, 1877–1882.

Jackson, D. R., and Dowling, D. R. (1991). Phase conjugation in underwater acoustics. *J. Acoust. Soc. Am.* **89**, 171–181.

Kobelev, Y. A., and Sutin, A. M. (1980). Difference-frequency sound generation in a liquid containing bubbles of different sizes. *Sov. Phys. Acoust.* **26**, 485–487.

Kustov, L. M., Nazarov, V. E., and Sutin, A. M. (1985). Phase conjugation of an acoustic wave at a bubble layer. *Sov. Phys. Acoust.* **31**, 517–518.

Kustov, L. M., Nazarov, V. E., and Sutin, A. M. (1986). Nonlinear sound scattering by a bubble layer. *Sov. Phys. Acoust.* **32**, 500–503.

Kustov, L. M., Nazarov, V. E., and Sutin, A. M. (1987). Demodulation of acoustic signals in a bubble layer. *Sov. Phys. Acoust.* **33**, 161–164.

Lyamshev, L. M., and Sakov, P. V. (1988). Phase conjugation in nonlinear sound scattering by a pulsating sphere. *Sov. Phys. Acoust.* **34**, 68–72.

Marston, P. L. (1992). Geometrical and catastrophe optics methods in scattering. In *Physical Acoustics*, Vol. 21, A. D. Pierce and R. N. Thurston, eds. (Academic Press, Boston), pp. 1–234.

Marston, P. L., and Kargl, S. G. (1990). Wavefront reversal and difference frequency generation via three-wave mixing in a bubble layer. In *Frontiers of Nonlinear Acoustics: Proceedings of 12th ISNA*, M. F. Hamilton and D. T. Blackstock, eds. (Elsevier Science Publishers, London), pp. 514–519.

Nazarov, V. E., and Sutin, A. M. (1989). Theory of a parametric sound receiver using a nonlinear layer. *Sov. Phys. Acoust.* **35**, 510–512.

Nikoonahad, M., and Pusateri, T. L. (1989). Ultrasonic phase conjugation. *J. Appl. Phys.* **66**, 4512–4513.

Ohno, M., and Takagi, K. (1992). Schlieren visualization of acoustic phase conjugate waves generated by nonlinear electroacoustic interaction in LiNbO$_3$. *Appl. Phys. Lett.* **60**, 29–31.

Pepper, D. M. (1982). Nonlinear optical phase conjugation. *Optical Engineering* **21**, 156–183.

Pepper, D. M. (1986). Applications of optical phase conjugation. *Sci. Am.* **254**, 74–83.

Sato, T., Kataoka, H., Nakayama, T., and Yamakoshi, Y. (1990). Ultrasonic phase conjugator using micro particle suspended cell and its application. In *Acoustical Imaging*, Vol. 17, H. Shimizu, N. Chubachi, and J. Kushibiki, eds. (Plenum, New York), pp. 361–390.

Simpson, H. J. (1992). Interaction of sound with sound by novel mechanisms: Ultrasonic four-wave mixing mediated by a suspension and ultrasonic three-wave mixing at a free surface. Ph.D. dissertation, Washington State University. For abstract, see *J. Acoust. Soc. Am.* **94**, 1167 (1993).

Simpson, H. J., and Marston, P. L. (1993). Ultrasonic four-wave mixing mediated by a suspension. In *Advances in Nonlinear Acoustics: Proceedings of 13th ISNA*, H. Hobæk, ed. (World Scientific, Singapore), pp. 644–655.

Simpson, H. J., and Marston, P. L. (1995). Ultrasonic four-wave mixing mediated by an aqueous suspension of microspheres: Theoretical steady-state properties. *J. Acoust. Soc. Am.* **98**, 1731–1741.

Smith, P. W., Ashkin, A., and Tomlinson, W. J. (1981). Four-wave mixing in an artificial Kerr medium. *Opt. Lett.* **6**, 284–286.

Smith, P. W., Maloney, P., and Ashkin, A. (1982). Use of a liquid suspension of dielectric spheres as an artificial Kerr medium. *Opt. Lett.* **7**, 347–349.

Welsby, V. G., and Safar, M. H. (1970). Acoustic non-linearity due to micro-bubbles in water. *Acustica* **22**, 178–182.

Westervelt, P. J. (1963). Parametric acoustic array. *J. Acoust. Soc. Am.* **35**, 535–537.

Wu, J., and Zhu, Z. (1991). Measurements of the effective nonlinearity parameter B/A of water containing trapped cylindrical bubbles. *J. Acoust. Soc. Am.* **89**, 2634–2639.

Yosioka, K., and Kawasima, Y. (1955). Acoustic radiation pressure on a compressible sphere. *Acustica* **5**, 167–173.

Zabolotskaya, E. A. (1984). Phase conjugation of sound beams in connection with four-phonon interaction in a liquid containing gas bubbles. *Sov. Phys. Acoust.* **30**, 462–463.

Zabolotskaya, E. A., and Soluyan, S. I. (1973). Emission of harmonic and combination-frequency waves by air bubbles. *Sov. Phys. Acoust.* **18**, 396–398.

Zel'dovich, B. Y., Pilipetsky, N. F., and Shkunov, V. V. (1985). *Principles of Phase Conjugation* (Springer-Verlag, Berlin).

Chapter 15
Biomedical Applications

Edwin L. Carstensen and David R. Bacon

Contents

15.1 Introduction

Since its beginnings in the late 1930s, medical ultrasound has become an indispensable diagnostic and therapeutic tool and a multibillion-dollar industry. Ultrasound produces detailed, high resolution images of most of the soft tissues of the body,

E. L. Carstensen
Department of Electrical Engineering, and Rochester Center for Biomedical Ultrasound, University of Rochester, Rochester, NY, USA

D. R. Bacon
Science Technology and Environment, British Embassy Bonn, Bonn, Germany

© The Author(s) 2024
M. F. Hamilton, D. T. Blackstock (eds.), *Nonlinear Acoustics*,
https://doi.org/10.1007/978-3-031-58963-8_15

measures the flow of blood by the Doppler effect, and combines flow information with images in a technique that has come to be known as color-flow Doppler. In addition to its diagnostic applications, ultrasound is used in therapy, notably in physiotherapy to treat soft-tissue injuries. Because of its ability to produce localized heating of deep tissues, ultrasound has advantages for use in hyperthermic treatment of tumors. At higher intensities, ultrasound has been used as a surgical tool to modify or destroy selected regions in tissues such as the prostate, the brain, and the eye. Most of these applications are achieved with ultrasound at frequencies between 0.5 and 10 MHz. More recent applications of ultrasound for examining the surface layers of skin and the walls of blood vessels have involved frequencies in the range 10–40 MHz. At even higher frequencies (40–1000 MHz), acoustic microscopy in transmission and reflection modes provides detailed images, based on the mechanical (acoustic) characteristics of tissues, that complement the optical images of conventional microscopes. Shock waves with peak positive pressures up to 100 MPa are used in extracorporeal and endoscopic lithotripsy for the treatment of kidney stones and gallstones.

Until the early 1980s, the development of medical ultrasound proceeded under the tacit assumption that sound propagation is a linear process. Today, we realize that the effects of finite-amplitude propagation can be seen in almost every medical use of ultrasound. This was apparent once the expertise that had evolved in nonlinear acoustics was directed to problems in medical ultrasound (Muir and Carstensen, 1980; Carstensen et al., 1980). In addition, there has been a slow escalation over the years in the amplitude of the sound fields applied to patients through diagnostic procedures (Duck and Martin, 1991). The payoff has been greater depth of penetration at higher frequencies and greater resolution and information content in the images. Together with today's increased capabilities of medical ultrasound comes a gradual increase in awareness of the possibility for the production of potentially adverse biological effects. This brings a need not only to know the levels of the acoustic fields in patients but also to understand the physical processes involved in the propagation of finite-amplitude sound in tissue as well as in water. One potential hazard of ultrasound is tissue heating, and finite-amplitude effects can significantly increase this heating above the levels predicted on the basis of small-signal propagation. If one wishes to know the relevant levels in vivo, it is necessary to take account of the differences between the characteristics of the test medium (usually water) and those of tissue. These differences give rise to a nonlinear relationship between the reference measurements and the exposure levels experienced by the patient. The physical processes involved in ultrasonic hyperthermia are similar to those in diagnosis, although the beam sizes and absolute temperature rises may be different.

Since the 1980s, the widespread use of shock-wave lithotripters to destroy kidney stones has increased by orders of magnitude the pressure levels used in therapeutic procedures. To understand the operation of these devices, one must analyze nonlinear propagation effects. Piezoelectric and electrodynamic lithotripters rely on nonlinear propagation to form the shock waves. Electrohydraulic lithotripters create shock waves by discharge of a spark.

This review describes the acoustical properties of tissues and discusses the impact of nonlinear propagation in medical ultrasound. Important aspects of this discussion are the characteristics of medical equipment as measured in water, the relationship of measurements in water to the levels in tissues, and the implications of these fields for biological effects.

15.2 Acoustical Properties of Tissues

With the exceptions of lung, bone, and fat, the tissues of the body have acoustic impedances that differ by only a few percent from that of water. Their small-signal absorption coefficients are determined largely by their macromolecular composition. A very broad spectrum of macromolecular relaxation processes gives rise to absorption that increases approximately with the first power of frequency over most of the medically interesting range of frequencies (1–10 MHz). Representative soft tissues have absorption coefficients of approximately 5 (Np/m)/MHz (NCRP, 1992). The total attenuation of an acoustic wave includes energy losses (absorption) and losses from diversion of the wave from its path (scattering). For most soft tissues, scattering is relatively small and the attenuation and absorption coefficients are approximately equal. In contrast, lung has the highest attenuation coefficient of any of the tissues of the body, and the attenuation appears to come almost entirely from scattering (Dunn, 1974; Pedersen and Ozcan, 1986; Hartman et al., 1992). Bone has the highest true absorption of the body tissues (>100 Np/m at 1 MHz).

15.2.1 Nonlinear Properties of Tissues

Techniques for the measurement of the nonlinearity parameter B/A of materials are described in Sect. 2.4, and a summary of the available data for tissues is provided in Table 2.3. Most of the macromolecular constituents of tissues have a somewhat higher nonlinearity parameter B/A than water does. For the purposes of this summary, it is possible to characterize the nonlinearity of tissues by a linear mixture equation. Figure 15.1 is a straight line connecting the value $B/A = 5.2$ for water at 30 °C to $B/A = 11$ for fat, a tissue with low water content. Tissues with high water content fall between these limits. Most measurements to date have been performed on excised tissue samples. However, in one study, Zhang and Dunn (1987) measured cat liver in vivo, and after excision found less than 2% difference between the preparations.

Fig. 15.1 The nonlinearity parameter B/A of representative soft tissues, blood, and water. To a first approximation, the values can be represented as a linear mixture of macromolecular constituents and water. Circles indicate scatter of data. Solutions of macromolecules such as hemoglobin and albumin also follow this relationship.

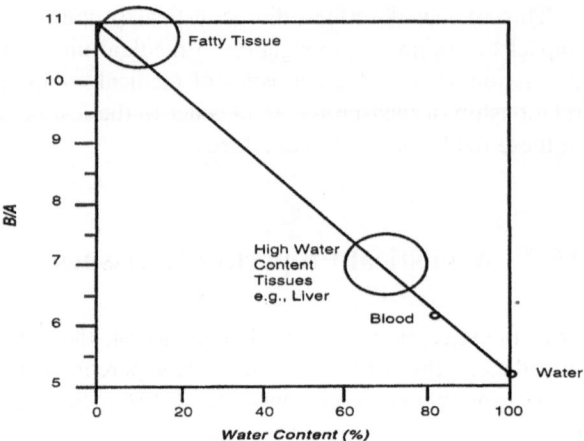

15.2.2 Microbubbles in Tissues

Very small concentrations of gas dispersed in the form of micrometer-sized gas bodies can have a dramatic effect on the acoustical properties of the medium. Bubbles near resonance size for the frequency of the sound can have absorption and scattering cross sections that are many times greater than their physical size. Furthermore, B/A for such a medium can be many times greater than its value for the same medium in the absence of bubbles [Eq. (5.66)]. No systematic studies of tissues from this point of view have been reported. It is known that certain specialized tissues, such as insect larvae and the leaves of aquatic plants, contain stabilized gas bodies associated with their respiration. Except in lung and intestine, microbubbles appear to be rare in mammalian tissues (Dalecki et al., 1997c). But it is reasonably well established that transient populations of bubbles can be created in the blood and tissues of mammals exposed to lithotripter fields (Delius, 1990). Ultrasound contrast agents are deliberately introduced into the blood of patients to increase scattering. Most of these agents use stabilized bubbles to achieve their effect (Nanda et al., 1997).

15.2.3 Tissue Characterization

The first "B-Scan" tomographic images made with ultrasound showed the outlines of tissue structures resulting from the relatively large reflections at acoustic impedance discontinuities. With the addition of a gray scale to the image, a qualitative indication of scatter from within organs provided the beginnings of ultrasonic tissue characterization. More sophisticated systems today provide quantitative measures of the scattering coefficients of tissue regions. Color Doppler gives a color-coded image of flow. Tissue characterization based on the attenuation, the sound speed, and the nonlinearity parameter B/A of tissues, and combinations of these parameters, have been explored (e.g., Jiang et al., 1991). Techniques have been proposed in which an acoustic pump wave of finite amplitude is used to modify the properties of tissues so that a probe beam can detect the nonlinear properties of the tissues through phase changes or from the production of difference or harmonic frequencies (Ichida et al., 1983).

15.3   High-Amplitude, Focused Fields of Medical Equipment

The ultrasonic echoes from most tissues are small relative to the transmitted signal amplitude. As a result, imaging equipment requires high output pressure levels to achieve a reasonable signal-to-noise ratio in the image. The time-average intensity is relatively low at any specific site in the tissue (on the order of a few milliwatts per centimeter squared), but nonlinear distortion of the wave can still be significant, as a result of the high amplitude ($>1000 \, \text{W/cm}^2$, or 5 MPa maximum negative pressure) of the individual pulses. At the other extreme, equipment used for physical therapy employs time-average intensities up to $3 \, \text{W/cm}^2$, but the relatively low frequency (typically 1 MHz), low amplitude, and short propagation distance mean that nonlinear propagation is unlikely to be important.

15.3.1   The Shock Parameter

In developing an understanding of nonlinear effects across the wide range of medical applications, it is useful to have a means of characterizing the degree of nonlinear distortion of the wave at any given position in the field. The dimensionless quantity σ used in weak shock theory (Blackstock, 1966), referred to here as the shock parameter, combines source amplitude and frequency, propagation distance, and properties of the medium in a single number that is correlated with the degree of shock formation over a wide range of conditions. For initially sinusoidal plane waves, its definition is [Eq. (4.23)]

$$\sigma = \beta \varepsilon k z, \tag{15.1}$$

where $\beta = 1 + B/2A$ is the coefficient of nonlinearity, $k = 2\pi/\lambda$ is the wavenumber, λ is the acoustic wavelength, ε is the ratio of the particle velocity amplitude at the source to the small-signal sound speed, and z is the distance from the source. For spherically converging waves, Eq. (15.1) is replaced by

$$\sigma = \beta \varepsilon k d \ln[1/(1 - z/d)], \tag{15.2}$$

where d is the radius of curvature of the source (i.e., focal length). On the axis of a focused Gaussian beam (see Sect. 8.3.3), a corresponding shock parameter can be defined (Dalecki et al., 1991a),

$$\sigma = \frac{\beta \varepsilon k d}{\sqrt{1 - G^{-2}}} \ln[(G + \sqrt{G^2 - 1})(R + \sqrt{R^2 + 1})], \tag{15.3}$$

where G is the focal gain (the ratio of focal to source amplitudes) and $R = -(1 - z/d)(G^2 - 1)^{1/2}$. Equation (15.2) is based on ray acoustics and has a singularity at $z = d$, but it can provide useful information in the prefocal region, $z < d$. Equation (15.3) takes some account of diffraction and can be used at the focus of realistic sound fields.

As shown in Eqs. (4.50)–(4.53), for nondiffracting waves, sound propagation is essentially linear for σ somewhat less than unity; a sinusoidal waveform at the source remains approximately sinusoidal as it propagates through the medium. As σ approaches unity, nonlinear distortion of the waveform becomes significant as energy is transferred from the fundamental component into higher harmonics. At $\sigma = 1$, the wave begins shock formation and by $\sigma = 3$, a full sawtooth wave has developed (Sect. 4.4.3.2).

In the absence of diffraction, the intensity at an axial field point z can be expressed as [see Eq. (4.300)]

$$\mathbf{I}(z) = \mathbf{e}_z \frac{I_0}{(1 - z/d)^a} \sum_{n=1}^{\infty} B_n^2(\sigma), \tag{15.4}$$

where I_0 is the intensity at the source, $\mathbf{e}_z$ is the unit vector in the direction of propagation, and $a = 0, 1,$ and 2 for plane, cylindrical, and spherical waves, respectively. The harmonic amplitude coefficients B_n are given by [Blackstock, 1966; see also Eq. (4.183)]

$$B_n = \frac{2}{n\pi\sigma}\left[\Phi_{sh} + \int_{\Phi_{sh}}^{\pi} \cos n(\Phi - \sigma \sin \Phi)\, d\Phi\right], \tag{15.5}$$

where $\Phi_{sh} = 0$ for $\sigma \leq 1$, in which case $B_n = (2/n\sigma)J_n(n\sigma)$ is obtained, where J_n is the Bessel function of the first kind. For $\sigma > 1$, Φ_{sh} is the smallest positive root of the transcendental equation $\Phi_{sh} = \sigma \sin \Phi_{sh}$, and Eq. (15.5) must be integrated numerically. For $\sigma > 3$, the asymptotic value $\Phi_{sh} = \pi\sigma/(1 + \sigma)$ is approached, corresponding to $B_n = 2/n(1 + \sigma)$.[1]

---

[1] Equation (15.4) can also be written in the equivalent time-domain form (Blackstock, 1990)

$$\mathbf{I}(z) = \mathbf{e}_z \frac{I_0}{(1 - z/d)^a}\left[1 - \frac{P_{sh}}{\pi}(\sigma - \cos \sigma P_{sh} - \tfrac{2}{3}\sigma P_{sh}^2)\right], \tag{15.4a}$$

Fig. 15.2 Dependence of intensity on the shock parameter σ, based on the summation in Eq. (15.4) (Dalecki et al., 1991a). Given is the ratio of the true intensity to the value that it would have if nonlinear losses were ignored. The theory (Blackstock, 1966) assumes negligible small-signal attenuation.

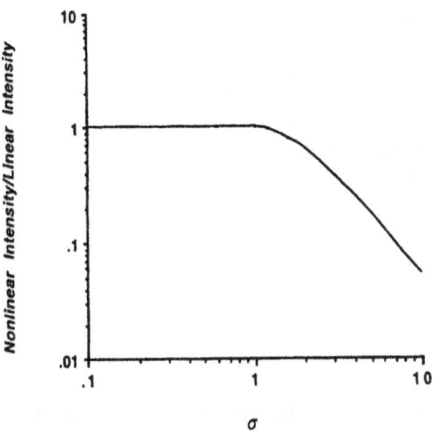

As illustrated in Fig. 15.2, even though the waveform gradually becomes modified as the shock parameter increases, losses are negligible up to $\sigma = 1$, but beyond $\sigma = 3$ losses eventually become so great that the field intensity becomes nearly independent of the source intensity (acoustical saturation). See Sect. 4.4.3.3 for related discussion based on Eq. (15.5).

15.3.2 Weak Shock Absorption

Although more general models to evaluate nonlinear heating rates numerically in specific cases are now available, weak shock theory (Blackstock, 1966, 1990; and Sect. 4.4) provides an adequate framework to model the general characteristics of the losses that are dominated by shock formation and are relatively independent of the magnitude of the small-signal absorption coefficient of the acoustic medium (Dalecki et al., 1991a). See also Sect. 4.6.2.

Application of the basic definition of the absorption coefficient,

$$\alpha = -\frac{\nabla \cdot \mathbf{I}}{2I},$$

$$(15.6)$$

to Eq. (15.4) yields a finite amplitude, "weak shock absorption coefficient" (Dalecki et al., 1991a)

$$\alpha_f = -\frac{\partial \sigma}{\partial s} \frac{(\partial/\partial\sigma) \sum B_n^2(\sigma)}{2 \sum B_n^2(\sigma)},$$

$$(15.7)$$

where $P_{sh} = \Phi_{sh}/\sigma$ is the dimensionless amplitude of the shock, and Φ_{sh} is defined following Eq. (15.5); see Eq. (4.303).

Fig. 15.3 The weak shock
absorption parameter α_f
(multiplied by the distance
function F) as a function of
the shock parameter (adapted
from Fig. 2 of Dalecki et al.,
1991a).

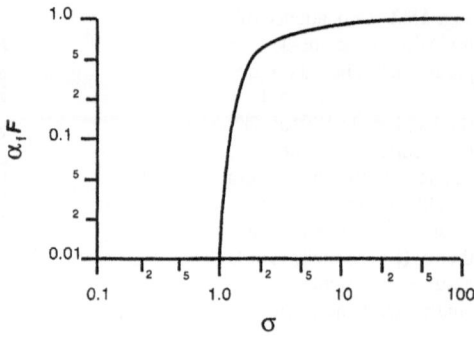

where s is the spatial variable along the direction of propagation of the wave.[2]
The expression on the right is determined almost entirely by the shock parameter
σ, which in turn is a function of the amplitude and frequency of the source, the
nonlinearity parameter of the medium, and the geometry of the field. There is thus
a direct relationship between weak shock absorption and the shock parameter at a
given point in the field.

Weak shock losses occur as a result of strong irreversible processes resulting
mainly from the accentuated influences of viscosity, heat conduction, and relaxation
at the discontinuity, and they depend upon the strength of the shock and the shape
of the waveform on either side of the shock. Figure 15.3 displays the losses in terms
of the weak shock absorption parameter, where the function $F = \sigma (\partial \sigma / \partial s)^{-1}$
depends on the geometry of the beam and has the dimensions of distance (Dalecki
et al., 1991a). The asymptotic value $\alpha_f F = 1$ is approached for $\sigma \gg 1$. With
the relation $F = z$ obtained for a plane wave, the limiting value of the weak
shock absorption parameter in this case is simply the reciprocal of the propagation
distance, $\alpha_f = 1/z$. Much larger values of α_f can be achieved in focused fields.
These weak shock losses are completely foreign to linear theory of acoustics.
They are essentially independent of the magnitude of the small-signal absorption
coefficient of the medium as long as the small-signal absorption is small. To a
first approximation, the total absorption parameter for a real but weakly absorbing
medium is the sum of the small-signal absorption coefficient and the weak shock
absorption parameter.

In evaluating the fields of clinical instruments, it is frequently difficult to
measure the (dimensionless) source amplitude ε directly. Bacon (1984) developed
an approximate method for determining the shock parameter from hydrophone

---

[2] The time-domain form of Eq. (15.7) is [see Eq. (4.306)]

$$\alpha_f = \frac{\partial \sigma}{\partial s} \frac{\frac{2}{3}\sigma P_{sh}^3}{\pi - P_{sh}(\sigma - \cos \sigma \, P_{sh} - \frac{2}{3}\sigma P_{sh}^2)}. \tag{15.7a}$$

Fig. 15.4 Weak shock
heating rate for a plane wave
at 4 MHz in water. The dotted
line shows the heating rate
that would be computed for
$z = 10$ cm according to linear
theory based on the
small-signal absorption
coefficient for water.

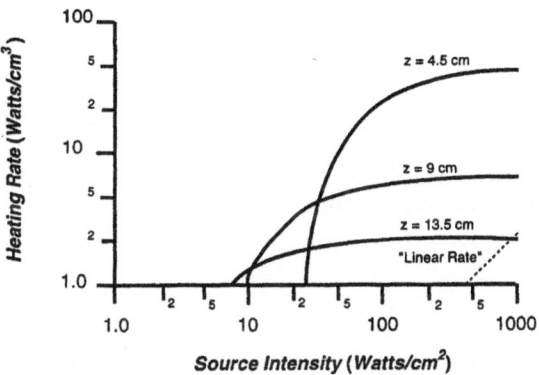

measurements made in the focal region that provides useful information for most
medically interesting output levels.

15.3.3 Heating

Absorption of sound leads to heat generation in the acoustic medium. Specifically,
the rate of heating (in mechanical units) is equal to the negative divergence of the
local acoustic intensity, which for a small-signal plane wave is proportional to the
product of the absorption coefficient of the medium and the local intensity [recall
Eq. (15.6)]. Figure 15.4 gives the heating rate of water in a plane-wave field as a
function of source intensity. Weak shock absorption is completely dominant in this
example. For perspective, the heating predicted for water by linear theory is shown
by the dotted line in the lower right-hand corner of the plot. Instead of a heating rate
that is directly proportional to source intensity, as predicted by linear theory, heating
rises rapidly (in the range $1 < \sigma < 3$), reaches a limiting value, and becomes
independent of source intensity thereafter. The limiting value of the heating rate
decreases as distance from the source increases. However, shocks develop at lower
source intensities as distance from the source increases. Similar phenomena occur
in focused fields (Dalecki et al., 1991a).

Description of sound fields in terms of the shock parameter provides insight
into the general phenomenon of nonlinear propagation and gives semiquantitative
predictions of weak shock absorption and heating rates for tissues with relatively
low small-signal attenuation (Dalecki et al., 1991a). However, weak shock theory
is not adequate to deal quantitatively with propagation in tissues possessing small-
signal absorption coefficients as large as those of most of the soft tissues of the
body. For this purpose, more powerful numerical methods are required (e.g., those
discussed in Chap. 11). Examples of such computations are given in Sect. 15.4.

At small-signal levels where the waveforms are sinusoidal, the amplitude of
the fundamental frequency component and the peak negative pressure amplitude

Fig. 15.5 An asymmetric, nonlinearly distorted waveform in a diffracting beam, typical of those produced by high-power diagnostic ultrasound devices (adapted from Fig. 3 of Cleveland et al., 1996).

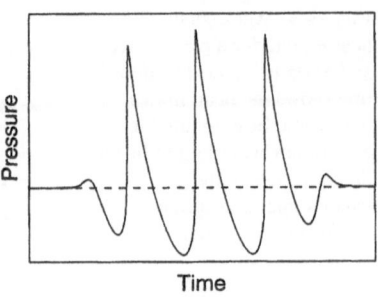

of the waveform are identical. In the case of a plane wave of infinite extent in a weakly attenuating medium such as water, as the amplitude of the wave increases, the wave front steepens symmetrically and, in the high-intensity limit, harmonics with amplitudes inversely proportional to their number are generated, as noted following Eq. (15.5), and the waveform assumes a sawtooth shape. In realistic medical ultrasound fields, diffraction and dispersion effects combine to yield an asymmetric wave, as shown in Fig. 15.5, where the amplitude of the negative phase of the wave is seen to be smaller than that of the positive phase [see also Fig. 8.8c]. Depending upon the phase relationships among the harmonics, the negative pressure can be either greater or smaller than the amplitude of the fundamental component alone, but frequently they are of comparable magnitude.

15.4   Predicting Fields in Tissues: The "Derating" Problem

Characterization of the fields of medical ultrasound devices usually is based on measurements made with hydrophones in water. Data obtained in this way can be used to estimate fields in tissues provided (1) that an appropriate model of the tissue as a propagation path is available and (2) that signals are small enough that linear theory applies. The use of measurements made in water to predict fields in tissue has come to be called "derating" in the medical device community. Two general tissue models have been proposed for this purpose (AIUM, 1988; NCRP, 1992). The most generally useful model requires a homogeneous tissue path from source to target. The other model, proposed originally by Carson et al. (1989), is intended for the special case in obstetrics in which most of the propagation path is through weakly attenuating fluids and most of the attenuation of the signal before it reaches the fetus is limited to a layer of skin and abdominal muscle near the face of the transducer.

  As currently defined, these derating models are based on linear acoustical theory. However, most diagnostic devices can operate at levels high enough to create shock waves in water (Duck et al., 1985; Duck and Martin, 1991). Under these conditions, extrapolation of water data to tissues becomes much more complex, and, as acoustical saturation is approached, the data obtained in water provide only very limited information about source amplitudes and tissue fields. Some of the problems of derating under the conditions of finite-amplitude propagation have been addressed (Bacon, 1989; Dalecki et al., 1991a; NCRP, 1992).

15.4.1 Intensity

Christopher recently developed a numerical propagation model that takes into account realistic tissue properties and includes the combined effects of nonlinearity, diffraction, absorption, and dispersion (Christopher and Parker, 1991; Christopher, 1993, 1994; see also Sect. 11.3.3). This model has been employed to evaluate the derating problem under nonlinear propagation conditions (Christopher and Carstensen, 1996). For illustration, consider a 3 MHz, 1-cm-radius source with a focal length of 8 cm. Figure 15.6 gives the focal intensities as a function of source intensity for propagation through water, liver, and fatty tissue. The small-signal absorption coefficient for liver is assumed to be 0.13 Np/cm in this example. For perspective, commercial diagnostic ultrasound devices may produce focal intensities as high as 400 W/cm^2 at this frequency in tissues such as liver. Dashed lines in the figure are linear extrapolations of the focal fields, for comparison.

In a general sense, highly attenuating media such as tissues reduce the amplitude of the wave by small-signal absorption and, in this way, minimize nonlinear effects of propagation. Note that acoustical saturation is approached at lower source intensities in water than in tissues. This makes it difficult to use measurements in water to estimate fields in tissues. Using the data in Fig. 15.6 for illustration, we see that, with a source intensity of 100 W/cm^2, the focal intensity in water is approximately 1000 W/cm^2. The waveform is highly distorted, and it is obvious in a qualitative sense that nonlinearity is important. However, if nonlinearity is ignored and the ratio of the small-signal absorption in water to that in liver is applied, the derated intensity will be 60 W/cm^2 instead of ~300 W/cm^2, the latter being the actual value in Fig. 15.6 that we find for liver by taking nonlinearity into account. In general, the closer sound fields in water approach saturation, the more difficult

Fig. 15.6 Focal intensities produced by a source of radius 1 cm, focal length 8 cm, radiating at 3 MHz (Christopher and Carstensen, 1996). Dotted lines are linear extrapolations corresponding to the source intensities given on the abscissa.

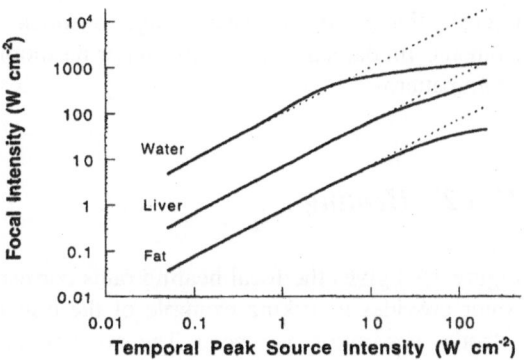

Fig. 15.7 Effects of
nonlinear propagation on
focal intensity (Christopher
and Carstensen, 1996). The
ordinate gives the ratio of the
solid to the dashed curves in
Fig. 15.6.

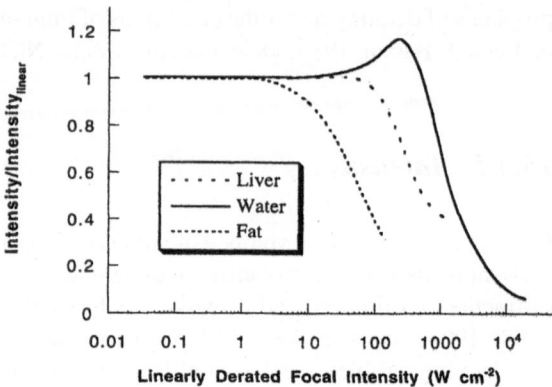

it becomes to use data obtained for sound in water to predict fields in any other medium. However, simple linear extrapolation of the small-signal derated fields, in general, overestimates the true focal fields.

Recognizing this problem, the World Federation of Ultrasound in Medicine and Biology (WFUMB, 1998) cautioned that "the common practice of linear derating from high level measurements in water is undesirable because it can lead to substantial underestimates of tissue fields" because of nonlinear propagation. To avoid this problem, the WFUMB (1998) recommended that "estimates of tissue field parameters at the point of interest should be based on derated values calculated according to an appropriate specified model and be extrapolated linearly from small signal characterization of source-field relationships." As noted above, this practice errs in the direction of overestimating true fields in vivo at high source intensities. When greater accuracy is desired, corrections for the effects of nonlinear propagation can be made theoretically. The data in Fig. 15.6 are replotted in Fig. 15.7 using the linearly extrapolated focal intensities as the independent variable and the nonlinear correction as the ordinate of the graph. Note that linear derating for liver according to the WFUMB procedure overestimates the true focal intensities by more than a factor of 2 near the upper limits of the outputs of diagnostic ultrasound devices. Figure 15.7 shows for water the nonlinear enhancement of the axial focal intensity that occurs in the early stages of shock formation when power is transferred from the fundamental beam to higher-harmonic components that have narrower beam patterns.

15.4.2 Heating

Figure 15.8 gives the focal heating rates corresponding to Fig. 15.6. The curve for water provides a striking example of the effects of nonlinear propagation on the effective absorption parameter. The small-signal absorption coefficient for water is much smaller than that for either liver or fat, which is evident in the heating rate

Fig. 15.8 Focal heating rates for water, liver, and fatty tissue, corresponding to Fig. 15.6 (Christopher and Carstensen, 1996).

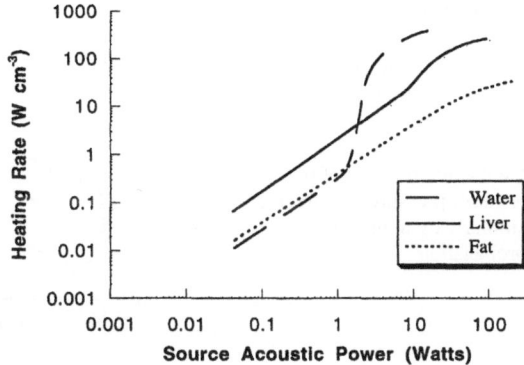

in water at low intensities. However, as shock waves begin to form, the effective absorption parameter for water increases dramatically, and, in this example, the focal heating rate in water substantially exceeds that in liver. Shock-wave absorption has been proposed as a mechanism for localized enhancement of tissue heating for hyperthermia and for ultrasonic thermal surgery (Carstensen and Muir, 1986; Hynynen, 1987). Even though nonlinear propagation effects strongly enhance focal heating rates in liver at the highest intensities, the steady-state temperature increment differs very little from that which would be predicted on the basis of linear theory. The explanation lies in the effects of nonlinear propagation on the heating patterns that characterize the sound field. For values of σ in the range between 1 and 3 at the focus of a sound beam, the heating rate due to nonlinear propagation effects is increased more strongly on- than off-axis, and the heating pattern is narrowed relative to the small-signal beam pattern. Enhanced heat diffusion from the narrower beam largely disperses the excess nonlinearly enhanced heating on the axis.

15.4.3   Acoustic Pressures

Figure 15.9 gives the focal pressures corresponding to Fig. 15.6. At high output levels, the combination of nonlinear generation of harmonic frequencies with diffraction and dispersion produces highly asymmetric focal waveforms, such as those in Fig. 15.5, and it is somewhat more arbitrary to characterize the field by pressure rather than by intensity. The positive pressure spike is difficult to measure and of questionable relevance to known biological effects. The fundamental component of the wave and the negative pressure, shown in Fig. 15.9 (solid and dashed curves, respectively), are affected similarly by nonlinear propagation. Figure 15.9 also shows that because of acoustical saturation, it is difficult in practice, regardless of output levels, to achieve fundamental or negative pressures greater than 3 or 4 MPa at distances of 8 cm or more at the frequencies that are commonly used in diagnosis (>2 MHz). Of course, much higher pressures can be achieved at

Fig. 15.9 Focal acoustic pressures for the conditions of Fig. 15.6 (Christopher and Carstensen, 1996). The amplitude of the fundamental component of the wave and the maximum value of the negative pressure are affected similarly by nonlinear propagation.

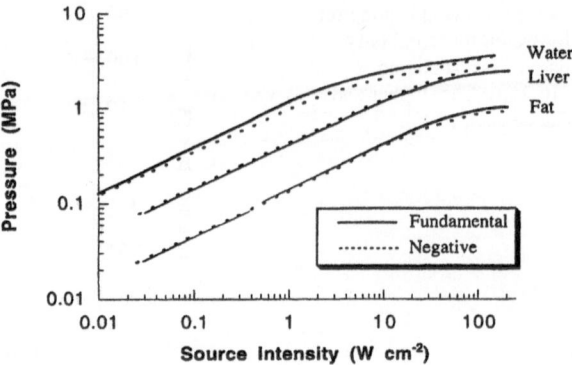

shorter focal lengths before saturation occurs because of the shorter high-intensity propagation path.

15.5 Implications of Nonlinear Contributions to Radiation Forces and Acoustic Streaming

In fluids, transfer of acoustic momentum to the medium via absorption results in macroscopic acoustic streaming in the direction of sound propagation. Acoustic streaming is readily observed in liquids. The second-order unidirectional body force exerted on a fluid by a sound field is equal to the divergence of the radiation pressure. This force is a function of the product of the local intensity and the absorption parameter of the fluid. Because nonlinear propagation can increase the effective absorption parameter of a medium such as water manyfold, streaming can be very much greater in a pulsed field than in a cw field of the same temporal average intensity. The problem that this causes for the use of a steel sphere radiometer for the measurement of the total intensity of a finite-amplitude wave was described by Carstensen et al. (1980). Mitome (1990) discussed the streaming phenomenon, and Starritt et al. (1991) reported the enhanced streaming observed with certain diagnostic devices operating with sound waves containing shocks. The general phenomenon of acoustic streaming and certain of its medical applications are discussed in detail in Chap. 7.

Lehman and Krusen (1955) observed that transport through membranes in vitro could be enhanced by irradiation with ultrasound. They attributed the effect to acoustic streaming at the liquid-tissue interface, which minimizes concentration gradients that otherwise would develop in a stagnant medium as solutes diffuse across the liquid-tissue boundary. Similar conclusions were reached by Howkins (1969) and by Mortimer et al. (1988). If this mechanistic explanation is correct, it should be possible to enhance the effect for a given temporal average intensity by using high-peak-intensity, pulsed ultrasound.

From general knowledge of the frequency response of nerves to mechanical and electrical stimuli, it is unlikely that peripheral receptors respond to the ac pressure changes in medical applications of ultrasound. Thresholds for perception of vibratory stimuli are in the range of 20–200 Pa at 200 Hz, depending upon exposed area and site (Moore, 1968). The thresholds for tactile perception rise rapidly for frequencies greater than 300 Hz. Radiation forces are another matter, however (Dalecki et al., 1995a). If all of the ultrasound were absorbed near the surface of the skin, intensities of 3–30 W/cm^2 would exert radiation pressures of 20–200 Pa. As discussed above in connection with heating, nonlinear enhancement of absorption increases the fraction of the momentum of the field that is transferred to the tissue and brings the thresholds closer to the values corresponding to complete absorption.

In lithotripsy, nonlinear absorption may contribute to perception in a more subtle way. Acoustic pressures used in lithotripter fields are very large, but durations of the pulses are typically only a few microseconds. The corresponding radiation force impulse has the same temporal characteristics as the envelope of the acoustic signal. However, because of nonlinearly enhanced absorption in the water, the lithotripter field causes fluid flow in the coupling medium. In this way, flows lasting for many milliseconds are directed toward the surface of the skin. Stimuli of this duration are readily perceived (Dalecki et al., 1995a).

15.6 Lithotripsy

More than a decade ago, it was discovered that acoustic shock waves could break kidney stones into fragments small enough to be eliminated through the urinary tract without surgery (Chaussy, 1986). Today, lithotripsy has become the technique of choice for the destruction of kidney stones. More than 100,000 cases of kidney stone disease are treated this way in the United States each year. Gallstones present a greater technical challenge for lithotripsy than kidney stones.

In most of these procedures, the wave is generated outside the body and propagates for distances of 20–50 cm through an aqueous coupling medium and the patient's tissues. In electromagnetic and piezoelectric lithotripters, the waves begin as low-amplitude pulses and become shocks by virtue of nonlinear propagation through the coupling medium. Electrohydraulic lithotripters employ spark sources, and the high amplitude acoustic shock created by the expanding plasma is collected by an ellipsoidal reflector and reconstituted at the other focus of the ellipsoid. At the focus inside the patient, the waveform typically rises to peak positive pressures ranging from 30 to 100 MPa in times typically shorter than can be measured by conventional hydrophones, i.e., less than $\sim$10 ns (Coleman and Saunders, 1989). The positive pressure decays to zero in roughly 1 μs, followed by a negative pressure excursion of 2–10 MPa that may last for several microseconds.

Until the advent of lithotripsy, the existence of nonthermal mechanisms in medical applications of ultrasound was open to question. But now it is clear not only that the destruction of stones is nonthermal but also that there are many side effects on nearby tissues that are caused by nonthermal mechanisms. Probably the most important of these mechanisms is acoustic cavitation.

15.6.1   Mechanisms of Stone Destruction

In spite of the success of extracorporeal shock-wave lithotripsy in eliminating the need for kidney stone surgery in the majority of cases, the details of how it works are still not clear. The "classic" explanation relied on purely mechanical forces. In particular, the shock wave was thought to enter the stone as a positive pulse but reflect from the back surface of the stone as a negative pulse, thus creating tensile stress that exceeds the tensile strength of the material (spalling) (Nasr, 1986; Whelan and Finlayson, 1988). Evidence for a role of cavitation in stone destruction has been accumulating gradually (Coleman et al., 1987; Delius et al., 1988a; Delius, 1990; Sass et al., 1991; Holmer et al., 1991). Probably both mechanisms contribute in some degree to stone destruction.

15.6.2   Side Effects of Lithotripsy

In contrast to cavitation produced by low-amplitude cw ultrasound, the bubbles produced by the acoustic pressures used in lithotripsy are large enough (Church, 1989) to be visualized easily in water and transparent gels by fast photography, and in tissues by transient excess scattering of ultrasound. Several investigators have presented strong evidence that cavitation is produced in tissues (Delius et al., 1988b; Delius, 1990; Delius and Brendel, 1988; Coleman et al., 1995). Hemorrhage is a common acute side effect of lithotripsy in kidney (Delius et al., 1988b, 1990c; Delius, 1990) and liver (Delius, 1990; Delius et al., 1990b; Albert et al., 1991; Prat et al., 1991; Rawat et al., 1991). Clear evidence for long-term complications, including hypertension, is more difficult to document (Williams et al., 1988; Wolfson et al., 1992). When mice are exposed to ten spherically diverging, spark-generated shock waves, the threshold for kidney hemorrhage appears to be in the range of 3–5 MPa peak positive pressure (Mayer et al., 1990). The negative pressures in those studies were immeasurably small. These exposures can be compared with clinical treatments for kidney stones, which employ 500–3000 individual shocks at 20–50 MPa positive pressure and 5–10 MPa negative pressure. However, when 1 and 4 MHz focused, pulsed ultrasound fields with peak positive pressures in excess of 10 MPa were delivered to murine kidneys, no hemorrhage was seen even when the total numbers of individual shocks exceeded one million (Carstensen et al., 1990b).

Early reports of the use of shock waves for the treatment of kidney stones noted side effects on lung (Konrad et al., 1979; Delius et al., 1987). Lung is particularly susceptible to damage by acoustic fields. With twenty, spherically diverging, spark-generated shock waves, the threshold for lung hemorrhage in mice is approximately 1.5 MPa peak positive pressure (Hartman et al., 1990a).

The intestine, which contains microscopic as well as macroscopic gas bodies, is susceptible to hemorrhage when exposed to lithotripter fields. In mice exposed to the fields of a piezoelectric lithotripter, the threshold for hemorrhage is approximately 2 MPa (Dalecki et al., 1995b,c). Most of the evidence points to cavitation as the responsible physical mechanism for intestinal hemorrhage.

With the exception of lung and intestine, most mammalian tissues are remarkably tolerant to lithotripter exposures. The development of lithotripsy has been possible because of this fortunate fact. The addition of microbubbles to the blood of mice dramatically decreases the threshold for hemorrhage in almost all tissues except lung and intestine (Dalecki et al., 1997d). This observation strongly supports the hypothesis that inertial cavitation can take place in vivo at acoustic pressures that are very small in comparison with those required to break kidney stones. The absence of comparable damage in normal mammals, even at much higher acoustic pressures, indicates that effective cavitation nuclei must be rare under ordinary physiological conditions.

Near term, fetal mice are particularly sensitive to lithotripter exposures near bony structures. Pregnant mice were exposed on the 18th day of gestation to 200 individual pulses from a piezoelectric lithotripter. Thresholds for hemorrhage at these sites are of the order of 1 MPa (Dalecki et al., 1997e). No similar, bone-related damage has been reported in adult animals. In the absence of evidence that cavitation nuclei are selectively associated with bony structures, there does not appear to be any basis for implicating cavitation in the damage. Possibly related to these observations is the report from Ohmori et al. (1994) that killing of fetuses exposed to lithotripter fields increased with gestational age.

During diastole, shock waves from a piezoelectric lithotripter stimulate heart muscle, resulting in premature contractions (Dalecki et al., 1991b). These acute effects appear to have thresholds between 5 and 10 MPa. Preliminary observations show that it is possible to pace the frog heart by shock-wave stimulation.

Figure 15.10 summarizes the available threshold data for biological effects in vivo that result from exposure to lithotripter fields. The numbers of individual acoustic pulses (10–200) used in these experimental studies are small in comparison with those of a typical clinical treatment for stone disease. In the case of murine lung, the experimentally determined threshold decreased with pulse number up to the maximum of 20 shocks used in the study. No systematic study of the relationship between pulse number and threshold pressure has been carried out.

Several independent investigating teams have reported qualified success in the treatment of tumors with lithotripter fields (e.g., Russo et al., 1986). Among these, the study by Weiss et al. (1990) is particularly notable in that striking success was achieved in tumor killing when the target tissues were backed by air.

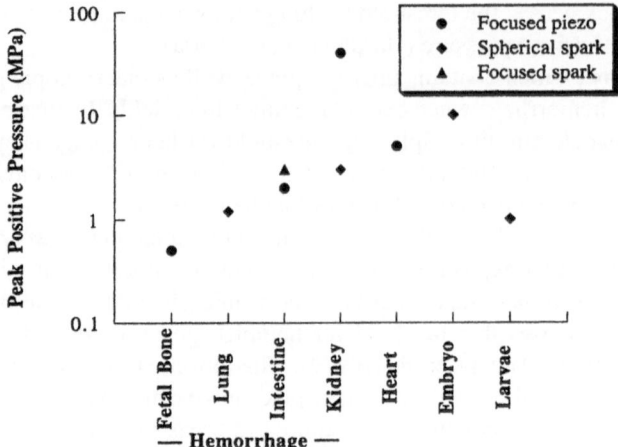

Fig. 15.10 Thresholds for biological effects of lithotripter fields. Hemorrhage has been observed near fetal bone (Dalecki et al., 1997e), adult murine lung (Hartman et al., 1990a), adult murine intestine (focused spark, Miller and Thomas, 1995; piezoelectric, Dalecki et al., 1995c), and murine kidney (spherically diverging spark, Mayer et al., 1990; piezoelectric, Raeman et al., 1994). Arrhythmias in frog heart (Dalecki et al., 1991b), killing of larvae (Carstensen et al., 1990a), and malformations and killing of chick embryos (Hartman et al., 1990b) have been observed. When microbubbles are added to the blood of mice, almost all of the tissues can be hemorrhaged at pressures less than 2 MPa (Dalecki et al., 1997d).

15.6.3   Endoscopic Lithotripsy

The shock amplitudes very near plasmas created by sparks or laser pulses can be greater than the focal fields of extracorporeal lithotripters. Although technically less sophisticated, endoscopic lithotripsy is an effective means for the treatment of kidney stones. In these procedures, a probe is inserted through the urinary tract and positioned adjacent to the stone. The general characteristics of the bubbles and the shock fields created by lasers and sparks are much the same (Vogel and Lauterborn, 1988; Doukas et al., 1991; Campbell et al., 1991). In representative clinical devices, the pressure in the initial shock rises very rapidly (in nanoseconds) and decays on the order of a microsecond. Doukas and colleagues report peak pressures well in excess of 100 MPa at distances a few tenths of a millimeter from the plasma created by a 3 mJ laser source. The plasma-generated bubbles continue to expand, reaching maximum diameters (in water) of the order of 1 cm. Collapse of the bubbles occurs at times of the order of 1 ms. Upon rebound, another shock wave is generated that is comparable to the initial shock in terms of peak pressure and total energy. Typically, the rise time of the shock due to rebound of the bubble is not as short as that of the initial shock. Weaker shocks are generated as the damped bubbles continue to oscillate.

In view of the violence of the mechanical processes that take place near these bubbles, it is not surprising that damage to soft tissues occurs in addition to that inflicted upon the target stone. In addition to the shock waves that are produced by radial rebound of bubbles, high-velocity fluid jets are produced when bubbles collapse asymmetrically (Vogel and Lauterborn, 1988; Crum, 1988). Although heating and chemical processes have been suggested as sources of tissue damage in endoscopic lithotripsy, the mechanical stresses associated with the plasma bubble appear to be the primary cause, and specific suggestions for mitigation of these problems have been proposed (Mayer et al., 1991).

15.7 Pulsed Ultrasound

The large-amplitude fields that are employed frequently in diagnostic ultrasound have waveforms that have been distorted by nonlinear propagation effects. A recent survey found that 90% of Doppler devices on the market had shock parameters greater than unity (Starritt and Duck, 1991). Effects of exposure to spark-generated shock waves and piezoelectric lithotripter fields may suggest directions for research, but at present we do not know how to extrapolate directly from those results to pulsed ultrasound exposure conditions in a meaningful way. Direct determinations of thresholds for biological effects of pulsed ultrasound in mammals are summarized in Fig. 15.11. In all cases, the tissues were exposed to diagnostically relevant

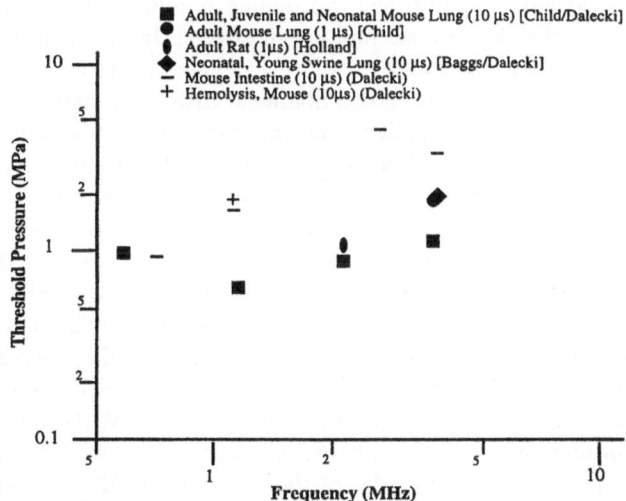

Fig. 15.11 Thresholds for biological effects of diagnostically relevant, pulsed ultrasound in vivo. Thresholds for lung hemorrhage are remarkably uniform among species, age, and between laboratories (Child et al., 1990; Dalecki, 1997a, 1997b; Holland et al., 1996; Baggs et al., 1996). Data on murine intestine are from Dalecki et al. (1995c). Hemolysis was observed in mice only after microbubbles were added to the blood (Dalecki et al., 1997c). No effects of pulsed ultrasound were observed on mouse kidney (Carstensen et al., 1990b). Acute effects on frog and murine heart required pulse lengths greater than 1 ms, and therefore the heart observations are not relevant to diagnostic ultrasound as currently practiced (Dalecki et al., 1993; McRobbie et al., 1997).

pulsed ultrasound. Pulse durations were less than 10 µs, and repetition frequencies were low enough that heating can be ruled out as a mechanism for the action of ultrasound. The effects include hemorrhage of lung and intestine, and hemolysis. Lung hemorrhage has been observed in mice, rats, rabbits, swine, and monkeys. Although lung owes its extreme sensitivity to acoustic stresses to the presence of air in the alveoli (Hartman et al., 1990a), it is not clear that the physical mechanism responsible for damage is cavitation (Raeman et al., 1997).

The large-amplitude fields that are employed frequently in biomedical ultrasound tend to involve waveforms that have been distorted by nonlinear propagation effects. Although, in a general sense, large acoustic pressures lead to cavitation, the nonlinear distortion per se has little effect on the behavior of bubbles except insofar as it makes the amplitude of the wave more difficult to characterize.

Comparison of Figs. 15.10 and 15.11 shows that the pressure thresholds for damage to lung are approximately the same for lithotripter fields and pulsed ultrasound. As mentioned above, hemorrhage in kidney occurs at a somewhat lower level with spherically diverging, positive shock waves than with a focused piezoelectric lithotripter or pulsed ultrasound with source frequencies in the 1–4 MHz frequency range. Acute effects on the frog heart occur at similar peak positive pressure levels with 1–4 MHz pulsed ultrasound and with a piezoelectric lithotripter. However, pulse durations in excess of 1 ms were required to produce the effects.

Thus far, the studies that have been conducted using either lithotripter fields or pulsed ultrasound have demonstrated no unique relationship between the shock characteristics of the waves and biological effects. In the murine lung study (Child et al., 1990), the threshold waveforms radiated by unfocused sources and detected at the subject showed marked nonlinear distortion, particularly at higher frequencies, whereas the threshold waveforms radiated by focused sources were nearly sinusoidal. Yet, threshold pressures for focused and unfocused fields were the same. Of several parameters that were used to characterize the exposures, the fundamental component of the wave provided the best correlation with the observed biological effects. A similar conclusion was reached in a study of the killing of Drosophila larvae by pulsed ultrasound (Aymé and Carstensen, 1988).

The bulk of the data available at the present time suggest that it is the pressure amplitude rather than the shock characteristics of the wave that correlate best with biological effects. There is an exception that should be noted before resting with that conclusion, however. Using laser-generated stress waves, McAuliffe et al. (1996) found that transient changes in membrane permeability in cells exposed to 100 ns positive stress pulses with amplitudes of the order of 50 MPa were more directly correlated with the rate of rise of the pressure than with the amplitude of the pressure pulse.

Acknowledgment The preparation of this review was supported in part by U.S.P.H.S. Grant No. DK39796.

References

Albert, M. B., Zook, B. C., Fromm, H., Hinson, D., Drelich, D., and Shehan, C. M. (1991). The effect of piezoelectric extracorporeal shock wave gallstone lithotripsy on dogs. *J. Lithotripsy Stone Dis.* **3**, 344–352.

American Institute of Ultrasound in Medicine (AIUM), Bioeffects Committee (1988). Bioeffects considerations for the safety of diagnostic ultrasound. *J. Ultrasound Med.* **7**, Supp. 9, S1–S38.

Aymé, E. J., and Carstensen, E. L. (1988). Occurrence of transient cavitation in pulsed sawtooth ultrasonic fields. *J. Acoust. Soc. Am.* **84**, 1598–1605.

Bacon, D. R. (1984). Finite amplitude distortion of the pulsed fields used in diagnostic ultrasound. *Ultrasound Med. Biol.* **10**, 189–195.

Bacon, D. R. (1989). Prediction of *in situ* exposure to ultrasound: An improved method. *Ultrasound Med. Biol.* **15**, 355–361.

Baggs, R. D., Penney, P., Cox, C., Child, S. Z., Raeman, C. H., Dalecki, D., and Carstensen, E. L. (1996). Thresholds for ultrasonically induced lung hemorrhage in neonatal swine. *Ultrasound Med. Biol.* **22**, 119–128.

Blackstock, D. T. (1966). Connection between the Fay and Fubini solutions for plane sound waves of finite amplitude. *J. Acoust. Soc. Am.* **39**, 1019–1026.

Blackstock, D. T. (1990). On the absorption of finite-amplitude sound. In *Frontiers of Nonlinear Acoustics: 12th ISNA*, M. F. Hamilton and D. T. Blackstock, eds. (Elsevier Science Publishers, London), pp. 119–124.

Campbell, D. S., Flynn, H. G., Blackstock, D. T., Linke, C. and Carstensen, E. L. (1991). The acoustic fields of the Wolf electrohydraulic lithotripter. *J. Lithotripsy Stone Dis.* **3**, 147–156.

Carson, P. L., Rubin, J. M., and Chiang, E. H. (1989). Constant soft tissue distance model in pregnancy. *Ultrasound Med. Biol.* **15**, Suppl. 1, 27–29.

Carstensen, E. L., Campbell, D. S., Hoffman, D., Child, S. Z., and Aymé-Bellegarda, E. J. (1990a). Killing of *Drosophila* larvae by the fields of an electrohydraulic lithotripter. *Ultrasound Med. Biol.* **16**, 687–698.

Carstensen, E. L., Hartman, C., Child, S. Z., Cox, C. A., Mayer, R., and Schenk, E. (1990b). Test for kidney hemorrhage following exposure to intense, pulsed ultrasound. *Ultrasound Med. Biol.* **16**, 681–685.

Carstensen, E. L., Law, W. K., McKay, N. D., and Muir, T. G. (1980). Demonstration of nonlinear acoustic effects at biomedical frequencies and intensities. *Ultrasound Med. Biol.* **6**, 359–368.

Carstensen, E. L., and Muir, T. G. (1986). The role of nonlinear acoustics in biomedical ultrasound. In *Tissue Characterization with Ultrasound*, J. G. Greenleaf, ed., Vol. I (CRC Press, Boca Raton), pp. 57–79.

Chaussy, C. (1986). *Extracorporeal Shock Wave Lithotripsy* (Karger, New York).

Child, S. Z., Hartman, C. L., McHale, L. A., and Carstensen, E. L. (1990). Lung damage from exposure to pulsed ultrasound. *Ultrasound Med. Biol.* **16**, 817–825.

Christopher, T. (1993). Modeling biomedical ultrasonic wave propagations. Ph.D. thesis, University of Rochester.

Christopher, T. (1994). Modeling the Dornier HM3 lithotripter. *J. Acoust. Soc. Am.* **96**, 3088–3095.

Christopher, T., and Carstensen, E. L. (1996). Finite amplitude distortion and its relationship to linear derating formulae for diagnostic ultrasound systems. *Ultrasound Med. Biol.* **22**, 1113–1116.

Christopher, P. T., and Parker, K. J. (1991). New approaches to nonlinear diffractive field propagation. *J. Acoust. Soc. Am.* **90**, 488–499.

Church, C. (1989). A theoretical study of cavitation generated by an extracorporeal shock wave lithotripter. *J. Acoust. Soc. Am.* **86**, 215–227.

Cleveland, R. O., Hamilton, M. F., and Blackstock, D. T. (1996). Time-domain modeling of finite-amplitude sound in relaxing fluids. *J. Acoust. Soc. Am.* **99**, 3312–3318.

Coleman, A. J., Kodama, T., Choi, M. J., Adams, T., and Saunders, J. E. (1995). The cavitation threshold of human tissue exposed to 0.2 MHz pulsed ultrasound: Preliminary measurements based on a study of clinical lithotripsy. *Ultrasound Med. Biol.* **21**, 405–417.

Coleman, A. J., and Saunders, J. E. (1989). A survey of the acoustic output of commercial extracorporeal shock wave lithotripters. *Ultrasound Med. Biol.* **15**, 213–227.

Coleman, A. J., Saunders, J. E., Crum, L. A., and Dyson, M. (1987). Acoustic cavitation generated by an extracorporeal shockwave lithotripter. *Ultrasound Med. Biol.* **13**, 69–76.

Crum, L. A. (1988). Cavitation microjets as a contributory mechanism for renal calculi disintegration in ESWL. *J. Urol.* **140**, 1587–1590.

Dalecki, D., Carstensen, E. L., Parker, K. J., and Bacon, D. R. (1991a). Absorption of finite amplitude focused ultrasound. *J. Acoust. Soc. Am.* **89**, 2435–2447.

Dalecki, D., Child, S. Z., Raeman, C. H., and Carstensen, E. L. (1995a). Tactile perception of ultrasound. *J. Acoust. Soc. Am.* **97**, 3165–3170.

Dalecki, D., Child, S. Z., Raeman, C. H., and Carstensen, E. L. (1997a). Age dependence of ultrasonically-induced lung hemorrhage in mice. *Ultrasound Med. Biol.*, **23**, 767–776.

Dalecki, D., Child, S. Z., Raeman, C. H., Cox, C., and Carstensen, E. L. (1997b). Ultrasonically induced lung hemorrhage in young swine. *Ultrasound Med. Biol.*, **23**, 777–781.

Dalecki, D., Child, S. Z., Raeman, C. H., Francis, C. W., Meltzer, R. S., and Carstensen, E. L. (1997c). Hemolysis in vivo from exposure to pulsed ultrasound. *Ultrasound Med. Biol.* **23**, 307–313.

Dalecki, D., Raeman, C. H., Child, S. Z., Penney, D. P., Mayer, R., and Carstensen, E. L. (1997d). The influence of contrast agents on hemorrhage produced by lithotripter fields. *Ultrasound Med. Biol.* **23**, 1435–1439.

Dalecki, D., Child, S. Z., Raeman, C. H., Penney, D. P., Mayer, R., Cox, C., and Carstensen, E. L. (1997e). Thresholds for fetal hemorrhage produced by lithotripter fields. *Ultrasound Med. Biol.* **23**, 287–297.

Dalecki, D., Keller, B. B., Carstensen, E. L., Neel, D. S., Palladino, J. L., and Noordergraaf, A. (1991b). Thresholds for premature ventricular contractions in frog hearts exposed to lithotripter fields. *Ultrasound Med. Biol.* **17**, 341–346.

Dalecki, D., Keller, B. B., Raeman, C. H., and Carstensen, E. L. (1993). Effects of pulsed ultrasound on the frog heart: I. Thresholds for changes in cardiac rhythm and aortic pressure. *Ultrasound Med. Biol.* **19**, 385–390.

Dalecki, D., Raeman, C. H., Child, S. Z., and Carstensen, E. L. (1995b). Intestinal hemorrhage from exposure to pulsed ultrasound. *Ultrasound Med. Biol.* **21**, 1067–1072.

Dalecki, D., Raeman, C. H., Child, S. Z., and Carstensen, E. L. (1995c). Thresholds for intestinal hemorrhage in mice exposed to a piezoelectric lithotripter. *Ultrasound Med. Biol.* **21**, 1239–1246.

Delius, M. (1990). Effects of lithotripter shock waves on tissues and materials. In *Frontiers of Nonlinear Acoustics: 12th ISNA*, M. F. Hamilton and D. T. Blackstock, eds. (Elsevier Science Publishers, London), pp. 31–46.

Delius, M., and Brendel, W. (1988). A model of extracorporeal shock wave action: Tandem action of shock waves. *Ultrasound Med. Biol.* **14**, 515–518.

Delius, M., Denk, R., Berding, C., Liebich, H.-G., Jordan, M., and Brendel, W. (1990c). Biological effects of shock waves: Cavitation by shock waves in piglet liver. *Ultrasound Med. Biol.* **16**, 467–472.

Delius, M., Enders, G., Heine, G., Stark, J., Remberger, K., and Brendel, W. (1987). Biological effects of shock waves: Lung hemorrhage by shock waves in dogs—Pressure dependence. *Ultrasound Med. Biol.* **13**, 61–67.

Delius, M., Heine, G., and Brendel, W. (1988a). A mechanism of gallstone destruction by extracorporeal shock wave. *Naturwissenschaften* **75**, 200–201.

Delius, M., Jordan, M., Eizenhoefer, H., Marlinghaus, E., Heine, G., Liebich, H., and Brendel, W. (1988b). Biological effects of shock waves: Kidney hemorrhage by shock waves in dogs—Administration rate dependence. *Ultrasound Med. Biol.* **14**, 689–694.

Delius, M., Jordan, M., Liebich, H.-G., and Brendel, W. (1990b). Biological effects of shock waves: Effect of shock waves on the liver and gallbladder wall of dogs—Administration rate dependence. *Ultrasound Med. Biol.* **16**, 459–466.

Delius, M., Mueller, W., Goetz, A., Liebich, H. G., and Brendel, W. (1990c). Biological effects of shock waves: Kidney hemorrhage in dogs at a fast shock wave administration rate of 15 Hz. *J. Lithotripsy Stone Dis.* **2**, 103–110.

Doukas, A. G., Zweig, A. D., Frisoli, J. K., Birngruber, R., and Deutsch, T. F. (1991). Noninvasive determination of shock wave pressure generated by optical breakdown. *Appl. Phys.* **B53**, 237–245.

Duck, F. A., and Martin, K. (1991). Trends in diagnostic ultrasound exposure. *Phys. Med. Biol.* **36**, 1423–1432.

Duck, F. A., Starritt, H. C., Aindow, J. D., Perkins, M. A., and Hawkins, A. J. (1985). The output of pulse-echo ultrasound equipment: A survey of power, pressure and intensities. *Br. J. Radiol.* **58**, 989–1001.

Dunn, F. (1974). Attenuation and speed of ultrasound in lung. *J. Acoust. Soc. Am.* **56**, 1638–1639.

Hartman, C., Child, S. Z., Mayer, R., Schenk, E., and Carstensen, E. L. (1990a). Lung damage from exposure to the fields of an electrohydraulic lithotripter. *Ultrasound Med. Biol.* **16**, 675–679.

Hartman, C. L., Child, S. Z., Penney, D. P., and Carstensen, E. L. (1992). Ultrasonic heating of lung tissue. *J. Acoust. Soc. Am.* **91**, 513–516.

Hartman, C., Cox, C. A., Brewer, L., Child, S. Z., Cox, C. F., and Carstensen, E. L. (1990b). Effects of lithotripter fields on development of chick embryos. *Ultrasound Med. Biol.* **16**, 581–585.

Holland, C. K., Dang, C. X., Apfel, R. E., Alderman, J. L., Fernandez, L. A., and Taylor, K. J. W. (1996). Direct evidence of cavitation in vivo from diagnostic ultrasound. *Ultrasound Med. Biol.* **22**, 917–925.

Holmer, N.-G., Almquist, L.-O., Hertz, T. G., Holm, A., Lindstedt, E., Persson, H. W., and Hertz, C. H. (1991). On the mechanism of kidney stone disintegration by acoustic shock waves. *Ultrasound Med. Biol.* **17**, 479–489.

Howkins, S. D. (1969). Diffusion rates and the effect of ultrasound. *Ultrasonics* **8**, 129–130.

Hynynen, K. (1987). Demonstration of enhanced temperature elevation due to nonlinear propagation of focussed ultrasound in dog's thigh in vivo. *Ultrasound Med. Biol.* **13**, 85–91.

Ichida, N., Sato, T., and Linzer, M. (1983). Imaging the nonlinear ultrasonic parameter. *Ultrason. Imag.* **9**, 295.

Jiang, P., Everbach, E. C., and Apfel, R. E. (1991). Application of mixture laws for predicting the composition of tissue phantoms. *Ultrasound Med. Biol.* **17**, 829–838.

Konrad, G., Ziegler, M., Haeusler, E., Kaspar-Fersch, U., Wurster, H., and Krauss, W.. (1979). Fokussierte Stowellen zur Berhrungsfreien Nierensteinzertrümmerung an der Freigelegten Niere. *Urologe* **A 18**, 289–293.

Lehman, J., and Krusen, F. H. (1955). Effects of pulsed and continuous application of ultrasound on active transport of ions through biological membranes. *Arch. Phys. Med. Rehab.* **35**, 20–23.

Mayer, R., Hartman, C., Child, S. Z., Dalecki, D., Schenk, E., and Carstensen, E. L. (1991). Tissue damage associated with the collapsing bubble of an endoscopic electrohydraulic lithotripter. *J. Lithotripsy Stone Dis.* **3**, 311–318.

Mayer, R., Schenk, E., Child, S., Norton, S., Cox, C., Hartman, C., Cox, C., and Carstensen, E. (1990). Pressure threshold for shock wave induced renal hemorrhage. *J. Urol.* **144**, 1505–1509.

McAuliffe, D. J., Lee, S., Flotte, T. J., and Doukas, A. G. (1996). Stress-wave-assisted transport through the plasma membrane in vitro. *Lasers Surg. Med.* **20**, 216–222.

McRobbie, A. G., Raeman, C. H., Child, S. Z., and Dalecki, D. (1997). Thresholds for premature contractions in murine hearts exposed to pulsed ultrasound. *Ultrasound Med. Biol.* **23**, 761–765.

Miller, D. L., and Thomas, R. M. (1995). Thresholds for hemorrhage in mouse skin and intestine induced by lithotripter shock waves, *Ultrasound Med. Biol.* **21**, 249–257.

Mitome, H. (1990). Acoustic streaming induced by tone burst waves in water. In *Frontiers of Nonlinear Acoustics: 12th ISNA*, M. F. Hamilton and D. T. Blackstock, eds. (Elsevier Science Publishers, London), pp. 353–358.

Moore, T. J. (1968). Vibratory stimulation of the skin by electrostatic field: Effects of size of electrode and site of stimulation on thresholds. *Am. J. Psychol.* **81**, 235–240.

Mortimer, A. J., Trollope, B. J., Villeneuve, E. J., and Roy, O. Z. (1988). Ultrasound-enhanced diffusion through isolated frog skin. *Ultrasonics* **26**, 348–351.

Muir, T. G., and Carstensen, E. L. (1980). Prediction of nonlinear acoustic effects at biomedical frequencies and intensities. *Ultrasound Med. Biol.* **6**, 345–357.

Nanda, N. C., Schlief, R., and Goldberg, B. B., eds. (1997). *Advances in Echo Imaging Using Contrast Enhancement*, 2nd ed. (Kluwer Academic Publishers, Dordrecht, The Netherlands).

Nasr, M. E. (1986). The evolution and dynamics of spark generated shock waves and their focusing by ellipsoidal reflectors in lithotripsy. Ph.D. thesis, University of Florida.

National Council for Radiation Protection and Measurements (NCRP) (1992). Exposure criteria for diagnostic ultrasound. Part I. Criteria based on thermal mechanisms (National Council for Radiation Protection and Measurements, Bethesda, MD).

Ohmori, K., Matsuda, T., Horii, Y., and Yoshida, O. (1994). Effects of shock waves on the mouse fetus. *J. Urology* **151**, 255–258.

Pedersen, P. C., and Ozcan, H. S. (1986). Ultrasound properties of lung tissue and their measurements. *Ultrasound Med. Biol.* **12**, 483–499.

Prat, F., Ponchon, T., Berger, F., Chapelon, J. Y., Gagnon, P., and Cathignol, D. (1991). Hepatic lesions in the rabbit induced by acoustic cavitation. *Gastroenterology* **100**, 1345–1350.

Raeman, C. H., Child, S. Z., Dalecki, D., Mayer, R., Parker, K. J., and Carstensen, E. L. (1994). Damage to murine kidney and intestine from exposure to the fields of a piezoelectric lithotripter. *Ultrasound Med. Biol.* **20**, 589–594.

Raeman, C. H., Dalecki, D., Child, S. Z., Meltzer, R. S., and Carstensen, E. L. (1997). Albunex® does not increase the sensitivity of the lung to pulsed ultrasound. *Echocardiography* **14**, 553–557.

Rawat, B., Riddler, C., Fache, J. S., and Burhenne, H. J. (1991). Hemoglobinuria and hematuria following biliary extracorporeal cholecystolithotripsy. *J. Lithotripsy Stone Dis.* **3**, 357–361.

Russo, P., Stephenson, R. A., Mies, C., Huryk, R., Heston, W. D. W., Melamed, M. R., and Fair, W. R. (1986). High energy shock waves suppress tumor growth in vitro and in vivo. *J. Urol.* **135**, 626–628.

Sass, W., Bräunlich, M., Dreyer, H.-P., Matura, E., Folberth, W., Priesmeyer, H.-G., and Seifert, J. (1991). The mechanisms of stone disintegration by shock waves. *Ultrasound Med. Biol.* **17**, 239–243.

Starritt, H. C., and Duck, F. (1991). Quantification of acoustic shock in routine exposure measurement. WFUMB Conference, Copenhagen, September, 1991.

Starritt, H. C., Duck, F., and Humphrey, V. F. (1991). Forces acting in the direction of propagation in pulsed ultrasound fields. *Phys. Med. Biol.* **36**, 1465–1474.

Vogel, A., and Lauterborn, W. (1988). Acoustic transient generation by laser produced cavitation bubbles near solid boundaries. *J. Acoust. Soc. Am.* **84**, 719–731.

Weiss, N., Delius, M., Gambihler, S., Dirschedl, P., Goetz, A., and Brendel, W. (1990). Influence of the shock wave application mode on the growth of A-Me1 3 and SSK2 tumors in vivo. *Ultrasound Med. Biol.* **16**, 595–605.

Whelan, J. P., and Finlayson, B. (1988). An experimental model for the systematic investigation of stone fracture by extracorporeal shock wave lithotripsy. *J. Urol.* **140**, 395–400.

Williams, C. M., Kaude, J. V., Newman, R. C., Peterson, J. C., and Thomas, W. C. (1988). Extracorporeal shock-wave lithotripsy: Long-term complications. *Am. J. Rad.* **150**, 311–315.

Wolfson, B. A., Fuchs, G. J., David, R. D., and Barbaric, Z. (1992). Creating of an animal model to investigate the bioeffects of alternative treatments for urinary stones. *J. Stone Dis.* **4**, 27–33.

World Federation of Ultrasound in Medicine and Biology (1998). Thresholds for nonthermal bioeffects: Theoretical and experimental basis for a threshold index. *Ultrasound Med. Biol.* **24**, Suppl. 1, S41–S49.

Zhang, J., and Dunn, F. (1987). In vivo B/A determination in a mammalian organ. *J. Acoust. Soc. Am.* **81**, 1635–1637.

Index